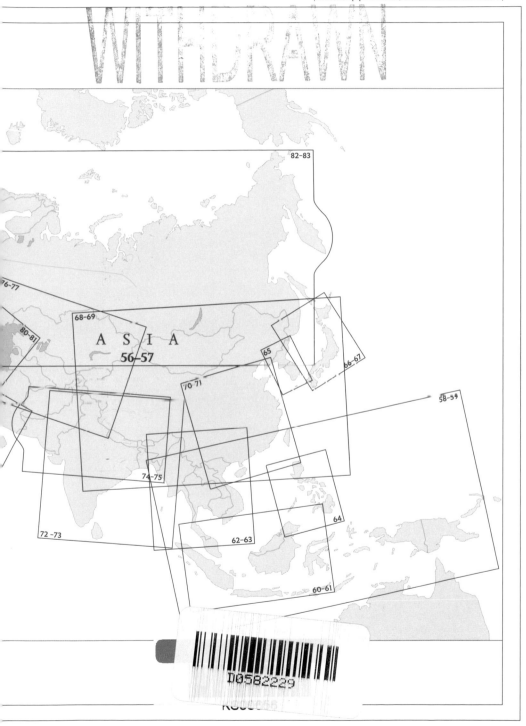

WITHDRAWN

82-83

76-77

68-69

80-81

A S I A
56-57

65

66-67

70 71

58-59

74-75

72 -73

64

62-63

60-61

D0582229

ATLAS OF THE WORLD

Times Books, 77-85 Fulham Palace Road,
London W6 8JB

The Times is a registered trademark of
Times Newspapers Ltd

First published 1994
Second Edition 2000
Third Edition 2004
Fourth Edition 2007

Fifth Edition 2010

Printed in Singapore

British Library Cataloguing in Publication Data.
A catalogue record for this book is
available from the British Library.

ISBN 978-0-00-731817-9

Imp 001

All mapping in this atlas is generated from Collins Bartholomew digital databases.
Collins Bartholomew, the UK's leading independent geographical information supplier,
can provide a digital, custom, and premium mapping service to a variety of markets.

For further information:
Tel: +44 (0) 141 306 3752
e-mail: collinsbartholomew@harpercollins.co.uk
or visit our website at: www.collinsbartholomew.com

www.timesatlas.com
The world's most authoritative and prestigious world atlases.

THE TIMES COMPACT
ATLAS OF THE WORLD

CALDAY GRANGE
GRAMMAR SCHOOL
LIBRARY

TIMES BOOKS
LONDON

CONTENTS

All independent countries and populated dependent and disputed territories are included in this list of the states and territories of the world; the list is arranged in alphabetical order by the conventional name form. For independent states, the full name is given below the conventional name, if this is different; for territories, the status is given. The capital city name is the same form as shown on the reference maps.

The statistics used for the area and population are the latest available and include estimates. The information on languages and religions is based on the latest information on 'de facto' speakers of the language or 'de facto' adherents to the religion. The information available on languages and religions varies greatly from country to country. Some countries include questions in censuses, others do not, in which case best estimates are used. The order of the languages and religions reflect their relative importance within the country; generally, languages or religions are included when more than one per cent of the population are estimated to be speakers or adherents.

Membership of selected international organizations is shown for each independent country. Territories are not shown as having separate memberships of these organizations.

ABBREVIATIONS

CURRENCIES

| CFA | Communauté Financière Africaine |
| CFP | Comptoirs Français du Pacifique |

ORGANIZATIONS

APEC	Asia-Pacific Economic Cooperation
ASEAN	Association of Southeast Asian Nations
CARICOM	Caribbean Community
CIS	Commonwealth of Independent States
Comm.	The Commonwealth
EU	European Union
OECD	Organization of Economic Co-operation and Development
OPEC	Organization of Petroleum Exporting Countries
SADC	Southern African Development Community
UN	United Nations

AFGHANISTAN
Islamic State of Afghanistan

Area Sq Km	652 225	Religions	Sunni Muslim,
Area Sq Miles	251 825		Shi'a Muslim
Population	28 150 000	Currency	Afghani
Capital	Kābul	Organizations	UN
Languages	Dari, Pushtu, Uzbek,Turkmen	Map page	76–77

ALBANIA
Republic of Albania

Area Sq Km	28 748	Religions	Sunni Muslim,
Area Sq Miles	11 100		Albanian Orthodo
Population	3 155 000		Roman Catholic
Capital	Tirana (Tiranë)	Currency	Lek
Languages	Albanian, Greek	Organizations	UN
		Map page	109

ALGERIA
People's Democratic Republic of Algeria

Area Sq Km	2 381 741	Religions	Sunni Muslim
Area Sq Miles	919 595	Currency	Algerian dinar
Population	34 895 000	Organizations	OPEC, UN
Capital	Algiers (Alger)	Map page	114–115
Languages	Arabic, French, Berber		

American Samoa
United States Unincorporated Territory

Area Sq Km	197	Religions	Protestant, Roma
Area Sq Miles	76		Catholic
Population	67 000	Currency	United States do
Capital	Fagatogo	Map page	49
Languages	Samoan, English		

ANDORRA
Principality of Andorra

Area Sq Km	465	Religions	Roman Catholic
Area Sq Miles	180	Currency	Euro
Population	86 000	Organizations	UN
Capital	Andorra la Vella	Map page	104
Languages	Spanish, Catalan, French		

ANGOLA
Republic of Angola

Area Sq Km	1 246 700	Religions	Roman Catholic,
Area Sq Miles	481 354		Protestant,
Population	18 498 000		traditional belief
Capital	Luanda	Currency	Kwanza
Languages	Portuguese, Bantu, local languages	Organizations	OPEC, SADC, UN
		Map page	120

Anguilla
United Kingdom Overseas Territory

Area Sq Km	155	Religions	Protestant, Roman Catholic
Area Sq Miles	60		
Population	15 000	Currency	East Caribbean dollar
Capital	The Valley		
Languages	English	Map page	147

ANTIGUA AND BARBUDA

Area Sq Km	442	Religions	Protestant, Roman Catholic
Area Sq Miles	171		
Population	88 000	Currency	East Caribbean dollar
Capital	St John's		
Languages	English, creole	Organizations	CARICOM, Comm., UN
		Map page	147

ARGENTINA
Argentine Republic

Area Sq Km	2 766 889	Religions	Roman Catholic, Protestant
Area Sq Miles	1 068 302		
Population	40 276 000	Currency	Argentinian peso
Capital	Buenos Aires	Organizations	UN
Languages	Spanish, Italian, Amerindian languages	Map page	152–153

ARMENIA
Republic of Armenia

Area Sq Km	29 800	Religions	Armenian Orthodox
Area Sq Miles	11 506	Currency	Dram
Population	3 083 000	Organizations	CIS, UN
Capital	Yerevan (Erevan)	Map page	81
Languages	Armenian, Azeri		

Aruba
Self-governing Netherlands Territory

Area Sq Km	193	Religions	Roman Catholic, Protestant
Area Sq Miles	75		
Population	107 000	Currency	Aruban florin
Capital	Oranjestad	Map page	147
Languages	Papiamento, Dutch, English		

Ascension
Dependency of St Helena

Area Sq Km	88	Religions	Protestant, Roman Catholic
Area Sq Miles	34		
Population	1 100	Currency	Pound sterling
Capital	Georgetown	Map page	113
Languages	English		

AUSTRALIA
Commonwealth of Australia

Area Sq Km	7 692 024	Religions	Protestant, Roman Catholic, Orthodox
Area Sq Miles	2 969 907		
Population	21 293 000	Currency	Australian dollar
Capital	Canberra	Organizations	APEC, Comm., OECD, UN
Languages	English, Italian, Greek	Map page	50–51

Australian Capital Territory (Federal Territory)

Area Sq Km	2 358	Population	346 400
Area Sq Miles	910	Capital	Canberra

Jervis Bay Territory (Territory)

Area Sq Km	73	Population	611
Area Sq Miles	28		

New South Wales (State)

Area Sq Km	800 642	Population	7 017 100
Area Sq Miles	309 130	Capital	Sydney

Northern Territory (Territory)

Area Sq Km	1 349 129	Population	221 100
Area Sq Miles	520 902	Capital	Darwin

Queensland (State)

Area Sq Km	1 730 648	Population	4 320 100
Area Sq Miles	668 207	Capital	Brisbane

South Australia (State)

Area Sq Km	983 482	Population	1 607 700
Area Sq Miles	379 725	Capital	Adelaide

Tasmania (State)

Area Sq Km	68 401	Population	498 900
Area Sq Miles	26 410	Capital	Hobart

Victoria (State)

Area Sq Km	227 416	Population	5 340 300
Area Sq Miles	87 806	Capital	Melbourne

Western Australia (State)

Area Sq Km	2 529 875	Population	2 188 500
Area Sq Miles	976 790	Capital	Perth

AUSTRIA
Republic of Austria

Area Sq Km	83 855	Religions	Roman Catholic, Protestant
Area Sq Miles	32 377		
Population	8 364 000	Currency	Euro
Capital	Vienna (Wien)	Organizations	EU, OECD, UN
Languages	German, Croatian, Turkish	Map page	102–103

AZERBAIJAN
Republic of Azerbaijan

Area Sq Km	86 600	Religions	Shi'a Muslim, Sunni Muslim, Russian and Armenian Orthodox
Area Sq Miles	33 436		
Population	8 832 000		
Capital	Baku (Bakı)	Currency	Azerbaijani manat
Languages	Azeri, Armenian, Russian, Lezgian	Organizations	CIS, UN
		Map page	81

Azores (Arquipélago dos Açores)
Autonomous Region of Portugal

Area Sq Km	2 300	Religions	Roman Catholic,
Area Sq Miles	888		Protestant
Population	244 000	Currency	Euro
Capital	Ponta Delgada	Map page	112
Languages	Portuguese		

THE BAHAMAS
Commonwealth of The Bahamas

Area Sq Km	13 939	Religions	Protestant,
Area Sq Miles	5 382		Roman Catholic
Population	342 000	Currency	Bahamian dollar
Capital	Nassau	Organizations	CARICOM,
Languages	English, creole		Comm., UN
		Map page	146–147

BAHRAIN
Kingdom of Bahrain

Area Sq Km	691	Religions	Shi'a Muslim,
Area Sq Miles	267		Sunni Muslim,
Population	791 000		Christian
Capital	Manama	Currency	Bahraini dinar
	(Al Manāmah)	Organizations	UN
Languages	Arabic, English	Map page	79

BANGLADESH
People's Republic of Bangladesh

Area Sq Km	143 998	Religions	Sunni Muslim,
Area Sq Miles	55 598		Hindu
Population	162 221 000	Currency	Taka
Capital	Dhaka (Dacca)	Organizations	Comm., UN
Languages	Bengali, English	Map page	75

BARBADOS

Area Sq Km	430	Religions	Protestant,
Area Sq Miles	166		Roman Catholic
Population	256 000	Currency	Barbados dollar
Capital	Bridgetown	Organizations	CARICOM,
Languages	English, creole		Comm., UN
		Map page	147

BELARUS
Republic of Belarus

Area Sq Km	207 600	Religions	Belorussian
Area Sq Miles	80 155		Orthodox,
Population	9 634 000		Roman Catholic
Capital	Minsk	Currency	Belarus rouble
Languages	Belorussian,	Organizations	CIS, UN
	Russian	Map page	88–89

BELGIUM
Kingdom of Belgium

Area Sq Km	30 520	Religions	Roman Catholic,
Area Sq Miles	11 784		Protestant
Population	10 647 000	Currency	Euro
Capital	Brussels (Bruxelles)	Organizations	EU, OECD, UN
Languages	Dutch (Flemish),	Map page	100
	French (Walloon),		
	German		

BELIZE

Area Sq Km	22 965	Religions	Roman Catholic,
Area Sq Miles	8 867		Protestant
Population	307 000	Currency	Belize dollar
Capital	Belmopan	Organizations	CARICOM,
Languages	English, Spanish,		Comm., UN
	Mayan, creole	Map page	147

BENIN
Republic of Benin

Area Sq Km	112 620	Religions	Traditional beliefs
Area Sq Miles	43 483		Roman Catholic,
Population	8 935 000		Sunni Muslim
Capital	Porto-Novo	Currency	CFA franc
Languages	French, Fon,	Organization	UN
	Yoruba, Adja,	Map page	114
	local languages		

Bermuda
United Kingdom Overseas Territory

Area Sq Km	54	Religions	Protestant,
Area Sq Miles	21		Roman Catholic
Population	65 000	Currency	Bermuda dollar
Capital	Hamilton	Map page	125
Languages	English		

BHUTAN
Kingdom of Bhutan

Area Sq Km	46 620	Religions	Buddhist, Hindu
Area Sq Miles	18 000	Currency	Ngultrum,
Population	697 000		Indian rupee
Capital	Thimphu	Organizations	UN
Languages	Dzongkha,	Map page	75
	Nepali, Assamese		

BOLIVIA
Republic of Bolivia

Area Sq Km	1 098 581	Religions	Roman Catholic,
Area Sq Miles	424 164		Protestant, Baha'i
Population	9 863 000	Currency	Boliviano
Capital	La Paz/Sucre	Organizations	UN
Languages	Spanish, Quechua,	Map page	152
	Aymara		

Bonaire
part of Netherlands Antilles

Area Sq Km	288	Religions	Roman Catholic,
Area Sq Miles	111		Protestant
Population	12 103	Currency	Netherlands Antilles
Capital	Kralendijk		guilder
Languages	Dutch, Papiamento	Map page	147

Bonin Islands (Ogasawara-shotō)
part of Japan

Area Sq Km	104	Religions	Shintoist, Buddhist,
Area Sq Miles	40		Christian
Population	2 772	Currency	Yen
Capital	Ōmura	Map page	69
Languages	Japanese		

BOSNIA-HERZEGOVINA
Republic of Bosnia and Herzegovina

Area Sq Km	51 130	Religions	Sunni Muslim,
Area Sq Miles	19 741		Serbian Orthodox,
Population	3 767 000		Roman Catholic,
Capital	Sarajevo		Protestant
Languages	Bosnian, Serbian,	Currency	Marka
	Croatian	Organizations	UN
		Map page	109

BOTSWANA
Republic of Botswana

Area Sq Km	581 370	Religions	Traditional beliefs,
Area Sq Miles	224 468		Protestant,
Population	1 950 000		Roman Catholic
Capital	Gaborone	Currency	Pula
Languages	English, Setswana,	Organizations	Comm., SADC, UN
	Shona, local	Map page	120
	languages		

BRAZIL
Federative Republic of Brazil

Area Sq Km	8 514 879	Religions	Roman Catholic,
Area Sq Miles	3 287 613		Protestant
Population	193 734 000	Currency	Real
Capital	Brasília	Organizations	UN
Languages	Portuguese	Map page	150–151

BRUNEI
State of Brunei Darussalam

Area Sq Km	5 765	Religions	Sunni Muslim,
Area Sq Miles	2 226		Buddhist, Christian
Population	400 000	Currency	Brunei dollar
Capital	Bandar Seri Begawan	Organizations	APEC, ASEAN,
Languages	Malay, English,		Comm., UN
	Chinese	Map page	61

BULGARIA
Republic of Bulgaria

Area Sq Km	110 994	Religions	Bulgarian
Area Sq Miles	42 855		Orthodox,
Population	7 545 000		Sunni Muslim
Capital	Sofia (Sofiya)	Currency	Lev
Languages	Bulgarian, Turkish,	Organizations	EU, UN
	Romany,	Map page	110
	Macedonian		

BURKINA
Democratic Republic of Burkina Faso

Area Sq Km	274 200	Religions	Sunni Muslim,
Area Sq Miles	105 869		traditional beliefs,
Population	15 757 000		Roman Catholic
Capital	Ouagadougou	Currency	CFA franc
Languages	French, Moore	Organizations	UN
	(Mossi), Fulani, local	Map page	114
	languages		

BURUNDI
Republic of Burundi

Area Sq Km	27 835	Religions	Roman Catholic,
Area Sq Miles	10 747		traditional beliefs,
Population	8 303 000		Protestant
Capital	Bujumbura	Currency	Burundian franc
Languages	Kirundi (Hutu,	Organizations	UN
	Tutsi), French	Map page	119

CAMBODIA
Kingdom of Cambodia

Area Sq Km	181 000	Religions	Buddhist, Roman
Area Sq Miles	69 884		Catholic, Sunni
Population	14 805 000		Muslim
Capital	Phnom Penh	Currency	Riel
Languages	Khmer, Vietnamese	Organizations	ASEAN, UN
		Map page	63

CAMEROON
Republic of Cameroon

Area Sq Km	475 442	Religions	Roman Catholic,
Area Sq Miles	183 569		traditional beliefs,
Population	19 522 000		Sunni Muslim,
Capital	Yaoundé		Protestant
Languages	French, English,	Currency	CFA franc
	Fang, Bamileke,	Organizations	Comm., UN
	local languages	Map page	118

CANADA

Area Sq Km	9 984 670	Religions	Roman Catholic,
Area Sq Miles	3 855 103		Protestant, Eastern
Population	33 573 000		Orthodox, Jewish
Capital	Ottawa	Currency	Canadian dollar
Languages	English, French,	Organizations	APEC, Comm.,
	local languages		OECD, UN
		Map page	126–127

Alberta (Province)

Area Sq Km	661 848	Population	3 610 782
Area Sq Miles	255 541	Capital	Edmonton

British Columbia (Province)

Area Sq Km	944 735	Population	4 405 534
Area Sq Miles	364 764	Capital	Victoria

Manitoba (Province)

Area Sq Km	647 797	Population	1 210 547
Area Sq Miles	250 116	Capital	Winnipeg

New Brunswick (Province)

Area Sq Km	72 908	Population	747 790
Area Sq Miles	28 150	Capital	Fredericton

Newfoundland and Labrador (Province)

Area Sq Km	405 212	Population	508 944
Area Sq Miles	156 453	Capital	St John's

Northwest Territories (Territory)

Area Sq Km	1 346 106	Population	43 151
Area Sq Miles	519 734	Capital	Yellowknife

CANADA

Nova Scotia (Province)

Area Sq Km	55 284	Population	939 125
Area Sq Miles	21 345	Capital	Halifax

Nunavut (Territory)

Area Sq Km	2 093 190	Population	31 522
Area Sq Miles	808 185	Capital	Iqaluit (Frobisher Bay)

Ontario (Province)

Area Sq Km	1 076 395	Population	12 977 059
Area Sq Miles	415 598	Capital	Toronto

Prince Edward Island (Province)

Area Sq Km	5 660	Population	140 750
Area Sq Miles	2 185	Capital	Charlottetown

Québec (Province)

Area Sq Km	1 542 056	Population	7 771 854
Area Sq Miles	595 391	Capital	Québec

Saskatchewan (Province)

Area Sq Km	651 036	Population	1 020 847
Area Sq Miles	251 366	Capital	Regina

Yukon (Territory)

Area Sq Km	482 443	Population	33 372
Area Sq Miles	186 272	Capital	Whitehorse

 Canary Islands (Islas Canarias)
Autonomous Community of Spain

Area Sq Km	7 447	Religions	Roman Catholic
Area Sq Miles	2 875	Currency	Euro
Population	2 075 968	Map page	114
Capital	Santa Cruz de Tenerife/Las Palmas		
Languages	Spanish		

 CAPE VERDE
Republic of Cape Verde

Area Sq Km	4 033	Religions	Roman Catholic,
Area Sq Miles	1 557		Protestant
Population	506 000	Currency	Cape Verde escudo
Capital	Praia	Organizations	UN
Languages	Portuguese, creole	Map page	46

 Cayman Islands
United Kingdom Overseas Territory

Area Sq Km	259	Religions	Protestant, Roman
Area Sq Miles	100		Catholic
Population	56 000	Currency	Cayman Islands
Capital	George Town		dollar
Languages	English	Map page	146

 CENTRAL AFRICAN REPUBLIC

Area Sq Km	622 436	Religions	Protestant,
Area Sq Miles	240 324		Roman Catholic,
Population	4 422 000		traditional beliefs,
Capital	Bangui		Sunni Muslim
Languages	French, Sango,	Currency	CFA franc
	Banda, Baya, local	Organizations	UN
	languages	Map page	118

 Ceuta
Autonomous Community of Spain

Area Sq Km	19	Religions	Roman Catholic,
Area Sq Miles	7		Muslim
Population	77 389	Currency	Euro
Capital	Ceuta	Map page	106
Languages	Spanish, Arabic		

 CHAD
Republic of Chad

Area Sq Km	1 284 000	Religions	Sunni Muslim,
Area Sq Miles	495 755		Roman Catholic,
Population	11 206 000		Protestant,
Capital	Ndjamena		traditional beliefs
Languages	Arabic, French,Sara,	Currency	CFA franc
	local languages	Organizations	UN
		Map page	115

 Chatham Islands
part of New Zealand

Area Sq Km	963	Religions	Protestant
Area Sq Miles	372	Currency	New Zealand doll
Population	640	Map page	49
Capital	Waitangi		
Languages	English		

 CHILE
Republic of Chile

Area Sq Km	756 945	Religions	Roman Catholic,
Area Sq Miles	292 258		Protestant
Population	16 970 000	Currency	Chilean peso
Capital	Santiago	Organizations	APEC, UN
Languages	Spanish, Amerindian	Map page	152–153
	languages		

CHINA
People's Republic of China

Area Sq Km	9 584 492	Religions	Confucian, Taoist
Area Sq Miles	3 700 593		Buddhist, Christia
Population	1 330 265 000		Sunni Muslim
Capital	Beijing (Peking)	Currency	Yuan, Hong Kong
Languages	Mandarin, Wu,		dollar, Macao pata
	Cantonese, Hsiang,	Organizations	APEC, UN
	regional languages	Map page	68–69

Anhui (Province)

Area Sq Km	139 000	Population	61 180 000
Area Sq Miles	53 668	Capital	Hefei

Bejing (Municipality)

Area Sq Km	16 800	Population	16 330 000
Area Sq Miles	6 487	Capital	Beijing (Peking)

Chongqing (Municipality)

Area Sq Km	23 000	Population	28 160 000
Area Sq Miles	8 880	Capital	Chongqing

Fujian (Province)

Area Sq Km	121 400	Population	35 810 000
Area Sq Miles	46 873	Capital	Fuzhou

ansu (Province)

Area Sq Km	453 700	Population	26 170 000
rea Sq Miles	175 175	Capital	Lanzhou

uangdong (Province)

Area Sq Km	178 000	Population	94 490 000
rea Sq Miles	68 726	Capital	Guangzhou (Canton)

uangxi Zhuangzu Zizhiqu (Autonomous Region)

Area Sq Km	236 000	Population	47 680 000
rea Sq Miles	91 120	Capital	Nanning

uizhou (Province)

Area Sq Km	176 000	Population	37 620 000
rea Sq Miles	67 954	Capital	Guiyang

ainan (Province)

Area Sq Km	34 000	Population	8 450 000
rea Sq Miles	13 127	Capital	Haikou

ebei (Province)

Area Sq Km	187 700	Population	69 430 000
rea Sq Miles	72 471	Capital	Shijiazhuang

eilongjiang (Province)

Area Sq Km	454 600	Population	38 240 000
rea Sq Miles	175 522	Capital	Harbin

enan (Province)

Area Sq Km	167 000	Population	93 600 000
rea Sq Miles	64 479	Capital	Zhengzhou

ong Kong (Special Administrative Region)

Area Sq Km	1 075	Population	6 926 000
rea Sq Miles	415	Capital	Hong Kong

ubei (Province)

Area Sq Km	185 900	Population	56 990 000
rea Sq Miles	71 776	Capital	Wuhan

unan (Province)

Area Sq Km	210 000	Population	63 550 000
rea Sq Miles	81 081	Capital	Changsha

angsu (Province)

Area Sq Km	102 600	Population	76 250 000
rea Sq Miles	39 614	Capital	Nanjing

angxi (Province)

Area Sq Km	166 900	Population	43 680 000
rea Sq Miles	64 440	Capital	Nanchang

lin (Province)

Area Sq Km	187 000	Population	27 300 000
rea Sq Miles	72 201	Capital	Changchun

aoning (Province)

Area Sq Km	147 400	Population	42 980 000
rea Sq Miles	56 911	Capital	Shenyang

Macao (Special Administrative Region)

Area Sq Km	17	Population	526 000
Area Sq Mile	7		

Nei Mongol Zizhiqu (Inner Mongolia) (Autonomous Region)

Area Sq Km	1 183 000	Population	24 050 000
Area Sq Miles	456 759	Capital	Hohhot

Ningxia Huizu Zizhiqu (Autonomous Region)

Area Sq Km	66 400	Population	6 100 000
Area Sq Miles	25 637	Capital	Yinchuan

Qinghai (Province)

Area Sq Km	721 000	Population	5 520 000
Area Sq Miles	278 380	Capital	Xining

Shaanxi (Province)

Area Sq Km	205 600	Population	37 480 000
Area Sq Miles	79 383	Capital	Xi'an

Shandong (Province)

Area Sq Km	153 300	Population	93 670 000
Area Sq Miles	59 189	Capital	Jinan

Shanghai (Municipality)

Area Sq Km	6 300	Population	18 580 000
Area Sq Miles	2 432	Capital	Shanghai

Shanxi (Province)

Area Sq Km	156 300	Population	33 930 000
Area Sq Miles	60 348	Capital	Taiyuan

Sichuan (Province)

Area Sq Km	569 000	Population	81 270 000
Area Sq Miles	219 692	Capital	Chengdu

Tianjin (Municipality)

Area Sq Km	11 300	Population	11 150 000
Area Sq Miles	4 363	Capital	Tianjin

Xinjiang Uygur Zizhiqu (Sinkiang) (Autonomous Region)

Area Sq Km	1 600 000	Population	20 950 000
Area Sq Miles	617 763	Capital	Ürümqi

Xizang Zizhiqu (Tibet) (Autonomous Region)

Area Sq Km	1 228 400	Population	2 840 000
Area Sq Miles	474 288	Capital	Lhasa

Yunnan (Province)

Area Sq Km	394 000	Population	45 140 000
Area Sq Miles	152 124	Capital	Kunming

Zhejiang (Province)

Area Sq Km	101 800	Population	50 600 000
Area Sq Miles	39 305	Capital	Hangzhou

Christmas Island
Australian External Territory

Area Sq Km	135	**Religions**	Buddhist, Sunni
Area Sq Miles	52		Muslim, Protestant,
Population	1 351		Roman Catholic
Capital	The Settlement	**Currency**	Australian dollar
Languages	English	**Map page**	58

Cocos Islands (Keeling Islands)
Australian External Territory

Area Sq Km	14	**Religions**	Sunni Muslim,
Area Sq Miles	5		Christian
Population	621	**Currency**	Australian dollar
Capital	West Island	**Map page**	58
Languages	English		

COLOMBIA
Republic of Colombia

Area Sq Km	1 141 748	**Religions**	Roman Catholic,
Area Sq Miles	440 831		Protestant
Population	45 660 000	**Currency**	Colombian peso
Capital	Bogotá	**Organizations**	UN
Languages	Spanish, Amerindian	**Map page**	150
	languages		

COMOROS
Union of the Comoros

Area Sq Km	1 862	**Religions**	Sunni Muslim,
Area Sq Miles	719		Roman Catholic
Population	676 000	**Currency**	Comoros franc
Capital	Moroni	**Organizations**	UN
Languages	Comorian, French,	**Map page**	121
	Arabic		

CONGO
Republic of the Congo

Area Sq Km	342 000	**Religions**	Roman Catholic,
Area Sq Miles	132 047		Protestant,
Population	3 683 000		traditional beliefs,
Capital	Brazzaville		Sunni Muslim
Languages	French, Kongo,	**Currency**	CFA franc
	Monokutuba, local	**Organizations**	UN
	languages	**Map page**	118

CONGO, DEMOCRATIC REPUBLIC OF THE

Area Sq Km	2 345 410	**Religions**	Christian, Sunni
Area Sq Miles	905 568		Muslim
Population	66 020 000	**Currency**	Congolese franc
Capital	Kinshasa	**Organizations**	SADC, UN
Languages	French, Lingala,	**Map page**	118–119
	Swahili, Kongo,		
	local languages		

Cook Islands
Self-governing New Zealand Overseas Territory

Area Sq Km	293	**Religions**	Protestant, Roman
Area Sq Miles	113		Catholic
Population	20 000	**Currency**	New Zealand dollar
Capital	Avarua	**Map page**	49
Languages	English, Maori		

COSTA RICA
Republic of Costa Rica

Area Sq Km	51 100	**Religions**	Roman Catholic,
Area Sq Miles	19 730		Protestant
Population	4 579 000	**Currency**	Costa Rican colón
Capital	San José	**Organizations**	UN
Languages	Spanish	**Map page**	146

CÔTE D'IVOIRE
Republic of Côte d'Ivoire

Area Sq Km	322 463	**Religions**	Sunni Muslim,
Area Sq Miles	124 504		Roman Catholic,
Population	21 075 000		traditonal beliefs,
Capital	Yamoussoukro		Protestant
Languages	French, creole, Akan,	**Currency**	CFA franc
	local languages	**Organizations**	UN
		Map page	114

CROATIA
Republic of Croatia

Area Sq Km	56 538	**Religions**	Roman Catholic,
Area Sq Miles	21 829		Serbian Orthodox,
Population	4 416 000		Sunni Muslim
Capital	Zagreb	**Currency**	Kuna
Languages	Croatian, Serbian	**Organizations**	UN
		Map page	109

CUBA
Republic of Cuba

Area Sq Km	110 860	**Religions**	Roman Catholic,
Area Sq Miles	42 803		Protestant
Population	11 204 000	**Currency**	Cuban peso
Capital	Havana (La Habana)	**Organizations**	UN
Languages	Spanish	**Map page**	146

Curaçao
part of Netherlands Antilles

Area Sq Km	444	**Religions**	Roman Catholic,
Area Sq Miles	171		Protestant
Population	140 796	**Currency**	Netherlands
Capital	Willemstad		Antilles guilder
Languages	Dutch, Papiamento	**Map page**	147

CYPRUS
Republic of Cyprus

Area Sq Km	9 251	**Religions**	Greek Orthodox,
Area Sq Miles	3 572		Sunni Muslim
Population	871 000	**Currency**	Euro
Capital	Nicosia (Lefkosia)	**Organizations**	Comm., EU, UN
Languages	Greek, Turkish,	**Map page**	80
	English		

CZECH REPUBLIC

Area Sq Km	78 864	**Religions**	Roman Catholic,
Area Sq Miles	30 450		Protestant
Population	10 369 000	**Currency**	Koruna
Capital	Prague (Praha)	**Organizations**	EU, OECD, UN
Languages	Czech, Moravian,	**Map page**	102–103
	Slovakian		

DENMARK
Kingdom of Denmark

Area Sq Km	43 075	**Religions**	Protestant
Area Sq Miles	16 631	**Currency**	Danish krone
Population	5 470 000	**Organizations**	EU, OECD, UN
Capital	Copenhagen	**Map page**	93
	(København)		
Languages	Danish		

DJIBOUTI
Republic of Djibouti

Area Sq Km	23 200	**Religions**	Sunni Muslim,
Area Sq Miles	8 958		Christian
Population	864 000	**Currency**	Djibouti franc
Capital	Djibouti	**Organizations**	UN
Languages	Somali, Afar, French,	**Map page**	117
	Arabic		

DOMINICA
Commonwealth of Dominica

Area Sq Km	750	**Religions**	Roman Catholic,
Area Sq Miles	290		Protestant
Population	67 000	**Currency**	East Caribbean
Capital	Roseau		dollar
Languages	English, creole	**Organizations**	CARICOM, Comm.,
			UN
		Map page	147

DOMINICAN REPUBLIC

Area Sq Km	48 442	**Religions**	Roman Catholic,
Area Sq Miles	18 704		Protestant
Population	10 090 000	**Currency**	Dominican peso
Capital	Santo Domingo	**Organizations**	UN
Languages	Spanish, creole	**Map page**	147

Easter Island (Isla de Pascua)
part of Chile

Area Sq Km	171	**Religions**	Roman Catholic
Area Sq Miles	66	**Currency**	Chilean peso
Population	4 781	**Map page**	157
Capital	Hanga Roa		
Languages	Spanish		

EAST TIMOR
Democratic Republic of Timor-Leste

Area Sq Km	14 874	**Religions**	Roman Catholic
Area Sq Miles	5 743	**Currency**	United States dollar
Population	1 134 000	**Organisations**	UN
Capital	Dili	**Map page**	59
Languages	Portuguese, Tetun,		
	English		

ECUADOR
Republic of Ecuador

Area Sq Km	272 045	**Religions**	Roman Catholic
Area Sq Miles	105 037	**Currency**	United States dollar
Population	13 625 000	**Organizations**	OPEC, UN
Capital	Quito	**Map page**	150
Languages	Spanish, Quechua,		
	Amerindian		
	languages		

EGYPT
Arab Republic of Egypt

Area Sq Km	1 000 250	**Religions**	Sunni Muslim,
Area Sq Miles	386 199		Coptic Christian
Population	82 999 000	**Currency**	Egyptian pound
Capital	Cairo (Al Qāhirah)	**Organizations**	UN
Languages	Arabic	**Map page**	116

EL SALVADOR
Republic of El Salvador

Area Sq Km	21 041	**Religions**	Roman Catholic,
Area Sq Miles	8 124		Protestant
Population	6 163 000	**Currency**	El Salvador colón,
Capital	San Salvador		United States dollar
Languages	Spanish	**Organizations**	UN
		Map page	146

EQUATORIAL GUINEA
Republic of Equatorial Guinea

Area Sq Km	28 051	**Religions**	Roman Catholic,
Area Sq Miles	10 831		traditional beliefs
Population	676 000	**Currency**	CFA franc
Capital	Malabo	**Organizations**	UN
Languages	Spanish, French,	**Map page**	118
	Fang		

ERITREA
State of Eritrea

Area Sq Km	117 400	**Religions**	Sunni Muslim,
Area Sq Miles	45 328		Coptic Christian
Population	5 073 000	**Currency**	Nakfa
Capital	Asmara	**Organizations**	UN
Languages	Tigrinya, Tigre	**Map page**	116

ESTONIA
Republic of Estonia

Area Sq Km	45 200	**Religions**	Protestant, Estonian
Area Sq Miles	17 452		and Russian
Population	1 340 000		Orthodox
Capital	Tallinn	**Currency**	Kroon
Languages	Estonian, Russian	**Organizations**	EU, UN
		Map page	88

ETHIOPIA
Federal Democratic Republic of Ethiopia

Area Sq Km	1 133 880	**Religions**	Ethiopian
Area Sq Miles	437 794		Orthodox,
Population	82 825 000		Sunni Muslim,
Capital	Addis Ababa		traditional beliefs
	(Ādīs Ābeba)	**Currency**	Birr
Languages	Oromo, Amharic,	**Organizations**	UN
	Tigrinya, local	**Map page**	117
	languages		

Falkland Islands
United Kingdom Overseas Territory

Area Sq Km	12 170	Religions	Protestant,
Area Sq Miles	4 699		Roman Catholic
Population	2 955	Currency	Falkland Islands
Capital	Stanley		pound
Languages	English	Map page	153

Faroe Islands
Self-governing Danish Territory

Area Sq Km	1 399	Religions	Protestant
Area Sq Miles	540	Currency	Danish krone
Population	50 000	Map page	94
Capital	Tórshavn		
Languages	Faroese, Danish		

FIJI
Republic of the Fiji Islands

Area Sq Km	18 330	Religions	Christian, Hindu,
Area Sq Miles	7 077		Sunni Muslim
Population	849 000	Currency	Fiji dollar
Capital	Suva	Organizations	UN, Comm.
Languages	English, Fijian,	Map page	49
	Hindi		

FINLAND
Republic of Finland

Area Sq Km	338 145	Religions	Protestant, Greek
Area Sq Miles	130 559		Orthodox
Population	5 326 000	Currency	Euro
Capital	Helsinki	Organizations	EU, OECD, UN
	(Helsingfors)	Map page	92–93
Languages	Finnish, Swedish		

FRANCE
French Republic

Area Sq Km	543 965	Religions	Roman Catholic,
Area Sq Miles	210 026		Protestant, Sunni
Population	62 343 000		Muslim
Capital	Paris	Currency	Euro
Languages	French, Arabic	Organizations	EU, OECD, UN
		Map page	104–105

French Guiana
French Overseas Department

Area Sq Km	90 000	Religions	Roman Catholic
Area Sq Miles	34 749	Currency	Euro
Population	226 000	Map page	151
Capital	Cayenne		
Languages	French, creole		

French Polynesia
French Overseas Country

Area Sq Km	3 265	Religions	Protestant, Roman
Area Sq Miles	1 261		Catholic
Population	269 000	Currency	CFP franc
Capital	Papeete	Map page	49
Languages	French, Tahitian,		
	Polynesian languages		

GABON
Gabonese Republic

Area Sq Km	267 667	Religions	Roman Catholic,
Area Sq Miles	103 347		Protestant,
Population	1 475 000		traditonal beliefs
Capital	Libreville	Currency	CFA franc
Languages	French, Fang, local	Organizations	UN
	languages	Map page	118

Galapagos Islands (Islas Galápagos)
part of Ecuador

Area Sq Km	8 010	Religions	Roman Catholic
Area Sq Miles	3 093	Currency	United States dol
Population	19 184	Map page	125
Capital	Puerto Baquerizo		
	Moreno		
Languages	Spanish		

THE GAMBIA
Republic of The Gambia

Area Sq Km	11 295	Religions	Sunni Muslim,
Area Sq Miles	4 361		Protestant
Population	1 705 000	Currency	Dalasi
Capital	Banjul	Organizations	Comm., UN
Languages	English, Malinke,	Map page	114
	Fulani, Wolof		

Gaza
Semi-autonomous region

Area Sq Km	363	Religions	Sunni Muslim,
Area Sq Miles	140		Shi'a Muslim
Population	1 486 816	Currency	Israeli shekel
Capital	Gaza	Map page	80
Languages	Arabic		

GEORGIA
Republic of Georgia

Area Sq Km	69 700	Religions	Georgian Orthod
Area Sq Miles	26 911		Russian Orthodo
Population	4 260 000		Sunni Muslim
Capital	T'bilisi	Currency	Lari
Languages	Georgian, Russian,	Organizations	CIS, UN
	Armenian, Azeri,	Map page	81
	Ossetian, Abkhaz		

GERMANY
Federal Republic of Germany

Area Sq Km	357 022	Religions	Protestant, Roma
Area Sq Miles	137 847		Catholic
Population	82 167 000	Currency	Euro
Capital	Berlin	Organizations	EU, OECD, UN
Languages	German, Turkish	Map page	102

GHANA
Republic of Ghana

Area Sq Km	238 537	Religions	Christian, Sunni
Area Sq Miles	92 100		Muslim, traditio
Population	23 837 000		beliefs
Capital	Accra	Currency	Cedi
Languages	English, Hausa,	Organizations	Comm., UN
	Akan, local	Map page	114
	languages		

Gibraltar
United Kingdom Overseas Territory

Area Sq Km	7	**Religions**	Roman Catholic,
Area Sq Miles	3		Protestant, Sunni
Population	31 000		Muslim
Capital	Gibraltar	**Currency**	Gibraltar pound
Languages	English, Spanish	**Map page**	106

GREECE
Hellenic Republic

Area Sq Km	131 957	**Religions**	Greek Orthodox,
Area Sq Miles	50 949		Sunni Muslim
Population	11 161 000	**Currency**	Euro
Capital	Athens (Athina)	**Organizations**	EU, OECD, UN
Languages	Greek	**Map page**	111

Greenland
Self-governing Danish Territory

Area Sq Km	2 175 600	**Religions**	Protestant
Area Sq Miles	840 004	**Currency**	Danish krone
Population	57 000	**Map page**	127
Capital	Nuuk (Godthåb)		
Languages	Greenlandic, Danish		

GRENADA

Area Sq Km	378	**Religions**	Roman Catholic,
Area Sq Miles	146		Protestant
Population	104 000	**Currency**	East Caribbean
Capital	St George's		dollar
Languages	English, creole	**Organizations**	CARICOM, Comm.,
			UN
		Map page	147

Guadeloupe
French Overseas Department

Area Sq Km	1 780	**Religions**	Roman Catholic
Area Sq Miles	687	**Currency**	Euro
Population	465 000	**Map page**	147
Capital	Basse-Terre		
Languages	French, creole		

Guam
United States Unincorporated Territory

Area Sq Km	541	**Religions**	Roman Catholic
Area Sq Miles	209	**Currency**	United States dollar
Population	178 000	**Map page**	59
Capital	Hagåtña		
Languages	Chamorro, English,		
	Tagalog		

GUATEMALA
Republic of Guatemala

Area Sq Km	108 890	**Religion**	Roman Catholic,
Area Sq Miles	42 043		Protestant
Population	14 027 000	**Currency**	Quetzal, United
Capital	Guatemala City		States dollar
Languages	Spanish, Mayan	**Organizations**	UN
	languages	**Map page**	146

Guernsey
United Kingdom Crown Dependency

Area Sq Km	78	**Religions**	Protestant, Roman
Area Sq Miles	30		Catholic
Population	64 801	**Currency**	Pound sterling
Capital	St Peter Port	**Map page**	95
Languages	English, French		

GUINEA
Republic of Guinea

Area Sq Km	245 857	**Religions**	Sunni Muslim,
Area Sq Miles	94 926		traditional beliefs,
Population	10 069 000		Christian
Capital	Conakry	**Currency**	Guinea franc
Languages	French, Fulani,	**Organizations**	UN
	Malinke, local	**Map page**	114
	languages		

GUINEA-BISSAU
Republic of Guinea-Bissau

Area Sq Km	36 125	**Religions**	Traditional beliefs,
Area Sq Miles	13 948		Sunni Muslim,
Population	1 611 000		Christian
Capital	Bissau	**Currency**	CFA franc
Languages	Portuguese, crioulo,	**Organizations**	UN
	local languages	**Map page**	114

GUYANA
Co-operative Republic of Guyana

Area Sq Km	214 969	**Religions**	Protestant, Hindu,
Area Sq Miles	83 000		Roman Catholic,
Population	762 000		Sunni Muslim
Capital	Georgetown	**Currency**	Guyana dollar
Languages	English, creole,	**Organizations**	CARICOM, Comm.,
	Amerindian		UN
	languages	**Map page**	150

HAITI
Republic of Haiti

Area Sq Km	27 750	**Religions**	Roman Catholic,
Area Sq Miles	10 714		Protestant, Voodoo
Population	10 033 000	**Currency**	Gourde
Capital	Port-au-Prince	**Organizations**	CARICOM, UN
Languages	French, creole	**Map page**	147

HONDURAS
Republic of Honduras

Area Sq Km	112 088	**Religions**	Roman Catholic,
Area Sq Miles	43 277		Protestant
Population	7 466 000	**Currency**	Lempira
Capital	Tegucigalpa	**Organizations**	UN
Languages	Spanish, Amerindian	**Map page**	147
	languages		

HUNGARY
Republic of Hungary

Area Sq Km	93 030	**Religions**	Roman Catholic,
Area Sq Miles	35 919		Protestant
Population	9 993 000	**Currency**	Forint
Capital	Budapest	**Organizations**	EU, OECD, UN
Languages	Hungarian	**Map page**	103

ICELAND
Republic of Iceland

Area Sq Km	102 820	Religions	Protestant
Area Sq Miles	39 699	Currency	Icelandic króna
Population	323 000	Organizations	OECD, UN
Capital	Reykjavík	Map page	92
Languages	Icelandic		

INDIA
Republic of India

Area Sq Km	3 064 898	Religions	Hindu,
Area Sq Miles	1 183 364		Sunni Muslim,
Population	1 198 003 000		Shi'a Muslim,
Capital	New Delhi		Sikh, Christian
Languages	Hindi, English, many	Currency	Indian rupee
	regional languages	Organizations	Comm., UN
		Map page	72–73

INDONESIA
Republic of Indonesia

Area Sq Km	1 919 445	Religions	Sunni Muslim,
Area Sq Miles	741 102		Protestant, Roman
Population	229 965 000		Catholic, Hindu,
Capital	Jakarta		Buddhist
Languages	Indonesian, local	Currency	Rupiah
	languages	Organizations	APEC, ASEAN,
			OPEC, UN
		Map page	58–59

IRAN
Islamic Republic of Iran

Area Sq Km	1 648 000	Religions	Shi'a Muslim,
Area Sq Miles	636 296		Sunni Muslim
Population	74 196 000	Currency	Iranian rial
Capital	Tehrān	Organizations	OPEC, UN
Languages	Farsi, Azeri, Kurdish,	Map page	81
	regional languages		

IRAQ
Republic of Iraq

Area Sq Km	438 317	Religions	Shi'a Muslim,
Area Sq Miles	169 235		Sunni Muslim,
Population	30 747 000		Christian
Capital	Baghdād	Currency	Iraqi dinar
Languages	Arabic, Kurdish,	Organizations	OPEC, UN
	Turkmen	Map page	81

IRELAND
Republic of Ireland

Area Sq Km	70 282	Religions	Roman Catholic,
Area Sq Miles	27 136		Protestant
Population	4 515 000	Currency	Euro
Capital	Dublin	Organizations	EU, OECD, UN
	(Baile Átha Cliath)	Map page	97
Languages	English, Irish		

Isle of Man
United Kingdom Crown Dependency

Area Sq Km	572	Religions	Protestant, Roman
Area Sq Miles	221		Catholic
Population	80 000	Currency	Pound sterling
Capital	Douglas	Map page	98
Languages	English		

ISRAEL
State of Israel

Area Sq Km	20 770	Religions	Jewish, Sunni
Area Sq Miles	8 019		Muslim, Christian
Population	7 170 000		Druze
Capital	Jerusalem*	Currency	Shekel
	(Yerushalayim)	Organizations	UN
	(El Quds)	Map page	80
Languages	Hebrew, Arabic		

*De facto capital. Disputed.

ITALY
Italian Republic

Area Sq Km	301 245	Religions	Roman Catholic
Area Sq Miles	116 311	Currency	Euro
Population	59 870 000	Organizations	EU, OECD, UN
Capital	Rome (Roma)	Map page	108–109
Languages	Italian		

JAMAICA

Area Sq Km	10 991	Religions	Protestant, Roman
Area Sq Miles	4 244		Catholic
Population	2 719 000	Currency	Jamaican dollar
Capital	Kingston	Organizations	CARICOM, Comm.,
Languages	English, creole		UN
		Map page	146

Jammu and Kashmir
Disputed territory (India/Pakistan/China)

Area Sq Km	222 236	Map page	74–75
Area Sq Miles	85 806		
Population	13 000 000		
Capital	Srinagar		

JAPAN

Area Sq Km	377 727	Religions	Shintoist, Buddhist,
Area Sq Miles	145 841		Christian
Population	127 156 000	Currency	Yen
Capital	Tōkyō	Organizations	APEC, OECD, UN
Languages	Japanese	Map page	66–67

Jersey
United Kingdom Crown Dependency

Area Sq Km	116	Religions	Protestant, Roman
Area Sq Miles	45		Catholic
Population	90 800	Currency	Pound sterling
Capital	St Helier	Map page	95
Languages	English, French		

JORDAN
Hashemite Kingdom of Jordan

Area Sq Km	89 206	Religions	Sunni Muslim,
Area Sq Miles	34 443		Christian
Population	6 316 000	Currency	Jordanian dinar
Capital	'Ammān	Organizations	UN
Languages	Arabic	Map page	80

Juan Fernández Islands
part of Chile

Area Sq Km	179	Religions	Roman Catholic,
Area Sq Miles	69		Protestant
Population	817	Currency	Chilean peso
Capital	San Juan Bautista	Map page	157
Languages	Spanish, Amerindian languages		

KAZAKHSTAN
Republic of Kazakhstan

Area Sq Km	2 717 300	Religions	Sunni Muslim,
Area Sq Miles	1 049 155		Russian Orthodox,
Population	15 637 000		Protestant
Capital	Astana (Akmola)	Currency	Tenge
Languages	Kazakh, Russian, Ukrainian, German, Uzbek, Tatar	Organizations	CIS, UN
		Map page	76–77

KENYA
Republic of Kenya

Area Sq Km	582 646	Religions	Christian,
Area Sq Miles	224 961		traditional beliefs
Population	39 802 000	Currency	Kenyan shilling
Capital	Nairobi	Organizations	Comm., UN
Languages	Swahili, English, local languages	Map page	119

KIRIBATI
Republic of Kiribati

Area Sq Km	717	Religions	Roman Catholic,
Area Sq Miles	277		Protestant
Population	98 000	Currency	Australian dollar
Capital	Bairiki	Organizations	Comm., UN
Languages	Gilbertese, English	Map page	49

KOSOVO
Republic of Kosovo

Area Sq Km	10 908	Religions	Sunni Muslim,
Area Sq Miles	4 212		Serbian Orthodox
Population	2 153 139	Currency	Euro
Capital	Prishtinë (Priština)	Map page	109
Languages	Albanian, Serbian		

KUWAIT
State of Kuwait

Area Sq Km	17 818	Religions	Sunni Muslim,
Area Sq Miles	6 880		Shi'a Muslim,
Population	2 985 000		Christian, Hindu
Capital	Kuwait (Al Kuwayt)	Currency	Kuwaiti dinar
Languages	Arabic	Organizations	OPEC, UN
		Map page	78

KYRGYZSTAN
Kyrgyz Republic

Area Sq Km	198 500	Religions	Sunni Muslim,
Area Sq Miles	76 641		Russian Orthodox
Population	5 482 000	Currency	Kyrgyz som
Capital	Bishkek (Frunze)	Organizations	CIS, UN
Languages	Kyrgyz, Russian, Uzbek	Map page	77

LAOS
Lao People's Democratic Republic

Area Sq Km	236 800	Religions	Buddhist,
Area Sq Miles	91 429		traditional beliefs
Population	6 320 000	Currency	Kip
Capital	Vientiane (Viangchan)	Organizations	ASEAN, UN
Languages	Lao, local languages	Map page	62–63

LATVIA
Republic of Latvia

Area Sq Km	64 589	Religions	Protestant,
Area Sq Miles	24 938		Roman Catholic,
Population	2 249 000		Russian Orthodox
Capital	Rīga	Currency	Lats
Languages	Latvian, Russian	Organizations	EU, UN
		Map page	88

LEBANON
Republic of Lebanon

Area Sq Km	10 452	Religions	Shi'a Muslim, Sunni
Area Sq Miles	4 036		Muslim, Christian
Population	4 224 000	Currency	Lebanese pound
Capital	Beirut (Beyrouth)	Organizations	UN
Languages	Arabic, Armenian, French	Map page	80

LESOTHO
Kingdom of Lesotho

Area Sq Km	30 355	Religions	Christian,
Area Sq Miles	11 720		traditional beliefs
Population	2 067 000	Currency	Loti, South African
Capital	Maseru		rand
Languages	Sesotho, English, Zulu	Organizations	Comm., SADC, UN
		Map page	123

LIBERIA
Republic of Liberia

Area Sq Km	111 369	Religions	Traditional beliefs,
Area Sq Miles	43 000		Christian,
Population	3 955 000		Sunni Muslim
Capital	Monrovia	Currency	Liberian dollar
Languages	English, creole, local languages	Organizations	UN
		Map page	114

LIBYA
Great Socialist People's Libyan Arab Jamahiriya

Area Sq Km	1 759 540	Religions	Sunni Muslim
Area Sq Miles	679 362	Currency	Libyan dinar
Population	6 420 000	Organizations	OPEC, UN
Capital	Tripoli (Ṭarābulus)	Map page	115
Languages	Arabic, Berber		

LIECHTENSTEIN
Principality of Liechtenstein

Area Sq Km	160	Religions	Roman Catholic,
Area Sq Miles	62		Protestant
Population	36 000	Currency	Swiss franc
Capital	Vaduz	Organizations	UN
Languages	German	Map page	105

LITHUANIA
Republic of Lithuania

Area Sq Km	65 200	Religions	Roman Catholic,
Area Sq Miles	25 174		Protestant,
Population	3 287 000		Russian Orthodox
Capital	Vilnius	Currency	Litas
Languages	Lithuanian, Russian, Polish	Organizations	EU, UN
		Map page	88

Lord Howe Island
part of Australia

Area Sq Km	17	Religions	Protestant,
Area Sq Miles	6		Roman Catholic
Population	350	Currency	Australian dollar
Languages	English	Map page	51

LUXEMBOURG
Grand Duchy of Luxembourg

Area Sq Km	2 586	Religions	Roman Catholic
Area Sq Miles	998	Currency	Euro
Population	486 000	Organizations	EU, OECD, UN
Capital	Luxembourg	Map page	100
Languages	Letzeburgish,		
	German, French		

MACEDONIA (F.Y.R.O.M.)
Republic of Macedonia

Area Sq Km	25 713	Religions	Macedonian
Area Sq Miles	9 928		Orthodox,
Population	2 042 000		Sunni Muslim
Capital	Skopje	Currency	Macedonian denar
Languages	Macedonian,	Organizations	UN
	Albanian, Turkish	Map page	111

MADAGASCAR
Republic of Madagascar

Area Sq Km	587 041	Religions	Traditional beliefs,
Area Sq Miles	226 658		Christian, Sunni
Population	19 625 000		Muslim
Capital	Antananarivo	Currency	Malagasy ariary,
Languages	Malagasy, French		Malagasy franc
		Organizations	SADC, UN
		Map page	121

Madeira
Autonomous Region of Portugal

Area Sq Km	779	Religions	Roman Catholic,
Area Sq Miles	301		Protestant
Population	246 689	Currency	Euro
Capital	Funchal	Map page	114
Languages	Portuguese		

MALAWI
Republic of Malawi

Area Sq Km	118 484	Religions	Christian,
Area Sq Miles	45 747		traditional beliefs,
Population	15 263 000		Sunni Muslim
Capital	Lilongwe	Currency	Malawian kwacha
Languages	Chichewa, English,	Organizations	Comm., SADC, UN
	local languages	Map page	121

MALAYSIA
Federation of Malaysia

Area Sq Km	332 965	Religions	Sunni Muslim,
Area Sq Miles	128 559		Buddhist, Hindu,
Population	27 468 000		Christian,
Capital	Kuala Lumpur/		traditional beliefs
	Putrajaya	Currency	Ringgit
Languages	Malay, English,	Organizations	APEC, ASEAN,
	Chinese, Tamil,		Comm., UN
	local languages	Map page	60–61

MALDIVES
Republic of the Maldives

Area Sq Km	298	Religions	Sunni Muslim
Area Sq Miles	115	Currency	Rufiyaa
Population	309 000	Organizations	Comm., UN
Capital	Male	Map page	56
Languages	Divehi (Maldivian)		

MALI
Republic of Mali

Area Sq Km	1 240 140	Religions	Sunni Muslim,
Area Sq Miles	478 821		traditional belief
Population	13 010 000		Christian
Capital	Bamako	Currency	CFA franc
Languages	French, Bambara,	Organizations	UN
	local languages	Map page	114

MALTA
Republic of Malta

Area Sq Km	316	Religions	Roman Catholic
Area Sq Miles	122	Currency	Euro
Population	409 000	Organizations	Comm., EU, UN
Capital	Valletta	Map page	84
Languages	Maltese, English		

MARSHALL ISLANDS
Republic of the Marshall Islands

Area Sq Km	181	Religions	Protestant, Roma
Area Sq Miles	70		Catholic
Population	62 000	Currency	United States doll
Capital	Delap-Uliga-Djarrit	Organizations	UN
Languages	English, Marshallese	Map page	48

Martinique
French Overseas Department

Area Sq Km	1 079	Religions	Roman Catholic,
Area Sq Miles	417		traditional beliefs
Population	405 000	Currency	Euro
Capital	Fort-de-France	Map page	147
Languages	French, creole		

MAURITANIA
Islamic Arab and African Republic of Mauritania

Area Sq Km	1 030 700	Religions	Sunni Muslim
Area Sq Miles	397 955	Currency	Ouguiya
Population	3 291 000	Organizations	UN
Capital	Nouakchott	Map page	114
Languages	Arabic, French,		
	local languages		

MAURITIUS
Republic of Mauritius

Area Sq Km	2 040	Religions	Hindu, Roman
Area Sq Miles	788		Catholic, Sunni
Population	1 288 000		Muslim
Capital	Port Louis	Currency	Mauritius rupee
Languages	English, creole,	Organizations	Comm., SADC, UI
	Hindi, Bhojpuri,	Map page	113
	French		

Mayotte
French Departmental Collectivity

Area Sq Km	373	Religions	Sunni Muslim,
Area Sq Miles	144		Christian
Population	194 000	Currency	Euro
Capital	Dzaoudzi	Map page	121
Languages	French, Mahorian		

Melilla
Autonomous Community of Spain

Area Sq Km	13	Religions	Roman Catholic,
Area Sq Miles	5		Muslim
Population	71 448	Currency	Euro
Capital	Melilla	Map page	114
Languages	Spanish, Arabic		

MEXICO
United Mexican States

Area Sq Km	1 972 545	Religions	Roman Catholic,
Area Sq Miles	761 604		Protestant
Population	109 610 000	Currency	Mexican peso
Capital	Mexico City	Organizations	APEC, OECD, UN
Languages	Spanish, Amerindian languages	Map page	144–145

MICRONESIA, FEDERATED STATES OF

Area Sq Km	701	Religions	Roman Catholic,
Area Sq Miles	271		Protestant
Population	111 000	Currency	United States dollar
Capital	Palikir	Organizations	UN
Languages	English, Chuukese, Pohnpeian, local languages	Map page	48

MOLDOVA
Republic of Moldova

Area Sq Km	33 700	Religions	Romanian
Area Sq Miles	13 012		Orthodox,
Population	3 604 000		Russian Orthodox
Capital	Chişinău (Kishinev)	Currency	Moldovan leu
Languages	Romanian, Ukrainian, Gagauz, Russian	Organizations	CIS, UN
		Map page	90

MONACO
Principality of Monaco

Area Sq Km	2	Religions	Roman Catholic
Area Sq Miles	1	Currency	Euro
Population	33 000	Organizations	UN
Capital	Monaco-Ville	Map page	105
Languages	French, Monégasque, Italian		

MONGOLIA

Area Sq Km	1 565 000	Religions	Buddhist,
Area Sq Miles	604 250		Sunni Muslim
Population	2 671 000	Currency	Tugrik (tögrög)
Capital	Ulan Bator (Ulaanbaatar)	Organizations	UN
		Map page	68–69
Languages	Khalka (Mongolian), Kazakh, local languages		

MONTENEGRO

Area Sq Km	13 812	Religions	Montenegrin,
Area Sq Miles	5 333		Orthodox,
Population	624 000		Sunni Muslim
Capital	Podgorica	Currency	Euro
Languages	Serbian, (Montenegrin), Albanian	Organizations	UN
		Map page	109

Montserrat
United Kingdom Overseas Territory

Area Sq Km	100	Religions	Protestant, Roman
Area Sq Miles	39		Catholic
Population	4 655	Currency	East Caribbean
Capital	Brades*		dollar
Languages	English	Organizations	CARICOM
*Temporary capital		Map page	147

MOROCCO
Kingdom of Morocco

Area Sq Km	446 550	Religions	Sunni Muslim
Area Sq Miles	172 414	Currency	Moroccan dirham
Population	31 993 000	Organizations	UN
Capital	Rabat	Map page	114
Languages	Arabic, Berber, French		

MOZAMBIQUE
Republic of Mozambique

Area Sq Km	799 380	Religions	Traditional beliefs,
Area Sq Miles	308 642		Roman Catholic,
Population	22 894 000		Sunni Muslim
Capital	Maputo	Currency	Metical
Languages	Portuguese, Makua, Tsonga, local languages	Organizations	Comm., SADC, UN
		Map page	121

MYANMAR (Burma)
Union of Myanmar

Area Sq Km	676 577	Religions	Buddhist,
Area Sq Miles	261 228		Christian,
Population	50 020 000		Sunni Muslim
Capital	Nay Pyi Taw/ Rangoon (Yangôn)	Currency	Kyat
		Organizations	ASEAN, UN
Languages	Burmese, Shan, Karen, local languages	Map page	62–63

NAMIBIA
Republic of Namibia

Area Sq Km	824 292	Religions	Protestant,
Area Sq Miles	318 261		Roman Catholic
Population	2 171 000	Currency	Namibian dollar
Capital	Windhoek	Organizations	Comm., SADC, UN
Languages	English, Afrikaans, German, Ovambo, local languages	Map page	121

NAURU
Republic of Nauru

Area Sq Km	21	Religions	Protestant, Roman
Area Sq Miles	8		Catholic
Population	10 000	Currency	Australian dollar
Capital	Yaren	Organizations	Comm., UN
Languages	Nauruan, English	Map page	48

NEPAL
Federal Democratic Republic of Nepal

Area Sq Km	147 181	Religions	Hindu, Buddhist,
Area Sq Miles	56 827		Sunni Muslim
Population	29 331 000	Currency	Nepalese rupee
Capital	Kathmandu	Organizations	UN
Languages	Nepali, Maithili, Bhojpuri, English, local languages	Map page	75

NETHERLANDS
Kingdom of the Netherlands

Area Sq Km	41 526	Religions	Roman Catholic,
Area Sq Miles	16 033		Protestant, Sunni
Population	16 592 000		Muslim
Capital	Amsterdam/ The Hague ('s-Gravenhage)	Currency	Euro
		Organizations	EU, OECD, UN
		Map page	100
Languages	Dutch, Frisian		

Netherlands Antilles
Self-governing Netherlands Territory

Area Sq Km	800	Religions	Roman Catholic,
Area Sq Miles	309		Protestant
Population	198 000	Currency	Netherlands
Capital	Willemstad		Antilles guilder
Languages	Dutch, Papiamento, English	Map page	147

New Caledonia
French Overseas Collectivity

Area Sq Km	19 058	Religions	Roman Catholic,
Area Sq Miles	7 358		Protestant, Sunni
Population	250 000		Muslim
Capital	Nouméa	Currency	CFP franc
Languages	French, local languages	Map page	48

NEW ZEALAND

Area Sq Km	270 534	Religions	Protestant, Roman
Area Sq Miles	104 454		Catholic
Population	4 266 000	Currency	New Zealand dollar
Capital	Wellington	Organizations	APEC, Comm.,
Languages	English, Maori		OECD, UN
		Map page	54

NICARAGUA
Republic of Nicaragua

Area Sq Km	130 000	Religions	Roman Catholic,
Area Sq Miles	50 193		Protestant
Population	5 743 000	Currency	Córdoba
Capital	Managua	Organizations	UN
Languages	Spanish, Amerindian languages	Map page	146

NIGER
Republic of Niger

Area Sq Km	1 267 000	Religions	Sunni Muslim,
Area Sq Miles	489 191		traditional beliefs
Population	15 290 000	Currency	CFA franc
Capital	Niamey	Organizations	UN
Languages	French, Hausa, Fulani, local languages	Map page	115

NIGERIA
Federal Republic of Nigeria

Area Sq Km	923 768	Religions	Sunni Muslim,
Area Sq Miles	356 669		Christian,
Population	154 729 000		traditional beliefs
Capital	Abuja	Currency	Naira
Languages	English, Hausa, Yoruba, Ibo, Fulani, local languages	Organizations	Comm., OPEC, UN
		Map page	115

Niue
Self-governing New Zealand Overseas Territory

Area Sq Km	258	Religions	Christian
Area Sq Miles	100	Currency	New Zealand dollar
Population	1 625	Map page	48
Capital	Alofi		
Languages	English, Nivean		

Norfolk Island
Australian External Territory

Area Sq Km	35	Religions	Protestant, Roman
Area Sq Miles	14		Catholic
Population	2 523	Currency	Australian dollar
Capital	Kingston	Map page	48
Languages	English		

Northern Mariana Islands
United States Commonwealth

Area Sq Km	477	Religions	Roman Catholic
Area Sq Miles	184	Currency	United States dollar
Population	87 000	Map page	59
Capital	Capitol Hill		
Languages	English, Chamorro, local languages		

NORTH KOREA
Democratic People's Republic of Korea

Area Sq Km	120 538	Religions	Traditional beliefs,
Area Sq Miles	46 540		Chondoist,
Population	23 906 000		Buddhist
Capital	P'yŏngyang	Currency	North Korean won
Languages	Korean	Organizations	UN
		Map page	65

NORWAY
Kingdom of Norway

Area Sq Km	323 878	Religions	Protestant, Roman
Area Sq Miles	125 050		Catholic
Population	4 812 000	Currency	Norwegian krone
Capital	Oslo	Organizations	OECD, UN
Languages	Norwegian	Map page	92–93

OMAN
Sultanate of Oman

Area Sq Km	309 500	Religions	Ibadhi Muslim,
Area Sq Miles	119 499		Sunni Muslim
Population	2 845 000	Currency	Omani riyal
Capital	Muscat (Masqat)	Organizations	UN
Languages	Arabic, Baluchi,	Map page	79
	Indian languages		

PAKISTAN
Islamic Republic of Pakistan

Area Sq Km	803 940	Religions	Sunni Muslim,
Area Sq Miles	310 403		Shi'a Muslim,
Population	180 808 000		Christian, Hindu
Capital	Islamabad	Currency	Pakistani rupee
Languages	Urdu, Punjabi,	Organizations	Comm., UN
	Sindhi, Pushtu,	Map page	74
	English		

PALAU
Republic of Palau

Area Sq Km	497	Religions	Roman Catholic,
Area Sq Miles	192		Protestant,
Population	20 000		traditional beliefs
Capital	Melekeok	Currency	United States dollar
Languages	Palauan, English	Organizations	UN
		Map page	59

PANAMA
Republic of Panama

Area Sq Km	77 082	Religions	Roman Catholic,
Area Sq Miles	29 762		Protestant, Sunni
Population	3 454 000		Muslim
Capital	Panama City	Currency	Balboa
Languages	Spanish, English,	Organizations	UN
	Amerindian	Map page	146
	languages		

PAPUA NEW GUINEA
Independent State of Papua New Guinea

Area Sq Km	462 840	Religions	Protestant,
Area Sq Miles	178 704		Roman Catholic,
Population	6 732 000		traditional beliefs
Capital	Port Moresby	Currency	Kina
Languages	English, Tok Pisin	Organizations	APEC, Comm., UN
	(creole), local	Map page	59
	languages		

PARAGUAY
Republic of Paraguay

Area Sq Km	406 752	Religions	Roman Catholic,
Area Sq Miles	157 048		Protestant
Population	6 349 000	Currency	Guaraní
Capital	Asunción	Organizations	UN
Languages	Spanish, Guaraní	Map page	152

PERU
Republic of Peru

Area Sq Km	1 285 216	Religions	Roman Catholic,
Area Sq Miles	496 225		Protestant
Population	29 165 000	Currency	Sol
Capital	Lima	Organizations	APEC, UN
Languages	Spanish, Quechua,	Map page	150
	Aymara		

PHILIPPINES
Republic of the Philippines

Area Sq Km	300 000	Religions	Roman Catholic,
Area Sq Miles	115 831		Protestant, Sunni
Population	91 983 000		Muslim, Aglipayan
Capital	Manila	Currency	Philippine peso
Languages	English, Filipino,	Organizations	APEC, ASEAN, UN
	Tagalog, Cebuano,	Map page	64
	local languages		

Pitcairn Islands
United Kingdom Overseas Territory

Area Sq Km	45	Religions	Protestant
Area Sq Miles	17	Currency	New Zealand
Population	66		dollar
Capital	Adamstown	Map page	49
Languages	English		

POLAND
Polish Republic

Area Sq Km	312 683	Religions	Roman Catholic,
Area Sq Miles	120 728		Polish Orthodox
Population	38 074 000	Currency	Złoty
Capital	Warsaw (Warszawa)	Organizations	EU, OECD, UN
Languages	Polish, German	Map page	103

PORTUGAL
Portuguese Republic

Area Sq Km	88 940	Religions	Roman Catholic,
Area Sq Miles	34 340		Protestant
Population	10 707 000	Currency	Euro
Capital	Lisbon (Lisboa)	Organizations	EU, OECD, UN
Languages	Portuguese	Map page	106

Puerto Rico
United States Commonwealth

Area Sq Km	9 104	Religions	Roman Catholic,
Area Sq Miles	3 515		Protestant
Population	3 982 000	Currency	United States
Capital	San Juan		dollar
Languages	Spanish, English	Map page	147

QATAR
State of Qatar

Area Sq Km	11 437	Religions	Sunni Muslim
Area Sq Miles	4 416	Currency	Qatari riyal
Population	1 409 000	Organizations	OPEC, UN
Capital	Doha (Ad Dawḩah)	Map page	79
Languages	Arabic		

Réunion
French Overseas Department

Area Sq Km	2 551	Religions	Roman Catholic
Area Sq Miles	985	Currency	Euro
Population	827 000	Map page	113
Capital	St-Denis		
Languages	French, creole		

Rodrigues Island
part of Mauritius

Area Sq Km	104	Religions	Christian
Area Sq Miles	40	Currency	Rupee
Population	37 699	Map page	159
Capital	Port Mathurin		
Languages	English, creole		

ROMANIA

Area Sq Km	237 500	Religions	Romanian Orthodox,
Area Sq Miles	91 699		Protestant,
Population	21 275 000		Roman Catholic
Capital	Bucharest (Bucureşti)	Currency	Romanian leu
Languages	Romanian, Hungarian	Organizations	EU, UN
		Map page	110

RUSSIAN FEDERATION

Area Sq Km	17 075 400	Religions	Russian Orthodox,
Area Sq Miles	6 592 849		Sunni Muslim,
Population	140 874 000		Protestant
Capital	Moscow (Moskva)	Currency	Russian rouble
Languages	Russian, Tatar, Ukrainian, local languages	Organizations	APEC, CIS, UN
		Map page	82–83

RWANDA
Republic of Rwanda

Area Sq Km	26 338	Religions	Roman Catholic,
Area Sq Miles	10 169		traditional beliefs,
Population	9 998 000		Protestant
Capital	Kigali	Currency	Rwandan franc
Languages	Kinyarwanda, French, English	Organizations	UN
		Map page	119

Saba
part of Netherlands Antilles

Area Sq Km	13	Religions	Roman Catholic,
Area Sq Miles	5		Protestant
Population	1 524	Currency	Netherlands
Capital	Bottom		Antilles guilder
Languages	Dutch, English	Map page	147

St-Barthélémy
French Overseas Collectivity

Area Sq Km	21	Religions	Roman Catholic
Area Sq Miles	8	Currency	Euro
Population	8 450	Map page	147
Capital	Gustavia		
Languages	French		

St Helena and Dependencies
United Kingdom Overseas Territory

Area Sq Km	121	Religions	Protestant, Roman Catholic,
Area Sq Miles	47		
Population	4 255	Currency	St Helena pound
Capital	Jamestown	Map page	113
Languages	English		

ST KITTS AND NEVIS
Federation of St Kitts and Nevis

Area Sq Km	261	Religions	Protestant, Roman Catholic
Area Sq Miles	101		
Population	52 000	Currency	East Caribbean dollar
Capital	Basseterre		
Languages	English, creole	Organizations	CARICOM, Comm., UN
		Map page	147

ST LUCIA

Area Sq Km	616	Religions	Roman Catholic, Protestant
Area Sq Miles	238		
Population	172 000	Currency	East Caribbean dollar
Capital	Castries		
Languages	English, creole	Organizations	CARICOM, Comm., UN
		Map page	147

St-Martin
French Overseas Collectivity

Area Sq Km	54	Religions	Roman Catholic
Area Sq Miles	21	Currency	Euro
Population	35 692	Map page	147
Capital	Marigot		
Languages	French		

St Pierre and Miquelon
French Territorial Collectivity

Area Sq Km	242	Religions	Roman Catholic
Area Sq Miles	93	Currency	Euro
Population	6 125	Map page	131
Capital	St-Pierre		
Languages	French		

ST VINCENT AND THE GRENADINES

Area Sq Km	389	Religions	Protestant, Roman Catholic
Area Sq Miles	150		
Population	109 000	Currency	East Caribbean dollar
Capital	Kingstown		
Languages	English, creole	Organizations	CARICOM, Comm., UN
		Map page	147

SAMOA
Independent State of Samoa

Area Sq Km	2 831	Religions	Protestant, Roman Catholic
Area Sq Miles	1 093		
Population	179 000	Currency	Tala
Capital	Apia	Organizations	Comm., UN
Languages	Samoan, English	Map page	49

SAN MARINO
Republic of San Marino

Area Sq Km	61	Religions	Roman Catholic
Area Sq Miles	24	Currency	Euro
Population	31 000	Organizations	UN
Capital	San Marino	Map page	108
Languages	Italian		

SÃO TOMÉ AND PRÍNCIPE
Democratic Republic of São Tomé and Príncipe

Area Sq Km	964	Religions	Roman Catholic,
Area Sq Miles	372		Protestant
Population	163 000	Currency	Dobra
Capital	São Tomé	Organizations	UN
Languages	Portuguese, creole	Map page	113

SAUDI ARABIA
Kingdom of Saudi Arabia

Area Sq Km	2 200 000	Religions	Sunni Muslim,
Area Sq Miles	849 425		Shi'a Muslim
Population	25 721 000	Currency	Saudi Arabian riyal
Capital	Riyadh (Ar Riyāḍ)	Organizations	OPEC, UN
Languages	Arabic	Map page	78–79

SENEGAL
Republic of Senegal

Area Sq Km	196 720	Religions	Sunni Muslim,
Area Sq Miles	75 954		Roman Catholic,
Population	12 534 000		traditional beliefs
Capital	Dakar	Currency	CFA franc
Languages	French, Wolof,	Organizations	UN
	Fulani, local	Map page	114
	languages		

SERBIA
Republic of Serbia

Area Sq Km	77 453	Religions	Roman Catholic,
Area Sq Miles	29 904		Serbian Orthodox,
Population	9 850 000		Sunni Muslim
Capital	Belgrade (Beograd)	Currency	Serbian dinar
Languages	Serbian, Hungarian	Organizations	UN
		Map page	109

SEYCHELLES
Republic of the Seychelles

Area Sq Km	455	Religions	Roman Catholic,
Area Sq Miles	176		Protestant
Population	84 000	Currency	Seychelles rupee
Capital	Victoria	Organizations	Comm., SADC, UN
Languages	English, French,	Map page	113
	creole		

SIERRA LEONE
Republic of Sierra Leone

Area Sq Km	71 740	Religions	Sunni Muslim,
Area Sq Miles	27 699		traditional beliefs
Population	5 696 000	Currency	Leone
Capital	Freetown	Organizations	Comm., UN
Languages	English, creole,	Map page	114
	Mende, Temne,		
	local languages		

SINGAPORE
Republic of Singapore

Area Sq Km	639	Religions	Buddhist, Taoist,
Area Sq Miles	247		Sunni Muslim,
Population	4 737 000		Christian, Hindu
Capital	Singapore	Currency	Singapore dollar
Languages	Chinese, English,	Organizations	APEC, ASEAN,
	Malay, Tamil		Comm., UN
		Map page	60

Sint Eustatius
part of Netherlands Antilles

Area Sq Km	21	Religions	Protestant, Roman
Area Sq Miles	8		Catholic
Population	2 754	Currency	Netherlands
Capital	Oranjestad		Antilles guilder
Languages	Dutch, English	Map page	147

Sint Maarten
part of Netherlands Antilles

Area Sq Km	34	Religions	Protestant, Roman
Area Sq Miles	13		Catholic
Population	40 007	Currency	Netherlands
Capital	Philipsburg		Antilles guilder
Languages	Dutch, English	Map page	147

SLOVAKIA
Slovak Republic

Area Sq Km	49 035	Religions	Roman Catholic,
Area Sq Miles	18 933		Protestant,
Population	5 406 000		Orthodox
Capital	Bratislava	Currency	Euro
Languages	Slovak,	Organizations	EU, OECD, UN
	Hungarian, Czech	Map page	103

SLOVENIA
Republic of Slovenia

Area Sq Km	20 251	Religions	Roman Catholic,
Area Sq Miles	7 819		Protestant
Population	2 020 000	Currency	Euro
Capital	Ljubljana	Organizations	EU, UN
Languages	Slovene, Croatian,	Map page	108–109
	Serbian		

SOLOMON ISLANDS

Area Sq Km	28 370	Religions	Protestant, Roman
Area Sq Miles	10 954		Catholic
Population	523 000	Currency	Solomon Islands
Capital	Honiara		dollar
Languages	English, creole,	Organizations	Comm., UN
	local languages	Map page	48

SOMALIA
Somali Republic

Area Sq Km	637 657	**Religions**	Sunni Muslim
Area Sq Miles	246 201	**Currency**	Somali shilling
Population	9 133 000	**Organizations**	UN
Capital	Mogadishu (Muqdisho)	**Map page**	117
Languages	Somali, Arabic		

SOUTH AFRICA, REPUBLIC OF

Area Sq Km	1 219 080	**Religions**	Protestant, Roman Catholic, Sunni Muslim, Hindu
Area Sq Miles	470 689		
Population	50 110 000		
Capital	Pretoria (Tshwane)/ Cape Town	**Currency**	Rand
		Organizations	Comm., SADC, UN
Languages	Afrikaans, English, nine official local languages	**Map page**	122–123

SOUTH KOREA
Republic of Korea

Area Sq Km	99 274	**Religions**	Buddhist, Protestant, Roman Catholic
Area Sq Miles	38 330		
Population	48 333 000		
Capital	Seoul (Sŏul)	**Currency**	South Korean won
Languages	Korean	**Organizations**	APEC, OECD, UN
		Map page	65

SPAIN
Kingdom of Spain

Area Sq Km	504 782	**Religions**	Roman Catholic
Area Sq Miles	194 897	**Currency**	Euro
Population	44 904 000	**Organizations**	EU, OECD, UN
Capital	Madrid	**Map page**	106–107
Languages	Spanish, Castilian, Catalan, Galician, Basque		

SRI LANKA
Democratic Socialist Republic of Sri Lanka

Area Sq Km	65 610	**Religions**	Buddhist, Hindu, Sunni Muslim, Roman Catholic
Area Sq Miles	25 332		
Population	20 238 000		
Capital	Sri Jayewardenepura Kotte	**Currency**	Sri Lankan rupee
		Organizations	Comm., UN
Languages	Sinhalese, Tamil, English	**Map page**	73

SUDAN
Republic of the Sudan

Area Sq Km	2 505 813	**Religions**	Sunni Muslim, traditional beliefs, Christian
Area Sq Miles	967 500		
Population	42 272 000		
Capital	Khartoum	**Currency**	Sudanese pound (Sudani)
Languages	Arabic, Dinka, Nubian, Beja, Nuer, local languages	**Organizations**	UN
		Map page	116–117

SURINAME
Republic of Suriname

Area Sq Km	163 820	**Religions**	Hindu, Roman Catholic, Protestant, Sunni Muslim
Area Sq Miles	63 251		
Population	520 000		
Capital	Paramaribo	**Currency**	Suriname guilder
Languages	Dutch, Surinamese, English, Hindi	**Organizations**	CARICOM, UN
		Map page	151

Svalbard
part of Norway

Area Sq Km	61 229	**Religions**	Protestant
Area Sq Miles	23 641	**Currency**	Norwegian krone
Population	2 449	**Map page**	82
Capital	Longyearbyen		
Languages	Norwegian		

SWAZILAND
Kingdom of Swaziland

Area Sq Km	17 364	**Religions**	Christian, traditional beliefs
Area Sq Miles	6 704		
Population	1 185 000	**Currency**	Emalangeni, South African rand
Capital	Mbabane		
Languages	Swazi, English	**Organizations**	Comm., SADC, UN
		Map page	123

SWEDEN
Kingdom of Sweden

Area Sq Km	449 964	**Religions**	Protestant, Roman Catholic
Area Sq Miles	173 732		
Population	9 249 000	**Currency**	Swedish krona
Capital	Stockholm	**Organizations**	EU, OECD, UN
Languages	Swedish	**Map page**	92–93

SWITZERLAND
Swiss Confederation

Area Sq Km	41 293	**Religions**	Roman Catholic, Protestant
Area Sq Miles	15 943		
Population	7 568 000	**Currency**	Swiss franc
Capital	Bern	**Organizations**	OECD, UN
Languages	German, French, Italian, Romansch	**Map page**	105

SYRIA
Syrian Arab Republic

Area Sq Km	185 180	**Religions**	Sunni Muslim, Shia Muslim, Christian
Area Sq Miles	71 498		
Population	21 906 000	**Currency**	Syrian pound
Capital	Damascus (Dimashq)	**Organizations**	UN
Languages	Arabic, Kurdish, Armenian	**Map page**	80

TAIWAN
Republic of China

Area Sq Km	36 179	**Religions**	Buddhist, Taoist, Confucian, Christian
Area Sq Miles	13 969		
Population	23 046 000		
Capital	T'aipei	**Currency**	Taiwan dollar
Languages	Mandarin, Min, Hakka, local languages	**Organizations**	APEC
		Map page	71

The People's Republic of China claims Taiwan as its 23rd province

TAJIKISTAN
Republic of Tajikistan

Area Sq Km	143 100	Religions	Sunni Muslim
Area Sq Miles	55 251	Currency	Somoni
Population	6 952 000	Organizations	CIS, UN
Capital	Dushanbe	Map page	77
Languages	Tajik, Uzbek, Russian		

TANZANIA
United Republic of Tanzania

Area Sq Km	945 087	Religions	Shi'a Muslim, Sunni
Area Sq Miles	364 900		Muslim, traditional
Population	43 739 000		beliefs, Christian
Capital	Dodoma	Currency	Tanzanian shilling
Languages	Swahili, English,	Organizations	Comm., SADC, UN
	Nyamwezi, local	Map page	119
	languages		

THAILAND
Kingdom of Thailand

Area Sq Km	513 115	Religions	Buddhist, Sunni
Area Sq Miles	198 115		Muslim
Population	67 764 000	Currency	Baht
Capital	Bangkok	Organizations	APEC, ASEAN, UN
	(Krung Thep)	Map page	62–63
Languages	Thai, Lao, Chinese,		
	Malay, Mon-Khmer		
	languages		

TOGO

Republic of Togo

Area Sq Km	56 785	Religions	Traditional beliefs,
Area Sq Miles	21 925		Christian, Sunni
Population	6 619 000		Muslim
Capital	Lomé	Currency	CFA franc
Languages	French, Ewe, Kabre,	Organizations	UN
	local languages	Map page	114

Tokelau
New Zealand Overseas Territory

Area Sq Km	10	Religions	Christian
Area Sq Miles	4	Currency	New Zealand dollar
Population	1 466	Map page	49
Capital	none		
Languages	English, Tokelauan		

TONGA
Kingdom of Tonga

Area Sq Km	748	Religions	Protestant, Roman
Area Sq Miles	289		Catholic
Population	104 000	Currency	Pa'anga
Capital	Nuku'alofa	Organizations	Comm., UN
Languages	Tongan, English	Map page	49

TRINIDAD AND TOBAGO
Republic of Trinidad and Tobago

Area Sq Km	5 130	Religions	Roman Catholic,
Area Sq Miles	1 981		Hindu, Protestant,
Population	1 339 000		Sunni Muslim
Capital	Port of Spain	Currency	Trinidad and
Languages	English, creole,		Tobago dollar
	Hindi	Organizations	CARICOM,
			Comm., UN
		Map page	147

Tristan da Cunha
Dependency of St Helena

Area Sq Km	98	Religions	Protestant, Roman
Area Sq Miles	38		Catholic
Population	264	Currency	Pound sterling
Capital	Settlement of	Map page	113
	Edinburgh		
Languages	English		

TUNISIA
Tunisian Republic

Area Sq Km	164 150	Religions	Sunni Muslim
Area Sq Miles	63 379	Currency	Tunisian dinar
Population	10 272 000	Organizations	UN
Capital	Tunis	Map page	115
Languages	Arabic, French		

TURKEY
Republic of Turkey

Area Sq Km	779 452	Religions	Sunni Muslim,
Area Sq Miles	300 948		Shi'a Muslim
Population	74 816 000	Currency	Lira
Capital	Ankara	Organizations	OECD, UN
Languages	Turkish, Kurdish	Map page	80

TURKMENISTAN
Republic of Turkmenistan

Area Sq Km	488 100	Religions	Sunni Muslim,
Area Sq Miles	188 456		Russian Orthodox
Population	5 110 000	Currency	Turkmen manat
Capital	Aşgabat (Ashkhabad)	Organizations	UN
Languages	Turkmen, Uzbek,	Map page	76
	Russian		

Turks and Caicos Islands
United Kingdom Overseas Territory

Area Sq Km	430	Religions	Protestant
Area Sq Miles	166	Currency	United States
Population	33 000		dollar
Capital	Grand Turk	Map page	147
	(Cockburn Town)		
Languages	English		

TUVALU

Area Sq Km	25	Religions	Protestant
Area Sq Miles	10	Currency	Australian dollar
Population	10 000	Organizations	Comm., UN
Capital	Vaiaku	Map page	49
Languages	Tuvaluan, English		

UGANDA
Republic of Uganda

Area Sq Km	241 038	**Religions**	Roman Catholic,
Area Sq Miles	93 065		Protestant, Sunni
Population	32 710 000		Muslim, traditional
Capital	Kampala		beliefs
Languages	English, Swahili,	**Currency**	Ugandan shilling
	Luganda, local	**Organizations**	Comm., UN
	languages	**Map page**	119

UKRAINE
Republic of Ukraine

Area Sq Km	603 700	**Religions**	Ukrainian
Area Sq Miles	233 090		Orthodox,
Population	45 708 000		Ukrainian Catholic,
Capital	Kiev (Kyiv)		Roman Catholic
Languages	Ukrainian, Russian	**Currency**	Hryvnia
		Organizations	CIS, UN
		Map page	90–91

UNITED ARAB EMIRATES
Federation of Emirates

Area Sq Km	77 700	**Religions**	Sunni Muslim,
Area Sq Miles	30 000		Shi'a Muslim
Population	4 599 000	**Currency**	United Arab
Capital	Abu Dhabi		Emirates dirham
	(Abū Ẓabī)	**Organizations**	OPEC, UN
Languages	Arabic, English	**Map page**	79

Abu Dhabi (Abū Ẓabī) (Emirate)

Area Sq Km	67 340	**Population**	1 559 000
Area Sq Miles	26 000	**Capital**	Abu Dhabi
			(Abū Ẓabī)

Ajman (Emirate)

Area Sq Km	259	**Population**	237 000
Area Sq Miles	100	**Capital**	Ajman

Dubai (Emirate)

Area Sq Km	3 885	**Population**	1 596 000
Area Sq Miles	1 500	**Capital**	Dubai

Fujairah (Emirate)

Area Sq Km	1 165	**Population**	143 000
Area Sq Miles	450	**Capital**	Fujairah

Ra's al Khaymah (Emirate)

Area Sq Km	1 684	**Population**	231 000
Area Sq Miles	650	**Capital**	Ra's al Khaymah

Sharjah (Emirate)

Area Sq Km	2 590	**Population**	946 000
Area Sq Miles	1 000	**Capital**	Sharjah

Umm al Qaywayn (Emirate)

Area Sq Km	777	**Population**	53 000
Area Sq Miles	300	**Capital**	Umm al Qaywayn

UNITED KINGDOM
of Great Britain and Northern Ireland

Area Sq Km	243 609	**Religions**	Protestant, Roman
Area Sq Miles	94 058		Catholic, Muslim
Population	61 565 000	**Currency**	Pound sterling
Capital	London	**Organizations**	Comm., EU, OECD,
Languages	English, Welsh,		UN
	Gaelic	**Map page**	94–95

England (Constituent country)

Area Sq Km	130 433	**Population**	51 092 000
Area Sq Miles	50 360	**Capital**	London

Northern Ireland (Province)

Area Sq Km	13 576	**Population**	1 759 000
Area Sq Miles	5 242	**Capital**	Belfast

Scotland (Constituent country)

Area Sq Km	78 822	**Population**	5 144 200
Area Sq Miles	30 433	**Capital**	Edinburgh

Wales (Principality)

Area Sq Km	20 778	**Population**	2 980 000
Area Sq Miles	8 022	**Capital**	Cardiff

UNITED STATES OF AMERICA
Federal Republic

Area Sq Km	9 826 635	**Religions**	Protestant, Roman
Area Sq Miles	3 794 085		Catholic, Sunni
Population	314 659 000		Muslim, Jewish
Capital	Washington D.C.	**Currency**	United States dollar
Languages	English, Spanish	**Organizations**	APEC, OECD, UN
		Map page	132–133

Alabama (State)

Area Sq Km	135 765	**Population**	4 661 900
Area Sq Miles	52 419	**Capital**	Montgomery

Alaska (State)

Area Sq Km	1 717 854	**Population**	686 293
Area Sq Miles	663 267	**Capital**	Juneau

Arizona (State)

Area Sq Km	295 253	**Population**	6 500 180
Area Sq Miles	113 998	**Capital**	Phoenix

Arkansas (State)

Area Sq Km	137 733	**Population**	2 855 390
Area Sq Miles	53 179	**Capital**	Little Rock

California (State)

Area Sq Km	423 971	**Population**	36 756 666
Area Sq Miles	163 696	**Capital**	Sacramento

olorado (State)

Area Sq Km	269 602	Population	4 939 456
rea Sq Miles	104 094	Capital	Denver

onnecticut (State)

Area Sq Km	14 356	Population	3 501 252
rea Sq Miles	5 543	Capital	Hartford

elaware (State)

Area Sq Km	6 446	Population	873 092
rea Sq Miles	2 489	Capital	Dover

istrict of Columbia (District)

Area Sq Km	176	Population	591 833
rea Sq Miles	68	Capital	Washington

lorida (State)

Area Sq Km	170 305	Population	18 328 340
rea Sq Miles	65 755	Capital	Tallahassee

eorgia (State)

Area Sq Km	153 910	Population	9 685 744
rea Sq Miles	59 425	Capital	Atlanta

awaii (State)

Area Sq Km	28 311	Population	1 288 198
rea Sq Miles	10 931	Capital	Honolulu

daho (State)

Area Sq Km	216 445	Population	1 523 816
rea Sq Miles	83 570	Capital	Boise

linois (State)

Area Sq Km	149 997	Population	12 901 563
rea Sq Miles	57 914	Capital	Springfield

ndiana (State)

Area Sq Km	94 322	Population	6 376 792
rea Sq Miles	36 418	Capital	Indianapolis

owa (State)

Area Sq Km	145 744	Population	3 002 555
rea Sq Miles	56 272	Capital	Des Moines

ansas (State)

Area Sq Km	213 096	Population	2 802 134
rea Sq Miles	82 277	Capital	Topeka

entucky (State)

Area Sq Km	104 659	Population	4 269 245
rea Sq Miles	40 409	Capital	Frankfort

Louisiana (State)

Area Sq Km	134 265	Population	4 410 796
Area Sq Miles	51 840	Capital	Baton Rouge

Maine (State)

Area Sq Km	91 647	Population	1 316 456
Area Sq Miles	35 385	Capital	Augusta

Maryland (State)

Area Sq Km	32 134	Population	5 633 597
Area Sq Miles	12 407	Capital	Annapolis

Massachusetts (State)

Area Sq Km	27 337	Population	6 497 967
Area Sq Miles	10 555	Capital	Boston

Michigan (State)

Area Sq Km	250 493	Population	10 003 422
Area Sq Miles	96 716	Capital	Lansing

Minnesota (State)

Area Sq Km	225 171	Population	5 220 393
Area Sq Miles	86 939	Capital	St Paul

Mississippi (State)

Area Sq Km	125 433	Population	2 938 618
Area Sq Miles	48 430	Capital	Jackson

Missouri (State)

Area Sq Km	180 533	Population	5 911 605
Area Sq Miles	69 704	Capital	Jefferson City

Montana (State)

Area Sq Km	380 837	Population	967 440
Area Sq Miles	147 042	Capital	Helena

Nebraska (State)

Area Sq Km	200 346	Population	1 783 432
Area Sq Miles	77 354	Capital	Lincoln

Nevada (State)

Area Sq Km	286 352	Population	2 600 167
Area Sq Miles	110 561	Capital	Carson City

New Hampshire (State)

Area Sq Km	24 216	Population	1 315 809
Area Sq Miles	9 350	Capital	Concord

New Jersey (State)

Area Sq Km	22 587	Population	8 682 661
Area Sq Miles	8 721	Capital	Trenton

UNITED STATES OF AMERICA
Federal Republic

New Mexico (State)
Area Sq Km	314 914	**Population**	1 984 356
Area Sq Miles	121 589	**Capital**	Santa Fe

New York (State)
Area Sq Km	141 299	**Population**	19 490 297
Area Sq Miles	54 556	**Capital**	Albany

North Carolina (State)
Area Sq Km	139 391	**Population**	9 222 414
Area Sq Miles	53 819	**Capital**	Raleigh

North Dakota (State)
Area Sq Km	183 112	**Population**	641 481
Area Sq Miles	70 700	**Capital**	Bismarck

Ohio (State)
Area Sq Km	116 096	**Population**	11 485 910
Area Sq Miles	44 825	**Capital**	Columbus

Oklahoma (State)
Area Sq Km	181 035	**Population**	3 642 361
Area Sq Miles	69 898	**Capital**	Oklahoma City

Oregon (State)
Area Sq Km	254 806	**Population**	3 790 060
Area Sq Miles	98 381	**Capital**	Salem

Pennsylvania (State)
Area Sq Km	119 282	**Population**	12 448 279
Area Sq Miles	46 055	**Capital**	Harrisburg

Rhode Island (State)
Area Sq Km	4 002	**Population**	1 050 788
Area Sq Miles	1 545	**Capital**	Providence

South Carolina (State)
Area Sq Km	82 931	**Population**	4 479 800
Area Sq Miles	32 020	**Capital**	Columbia

South Dakota (State)
Area Sq Km	199 730	**Population**	804 194
Area Sq Miles	77 116	**Capital**	Pierre

Tennessee (State)
Area Sq Km	109 150	**Population**	6 214 888
Area Sq Miles	42 143	**Capital**	Nashville

Texas (State)
Area Sq Km	695 622	**Population**	24 326 974
Area Sq Miles	268 581	**Capital**	Austin

Utah (State)
Area Sq Km	219 887	**Population**	2 736 424
Area Sq Miles	84 899	**Capital**	Salt Lake City

Vermont (State)
Area Sq Km	24 900	**Population**	621 270
Area Sq Miles	9 614	**Capital**	Montpelier

Virginia (State)
Area Sq Km	110 784	**Population**	7 769 089
Area Sq Miles	42 774	**Capital**	Richmond

Washington (State)
Area Sq Km	184 666	**Population**	6 549 224
Area Sq Miles	71 300	**Capital**	Olympia

West Virginia (State)
Area Sq Km	62 755	**Population**	1 814 468
Area Sq Miles	24 230	**Capital**	Charleston

Wisconsin (State)
Area Sq Km	169 639	**Population**	5 627 967
Area Sq Miles	65 498	**Capital**	Madison

Wyoming (State)
Area Sq Km	253 337	**Population**	532 668
Area Sq Miles	97 814	**Capital**	Cheyenne

URUGUAY
Oriental Republic of Uruguay

Area Sq Km	176 215	**Religions**	Roman Catholic,
Area Sq Miles	68 037		Protestant, Jewish
Population	3 361 000	**Currency**	Uruguayan peso
Capital	Montevideo	**Organizations**	UN
Languages	Spanish	**Map page**	153

UZBEKISTAN
Republic of Uzbekistan

Area Sq Km	447 400	**Religions**	Sunni Muslim,
Area Sq Miles	172 742		Russian Orthodox
Population	27 488 000	**Currency**	Uzbek som
Capital	Tashkent	**Organizations**	CIS, UN
Languages	Uzbek, Russian,	**Map page**	76–77
	Tajik, Kazakh		

VANUATU
Republic of Vanuatu

Area Sq Km	12 190	**Religions**	Protestant,
Area Sq Miles	4 707		Roman Catholic,
Population	240 000		traditional beliefs
Capital	Port Vila	**Currency**	Vatu
Languages	English, Bislama	**Organizations**	Comm., UN
	(creole), French	**Map page**	48

VATICAN CITY
Vatican City State or Holy See

Area Sq Km	0.5	**Religions**	Roman Catholic
Area Sq Miles	0.2	**Currency**	Euro
Population	557	**Map page**	108
Capital	Vatican City		
Languages	Italian		

VENEZUELA
Bolivarian Republic of Venezuela

Area Sq Km	912 050	**Religions**	Roman Catholic,
Area Sq Miles	352 144		Protestant
Population	28 583 000	**Currency**	Bolívar fuerte
Capital	Caracas	**Organizations**	OPEC, UN
Languages	Spanish, Amerindian	**Map page**	150
	languages		

VIETNAM
Socialist Republic of Vietnam

Area Sq Km	329 565	**Religions**	Buddhist, Taoist,
Area Sq Miles	127 246		Roman Catholic,
Population	88 069 000		Cao Dai, Hoa Hoa
Capital	Ha Nôi (Hanoi)	**Currency**	Dong
Languages	Vietnamese, Thai,	**Organizations**	APEC, ASEAN, UN
	Khmer, Chinese,	**Map page**	62–63
	local languages		

Virgin Islands (U.K.)
United Kingdom Overseas Territory

Area Sq Km	153	**Religions**	Protestant, Roman
Area Sq Miles	59		Catholic
Population	23 000	**Currency**	United States dollar
Capital	Road Town	**Map page**	147
Languages	English		

Virgin Islands (U.S.)
United States Unincorporated Territory

Area Sq Km	352	**Religions**	Protestant,
Area Sq Miles	136		Roman Catholic
Population	110 000	**Currency**	United States dollar
Capital	Charlotte Amalie	**Map page**	147
Languages	English, Spanish		

Wallis and Futuna Islands
French Overseas Collectivity

Area Sq Km	274	**Religions**	Roman Catholic
Area Sq Miles	106	**Currency**	CFP franc
Population	15 000	**Map page**	49
Capital	Matā'utu		
Languages	French, Wallisian,		
	Futunian		

West Bank
Disputed Territory

Area Sq Km	5 860	**Religions**	Sunni Muslim,
Area Sq Miles	2 263		Jewish,
Population	2 448 433		Shi'a Muslim,
Capital	none		Christian
Languages	Arabic, Hebrew	**Currency**	Jordanian dinar,
			Israeli shekel
		Map page	80

Western Sahara
Disputed Territory (Morocco)

Area Sq Km	266 000	**Religions**	Sunni Muslim
Area Sq Miles	102 703	**Currency**	Moroccan dirham
Population	513 000	**Map page**	114
Capital	Laâyoune		
Languages	Arabic		

YEMEN
Republic of Yemen

Area Sq Km	527 968	**Religions**	Sunni Muslim,
Area Sq Miles	203 850		Shi'a Muslim
Population	23 580 000	**Currency**	Yemeni riyal
Capital	Şan'ā'	**Organizations**	UN
Languages	Arabic	**Map page**	78–79

ZAMBIA
Republic of Zambia

Area Sq Km	752 614	**Religions**	Christian,
Area Sq Miles	290 586		traditional beliefs
Population	12 935 000	**Currency**	Zambian kwacha
Capital	Lusaka	**Organizations**	Comm., SADC, UN
Languages	English, Bemba,	**Map page**	120–121
	Nyanja, Tonga,		
	local languages		

ZIMBABWE
Republic of Zimbabwe

Area Sq Km	390 759	**Religions**	Christian,
Area Sq Miles	150 873		traditional beliefs
Population	12 523 000	**Currency**	Zimbabwean dollar
Capital	Harare	**Organizations**	SADC, UN
Languages	English, Shona,	**Map page**	121
	Ndebele		

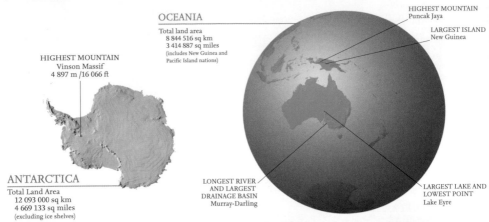

OCEANIA

Total land area
8 844 516 sq km
3 414 887 sq miles
(includes New Guinea and
Pacific Island nations)

HIGHEST MOUNTAIN
Puncak Jaya

LARGEST ISLAND
New Guinea

HIGHEST MOUNTAIN
Vinson Massif
4 897 m /16 066 ft

ANTARCTICA

Total Land Area
12 093 000 sq km
4 669 133 sq miles
(excluding ice shelves)

LONGEST RIVER
AND LARGEST
DRAINAGE BASIN
Murray-Darling

LARGEST LAKE AND
LOWEST POINT
Lake Eyre

HIGHEST MOUNTAINS	metres	feet	LARGEST ISLANDS	sq km	sq miles	LARGEST LAKES	sq km	sq miles	LONGEST RIVERS	km	miles
Puncak Jaya	5 030	16 502	New Guinea	808 510	312 167	Lake Eyre	0–8 900	0–3 436	Murray-Darling	3 672	2 282
Puncak Trikora	4 730	15 518	South Island	151 215	58 384	Lake Torrens	0–5 780	0–2 232	Darling	2 844	1 767
Puncak Mandala	4 700	15 420	North Island	115 777	44 701				Murray	2 375	1 476
Puncak Yamin	4 595	15 075	Tasmania	67 800	26 178				Murrumbidgee	1 485	923
Mt Wilhelm	4 509	14 793							Lachlan	1 339	832
Mt Kubor	4 359	14 301							Cooper Creek	1 113	692

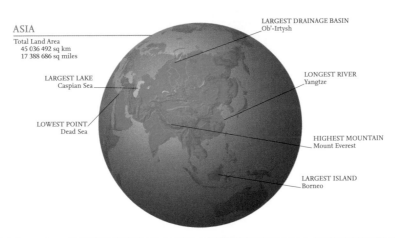

ASIA

Total Land Area
45 036 492 sq km
17 388 686 sq miles

LARGEST DRAINAGE BASIN
Ob'-Irtysh

LARGEST LAKE
Caspian Sea

LONGEST RIVER
Yangtze

LOWEST POINT
Dead Sea

HIGHEST MOUNTAIN
Mount Everest

LARGEST ISLAND
Borneo

HIGHEST MOUNTAINS	metres	feet	LARGEST ISLANDS	sq km	sq miles	LARGEST LAKES	sq km	sq miles	LONGEST RIVERS	km	miles
Mt Everest	8 848	29 028	Borneo	745 561	287 861	Caspian Sea	371 000	143 243	Yangtze	6 380	3 965
K2	8 611	28 251	Sumatra	473 606	182 859	Lake Baikal	30 500	11 776	Ob'-Irtysh	5 568	3 460
Kangchenjunga	8 586	28 169	Honshū	227 414	87 805	Lake Balkhash	17 400	6 718	Yenisey-Angara-Selenga	5 550	3 449
Lhotse	8 516	27 939	Celebes	189 216	73 056	Aral Sea	17 158	6 625	Yellow	5 464	3 395
Makalu	8 463	27 765	Java	132 188	51 038	Ysyk-Köl	6 200	2 394	Irtysh	4 440	2 759
Cho Oyu	8 201	26 906	Luzon	104 690	40 421						

EUROPE

Total Land Area
9 908 599 sq km
3 825 731 sq miles

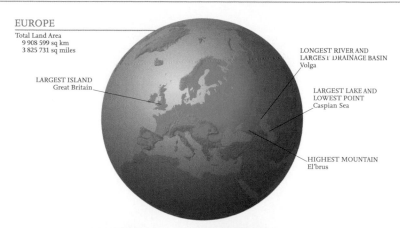

LARGEST ISLAND
Great Britain

LONGEST RIVER AND
LARGEST DRAINAGE BASIN
Volga

LARGEST LAKE AND
LOWEST POINT
Caspian Sea

HIGHEST MOUNTAIN
El'brus

HIGHEST MOUNTAINS	metres	feet	LARGEST ISLANDS	sq km	sq miles	LARGEST LAKES	sq km	sq miles	LONGEST RIVERS	km	miles
El'brus	5 642	18 510	Great Britain	218 476	84 354	Caspian Sea	371 000	143 243	Volga	3 688	2 292
Gora Dykh-Tau	5 204	17 073	Iceland	102 820	39 699	Lake Ladoga	18 390	7 100	Danube	2 850	1 771
Shkhara	5 201	17 063	Novaya Zemlya	90 650	35 000	Lake Onega	9 600	3 707	Dnieper	2 285	1 420
Kazbek	5 047	16 558	Ireland	83 045	32 064	Vänern	5 585	2 156	Kama	2 028	1 260
Mont Blanc	4 808	15 774	Spitsbergen	37 814	14 600	Rybinskoye Vodokhranilishche	5 180	2 000	Don	1 931	1 200
Dufourspitze	4 634	15 203	Sicily (Sicilia)	25 426	9 817				Pechora	1 802	1 120

AFRICA

Total Land Area
30 343 578 sq km
11 715 721 sq miles

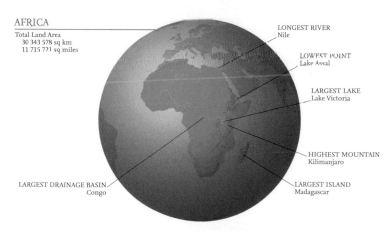

LONGEST RIVER
Nile

LOWEST POINT
Lake Assal

LARGEST LAKE
Lake Victoria

HIGHEST MOUNTAIN
Kilimanjaro

LARGEST DRAINAGE BASIN
Congo

LARGEST ISLAND
Madagascar

HIGHEST MOUNTAINS	metres	feet	LARGEST ISLANDS	sq km	sq miles	LARGEST LAKES	sq km	sq miles	LONGEST RIVERS	km	miles
Kilimanjaro	5 892	19 330	Madagascar	587 040	226 656	Lake Victoria	68 870	26 591	Nile	6 695	4 160
Mt Kenya	5 199	17 057				Lake Tanganyika	32 600	12 587	Congo	4 667	2 900
Margherita Peak	5 110	16 765				Lake Nyasa	29 500	11 390	Niger	4 184	2 600
Meru	4 565	14 977				Lake Volta	8 482	3 275	Zambezi	2 736	1 700
Ras Dejen	4 533	14 872				Lake Turkana	6 500	2 510	Webi Shabeelle	2 490	1 547
Mt Karisimbi	4 510	14 796				Lake Albert	5 600	2 162	Ubangi	2 250	1 398

NORTH AMERICA

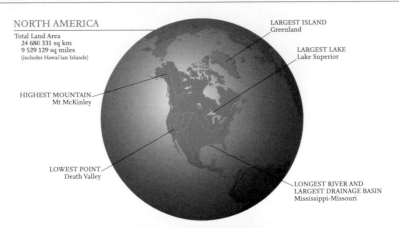

Total Land Area
24 680 331 sq km
9 529 129 sq miles
(includes Hawai'ian Islands)

LARGEST ISLAND
Greenland

LARGEST LAKE
Lake Superior

HIGHEST MOUNTAIN
Mt McKinley

LOWEST POINT
Death Valley

LONGEST RIVER AND
LARGEST DRAINAGE BASIN
Mississippi-Missouri

HIGHEST MOUNTAINS	metres	feet	LARGEST ISLANDS	sq km	sq miles	LARGEST LAKES	sq km	sq miles	LONGEST RIVERS	km	miles
Mt McKinley	6 194	20 321	Greenland	2 175 600	839 999	Lake Superior	82 100	31 699	Mississippi-Missouri	5 969	3 709
Mt Logan	5 959	19 550	Baffin Island	507 451	195 927	Lake Huron	59 600	23 012	Mackenzie-Peace-Finlay	4 241	2 635
Pico de Orizaba	5 610	18 405	Victoria Island	217 291	83 896	Lake Michigan	57 800	22 317	Missouri	4 086	2 539
Mt St Elias	5 489	18 008	Ellesmere Island	196 236	75 767	Great Bear Lake	31 328	12 096	Mississippi	3 765	2 340
Volcán Popocatépetl	5 452	17 887	Cuba	110 860	42 803	Great Slave Lake	28 568	11 030	Yukon	3 185	1 979
			Newfoundland	108 860	42 031	Lake Erie	25 700	9 923			

SOUTH AMERICA

Total Land Area
17 815 420 sq km
6 878 572 sq miles

LONGEST RIVER AND
LARGEST DRAINAGE BASIN
Amazon

LARGEST LAKE
Lake Titicaca

HIGHEST MOUNTAIN
Cerro Aconcagua

LOWEST POINT
Laguna del Carbón

LARGEST ISLAND
Isla Grande de Tierra del Fuego

HIGHEST MOUNTAINS	metres	feet	LARGEST ISLANDS	sq km	sq miles	LARGEST LAKES	sq km	sq miles	LONGEST RIVERS	km	miles
Cerro Aconcagua	6 959	22 831	Isla Grande de Tierra del Fuego	47 000	18 147	Lake Titicaca	8 340	3 220	Amazon	6 516	4 049
Nevado Ojos del Salado	6 908	22 664	Isla de Chiloé	8 394	3 241				Río de la Plata-Paraná	4 500	2 796
Cerro Bonete	6 872	22 546	East Falkland	6 760	2 610				Purus	3 218	2 000
Cerro Pissis	6 858	22 500	West Falkland	5 413	2 090				Madeira	3 200	1 988
Cerro Tupungato	6 800	22 309							São Francisco	2 900	1 802

CONTINENTS AND OCEANS

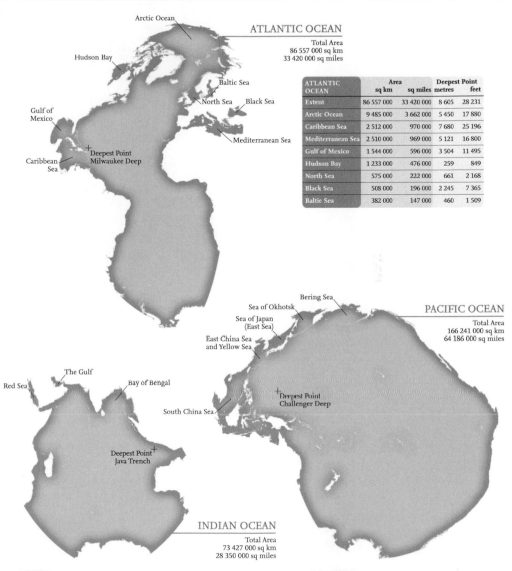

ATLANTIC OCEAN
Total Area
86 557 000 sq km
33 420 000 sq miles

ATLANTIC OCEAN	Area sq km	sq miles	Deepest Point metres	feet
Extent	86 557 000	33 420 000	8 605	28 231
Arctic Ocean	9 485 000	3 662 000	5 450	17 880
Caribbean Sea	2 512 000	970 000	7 680	25 196
Mediterranean Sea	2 510 000	969 000	5 121	16 800
Gulf of Mexico	1 544 000	596 000	3 504	11 495
Hudson Bay	1 233 000	476 000	259	849
North Sea	575 000	222 000	661	2 168
Black Sea	508 000	196 000	2 245	7 365
Baltic Sea	382 000	147 000	460	1 509

PACIFIC OCEAN
Total Area
166 241 000 sq km
64 186 000 sq miles

INDIAN OCEAN
Total Area
73 427 000 sq km
28 350 000 sq miles

INDIAN OCEAN	Area sq km	sq miles	Deepest Point metres	feet
Extent	73 427 000	28 350 000	7 125	23 376
Bay of Bengal	2 172 000	839 000	4 500	14 763
Red Sea	453 000	175 000	3 040	9 973
The Gulf	238 000	92 000	73	239

PACIFIC OCEAN	Area sq km	sq miles	Deepest Point metres	feet
Extent	166 241 000	64 186 000	10 920	35 826
South China Sea	2 590 000	1 000 000	5 514	18 090
Bering Sea	2 261 000	873 000	4 150	13 615
Sea of Okhotsk	1 392 000	537 000	3 363	11 033
Sea of Japan (East Sea)	1 013 000	391 000	3 743	12 280
East China Sea and Yellow Sea	1 202 000	464 000	2 717	8 913

© Collins Bartholomew Ltd

MAJOR CLIMATIC REGIONS AND SUB-TYPES 2006

Winkel Tripel Projection
1:155 000 000

Köppen classification system

Polar

| EF | Ice cap |
| ET | Tundra |

Dry

| BS | Steppe |
| BW | Desert |

Cooler humid

Dc Dd	Subarctic
Db	Continental cool summer
Da	Continental warm summer

Tropical humid

| Aw As | Savanna |
| Af Am | Rain forest |

Warmer humid

Cb Cc	Temperate
Ca	Humid subtropical
Cs	Mediterranean

o Weather extreme location

A Rainy climate with no winter: coolest month above 18°C (64.4°F).

B Dry climates; limits are defined by formulae based on rainfall effectiveness:
  BS Steppe or semi-arid climate.
  BW Desert or arid climate.

*C Rainy climates with mild winters: coolest month above 0°C (32°F), but below 18°C (64.4°F); warmest month above 10°C (50°F).

*D Rainy climates with severe winters: coldest month below 0°C (32°F); warmest month above 10°C (50°F).

E Polar climates with no warm season: warmest month below 10°C (50°F).
  ET Tundra climate: warmest month below 10°C (50°F) but above 0°C (32°F).
  EF Perpetual frost: all months below 0°C (32°F).

a Warmest month above 22°C (71.6°F).
b Warmest month below 22°C (71.6°F).
c Less than four months over 10°C (50°F).
d As 'c', but with severe cold: coldest month below -38°C (-36.4°F).
f Constantly moist rainfall throughout the year.
*h Warmer dry: all months above 0°C (32°F).
*k Cooler dry: at least one month below 0°C (32°F).
m Monsoon rain: short dry season, but is compensated by heavy rains during rest of the year.
n Frequent fog.
s Dry season in summer.
w Dry season in winter.

* Modification of Köppen definition

TRACKS OF TROPICAL STORMS

(wind speeds often over 160 km per hour)
1:300 000 000

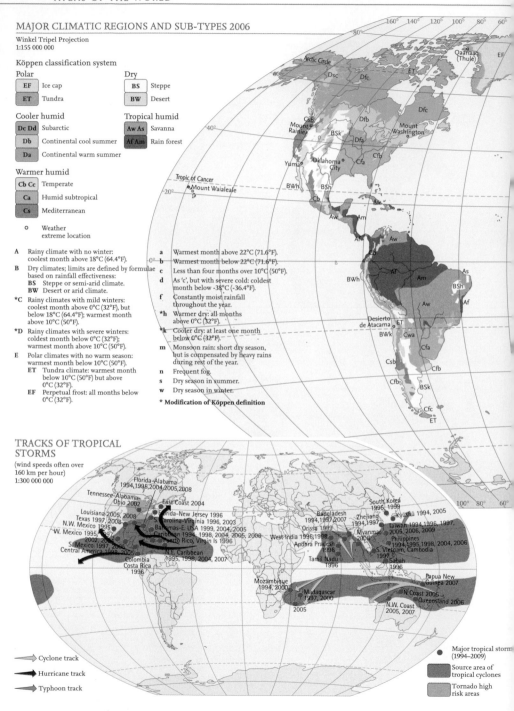

→ Cyclone track

→ Hurricane track

→ Typhoon track

● Major tropical storm (1994–2009)

Source area of tropical cyclones

Tornado high risk areas

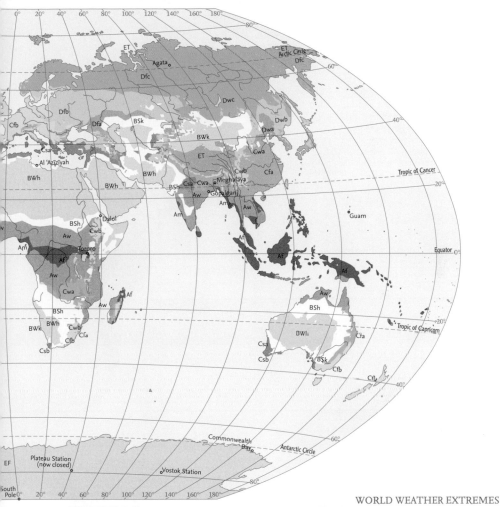

WORLD WEATHER EXTREMES

	Location		Location
Highest shade temperature	57.8°C/136°F Al ´Azīzīyah, Libya (13th September 1922)	**Highest surface wind speed**	
		High altitude	372 km per hour/231 miles per hour Mount Washington, New Hampshire, USA (12th April 1934)
Hottest place — Annual mean	34.4°C/93.9°F Dalol, Ethiopia		
Driest place — Annual mean	0.1 mm/0.004 inches Atacama Desert, Chile	Low altitude	333 km per hour/207 miles per hour Qaanaaq (Thule), Greenland (8th March 1972)
Most sunshine — Annual mean	90% Yuma, Arizona, USA (over 4 000 hours)	Tornado	512 km per hour/318 miles per hour Oklahoma City, Oklahoma, USA (3rd May 1999)
Least sunshine	Nil for 182 days each year, South Pole		
Lowest screen temperature	-89.2°C/-128.6°F Vostok Station, Antarctica (21st July 1983)	Greatest snowfall	31 102 mm/1 224.5 inches Mount Rainier, Washington, USA (19th February 1971 — 18th February 1972)
Coldest place — Annual mean	-56.6°C/-69.9°F Plateau Station, Antarctica		
Wettest place — Annual mean	11 873 mm/467.4 inches Meghalaya, India	Heaviest hailstones	1 kg/2.21 lb Gopalganj, Bangladesh (14th April 1986)
Most rainy days	Up to 350 per year Mount Waialeale, Hawaii, USA	Thunder-days average	251 days per year Tororo, Uganda
Windiest place	322 km per hour/200 miles per hour in gales, Commonwealth Bay, Antarctica	Highest barometric pressure	1 083.8 mb Agata, Siberia, Rus. Fed. (31st December 1968)
		Lowest barometric pressure	870 mb 483 km/300 miles west of Guam, Pacific Ocean (12th October 1979)

WORLD LAND COVER

Winkel Tripel Projection
1:155 000 000

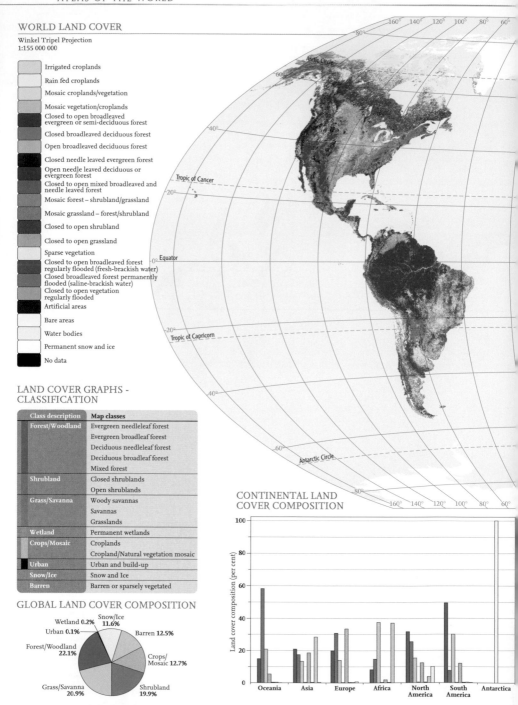

Irrigated croplands

Rain fed croplands

Mosaic croplands/vegetation

Mosaic vegetation/croplands

Closed to open broadleaved evergreen or semi-deciduous forest

Closed broadleaved deciduous forest

Open broadleaved deciduous forest

Closed needle leaved evergreen forest

Open needle leaved deciduous or evergreen forest

Closed to open mixed broadleaved and needle leaved forest

Mosaic forest – shrubland/grassland

Mosaic grassland – forest/shrubland

Closed to open shrubland

Closed to open grassland

Sparse vegetation

Closed to open broadleaved forest regularly flooded (fresh-brackish water)

Closed broadleaved forest permanently flooded (saline-brackish water)

Closed to open vegetation regularly flooded

Artificial areas

Bare areas

Water bodies

Permanent snow and ice

No data

LAND COVER GRAPHS - CLASSIFICATION

Class description	Map classes
Forest/Woodland	Evergreen needleleaf forest
	Evergreen broadleaf forest
	Deciduous needleleaf forest
	Deciduous broadleaf forest
	Mixed forest
Shrubland	Closed shrublands
	Open shrublands
Grass/Savanna	Woody savannas
	Savannas
	Grasslands
Wetland	Permanent wetlands
Crops/Mosaic	Croplands
	Cropland/Natural vegetation mosaic
Urban	Urban and build-up
Snow/Ice	Snow and Ice
Barren	Barren or sparsely vegetated

GLOBAL LAND COVER COMPOSITION

Wetland 0.2%
Urban 0.1%
Snow/Ice 11.6%
Barren 12.5%
Crops/Mosaic 12.7%
Shrubland 19.9%
Grass/Savanna 20.9%
Forest/Woodland 22.1%

CONTINENTAL LAND COVER COMPOSITION

Land cover composition (per cent)

Oceania, Asia, Europe, Africa, North America, South America, Antarctica

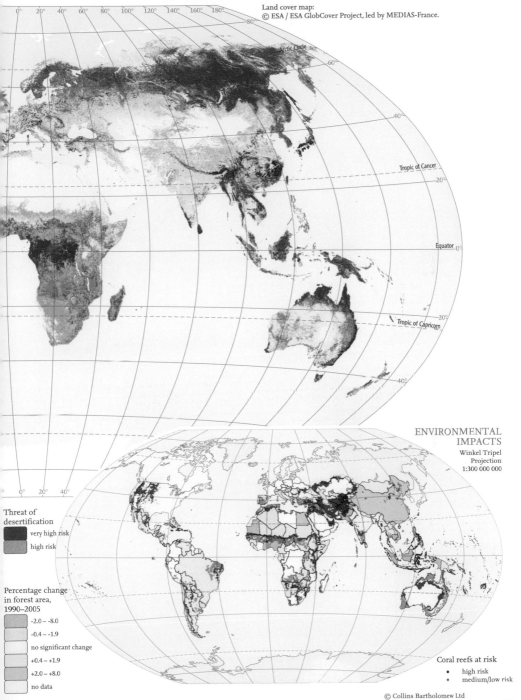

Land cover map:
© ESA / ESA GlobCover Project, led by MEDIAS-France.

Arctic Circle

Tropic of Cancer

Equator

Tropic of Capricorn

ENVIRONMENTAL
IMPACTS
Winkel Tripel
Projection
1:300 000 000

Threat of
desertification

very high risk

high risk

Percentage change
in forest area,
1990–2005

-2.0 – -8.0

-0.4 – -1.9

no significant change

+0.4 – +1.9

+2.0 – +8.0

no data

Coral reefs at risk

• high risk
• medium/low risk

© Collins Bartholomew Ltd

WORLD POPULATION DISTRIBUTION AND THE WORLD'S MAJOR CITIES

Winkel Tripel Projection
1:155 000 000

Major Urban Agglomerations

- over 20 million
- 10 million – 20 million
- 5 million – 10 million

Density of inhabitants

per sq km	per sq mile
500	1 250
100	250
25	62.5
1	2.5
0	0
	Uninhabited

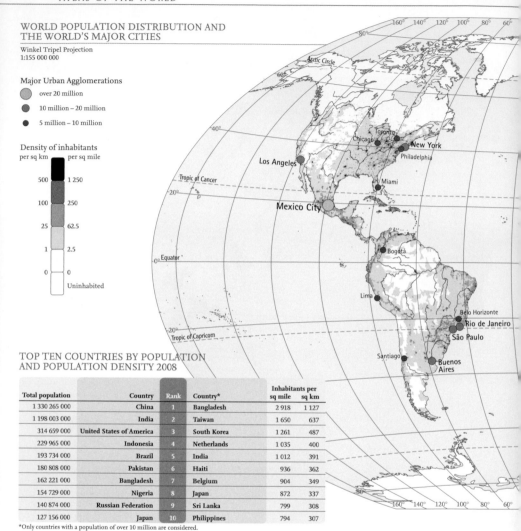

TOP TEN COUNTRIES BY POPULATION AND POPULATION DENSITY 2008

Total population	Country	Rank	Country*	Inhabitants per sq mile	sq km
1 330 265 000	China	1	Bangladesh	2 918	1 127
1 198 003 000	India	2	Taiwan	1 650	637
314 659 000	United States of America	3	South Korea	1 261	487
229 965 000	Indonesia	4	Netherlands	1 035	400
193 734 000	Brazil	5	India	1 012	391
180 808 000	Pakistan	6	Haiti	936	362
162 221 000	Bangladesh	7	Belgium	904	349
154 729 000	Nigeria	8	Japan	872	337
140 874 000	Russian Federation	9	Sri Lanka	799	308
127 156 000	Japan	10	Philippines	794	307

*Only countries with a population of over 10 million are considered.

KEY POPULATION STATISTICS FOR MAJOR REGIONS

	Population 2009 (millions)	Growth (per cent)	Infant mortality rate	Total fertility rate	Life expectancy (years)	% aged 60 and over	
						2010	2050
World	6 829	1.2	47	2.6	68	11	22
More developed regions	1 223	0.1	6	1.6	77	22	33
Less developed regions	5 596	1.4	52	2.7	66	9	20
Africa	1 010	2.4	83	4.6	54	5	11
Asia	4 121	1.2	42	2.4	69	10	24
Europe	732	-0.1	7	1.5	75	22	34
Latin America and the Caribbean	582	1.3	22	2.3	73	10	26
North America	348	0.6	6	2.0	79	18	28
Oceania	35	1.0	23	2.4	76	15	24

Except for population and % aged 60 and over figures, the data are annual averages projected for the period 2005–2010.

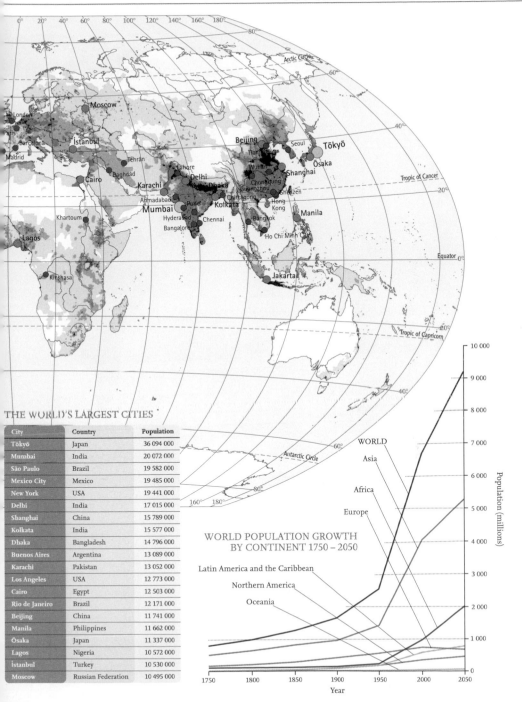

THE WORLD'S LARGEST CITIES

City	Country	Population
Tōkyō	Japan	36 094 000
Mumbai	India	20 072 000
São Paulo	Brazil	19 582 000
Mexico City	Mexico	19 485 000
New York	USA	19 441 000
Delhi	India	17 015 000
Shanghai	China	15 789 000
Kolkata	India	15 577 000
Dhaka	Bangladesh	14 796 000
Buenos Aires	Argentina	13 089 000
Karachi	Pakistan	13 052 000
Los Angeles	USA	12 773 000
Cairo	Egypt	12 503 000
Rio de Janeiro	Brazil	12 171 000
Beijing	China	11 741 000
Manila	Philippines	11 662 000
Ōsaka	Japan	11 337 000
Lagos	Nigeria	10 572 000
İstanbul	Turkey	10 530 000
Moscow	Russian Federation	10 495 000

WORLD POPULATION GROWTH
BY CONTINENT 1750 – 2050

© Collins Bartholomew Ltd

INTERNET USERS 2008

Winkel Tripel Projection
1:155 000 000

Internet users per
10 000 inhabitants 2008

- 4 000–9 999
- 2 000–3 999
- 700–1 999
- 200–699
- 0–199
- no data

Total internet users 2008
Top ten countries

China
298 000 000

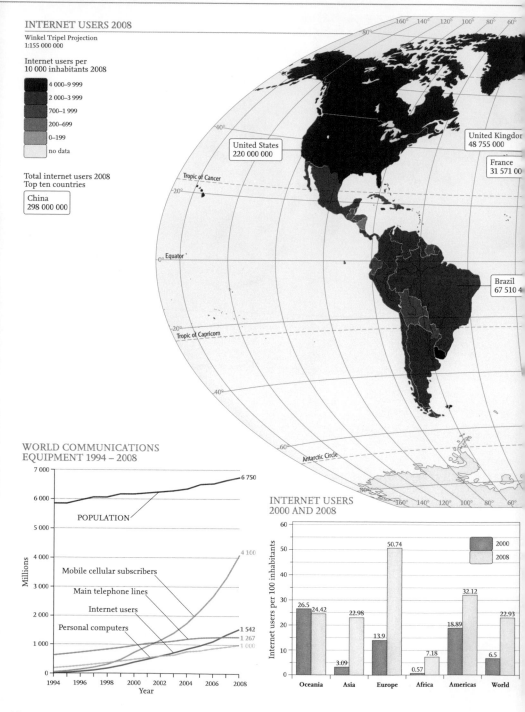

United States
220 000 000

United Kingdom
48 755 000

France
31 571 00

Brazil
67 510 4

Tropic of Cancer

Equator

Tropic of Capricorn

Antarctic Circle

WORLD COMMUNICATIONS
EQUIPMENT 1994 – 2008

POPULATION — 6 750

Mobile cellular subscribers — 4 100

Main telephone lines

Internet users — 1 542

Personal computers — 1 267

1 000

Millions

1994 1996 1998 2000 2002 2004 2006 2008
Year

INTERNET USERS
2000 AND 2008

Internet users per 100 inhabitants

- 2000
- 2008

Region	2000	2008
Oceania	26.5	24.42
Asia	3.09	22.98
Europe	13.9	50.74
Africa	0.57	7.18
Americas	18.89	32.12
World	6.5	22.93

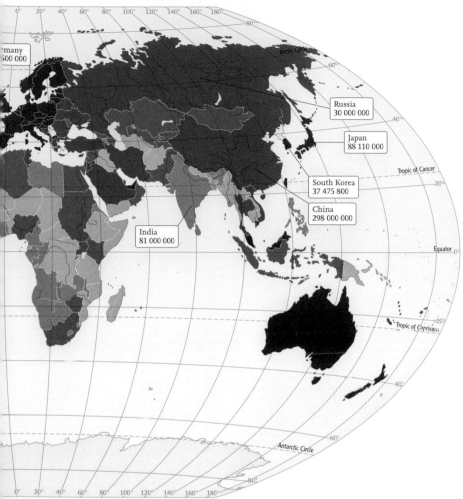

rmany
00 000

Russia
30 000 000

Japan
88 110 000

South Korea
37 475 800

China
298 000 000

India
81 000 000

Arctic Circle

Tropic of Cancer

Equator

Tropic of Capricorn

Antarctic Circle

TOP BROADBAND ECONOMIES 2008

Countries with the highest broadband penetration rate –
subscribers per 100 inhabitants

	Top Economies	Rate
1	Sweden	37.3
2	Denmark	36.8
3	Netherlands	35.0
4	Norway	34.0
5	Switzerland	33.0
6	Iceland	32.9
7	South Korea	32.0
8	Finland	30.6
9	Luxembourg	30.3
10	Canada	29.0
11	France	28.6
12	United Kingdom	28.3

INTERNET USERS

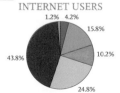

1.2% 4.2%
15.8%
10.2%
43.8%
24.8%

MOBILE CELLULAR SUBSCRIBERS

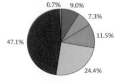

0.7% 9.0%
7.3%
11.5%
47.1%
24.4%

MAIN TELEPHONE LINES

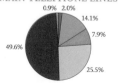

0.9% 2.0%
14.1%
7.9%
49.6%
25.5%

Telecommunications indicators
By region 2008

Africa
North America
Latin America and
the Caribbean*

Europe
Asia
Oceania

*Includes Mexico.

© Collins Bartholomew Ltd

41

MAP POLICIES

PLACE NAMES

The spelling of place names on maps has always been a matter of great complexity, because of the variety of the world's languages and the systems used to write them down. There is no standard way of spelling names or of converting them from one alphabet, or symbol set, to another. Instead, conventional ways of spelling have evolved in each of the world's major languages, and the results often differ significantly from the name as it is spelled in the original language. Familiar examples of English conventional names include Munich (München), Florence (Firenze) and Moscow (from the transliterated form, Moskva).

In this atlas, local name forms are used where these are in the Roman alphabet, though for major cities and main physical features, conventional English names are given first. The local forms are those which are officially recognized by the government of the country concerned, usually as represented by its official mapping agency. This is a basic principle laid down by the United Kingdom government's Permanent Committee on Geographical Names (PCGN) and the equivalent United States Board on Geographic Names (BGN). Prominent English-language and historic names are not neglected, however. These, and significant superseded names and alternate spellings, are included in brackets on the maps where space permits, and are cross-referenced in the index.

Country names are shown in conventional English form and include any recent changes promulgated by national governments and adopted by the United Nations. The names of continents, oceans, seas and under-water features in international waters also appear in English throughout the atlas, as do those of other international features where such an English form exists and is in common use. International features are defined as features crossing one or more international boundary.

BOUNDARIES

The status of nations, their names and their boundaries, are shown in this atlas as they are at the time of going to press, as far as can be ascertained. Where an international boundary symbol appears in the sea or ocean it does not necessarily infer a legal maritime boundary, but shows which offshore islands belong to which country. The extent of island nations is shown by a short boundary symbol at the extreme limits of the area of sea or ocean within which all land is part of that nation.

Where international boundaries are the subject of dispute it may be that no portrayal of them will meet with the approval of any of the countries involved, but it is not seen as the function of this atlas to try to adjudicate between the rights and wrongs of political issues. Although reference mapping at atlas scales is not the ideal medium for indicating the claims of many separatist and irredentist movements, every reasonable attempt is made to show where an active territorial dispute exists, and where there is an important difference between 'de facto' (existing in fact, on the ground) and 'de jure' (according to law) boundaries. This is done by the use of a different symbol where international boundaries are disputed, or where the alignment is unconfirmed, to that used for settled international boundaries. Ceasefire lines are also shown by a separate symbol. For clarity, disputed boundaries and areas are annotated where this is considered necessary. The atlas aims to take a strictly neutral viewpoint of all such cases, based on advice from expert consultants.

MAP PROJECTIONS

Map projections have been selected specifically for the area and scale of each map, or suite of maps. As the only way to show the Earth with absolute accuracy is on a globe, all map projections are compromises. Some projections seek to maintain correct area relationships (equal area projections), true distances and bearings from a point (equidistant projections) or correct angles and shapes (conformal projections); others attempt to achieve a balance between these properties. The choice of projections used in this atlas has been made on an individual continental and regional basis. Projections used, and their individual parameters, have been defined to minimize distortion and to reduce scale errors as much as possible. The projection used is indicated at the bottom left of each map page.

SCALE

In order to directly compare like with like throughout the world it would be necessary to maintain a single scale throughout the atlas. However, the desirability of mapping the more densely populated areas of the world at larger scales, and other geographical considerations, such as the need to fit a homogeneous physical region within a uniform rectangular page format, mean that a range of scales have been used. Scales for continental maps range between 1:25 000 000 and 1:55 000 000, depending on the size of the continental land mass being covered. Scales for regional maps are typically in the range of 1:15 000 000 to 1:25 000 000. Mapping for most countries is at scales between 1:6 000 000 and 1:12 000 000, although for the more densely populated areas of Europe the scale increases to 1:3 000 000.

ABBREVIATIONS

Arch.	Archipelago			Mts	Mountains		
B.	Bay				Monts	French	hills, mountains
	Bahia, Baía	Portuguese	bay	N.	North, Northern		
	Bahía	Spanish	bay	O.	Ostrov	Russian	island
	Baie	French	bay	Pt	Point		
C.	Cape			Pta	Punta	Italian, Spanish	cape, point
	Cabo	Portuguese, Spanish	cape, headland	R.	River		
	Cap	French	cape, headland		Rio	Portuguese	river
Co	Cerro	Spanish	hill, peak, summit		Río	Spanish	river
E.	East, Eastern				Rivière	French	river
Est.	Estrecho	Spanish	strait	Ra.	Range		
Gt	Great			S.	South, Southern		
I.	Island, Isle				Salar, Salina, Salinas	Spanish	saltpan, saltpans
	Ilha	Portuguese	island	Sa	Serra	Portuguese	mountain range
	Islas	Spanish	island		Sierra	Spanish	mountain range
Is	Islands, Isles			Sd	Sound		
	Islas	Spanish	islands	S.E.	Southeast, Southeastern		
Khr.	Khrebet	Russian	mountain range	St	Saint		
L.	Lake				Sankt	German	saint
	Loch	(Scotland)	lake		Sint	Dutch	saint
	Lough	(Ireland)	lake	Sta	Santa	Italian, Portuguese, Spanish	saint
	Lac	French	lake				
	Lago	Portuguese, Spanish	lake	Ste	Sainte	French	saint
M.	Mys	Russian	cape, point	Str.	Strait		
Mt	Mount			W.	West, Western		
	Mont	French	hill, mountain		Wadi, Wādī	Arabic	watercourse
Mt.	Mountain						

MAP SYMBOLS

LAND AND WATER FEATURES

Lake

Impermanent lake

Salt lake or lagoon

Impermanent salt lake

Dry salt lake or salt pan

—— River

---- Impermanent river

Ice cap / Glacier

$\overset{123}{\sim}$ Pass
Height in metres

∴ Site of special interest

◡ Oasis

⌒⌒⌒ Wall

TRANSPORT

═══ Motorway

—— Main road

--- Track

—— Main railway

⊥⊥⊥⊥ Canal

✈ Main airport

BOUNDARIES

▬▬▬ International boundary

-■-■- Disputed international boundary or alignment unconfirmed

Undefined international boundary in the sea.
All land within this boundary is part of state or territory named.

—— Administrative boundary
Shown for selected countries only.

●●●● Ceasefire line or other boundary described on the map

RELIEF

Contour intervals used in layer-colouring, for land height and sea depth

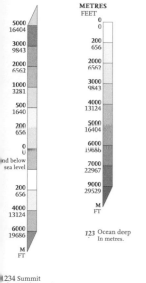

METRES FEET	
5000	16404
3000	9843
2000	6562
1000	3281
500	1640
200	656
0	0
land below sea level	
200	656
4000	13124
6000	19686
M	FT

Ocean pages

METRES FEET	
0	0
200	656
2000	6562
3000	9843
4000	13124
5000	16404
6000	19686
7000	22967
9000	29529
M	FT

123 Ocean deep In metres.

▲234 Summit
△ Height in metres

▲234 Volcano
▲ Height in metres

STYLES OF LETTERING

Cities and towns are explained separately

Country	**FRANCE**
Overseas Territory/Dependency	**Guadeloupe**
Disputed Territory	AKSAI CHIN
Administrative name Shown for selected countries only.	SCOTLAND
Area name	PATAGONIA

Physical features

Island	*Gran Canaria*
Lake	*Lake Erie*
Mountain	*Mt Blanc*
River	*Thames*
Region	*LAPPLAND*

CITIES AND TOWNS

Population	National Capital	Administrative Capital Shown for selected countries only	Other City or Town
over 10 million	**DHAKA** ▣	**Karachi** ⊙	**New York** ⊙
5 million to 10 million	**MADRID** ▣	**Toronto** ⊙	**Philadelphia** ⊙
1 million to 5 million	**KĀBUL** ☐	**Sydney** ○	**Koahsiung** ○
500 000 to 1 million	**BANGUI** ☐	Winnipeg ○	Jeddah ○
100 000 to 500 000	WELLINGTON ☐	Edinburgh ○	Apucarana ○
50 000 to 100 000	PORT OF SPAIN ☐	Bismarck ○	Invercargill ○
under 50 000	MALABO ▫	Charlottetown ○	Ceres ○

CONTINENTAL MAPS

BOUNDARIES —— International boundary ------ Disputed international boundary .... Ceasefire line

CITIES AND TOWNS National Capital **Beijing** ☐ Other City or Town **New York** ○

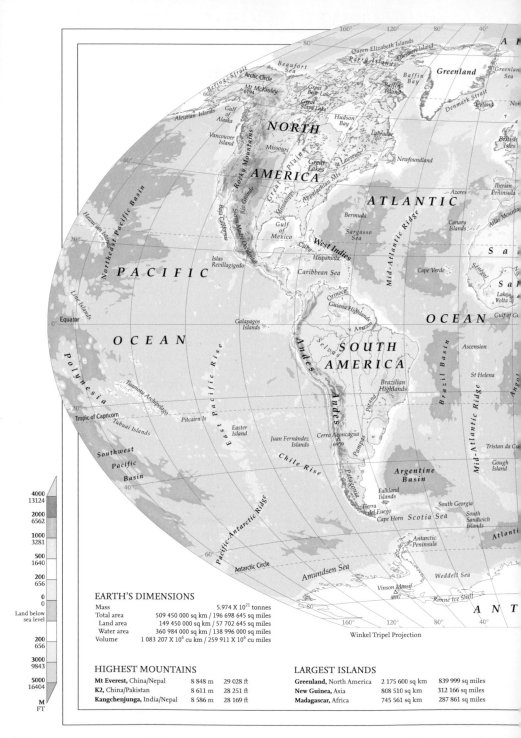

EARTH'S DIMENSIONS

Mass		5.974 X 10^{21} tonnes
Total area	509 450 000 sq km	196 698 645 sq miles
Land area	149 450 000 sq km	57 702 645 sq miles
Water area	360 984 000 sq km	138 996 000 sq miles
Volume	1 083 207 X 10^6 cu km	259 911 X 10^6 cu miles

Winkel Tripel Projection

HIGHEST MOUNTAINS

Mt Everest, China/Nepal	8 848 m	29 028 ft
K2, China/Pakistan	8 611 m	28 251 ft
Kangchenjunga, India/Nepal	8 586 m	28 169 ft

LARGEST ISLANDS

Greenland, North America	2 175 600 sq km	839 999 sq miles
New Guinea, Asia	808 510 sq km	312 166 sq miles
Madagascar, Africa	745 561 sq km	287 861 sq miles

Map legend (elevation scale):

M	FT
4000	13124
2000	6562
1000	3281
500	1640
200	656
0	0
Land below sea level	
200	656
3000	9843
5000	16404

C O C E A N

40° 80° 120° 160°

New Siberia
Islands

East Siberian
Sea

Novaya
Zemlya

Kara Sea

Barents
Sea

Arctic Circle

80°

Central

Siberian
Plateau

Verkhoyanskiy Khrebet

Lena

Bering
Sea

60°

Yenisey

Siberia

Kamchatka Pen.

Aleutian Is

Emperor Seamount Chain

West
Siberian
Plain

Sea of
Okhotsk

Ural Mountains

Irtysh

Lake
Baikal

Amur

40°

ROPE

El'brus
5642

Volga

Caspian Sea

Aral
Sea

Lake
Balkhash

Altai Mountains

Tien Shan

Gobi

ASIA

Manchurian
Plain

Hokkaidō

Danube

Black Sea

Turan
Lowland

Kunlun Shan

Qilian Shan

Sea of
Japan
(East Sea)

Honshū

ean Sea

Euphrates

Zagros Mts

Indus

Plateau of Tibet

Yangtze

East
China
Sea

PACIFIC

Tropic of Cancer

20°

Libyan
Desert

Nile

Arabian

Peninsula

Red Sea

The Gulf

Himalaya

Mt Everest △ 8848

Ganges

Deccan

Bonin
Islands

Mariana Trench

Mid-Pacific Mountains

OCEAN

Blue Nile

Gulf of Aden

Rub' al Khali

Arabian
Sea

Bay
of
Bengal

Mekong

South
China
Sea

Challenger
Deep
10920

Marshall Islands

Micronesia

CA

Ethiopian
Highlands

White Nile

Maldives

Sri Lanka

Peninsular
Malaysia

Philippines

Caroline Islands

Equator 0°

Congo
Basin

Lake
Victoria

Kilimanjaro △ 5892

Somali Basin

Sumatra

Borneo

Celebes

Puncak Jaya
△ 5030

New
Guinea

Melanesia

Solomon Is

Tuvalu

Great Rift Valley

Seychelles

INDIAN

Greater Sunda Islands

Laut Jawa

Java

Laut
Banda

Arafura
Sea

Coral
Sea

Fiji

Tonga Trench

Zambezi

Madagascar

Mauritius
Réunion

OCEAN

Timor
Sea

AUSTRALIA

Great Barrier Reef

Tropic of Capricorn

20°

alahari
Desert

Great
Victoria
Desert

Great
Australian
Bight

Norfolk I.

Lord Howe I.

Great Dividing Range

Darling

of
Hope

Crozet
Basin

Southeast Indian Ridge

Mt Kosciuszko
△ 2229

Murray

Tasman
Sea

New Zealand

North
Island

40°

Prince
Edward Is

Îles Kerguelen

Tasmania

Mt Cook △
3754

South
Island

ntarctic Basin

Australian-Antarctic Basin

Davis Sea

60°

C T I C A

40° 80° 120° 160°

Antarctic Circle

Antarctic Mountains

Ross Sea

80°

1: 126 000 000

Equatorial diameter	12 756 km / 7 927 miles	
Polar diameter	12 714 km / 7 901 miles	
Equatorial circumference	40 075 km / 24 903 miles	
Meridional circumference	40 008 km / 24 861 miles	

LARGEST LAKES

Caspian Sea, Asia / Europe	371 000 sq km	143 243 sq miles
Lake Superior, North America	82 100 sq km	31 366 sq miles
Lake Victoria, Africa	68 800 sq km	6 591 sq miles

LONGEST RIVERS

Nile, Africa	6 695 km	4 160 miles
Amazon, South America	6 516 km	4 049 miles
Yangtze, Asia	6 380 km	3 965 miles

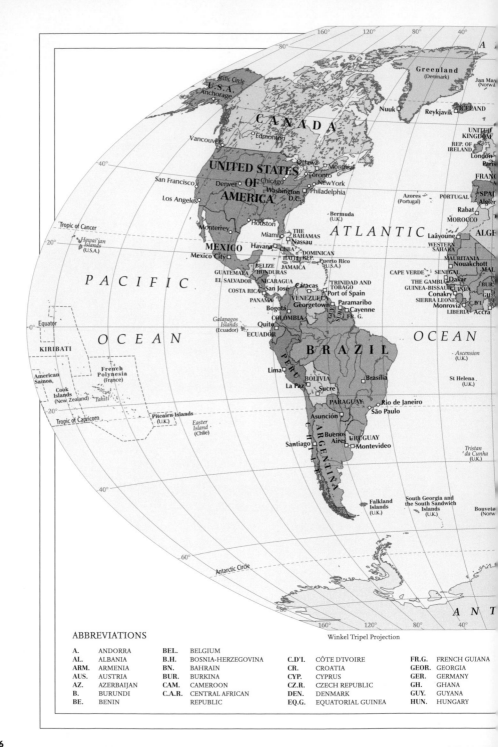

ABBREVIATIONS

A.	ANDORRA	BEL.	BELGIUM	
AL.	ALBANIA	B.H.	BOSNIA-HERZEGOVINA	C.D'I.	CÔTE D'IVOIRE	FR.G.	FRENCH GUIANA
ARM.	ARMENIA	BN.	BAHRAIN	CR.	CROATIA	GEOR.	GEORGIA
AUS.	AUSTRIA	BUR.	BURKINA	CYP.	CYPRUS	GER.	GERMANY
AZ.	AZERBAIJAN	CAM.	CAMEROON	CZ.R.	CZECH REPUBLIC	GH.	GHANA
B.	BURUNDI	C.A.R.	CENTRAL AFRICAN	DEN.	DENMARK	GUY.	GUYANA
BE.	BENIN		REPUBLIC	EQ.G.	EQUATORIAL GUINEA	HUN.	HUNGARY

Winkel Tripel Projection

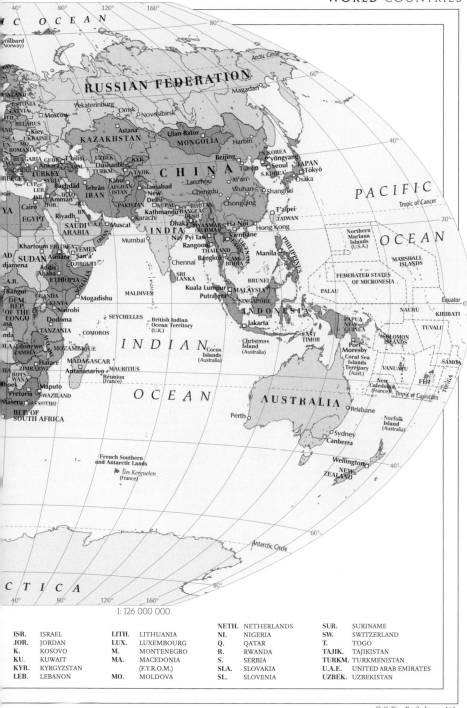

40° 80° 120° 160°

C OCEAN 80°

Svalbard (Norway)

Arctic Circle

RUSSIAN FEDERATION 60°

Magadan

Yekaterinburg

FINLAND Omsk Novosibirsk

ESTONIA LATVIA Moscow Astana

LITH. BELARUS Ulan Bator 40°

AND Kiev **KAZAKHSTAN** Harbin

'SLA. UKRAINE MONGOLIA

UN. MO. Beijing N.KOREA

N.ROMANIA UZBEK. KYR. P'yŏngyang

BULGARIA GEOR. T'bilisi Dushanbe Tianjin S.KOREA Seoul JAPAN

ISTANBUL Ankara ARM. TURKM. TAJIK. **C H I N A** Lanzhou Xi'an Tōkyō

TURKEY AZER. Baghdad Kabul. Xi'an Osaka

REECE CYP. LEB. Tehrān AFGHAN. Islamabad Chengdu Wuhan Shanghai

oli ISR. IRAQ New Delhi **P A C I F I C**

Amman JOR. **IRAN** PAKISTAN NEPAL BHUTAN Chongqing

YA Cairo Riyadh BN. Kathmandu BANGLA- T'aipei Tropic of Cancer

EGYPT Karachi DESH TAIWAN 20°

SAUDI U.A.E. Muscat **I N D I A** MYANMAR Ha Noi Hong Kong

ARABIA BURMA Nay Pyi Taw LAOS **O C E A N**

Khartoum ERITREA YEMEN Mumbai Rangoon Vientiane

AD **SUDAN** Asmara San'a' THAILAND CAM- Manila Northern

djamena Addis DJIBOUTI Bangkok BODIA **PHILIPPINES** Mariana

.A.R. Ababa **ETHIOPIA** Chennai Islands (U.S.A.)

Bangui **SOMALIA** SRI **MARSHALL**

DEM. UGANDA KENYA Mogadishu LANKA BRUNEI **ISLANDS**

REP. Nairobi Kuala Lumpur **MALAYSIA** **FEDERATED STATES**

OF THE SEYCHELLES **MALDIVES** Putrajaya SINGAPORE **OF MICRONESIA**

CONGO Dodoma PALAU Equator 0°

asa **TANZANIA** British Indian **I N D O N E S I A** NAURU KIRIBATI

nda COMOROS Ocean Territory Jakarta

(U.K.) EAST PAPUA TUVALU

I N D I A N MOZAMBIQUE Christmas TIMOR NEW

OLA ZAMBIA Island Port GUINEA SOLOMON

Harare MADAGASCAR Cocos (Australia) Moresby ISLANDS

ZIMBABWE Antananarivo Islands Coral Sea SAMOA

BOTS- MAURITIUS (Australia) Islands VANUATU

hoek WANA Réunion Territory New FIJI

Maputo (France) (Aust.) Caledonia TONGA

Pretoria SWAZILAND **O C E A N** **A U S T R A L I A** (France) Tropic of Capricorn

Maseru LESOTHO Brisbane

REP. OF Norfolk

SOUTH AFRICA Perth Island

(Australia)

French Southern Sydney

and Antarctic Lands Canberra

Îles Kerguelen Wellington 40°

(France) NEW

ZEALAND

60°

Antarctic Circle

C T I C A 80°

40° 80° 120° 160°

1: 126 000 000

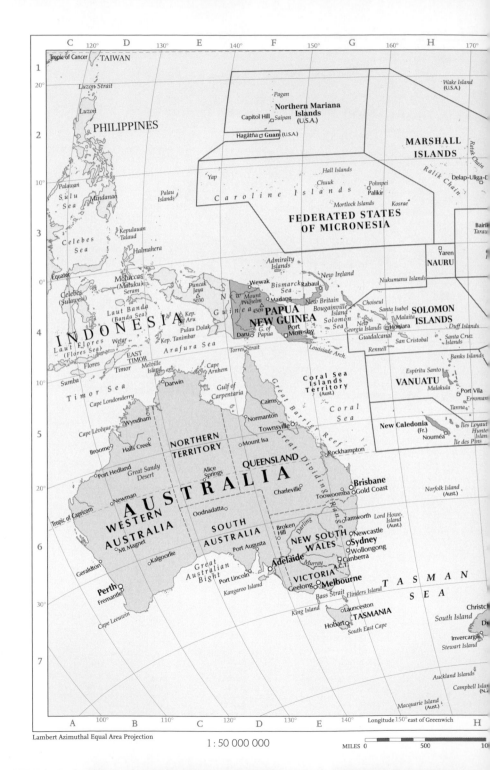

C 120° D 130° E 140° F 150° G 160° H 170°

Tropic of Cancer TAIWAN

20°

Luzon Strait

Luzon

PHILIPPINES

Wake Island
(U.S.A.)

Pagan

Northern Mariana
Islands
(U.S.A.)

Capitol Hill ▫ Saipan

Hagåtña ▫ Guam (U.S.A.)

MARSHALL
ISLANDS

Palawan

Sulu
Sea

Mindanao

10°

Palau
Islands

Yap

Hall Islands

Chuuk

Pohnpei
Palikir

Caroline Islands

Mortlock Islands Kosrae

Delap-Uliga-D

Ralik Chain

Ratak Chain

Kepulauan
Talaud

Celebes
Sea

Halmahera

FEDERATED STATES
OF MICRONESIA

Bairiki
Tarau

▫ Yaren

NAURU

Equator 0°

Celebes
(Sulawesi)

Moluccas
(Maluku)
Seram

Laut Banda
(Banda Sea)

Puncak
Jaya
5030

Admiralty
Islands

New Ireland

Wewak

Mount
Wilhelm
4509

Bismarck
Sea

Madang

PAPUA
NEW GUINEA

Rabaul

New Britain

Bougainville
Island

Solomon
Sea

Choiseul

Santa Isabel

Malaita

New
Georgia Islands ▫ Honiara

SOLOMON
ISLANDS

Nukumanu Islands

Duff Islands

INDONESIA

Laut Flores
(Flores Sea)

Wetar

Kep.
Aru

Pulau Dolak

Kep. Tanimbar

Port
▫ Moresby

Daru ▫ G. of
Papua

Guadalcanal

Rennell

San Cristobal

Santa Cruz
Islands

Banks Islands

EAST
TIMOR

Flores

Timor

Melville
Island

Cape
Arnhem

Arafura Sea

Torres Strait

Louisiade Arch.

Espíritu Santo

VANUATU

Sumba

Timor Sea

Cape Londonderry

Darwin

Gulf of
Carpentaria

Coral Sea
Islands
Territory
(Aust.)

Malakula

Port Vila

Erroman

Coral
Sea

Tanna

Cape Léveque

Wyndham

Cairns

Normanton

Townsville

Great Barrier Reef

Rockhampton

New Caledonia
(Fr.)

Nouméa

Iles Loyauté

Hunte

Islan

Ile des Pins

Broome

Halls Creek

NORTHERN
TERRITORY

Mount Isa

Norfolk Island
(Aust.)

Port Hedland

Newman

Great Sandy
Desert

Alice
Springs

QUEENSLAND

Great Dividing Range

Charleville

AUSTRALIA

Brisbane

Toowoomba Gold Coast

Lord Howe
Island
(Aust.)

Tamworth

Tropic of Capricorn

WESTERN
AUSTRALIA

Mt Magnet

Oodnadatta

SOUTH
AUSTRALIA

Broken
Hill

Darling

NEW SOUTH
WALES

Newcastle

Sydney

Wollongong

TASMAN

Geraldton

Kalgoorlie

Great
Australian
Bight

Port Augusta

Adelaide

Murray

A.C.T.

Canberra

SEA

Perth

Fremantle

Port Lincoln

Kangaroo Island

VICTORIA

Geelong

Melbourne

Bass Strait Flinders Island

Christc

South Island

D

Cape Leeuwin

King Island

Launceston

Hobart TASMANIA

South East Cape

Invercargill

Stewart Island

Auckland Islands

Campbell Islan
(N.Z

Macquarie Island
(Aust.)

A 100° B 110° C 120° D 130° E Longitude 150° east of Greenwich H

Lambert Azimuthal Equal Area Projection

1 : 50 000 000

MILES 0   500   10

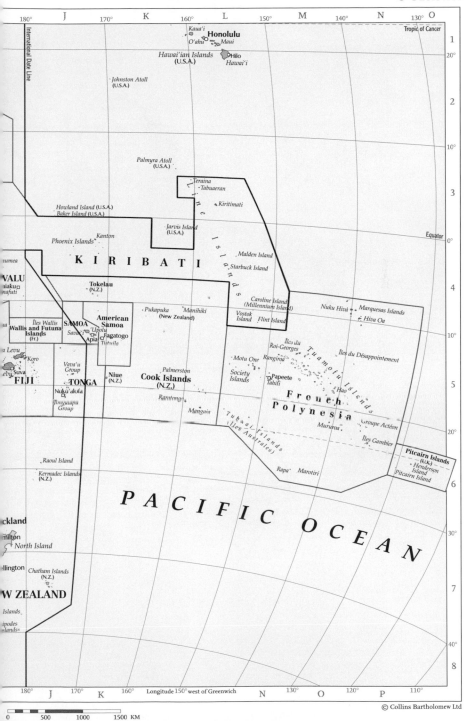

International Date Line

| 180° | J | 170° | K | 160° | L | 150° | M | 140° | N | 130° | O |

Tropic of Cancer

Kaua'i
O'ahu □ **Honolulu**
⊙*Maui*
Hawai'ian Islands
(U.S.A.)
○*Hilo*
Hawai'i

20°

· *Johnston Atoll*
(U.S.A.)

10°

· *Palmyra Atoll*
(U.S.A.)

·*Teraina*
· *Tabuaeran*

· *Kiritimati*

Equator 0°

· *Howland Island* (U.S.A.)
Baker Island (U.S.A.)

· *Jarvis Island*
(U.S.A.)

Kanton
Phoenix Islands·

K I R I B A T I

· *Malden Island*
· *Starbuck Island*

umea

VALU
aiaku □
nafuti

Tokelau
·
(N.Z.)

· *Caroline Island*
(Millennium Island)
Vostok
Island · *Flint Island*

· *Nuku Hiva* · *Marquesas Islands*
·· *Hiva Oa*

· *Pukapuka* · *Manihiki*
(New Zealand)

SAMOA
Savai'i
Upolu
□*Apia* □*Fagatogo*
Tutuila
Wallis and Futuna
Islands
(Fr.)
Îles Wallis
American
Samoa

Motu One ·
Society
Islands
□*Papeete*
Tahiti
Îles du
Roi-Georges
Rangiroa
· *Îles du Désappointement*

Tuamotu Islands

a Levu
Koro ·
·*Suva*
FIJI

Vava'u
Group
Niue
(N.Z.)
Cook Islands
(N.Z.)
Palmerston ·

Hao
· *Hao*

French
Polynesia

TONGA
Nuku'alofa□
Tongatapu
Group

Rarotonga ·

Mangaia ·

Tubuai Islands
(Îles Australes)
· *Mururoa*
· *Groupe Actéon*
· *Îles Gambier*

Pitcairn Islands
(U.K.)
· *Henderson*
Island
Pitcairn Island

· *Raoul Island*

Kermadec Islands
(N.Z.)

· *Rapa* · *Marotiri*

P A C I F I C   O C E A N

ckland
milton
⌀ *North Island*

llington
Chatham Islands
(N.Z.)

W ZEALAND

Islands ·
ipodes
slands ·

20°

30°

40°

© Collins Bartholomew Ltd

| 180° | J | 170° | K | 160° | Longitude 150° west of Greenwich | N | 130° | O | 120° | P | 110° |

0   500   1000   1500 KM

49

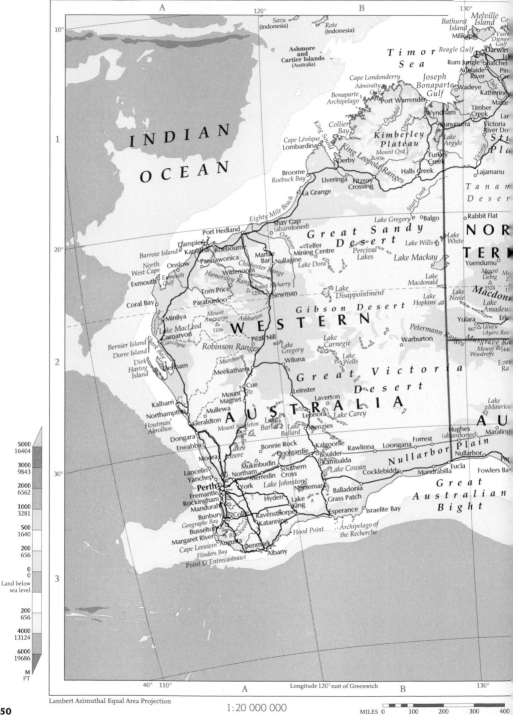

INDIAN

OCEAN

Timor
Sea

Savu
(Indonesia)   Rote
(Indonesia)

Ashmore
and
Cartier Islands
(Australia)

Bathurst   Melville
Island   Island   Co
Milikapiti   Van
Diemen
Gulf
Beagle Gulf   Darwin
Rum Jungle   Batchel
Adelaide   Pin
River
Wadeye   Katherine
Timber   Mataranka
Creek   Lar
Victoria   River Do
River

Cape Londonderry
Admiralty   Joseph
Bonaparte
Bonaparte   Gulf
Archipelago   Port Warrender
Collier   Wyndham
Bay   Kununurra   S
Cape Lévêque   Kimberley   Lake   Pl
Lombardina   Plateau   Argyle
Mount Ord   Turkey
936   Creek
Derby   Halls Creek   Lajamanu
Broome   Liveringa   Fitzroy
Roebuck Bay   Crossing   *Tan a*
La Grange   *Dese r*

Eighty Mile Beach   Lake Gregory   Balgo   Rabbit Flat

Port Hedland   Shay Gap   *Great Sandy*   NOR
(abandoned)   *Desert*   TER
Dampier   Roebourne   Telfer   Lake   Lake
Barrow Island   Karratha   Marble   Mining Centre   Lake Wills   White
North   Onslow   Pannawonica   Bar   Percival   Lake Mackay   Yuendumu   Me
West Cape   Wittenoom   Nullagine   Lakes   Liebig
Exmouth   Exmouth   Hamersley Range   Lake Dora   Lake   *Macdon*
Gulf   Tom Price   Mount Meharry   Newman   Macdonald   Lake   Lake
Coral Bay   Paraburdoo   1250   Lake   Neale   Amadeus
Minilya   Mount   Ashburton   *Gibson Desert*   Disappointment   Lake   Uluru   Erl
Lake MacLeod   Augustus   Hopkins   Yulara   (Ayers Roc
Carnarvon   1106   Lake   867   Musgrave Ra
Bernier Island   Gascoyne   *Robinson Ranges*   Peak Hill   Carnegie   Warburton   Petermann Ranges   Mount   1440
Dorre Island   Murchison   Lake   Lake   Woodroffe
Dirk   Wiluna   Gregory   Wells   Ever
Hartog   Denham   *Great Victoria*
Island   Meekatharra   Lake   Maurice
Cue   *Desert*   *A U*
Mount   Leinster   Laverton
Magnet   Leonora   Lake Carey   Maraling
Kalbarri   Mullewa   Lake   Menzies
Northampton   Geraldton   AUSTRALIA   Barlee   Hughes
Houtman   Mount Singleton   Ballard   Forrest   (abandoned)
Abrolhos   Dongara   Bonnie Rock   Kalgoorlie   Rawlinna   Loongana   Nullarbor
Eneabba   Lake   Coolgardie   Boulder   Nullarbor Plain
Moora   Moore   Mukinbudin   Kambalda   Cocklebiddy   Mundrabilla   Eucla   Fowlers Ba
Lancelin   Southern   Lake Cowan
Yanchep   Northam   Cross   *Great*
Perth   Merredin   Norseman   *Australian*
Fremantle   York   Lake Johnston   Balladonia   *Bight*
Rockingham   Hyden   Lake   Grass Patch
Mandurah   King   Esperance   Israelite Bay   Archipelago of
Bunbury   Collie   Ravensthorpe   the Recherche
Geographe Bay   Busselton   Katanning   Hood Point
Margaret River   Augusta   Denmark
Cape Leeuwin   Albany
Flinders Bay
Point D'Entrecasteaux

5000
16404
3000
9843
2000
6562
1000
3281
500
1640
200
656
0
0
Land below
sea level
200
656
4000
13124
6000
19686
M
FT

Lambert Azimuthal Equal Area Projection   1:20 000 000   MILES 0   100   200   300   400

Longitude 120° east of Greenwich

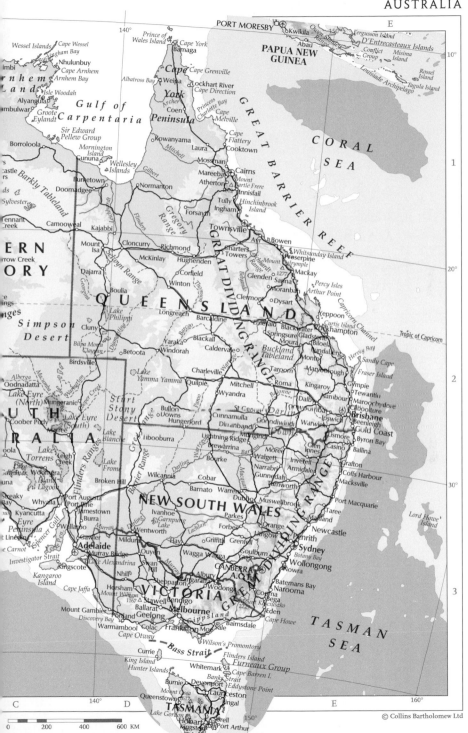

PORT MORESBY

Kwikila

PAPUA NEW GUINEA

Abau

Fergusson Island

D'Entrecasteaux Islands

Conflict Group

Misima Island

Rossel Island

Tagula Island

Louisiade Archipelago

Wessel Islands

Cape Wessel

Buckingham Bay

Nhulunbuy

Cape Arnhem

Arnhem Bay

Arnhem Land

Isle Woodah

Alyangula

Groote Eylandt

Gulf of Carpentaria

Sir Edward Pellew Group

Cape York

Prince of Wales Island

Bamaga

Cape York

Cape Grenville

Weipa

Albatross Bay

Lockhart River

Cape Direction

Princess Charlotte Bay

Cape Melville

Cape Flattery

CORAL SEA

Borroloola

Mornington Island

Gununa

Wellesley Islands

Kowanyama

York

Peninsula

Coen

Laura

Cooktown

Mossman

Cairns

Barkly Tableland

Burketown

Doomadgee

Normanton

Mareeba

Atherton

Mount Bartle Frere

Tully

Ingham

Innisfail

Hinchinbrook Island

Camooweal

Kajabbi

Forsayth

Gregory Range

Townsville

GREAT BARRIER REEF

Mount Isa

Cloncurry

Richmond

Hughenden

Charters Towers

Ayr

Bowen

Proserpine

Whitsunday Island

McKinlay

Corfield

Winton

Glenden

Sarina

Mackay

Percy Isles

Arthur Point

Clermont

Moranbah

Dysart

QUEENSLAND

Longreach

Barcaldine

Emerald

Blackwater

Springsure

Moura

Gladstone

Rockhampton

Curtis Island

Yeppoon

Capricorn Channel

Tropic of Capricorn

Simpson Desert

Boulia

Cluny

Yaraka

Blackall

Caldervale

Buckland Tableland

Tambo

Biloela

Monto

Bundaberg

Hervey Bay

Sandy Cape

Fraser Island

Birdsville

Windorah

Charleville

Tarnom

Roma

Maryborough

Gympie

SOUTH AUSTRALIA

Oodnadatta

Lake Eyre (North)

Simpson Desert

Quilpie

Mitchell

Wyandra

Kingaroy

Tewantin

Maroochydore

Coober Pedy

Lake Eyre (South)

Sturt Stony Desert

Bullo Downs

Hungerford

Cunnamulla

Dirranbandi

St George

Goondiwindi

Dalby

Nanango

Caboolture

Brisbane

Gold Coast

Beenleigh

Lake Torrens

Leigh Creek

Lake Blanche

Tibooburra

Lightning Ridge

Warwick

Ipswich

Byron Bay

Ballina

Lake Frome

Bourke

Brewarrina

Walgett

Moree

Inverell

Glen Innes

Casino

Lismore

Grafton

Broken Hill

Wilcannia

Cobar

Narrabri

Gunnedah

Armidale

Coffs Harbour

Macksville

Port Augusta

Port Pirie

Barnato

Warren

Dubbo

Muswellbrook

Tamworth

Port Macquarie

NEW SOUTH WALES

Taree

Whyalla

Jamestown

Burra

Ivanhoe

Parkes

Orange

Singleton

Maitland

Newcastle

Adelaide

Gawler

Mildura

Wentworth

Hay

Griffith

Forbes

Grenfell

Lithgow

Goulburn

Penrith

Sydney

Botany Bay

Wollongong

Murray Bridge

Swan Hill

Ouyen

Wagga Wagga

Canberra

A.C.T.

Nowra

Batemans Bay

Kangaroo Island

Kingscote

Nhill

Horsham

Shepparton

Wangaratta

Albury

Wodonga

Mount Kosciuszko

Cooma

Narooma

Bega

Eden

VICTORIA

Stawell

Bendigo

Ballarat

Geelong

Melbourne

Sale

Bairnsdale

Cape Howe

Mount Gambier

Portland

Warrnambool

Colac

Frankston

Morwell

Gippsland

Wilson's Promontory

TASMAN SEA

Lord Howe Island

Cape Otway

Bass Strait

Flinders Island

Furneaux Group

King Island

Hunter Islands

Currie

Whitemark

Banks Strait

Cape Barren I.

Eddystone Point

Burnie

Devonport

Queenstown

Launceston

TASMANIA

Hobart

Port Arthur

0   200   400   600 KM

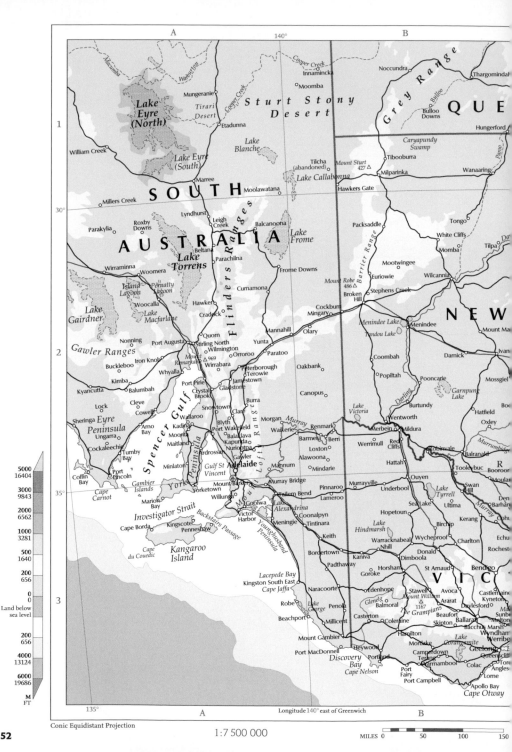

A   140°   B

Warburton   Cooper Creek
Macumba   Innamincka   Noccundra   Thargomindal
                    Mungeranie   Moomba
Lake   Tirari   Cooper Creek   Sturt Stony   Bulloo   QUE
Eyre   Desert   Desert   Downs   Hungerford
(North)   Etadunna   Caryapundy
                                                                                        Swamp
William Creek   Lake   Tilcha   Mount Sturt   Tibooburra
                    Lake Eyre   Blanche   (abandoned)   427 △   Milparinka   Wanaaring
                    (South)   Lake Callabonna
                    Marree   Moolawatana   Hawkers Gate
              Millers Creek   SOUTH   Packsaddle   Tongo
                                                                                        White Cliffs
30°   Lyndhurst   Leigh   Balcanoona   Momba   Tilpa   Da
Parakylia   Roxby   Creek   Mootwingee
              Downs   Lake   Euriowie   Wilcannia
AUSTRALIA   Beltana   Frome   Mount Robe
Wirraminna   Lake   Parachilna   486 △   Stephens Creek   NEW
              Torrens   Woomera   Frome Downs   Broken Hill
Island   Pernatty   Curnamona   Meninee Lake
Lagoon   Lagoon   Cockburn   Tandou Lake   Meninee   Mount Ma
Woocalla   Lake   Hawker   Mingary
Lake   Macfarlane   Cradock   Olary   Darnick   Ivan
Gairdner   Quorn   Mannahill   Coombah
Nonning   Port Augusta Stirling North Wilmington   Yunta   Popiltah   Pooncarie   Mossgiel
Gawler Ranges   Iron Knob   Mount   Orroroo   Paratoo   Garnpung
Buckleboo   Remarkable△ 969   Peterborough   Oakbank   Burtundy   Lake   Boe
Kimba   Balumbah   Whyalla   Wirrabara   Terowie   Hatfield   Oxley
Kyancutta   Port Pirie   Jamestown   Canopus   Lake   Darling
Lock   Cleve   Crystal   Gladstone   Victoria   Wentworth
Sheringa Eyre   Cowell   Brook   Burra   Merbein   Mildura   Hattah   Murrumbidg
Peninsula   Snowtown   Clare   Morgan Murray   Renmark   Red   Balranald   R
Ungarra   Arno   Wallaroo   Blyth   Waikerie   Cliffs   Tooleybuc   Booror
Cockaleechie   Bay   Kadina   Moonta   Port Wakefield   Barmera   Berri   Robinvale   Moula
Tumby   Maitland   Balaklava   Kapunda   Loxton   Werrimull   Den
Bay   Ardrossan   Nuriootpa   Alawoona   Ouyen   Swan
Coffin   Port   Minlaton   Gawler   Mannum   Murrayville   Sea Lake   Hill   Barha
Bay   Lincoln   Gulf St   Adelaide   Mindarie   Pinnaroo   Lake   Ultima   Kerang
Gambier   Vincent   Mount Barker   Murray Bridge   Lameroo   Tyrrell   Echu
Cape   Islands   York   Yorketown   Tailem Bend   Hopetoun   Birchip   Rochest
Carnot   Marion   Willunga   Coonalpyn   Lake   Wycheproof   Charlton
Bay   Kingscote   Goolwa Lake   Tintinara   Hindmarsh   Warracknabeal
Investigator Strait   Penneshaw   Victor   Alexandrina   Keith   Nhill   Donald   St Arnaud   Bendigo
Cape Borda   Harbor   Meningie   Bordertown   Kaniva   Dimboola   VIC
Backstairs   Youngh   Padthaway   Goroke   Horsham   Stawell   Avoca   Castlemaine
Cape   Passage   Kangaroo   Lacepede Bay   Mount William   Ararat   Daylesford   Kyneton
du Couedic   Island   Kingston South East   Naracoorte   Edenhope   1167 △   Beaufort   Ballarat   Sunb
Cape Jaffa   Penola   Glenelg   Balmoral   The Grampians   Skipton   Bacchus Marsh   Milton
Robe   Lake   Casterton   Coleraine   Geelong   Cliffs
Beachport   George   Millicent   Hamilton   Montake   Lake   Queens
                                                                        Tepko   Corangamite   Tore
Mount Gambier   Camperdown   Colac   Angles
Port MacDonnell   Heywood   Warrnambool   Lorne
Discovery   Portland   Port   Apollo Bay
Bay   Cape Nelson   Fairy   Port Campbell   Cape Otway

5000
16404

3000
9843

2000
6562

1000
3281

500
1640

200
656

0
0
Land below
sea level

200
656

4000
13124

6000
19686

M
FT

135°   A   Longitude 140° east of Greenwich   B

Conic Equidistant Projection

52   1:7 500 000   MILES 0   50   100   150

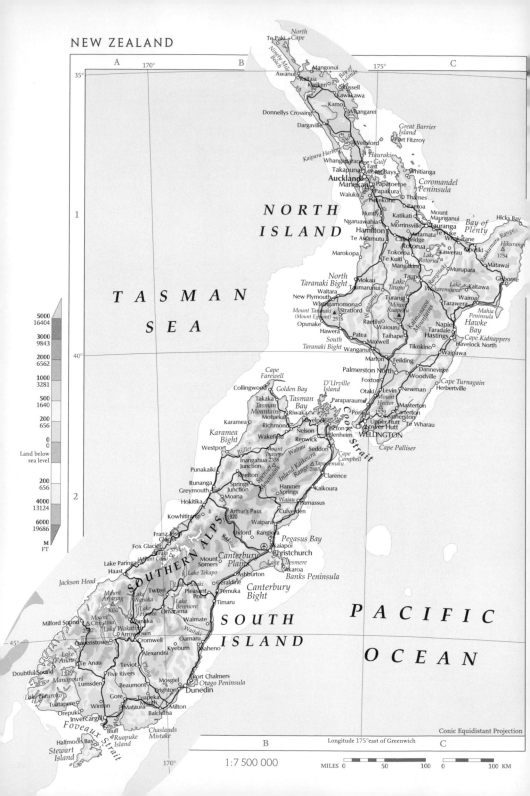

ANTARCTICA

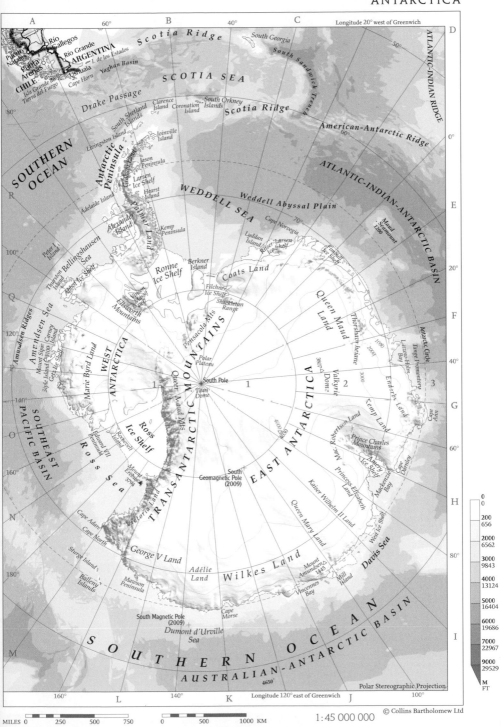

A
60°
B
40°
C
Longitude 20° west of Greenwich
50°
D

ATLANTIC-INDIAN RIDGE

Scotia Ridge
South Georgia
South Sandwich Trench

SCOTIA SEA

ARGENTINA
Río Gallegos
Río Grande
Puerto Natales
Punta Arenas
CHILE
Ushuaia
Cape Horn
I. de los Estados
Tierra del Fuego
Isla Grande
Yaghán Basin

Drake Passage

South Shetland Islands
Clarence Island
Coronation Island
South Orkney Islands
Scotia Ridge

American-Antarctic Ridge

SOUTHERN OCEAN

Livingston Island
Joinville Island

Antarctic Peninsula
Jason Peninsula
Larsen Ice Shelf
Hearst Island

Weddell Abyssal Plain

WEDDELL SEA

ATLANTIC-INDIAN-ANTARCTIC BASIN

Adelaide Island

Palmer Land
Kemp Peninsula

Cape Norvegia

Maud Seamount 1200

70°

Alexander Island

Peter I Island

Bellingshausen Sea
Thurston Island
Abbot Ice Shelf

Ronne Ice Shelf

Berkner Island

Lyddan Island
Kaiser Wilhelm II Land
Larsen Ice Shelf

Fimbul Ice Shelf

Coats Land

Queen Maud Land

Thorshavn heiane

Tinggi-Helm Bay

Antarctic Circle

Tange Promontory

Enderby Land

Amundsen Sea
Mount Sipte
Carney Island
Siple Island
Getz Ice Shelf

Ellsworth Mountains
Vinson Massif

Filchner Ice Shelf
Shackleton Range

Pensacola Mts

Polar Plateau

Queen Maud Land

Valkyrie Dome
3807

Kemp Land

WEST ANTARCTICA

Marie Byrd Land

South Pole

TRANSANTARCTIC MOUNTAINS

Titan Dome

EAST ANTARCTICA

1

2

3

Amundsen Ridges

Edward VII Peninsula

Roosevelt Island

Ross Ice Shelf

Queen Maud Mts

Mount Erebus 3794

South Geomagnetic Pole (2009)

Mawson
Robertson Land

Prince Charles Mountains

Princess Elizabeth Land

Amery Ice Shelf

Cape Darnley

Mackenzie Bay

SOUTHEAST PACIFIC BASIN

Ross Sea

Cape Adare
Cape North

Victoria Land

Kaiser Wilhelm II Land

Queen Mary Land

West Ice Shelf

Davis Sea

George V Land

Sturge Island

Balleny Islands

Mawson Peninsula

Adélie Land

Wilkes Land

Mount Amundsen 1445

Vincennes Bay

Mill Island

South Magnetic Pole (2009)

Dumont d'Urville Sea

Cape Morse

SOUTHERN OCEAN

AUSTRALIAN-ANTARCTIC BASIN

4650

Polar Stereographic Projection

160°
L
140°
K
Longitude 120° east of Greenwich
J
100°

0	
0	
200	656
2000	6562
3000	9843
4000	13124
5000	16404
6000	19686
7000	22967
9000	29529
M	
FT	

MILES 0   250   500   750

0   500   1000 KM

1:45 000 000

© Collins Bartholomew Ltd

55

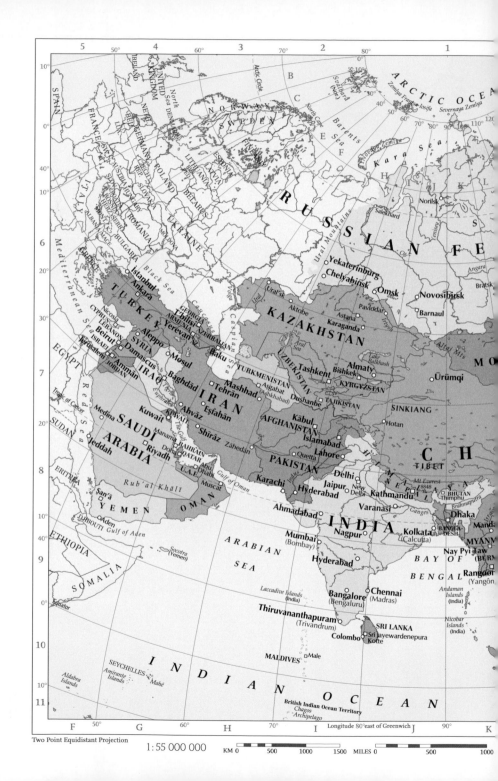

Two Point Equidistant Projection

1 : 55 000 000

KM 0   500   1000   1500   MILES 0   500   1000

Map Grid and Coordinates

2 4 50° 5 40° 6 30°

180° 70°

60°

BERING

SEA

T

S

R

East Siberian Sea

Wrangel Island

Arctic Circle

U.S.A.

Aleutian Islands (U.S.A.)

Aleutian Islands (U.S.A.)

170°

R A T I O N

Lena

Yakutsk

Kamchatka Peninsula

Magadan

Petropavlovsk-Kamchatskiy

Sea of Okhotsk

Midway Islands (U.S.A.)

Kure Atoll

7

180°

P A C I F I C

O C E A N

Sakhalin

Amur

Khabarovsk

Sapporo

Hokkaidō

Kuril Islands

Tropic of Cancer

20°

Wake Atoll (U.S.A.)

170°

Lake Baikal

Qiqihar

Harbin

Vladivostok

Ulan Bator

LIA

MONGOLIA

Changchun

Shenyang

NORTH KOREA

Sea of Japan (East Sea)

Honshū

Tokyo

J A P A N

8

Beijing

P'yŏngyang

Baotou

Dalian

SOUTH KOREA

Seoul

Ōsaka

Tianjin

Huang He

Yellow Sea

Fukuoka

Kyūshū

Bōnin Islands (Japan)

10°

160°

Taiyuan

Lanzhou

Xi'an

Shanghai

East China Sea

Volcano Islands (Japan)

Chengdu

N A

Wuhan

Nanjing

Yangtze

Ryukyu Islands

Northern Mariana Islands (U.S.A.)

9

gqing

Changsha

Fuzhou

T'aipei

TAIWAN

Guam (U.S.A.)

Kunming

Guangzhou

Hong Kong

Caroline Islands

Equator

0°

Nanning

Ha Nôi

Hainan

Luzon

Quezon City

PHILIPPINES

iane

VIETNAM

South China Sea

Manila

Melekeok

PALAU

LAND

Bangkok

Phnom Penh

CAMBODIA

Ho Chi Minh City

Mindanao

Palawan

Admiralty Island

New Britain

10

10°

Gulf of Thailand

Sulu Sea

Davao

Bandar Seri Begawan

BRUNEI

Celebes Sea

Halmahera

Jayapura

PAPUA

NEW GUINEA

New Guinea

Manado

Maluku (Moluccas)

Puncak Jaya 5030

an

MALAYSIA

Kuala Lumpur

SINGAPORE

Putrajaya

Kuching

Borneo

Balikpapan

Banjarmasin

Celebes (Sulawesi)

Seram

Kepulauan Aru

Pulau Dolak

10°

tera

Palembang

Makassar

Kepulauan Tanimbar

Arafura Sea

Cape Arnhem

11

I N D O N E S I A

Laut Jawa

Jakarta

Java

Surabaya

Sumbawa

Laut Banda

Dili

EAST TIMOR

Laut Sawu

Timor

AUSTRALIA

Bandung

Sumba

O 140° P 150°

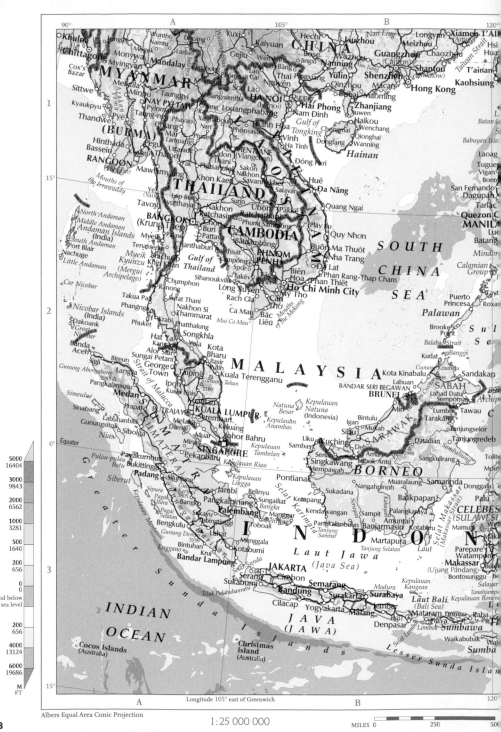

CHINA

Khulna   Jungle   Namtu   Lincang   Yuxi   Kaiyuan   Hechi   Liuzhou   Nan Ling   Longyan   Xiamen   T'AI

Chittagong   Monywa   Megok   Lashio   Gejiu   Nanning   Wuzhou   Guangzhou   Chaozhou   Amoy   Hsi

Cox's Bazar   Myingyan   Mandalay   Simao   Cao Bang   Qinzhou   Shantou   Shenzhen   Macau   (Canton)   Hua

MYANMAR

Sittwe   Meiktila   Minbu   NAY PYI TAW   Loungphabang   Hai Phong   Zhanjiang   Hong Kong   Kaohsiung

Kyaukpyu   Pyinmana   Chiang   Nam Dinh   Naikou   Wenchang

Thandwe   Pye   (BURMA)   Phayao   Ha Tinh   Qionghai   Wanning

(BURMA)

Hinthada   Pegu   VIENTIANE   Dong Hoi   Hainan

Bassein   RANGOON   Khon Kaen   Savannakhet   Huê

Mawlamyaing   Nakhon   Salavan   Da Nang

Mouths of the Irrawaddy   **THAILAND**   Surin   Ubon   Pakse   Quang Ngai

Tavoy   Ayutthaya   Ratchasima   Ratchathani   Phumi Samrong   Play Ku

North Andaman   **BANGKOK**   Chon   Buri   **CAMBODIA**   Quy Nhon

Middle Andaman   (Krung Thep)   Pattaya   Batdambang   Ma Thuôt

Andaman Islands   Myeik   Palaw   Chanthaburi   **PHNOM**   Nha Trang

South Andaman   Tenasserim   Kompong   **PENH**   Da Lat

Port Blair   Myeik   Gulf of   Kompong   Takev   Bien   Phan Rang-Thap Cham

Nachuge   Kyunzu   Thailand   Spoe   Hoa   Phan Thiet

Little Andaman   (Mergui   Sihanoukville   **Ho Chi Minh City**

Car Nicobar   Archipelago)   Chumphon   Long Xuyen

Ranong   Rach Gia   Can Tho   My Tho

Nicobar Islands   Takua Pa   Surat Thani   Ca Mau   Bac Lieu

(India)   Phangnga   Nakhon Si   Mui Ca Mau   Mouths of the Mekong

Dakoank   Krabi   Thammarat

Great Nicobar   Phuket   Phatthalung   **SOUTH**

Banda Aceh   Hat Yai   Songkhla   **CHINA**

Bireun   Yala   Kota   **SEA**

Sigli   Sungai Petani   Bharu   Pasir

Gunung Abongabong   Langsa   George   Putih   Palawan

Pangkalansusu   Town   Taiping   Kuala Terengganu   Puerto Princesa

Simeulue   Ipoh   Gunung Tahan   **MALAYSIA**   Kota Kinabalu   Sandakan

Medan   2189   Labuan   Gunung Kinabalu

Sinabang   Prapat   **KUALA LUMPUR**   BANDAR SERI BEGAWAN   **SABAH**   Lahad Datu

Labuhanbilik   Seremban   Kuantan   **BRUNEI**   Semporna

Gunungsitoli   Sibolga   Melaka   Tawau

Nias   Dumai   Muar   Keluang   Tanjungselor

Payakumbuh   Johor Bahru   Kepulauan   Bintulu   Mukah   Sibu   Tanjungredeb

Bukittinggi   Minas   Tambelan   **SINGAPORE**   Kuching   Tolit

Padang   Pekanbaru   Kepulauan   Sambas   Debak   Sangkulirang

Siberut   **SUMATERA**   Kepulauan Riau   Singkawang   Lubok Antu

Siberut   Jambi   Lingga   Mempawah   **BORNEO**   Samarinda

Sipura   Pangkalpinang   Belinyu   Pontianak   Balikpapan

Pagai Utara   Sekayu   Sungailiat   Sukadana   Muaralaung   Donggala

Pagai Selatan   Bangko   Bangka   Ketapang   Nangahpinoh   Sampit   Palu

INDONESIA

Bengkulu   Palembang   Manggar   Kendawangan   Pangkalanbun   Banjarmasin   Kotabaru   Mamuju

Tebingtinggi   Belitung   Amuntai

Bintuhan   Toboali   Parepare

Gunung Dempo   Menggala   Kepulauan   Tanjung   Martapura   Laut   Watampone

3159   Krui   Sambar   Makassar   **CELEBES**

Enggano   Bandar Lampung   Selat Karimata   Tanjung Selatan   (Ujung Pandang)   **(SULAWESI)**

Serang   **Laut Jawa**   Bontosungguu   Salayar

JAKARTA   Cirebon   (Java Sea)   Kepulauan Kangean

Sukabumi   Semarang   Madura   Tanjiampea

Bandung   Surakarta   **Surabaya**   Pamekasan   Lembar   **Laut Bali**   Kepulauan Bonera

Cilacap   Yogyakarta   Malang   (Bali Sea)   Le

Teluk Palabuhanratu   **JAVA**   Denpasar   Mataram   Lombok   Dompu   Raba

(JAWA)   Sumbawa

Sumba

INDIAN

OCEAN

Cocos Islands   Christmas Island   Waikabubak   Sumba

(Australia)   (Australia)   Greater Sunda Islands   Lesser Sunda Islands

Longitude 105° east of Greenwich
A   B   120°
15°
Albers Equal Area Conic Projection
58   1:25 000 000   MILES 0   250   500

M FT
5000 / 16404
3000 / 9843
2000 / 6562
1000 / 3281
500 / 1640
200 / 656
0 / 0
Land below sea level
200 / 656
4000 / 13124
6000 / 19686

C   135°   D   150°   E

Tropic of Cancer

WAN
eople's Republic of China
Taiwan as its 23rd Province.

PACIFIC

OCEAN

Philippine
Sea

on

PHILIPPINES

Catanduanes
Legaspi
Sorsogon
Catarman
s
Catalogan   Samar
oxas   Catbalogan
ays   Tacloban
Bacolod
Cebu   Surigao
I   *Bohol*   Butuan
Bohol Sea
Cagayan de Oro
Iligan   Mindanao
otabato   Davao
nboanga   Mati
la   *oro*
General Santos
ulf

es

Kepulauan
Talaud

Sangir

PACIFIC
OCEAN

Northern
Mariana
Islands
(U.S.A.)

Pagan

CAPITOL HILL   Saipan
Tinian

HAGÅTÑA   *Rota*
Guam
(U.S.A.)

Mariana Trench

FEDERATED STATES
OF MICRONESIA

Ulithi
Yap   *Fais*
Colonia
Faraulep
Ngulu   *Sorol*
Eauripik   Caroline
Islands
East Caroline
Basin

PALAU   *Babeldaob*
MELEKEOK

Kepulauan
Sangir   *Morotai*

anjung Manado
ahasa   *Tondano*   Tobelo
Kwandang   Ternate   Halmahera
Gorontalo   Sau-silu
uwuk   *Kahuna*   Waigeo
Peleng   Sofifi
Bacan
Todeli Mangole   *Obi*
alauan Molucas
Banggai Kepulauan   *(Maluku)*   Misool   Fakfak
nggai   *Sula*

Manui
Kendari   Namlea
Buton   Piru   Bula
Ambon   Seram
Kepulauan Banda
Kepulauan Wahtubela
Banda
Kepulauan Kai Kecil
Tukangbesi   Kai
Besar
Kepulauan Barat Daya   Aru
Kalabahi *Alor*   *Damar* Wuliaru   Trangan
Kepulauan Tanimbar
Dili *Wetar* Kisar   Tepa   Dobar
OCUSSI   *Leti* Manatuto
Maliana   Kepulauan Kepulauan Selaru
Saumlakki
EAST   *Suai*   *Kefamenanu*
TIMOR   Timor
Kupang   *Rote*

Equator

Pelleluhu   St Matthias
Islands   Group
Admiralty   Lorengau
Islands   Umbukul
Hermit Islands   Manus Island
Wuvulu   Aitape   Schouten Islands
Island   Vanimo
Manokwari   Biak
Selat Yapen
Nimfoor   Serui   Sarmi   Maprik   Wewak
Ransiki   Jayapura   Sepik
Teluk   Nabire   PAPUA
Cenderawasih   (IRIAN JAYA)   Maikd
Teluk Berau   Babo   Pegunungan Van Rees
Kaimana   Pegunungan Maoke
Adi
Amamapare   Pincak Jaya   Central Ra   Wilhelm
4750   Kiunga   Mendi   Hagen   Goroka
Timika   Enarotali   4500   Wau
Wokam   Tanjung Deyong   Balimo   Kikori
Kobroor   *Larat*   Digul   Merauke   Morehead   Daru
Tanjung Vals   Pulau   Kerema
Dolak   Thursday   Cape York
Island
Arafura   Sea   Prince of Wales   Bamaga
Island

Bismarck Archipelago
Ireland
Kavieng   New Hanover
Ysabel Channel
Bismarck Sea   Rabaul
Long   Ulamona   New Britain
Manam   Island
Madang   Kimbe
Bogia   Gasmata
NEW   Umbol
PAPUA   GUINEA   Morobe
Lae
NEW GUINEA   Wau
Gulf   Bereina   PORT
of Papua   MORESBY   Abau

Trobriand
Islands
Losuia

D'Entrecasteaux Is
Boluboldi
Goschen Strait
Alotau
KWIKILA
Samarai

150°

AUSTRALIA

Melville
Island   Croker Island
Milikapiti
Bathurst Island   *Van Diemen*
Gulf
Beagle Gulf   Jabiru
Batchelor   Darwin   Milingimbi
Adelaide River   Pine Creek   Maningrida
Arnhem
Land   *Alyangula*

Cape Wessel
Wessel Islands
Nhulunbuy
Cape Arnhem   Gulf
of
Carpentaria

Cape Grenville
Weipa   Lockhart River
Cape Melville
Coen   Cape
Flattery
Cape York
Peninsula   Cooktown
Laura

Timor Sea

C   135°   D

© Collins Bartholomew Ltd

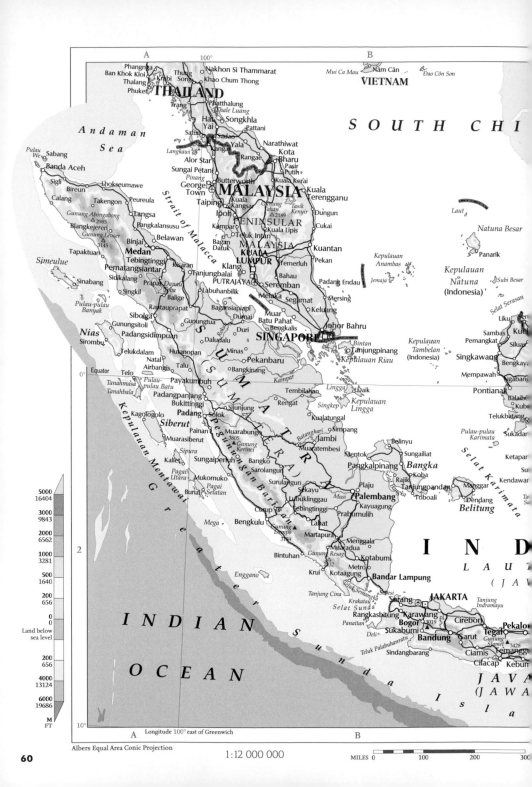

A   100°   B

Andaman
Sea

Phangnga
Ban Khok Kloi
Thalang
Phuket
Krabi
Thung
Song
Trang
Phatthalung
Thale Luang
Hat
Yai
Songkhla
Pattani
Sadao
Yala
Narathiwat
Kota
Bharu
Pasir
Putih
Kuala Kerai
Kuala
Terengganu
Dungun
Cukai

Nakhon Si Thammarat
Khao Chum Thong

THAILAND

Mui Ca Mau   Nam Căn   *Đao Côn Son*

VIETNAM

S O U T H   C H I

Pulau
Wě
Sabang
Banda Aceh
Sigli
Bireun
Calang
Takengon
Peureula
Lhokseumawe
Gunung Abongabong
△ 2985
Blangkejeren
Gunung Leuser
△ 3145
Binjai
Belawan
Medan
Tebingtinggi
Pematangsiantar
Sidikalang
Danau
Toba
Prapat
Balige
Rantauprapat
Gunungtua
Labuhanbilik
Bagansiapiapi
Dumai
Duri
Daludalu
Huktanopan
Minas
Pekanbaru
Talu
Bangkinang
Airbangis
Kampar

Langkawi
Alor Star
Kangar
Sungai Petani
Pinang
George
Town
Butterworth
Taiping
Ipoh
MALAYSIA
Kuala
Kangsar
PENINSULAR
Gunung
Tahan
△ 2189
Kampar
Teluk Intan
Bagan
Datuk
KUALA
LUMPUR
Klang
PUTRAJAYA
Seremban
Bahau
Meláka
Segamat
Muar
Batu Pahat
Bengkalis
SINGAPORE
Johor Bahru
Tasik
Kenyir
Kuala Lipis
Temerluh
Pekan
Kuantan
Padang Endau
Mersing
Keluang
Bintan
Tanjungpinang
Kepulauan
Riau

Laut

Natuna Besar

Panarik

Kepulauan
Anambas

Kepulauan
Natuna
(Indonesia)

Jemaja

Subi Besar

Selat Serasan

Liku
Sambas
Pemangkat
Singkawang
Kepulauan
Tambelan
(Indonesia)

Mempawah

Pontianak

Kuc
Siluas
Ngabar
Bengkaya

Simeulue

Sinabang
Singkil
Sibolga
Gunungsitoli
Padangsidimpuan
Nias
Sirombu
Telukdalam
Natal

Pulau-pulau
Banyak

Equator
Telo
Tanahmasa
Tanahbala
Pulau-
pulau Batu
Payakumbuh
Kagologolo
Bukittinggi
Padangpanjang
Padang
Solok
Muarabungo
△ *Gunung*
Kerinci
3805
Sungaipenuh
Bangko
Sarolangun
Surulangun
Sekayu
Lubuklinggau
Tebingtinggi
Curup
△ *Gunung*
Dempo
3159
Bengkulu
Martapura
Bintuhan
Krui
Kotaagung
Teluk Semangka

Siberut
Painan
Muarasiberut
Sipura
Kaliet
Mukomuko
Pagai
Utara
Pagai
Selatan
Buriat

0°

Sijunjung
Batanghari
Muaratembesi
Jambi

Tembilahan
Rengat
Kualatungal

Simpang
Mentok
Plaju
Rajik
Palembang
Kayuagung
Prabumulih
Lahat
Menggala
Muaradua
Metro
Kotabumi
Bandar Lampung

Lingga
Daik
Singkep
Kepulauan
Lingga

Balaibe
Kubi
Telukbatang
Ngabar

Belinyu
Sungailiat
Pangkalpinang
Koba
Tanjungpandan
Toboali
Bangka
Manggar
Dendang
Belitung

Pulau-pulau
Karimata

Ketapar
Su

Selat Karimata

Kendawar
Ta
Sa

I N D

L A U T

(J A V A

S U M A T E R A

Pegunungan Barisan

Greater

Kepulauan Mentawai

Enggano

I N D I A N

O C E A N

Sunda

Islā

Krakatau
Selat Sunda
Panaitan
Deli
Serang
Rangkasbitung
Bogor
△ *Gunung*
2019
Sukabumi
Sindangbarang
Teluk Palabuhanratu
JAKARTA
Karawang
Cirebon
Garut
Bandung
Ciamis
Cifacap
Tanjung
Indramayu
Teganggl
Pekalo
Gunung
Slamet
△ 3428
Kebun

J A V A
(J A W A

Tanjung Cina
Kotaagung

2

Land below
sea level

5000	16404
3000	9843
2000	6562
1000	3281
500	1640
200	656
0	0
200	656
4000	13124
6000	19686

M
FT

A   Longitude 100° east of Greenwich   B

Albers Equal Area Conic Projection

60

1:12 000 000

MILES 0   100   200   3

C   120°   D

Palawan   Rio Tuba

Balabac   Bugsuk

Balabac

SULU

SEA   Balabac Strait

Banggi

Roxas   Oroquieta

Liloy   Ozamis

Siocon   Pagadian   Iligan

Zamboanga   Cotabato

Peninsula   Datu Piang

Cagayan de   Tawi-Tawi

Kudat   Kanibongan

Kota Belud

Gunung   Kinabalu

Sulu   Zamboanga   *Moro*

Archipelago   Isabela   *Gulf*

Basilan   Lebak

Kota   △4095   Jolo   Iolo   **PHILIPPINES**

Kinabalu   Ranau   Sandakan

Gunung Trus Madi

Beaufort   △2649   Siasi

Labuan   Lamag   Tawi-Tawi

BANDAR SERI   Tenom   Tambisan

BEGAWAN   Lahad

BRUNEI   Kuamut   Datu   Balimbing

Kuala Belait   Tomani   **SABAH**   Sibutu

Lutong   Seria   Pensiangan

Miri   Tumbis   1

△Bukit Harden   Tawau

2136   Mensalong   *CELEBES*

Labang   Kubuang   *SEA*

Bintulu   Long   Tarakan

Akah

Igan   Mukah   Belaga

rik   Sibu   **SARAWAK**   Tanjungselor

kei   Kapit   Datadian

Saratok   *Rajang*   Tanjungredeb

Debak

afian   Sri Aman   △2988   Sepinang   Tolitoli   Kwandang

Lubok   Putusibau   Longwai   *Tanjung*   *Semenanjung Minahasa*

Antu   *Mahakam*   *Mangkalihat*   Gorontalo

Semitau   **BORNEO**   Sangkulirang   *Kepulauan*

au   Sintang   Longiram   Bontang   Moutong   *Togian*

hpinou   Sidoan   *Tanjung*   0°

Muaralaung   Tenggarong   *Teluk*   Togian   *Pangkalsiang*

Pegunungan Schwaner   Tewah   Samarinda   Tomali   *Tomini*   Batudaka

atayap   Rantaupanjang   Muarateweh   Donggala   Luwuk   *Peleng*

angtiti   **K A L I M A N T A N**   Samboja   Palu   Mapane   Poso   Tataba

Palangkaraya   Balikpapan   *Selat Makassar*   Uekuli   Banggai

gkalanbuun   Tanahgrogot   (Macassar Strait)   Tentena   Kolonedale   *Kepulauan*

Sampit   Tanjung   Babana   *Teluk Towori*   *Banggai*

Kualapembuang   Amuntai   Mamuju   **CELEBES**   Wotu   *Manui*

Kandangan   Bukit △3074   (SULAWESI)   Wowom

Tanjung   Kotabaru   *Gandadiwata*   Masamba

Puting   Banjarmasin   Sambo   Rantepao   Mulamela

Martapura   *Sebuku*   Majene   Makale   Palopo   Kendari

Pagatan   Polewali   Kolaka

Laut   Parepare   Anabanua

Tanjung   Majene   Singkang   Raha

Selatan   Watampone   *Buton*

O N E S I A   *Kepulauan*   Maros   Sinjai   *Kabaena*   Baubau   2

Laut Kecil   *Gunung Lompobattang*

A W A   **Makassar**   Bulukumba   *Batuata*

E A)   (Ujung Pandang)

pulau   *Masalembu*   Bontosunggu   *Salayar*

unjawa   *Bawean*   *Besar*   Benteng

Tanjung   *Kepulauan*   *Sabalana*   *Tanahjampea*   *Kalao*   *Kalaotoa*

Bugel   *Kangean*

Pati   *Madura*   Sumenep   *Kepulauan*   *Kepulauan Bonerate*

urwodadi   Bangkalan   Anjasa   *Tengah*   *Laut Flores*   *Kepulauan*

arta   Jombang   **Surabaya**   Genteng Raas   *Laut Bali*   (Flores Sea)   *Solor*

Madiun   *Selat Madura*   Situbondo   (Bali Sea)   Reo   *Flores*   Larantuka

akarta   Malang   Pasuruan   *Sumbawa*   Labuanbajo   Maumere   Labala

itan   Nganjuk   Banyuwangi   *Gunung*   *Gunung*   △2821   Ruteng   Bajawa   Ende

Lumajang   Jember   *Semeru*   *Raung*   Taliwang   Tambora   Raba

Singaraja   Mataram   Alas   Dompu

Barung   Glawar   Sumbawabesar   Plampang   *Laut Sawu*

Denpasar   Praya   *Selat Lombok*   *Selat Sumba*   Memboro   (Savu Sea)

Bali   *Lombok*   Taliwang   *Sumba*   Waingapu

S   Waikabubak   10°

C   120°   D

© Collins Bartholomew Ltd

0   200   400 KM

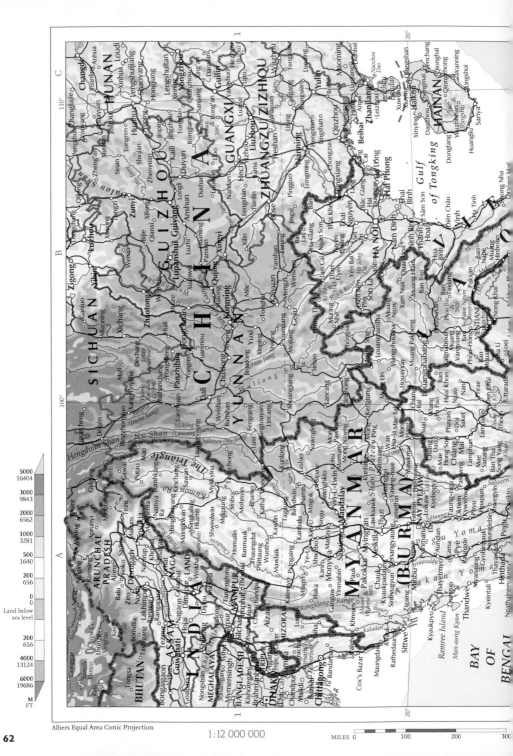

Albers Equal Area Conic Projection

1:12 000 000

MILES 0   100   200   300

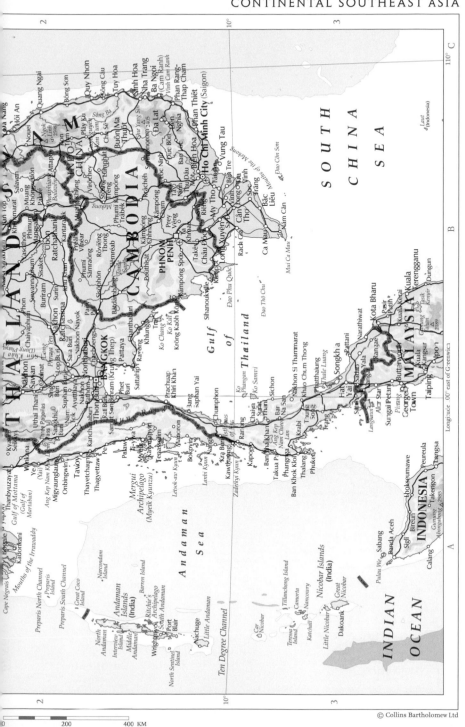

SOUTH CHINA SEA

V I E T N A M

L A O S

C A M B O D I A

T H A I L A N D

Gulf of Thailand

Andaman Sea

INDIAN OCEAN

INDONESIA

MALAYSIA

Gulf of Martaban (Gulf of Mottama)

Mouths of the Irrawaddy

Meryui Archipelago (Myeik Kyunzu)

Nicobar Islands (India)

Andaman Islands (India)

Preparis North Channel

Preparis South Channel

Ten Degree Channel

BANGKOK

PHNOM PENH

Ho Chi Minh City (Saigon)

Da Nang
Hoi An
Quang Ngai
Quy Nhon
Song Cau
Tuy Hoa
Nha Trang
Vinh Hoa
Ba Ngoi (Cam Ranh)
Phan Rang–Thap Cham
Phan Thiet
Vung Tau
Bien Hoa
Da Lat

Longitude 00° east of Greenwich

Da Nang

0   200   400 KM

© Collins Bartholomew Ltd

PHILIPPINES

A   120°   **B**

1

20°

Dongsha Qundao

Luzon Strait

Batan Islands
Itbayat Basco
Batan

Balintang Channel

Babuyan

Calayan *Babuyan Islands*

Fuga Camiguin

Babuyan Channel

Bangui
Laoag
San Vicente
Aparri

Bangued
Vigan   Tuguegarao
Mount Chico
Tagudin   *Sapocoy*   Ilagan
Bontoc   Palanan
San Fernando   Mount Pulog
La Trinidad   △ 2928
Baguio   Bayombong   Santiago
Dagupan
Lingayen   San Carlos
Tarlac   San Jose
Mount Pinatubo   Cabanatuan
Iba 1660
Angeles   Gapan
Olongapo   San Fernando
Balanga   Valenzuela
Tagaytay City   **Quezon City**
MANILA   Pasig
Santa Cruz
Tagaytay City   Lucena   Labo   Pandan
San Pablo   Lopez   Libmanan   Naga
Lubang Islands   Batangas   Calapan   Beac   Oas   Tabaco
Mamburao   Naujan   Legaspi   Sorsogon
Mount Halcon   Burias   Irosin
△ 2585   *Mindoro*   Roxas   *Sibuyan*   Catarman
San Jose   Romblon   Masbate   Calbayog
Busuanga   *Tablas*   *Masbate*   **Samar**
Calamian Group   Pandan   *Sibuyan Sea*   Catbalogan
Coron   Roxas   *Visayan Sea*   Tacloban
Culion   Culasi   **Panay**   Guiuan
El Nido   *Cuyo Islands*   Potolan   Cadiz   Ormoc
Dalanganem Islands   San Jose del   Iloilo   Bacolod   Cebu   **Leyte**   Dinagat
Tayta   Buenavista   △ 2450   Talisay   *Bohol*   Siargao
Roxas   *Dumaran*   **Negros**   Cebu   Maasin   Dapa
Cauayan   Tagbilaran   Surigao
Palawan   Puerto Princesa   Tanjay   *Bohol Sea*   Tandag
Quezon   Apurahuan   Bayawan   Siquijor   Mambajao
Aborlan   Dumaguete   Camiguin   Butuan
Mount Mantalingajan   Dipolog   Cagayan de Oro   Gingoog
△ 2054   Roxas   Oroquieta   Bislig
Brooke's Point   Liloy   Ozamiz   **MINDANAO**   Baganga
Bugsuk   Siocon   Pagadian   Mount Ragang   Malaybalay
Rio Tuba   Zamboanga Peninsula   ▲ 2815
Balabac   Cotabato   Mount Apo   Davao
Balabac   Zamboanga   Datu Piang   ▲ 2954   Digos   Mati
Balabac Strait   *Moro Gulf*   Lebak   *Davao Gulf*
Banggi   Isabela   Banga   General Santos
Kudat   *Cagayan de Tawi-Tawi*   *Basilan*   Kiamba   Batulaki
Kota Belud   Kanibongan   Jolo   *Sarangani Islands*
Turtle Islands (Philippines)   *Jolo*
Kota Kinabalu   *Gunung Kinabalu*   Sandakan   Siasi   *Sulu Archipelago*
Ranau   △ 4095   Tamag   Tambisan   *Tawi-Tawi*
△ Trus Madi   Balimbing   *Sibutu*
2649   Lahad Datu
MALAYSIA   Tenom   Kuamut
Lawas   **SABAH**   Semporna
Tomani   Pensiangan
Tawau
INDONESIA   Lumbis
Mensalong
Kubuang   Tarakan

PHILIPPINE SEA

PHILIPPINES

SOUTH CHINA SEA

Scarborough Shoal

Lubang Islands

Mindoro Strait

Palawan Passage

Cordillera Central

Catanduanes

Polillo Islands

LUZON

Cordillera Rimbing

SULU SEA

Crocker Range

CELEBES SEA

INDONESIA

Kepulauan Nanusa

Kepulauan Talaud

Karakelong   Pulutan

Sangir   Tahuna   Kaburuang

2

10°

3

5000
16404

3000
9843

2000
6562

1000
3281

500
1640

200
656

0
0
Land below
sea level

200
656

4000
13124

6000
19686

M
FT

A   Longitude 120° east of Greenwich   **B**

Albers Equal Area Conic Projection

1:12 000 000

KM 0   200

MILES 0   100   200

A 125° **B** 130° **C**

Siping
Yingchengxi
Huadian
Songhua Hu
Wangqing
Tumen
Hunchun
Kangping
Liaoyuan
Panshi
Laotougou
Yanji
R.U.S. FED.
Changtu
Meihekou
Baishanzhen
Zengfeng
Zhabino
Faku
Kaiyuan
Huinan
Jingyu
Fusong
Baihe
Sŏnbong
Zhangwu
Xinmin
Qingyuan
Baishan
Songjianghe
Baitou Shan
Helong
Musan
Puryŏng
Najin
Tonghua
Jinjiang
(Paektu-san)
Samjiyŏn
Ch'ŏngjin
CHINA
Linjiang
Changbai
Paegam
Kwanmo
Orang
Fushun
Hun He
Lao Ling
Chasŏng
Hyesan
Kapsan
Myŏnggan
LIAONING
Huanren
Laoling
Ji'an
Nanp'o
Puksubaek-
Kilchu
Shenyang
Guanshui
Kanggye
Sŏnggan
san
P'ungsan
Liaoyang
Benxi
Kuandian
Ch'osan
2522
Kimch'aek
Anshan
Maoku
Fengcheng
Sakchu
Changjin
Tanch'ŏn
Haicheng
Shan
Pukchin
Hongwŏn
Pukch'ŏng
Dashiqiao
1110
Uiju
Sinŭiju
Huich'ŏn
Hamhŭng
Sinp'o
Buyun
Dandong
NORTH
Kujiong
Chŏngp'yŏng
Hŭngnam
Shan
Gushan
Anju
KOREA
Zhuanghe
Dongang
Sidari
Sunch'ŏn
Yŏnghŭng
Wŏnsan
Sikuaishi
Korea Bay
P'yŏngsong
Yangdok
Anbyŏn
Sunan
Kangdong
Kosan
P'YŎNGYANG
Chinghwa
Hoeyang
Kosŏng
Namp'o
Kwail
Ch'angdo
Sŏngnim
P'yŏnggang
Ich'ŏn
Ch'ŏrwŏn
Sariwŏn
Sŏkch'o
Chaeryŏng
Changyŏn
Haeju
P'aro-ho
Sokch'o
Kaesŏng
Tongduch'ŏn
Ch'unch'ŏn
Ongjin
Soyang-ho
Kangnŭng
Ŭijŏngbu
Samch'ŏk
Inch'ŏn
SEOUL (Sŏul)
Anyang
Sŏngnam
Wŏnju
Tonghae
Ansan
Suwŏn
Chech'ŏn
P'yŏngt'aek
SOUTH
Sosan
Ch'ŏnan
T'aebaek
Ulchin
Yesan
KOREA
Ch'ŏngju
Andong
Yŏngdŏk
YELLOW
Poryŏng
Kongju
Sangju
Ŭisŏng
SEA
Nonsan
Taejŏn
Kimch'ŏn
P'ohang
(HUANG HAI)
Kunsan
Iksan
Muju
Kumi
Taegu
Kyŏngju
Chŏnju
Koryŏng
Miryang
Ulsan
Chŏngŭp
Ch'angsŏng
Namwŏn
Chiri-san
Ch'angwŏn
Kimhae
1915
Chinju
Masan
Pusan
Kwangju
Sach'on
Chinhae
Mokp'o
Sunch'ŏn
T'ongyŏng
Korea Strait

SEA

OF

JAPAN

(EAST SEA)

Ullŭng-do
(S. Korea)

Tsushima
Izuhara

Paengnyŏng-do
(S. Korea)

Kyŏnggi-man

Haeju-
man

Chin-do
Chindo
Haenam

Cheju-haehyŏp

Cheju-do
(S. Korea)
Cheju
Halla-san
1950
Taejŏng

Higashi-suidō
Iki-shima
Shimonoseki
Kita-Kyūshū
Fukuoka
Iizuka
Karatsu
Imari
Kurume
Sasebo
Saga
JAPAN

40°
1
2
35°
3

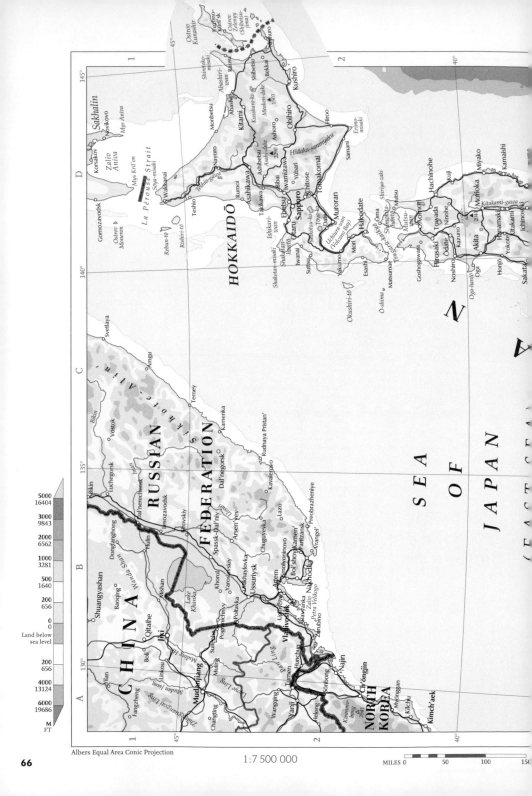

Sakhalin

Korsakov
Novikovo
Zaliv
Aniva
Mys Aniva

Mys Kril'on

La Pérouse Strait

Ostrov
Moneron

Gornozavodsk

Rebun-tō

Rishiri-tō

Ostrov Zelenyy
(Shibotsu-jima)
Ostrov
Yuzhno-Kuril'sk
Ostrov Kunashir

Shiretoko-misaki
Abashiri-wan
Nemuro
Rausu
Shibetsu
Bekka
Kushiro

45°

1

2

40°

145°

Shiretoko-misaki
Abashiri
Abashiri-wan
Kussharo-ko
Meakan-dake
1503
Kitami
Mombetsu
Akkeshi
Ashoro
Obihiro
Hiroo
Erimo-misaki

Teshio-gawa
Nayoro
Rumoi
Takikawa
Ashibetsu
Asahikawa
Bibai
Iwamizawa
Yubari
Chitose
Tomakomai
Samani
Hidaka-sanmyaku
Asahi-dake
2290

HOKKAIDŌ

Wakkanai

Teshio

Ishikari-gawa
Ebetsu
Sapporo
Otaru
Tomakomai
Tōya-ko
Date
Muroran
Uchiura-wan
(Volcano Bay)

Ishikari-wan
Shakotan-misaki
Iwanai
Suttsu
Yakumo
Mori
Hakodate

Okushiri-tō
Esashi
Matsumae

Shiriya-zaki
Ōma
Mutsu

Shimokita-
hantō
Tappi-zaki
Kitakami-gawa

Miyako

140°

N

SEA

OF

JAPAN

(EAST SEA)

Kamaishi

Miyako
Morioka
2041

Hachinohe
Towada-ko
Kuji
Aomori
Noheji
Goshogawara
Towada
Akita
Hanamaki
Ōdate
Kazuno
Yokote
Ichinoseki
Hirosaki
Ninohe
Honjō
Oga
Oga-hantō
Noshiro
Sakata

Svetlaya

Amgu

Terney

RUSSIAN

FEDERATION

Sikhote-Alin'

Bikin

Iman

Ussuri

Vostok

Kavalerovo
Kamenka
Rudnaya Pristan'
Dal'negorsk

Bikin
Luchegorsk
Dal'nerechensk
Vesozavodsk
Dongfanghong
Hulin

Mishan
Lake
Khanka

Smirnovskiy
Lazo
Preobrazheniye
Vrangel'

Kavalerovo
Arsen'yev
Chuguyevka
Yarosiavka
Mikhaylovka
Smolyaninovo
Kamen'
Bol'shoy Kamen'
Nakhodka
Partizansk

Shuangyashan

Baoqing

Qitaihe

135°

B

Khorol
Spassk-Dal'niy

Ussuriysk

Artem
Vladivostok
Zaliv
Petra Velikogo

Wanda Shan

Poltavka
Pogranichnyy

Ussuriysk
Razdol'noye
Slavyanka
Zarubino

Boli

Linkou

Ilki

Mulan Jiang

Suifenhe
Dongning

Hunchun

Wangqing

Tumen

Najin
Sŏnbong

Hún Jiang

Lake

Tumen

CHINA

Fangzheng
Yilan

Mudan Jiang

Mudanjiang

Changting

Zhangguangcai Ling

Helong

Yanji
Helong
Wangqing

Laoye Ling

Kwanmo-
bong
2541

NORTH
KOREA

Ch'ŏngjin

Musan
Myŏnggan
Kilchu
Kimch'aek

45°

1

2

40°

130°

A

5000
16404

3000
9843

2000
6562

1000
3281

500
1640

200
656

0
0

Land below
sea level

200
656

4000
13124

6000
19686

M
FT

1:7 500 000

MILES 0   50   100   150

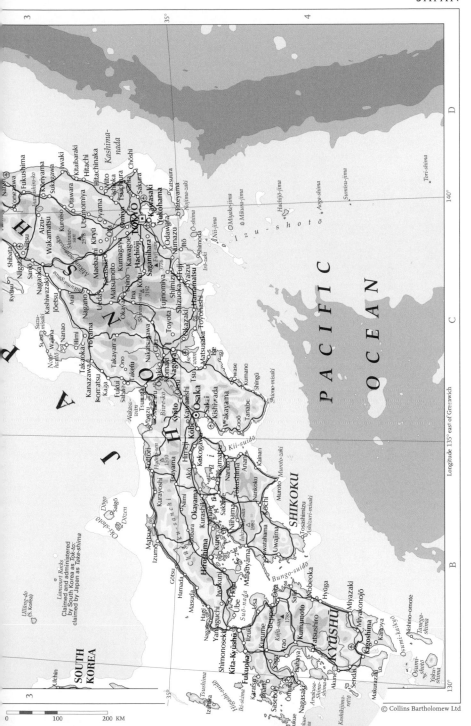

PACIFIC OCEAN

SHIKOKU

KYUSHU

SOUTH KOREA

© Collins Bartholomew Ltd

Longitude 135° east of Greenwich

Claimed and administered
by South Korea as Tok-to;
claimed by Japan as Take-shima

Liancourt Rocks

Ulleung-do
(S. Korea)

0   100   200 KM

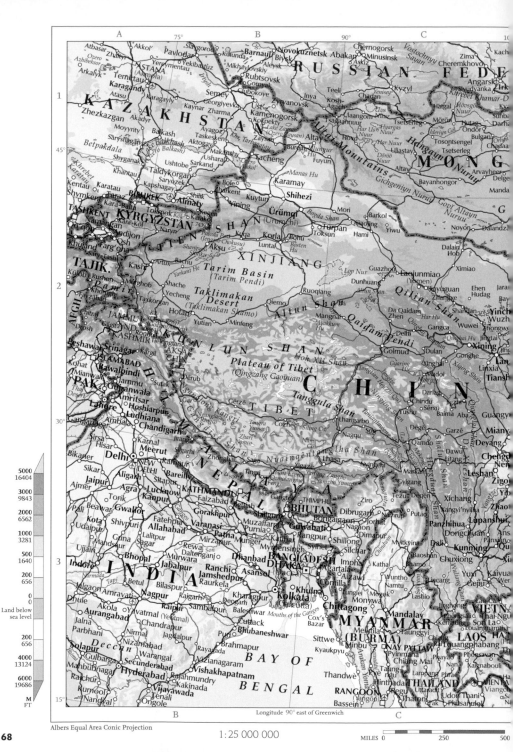

Albers Equal Area Conic Projection

Longitude 90° east of Greenwich

1 : 25 000 000

MILES 0   250   500

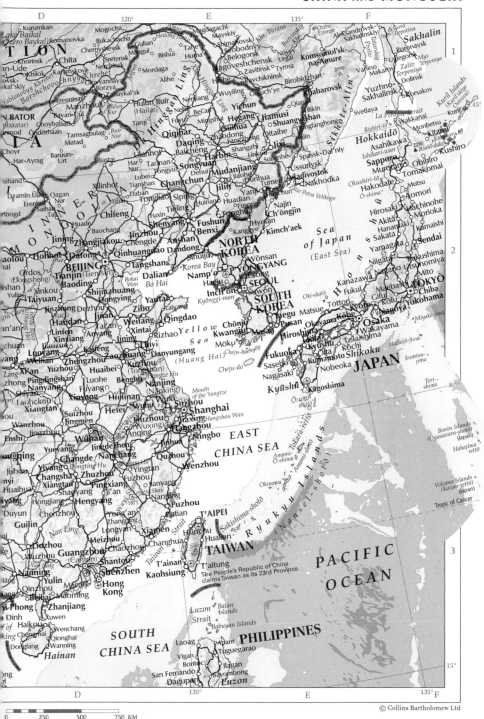

0   250   500   750 KM

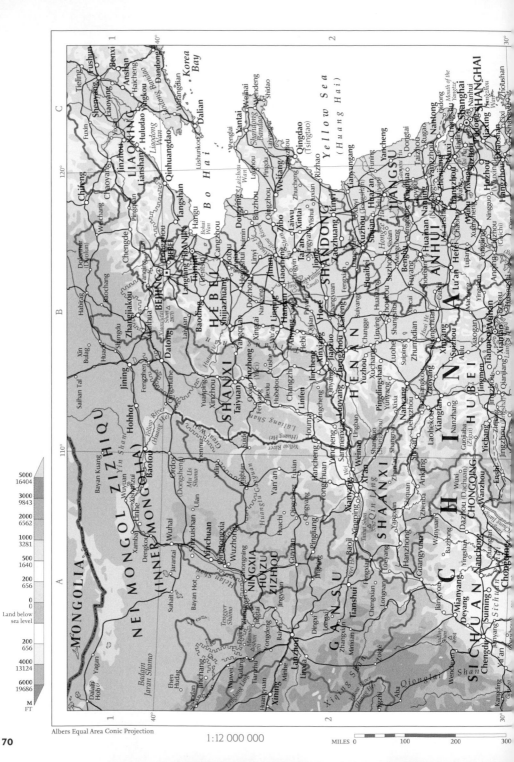

Albers Equal Area Conic Projection

1:12 000 000

MILES 0   100   200   300

EAST CHINA SEA

SOUTH CHINA SEA

TAIWAN

The People's Republic
of China claims Taiwan
as its 23rd Province

Luzon Strait

Bashi Channel

Balintang Channel

Babuyan Channel

PHILIPPINES

LUZON

ZHEJIANG

FUJIAN

JIANGXI

HUNAN

GUANGDONG

GUANGXI ZHUANGZU ZIZHIQU

GUIZHOU

YUNNAN

Hong Kong

Macao

HAINAN

Gulf of Tongking

VIETNAM

HANOI

LAOS

VIENTIANE

THAILAND

Taiwan Strait

Chungyang Shanmo

Matsu Tao (Taiwan)

Quemoy

Longitude 110° east of Greenwich

© Collins Bartholomew Ltd

0   200   400 KM

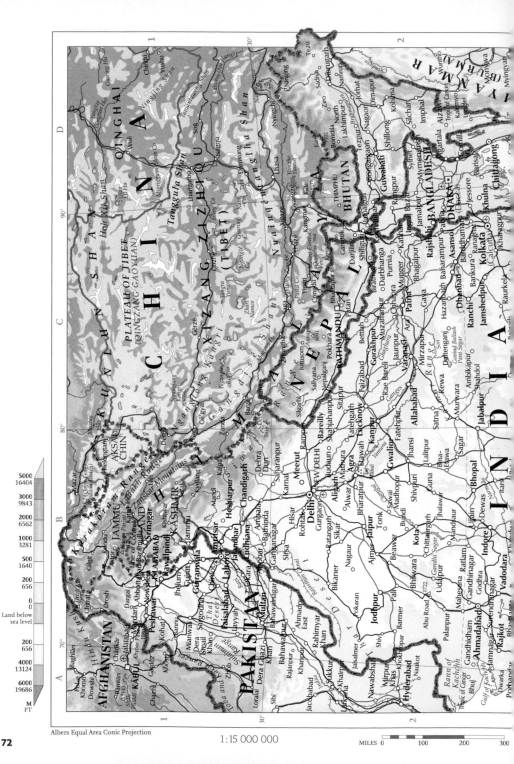

Albers Equal Area Conic Projection

1:15 000 000

MILES 0   100   200   300

5000
16404
3000
9843
2000
6562
1000
3281
500
1640
200
656
0
0
Land below
sea level
200
656
4000
13124
6000
19686
M
FT

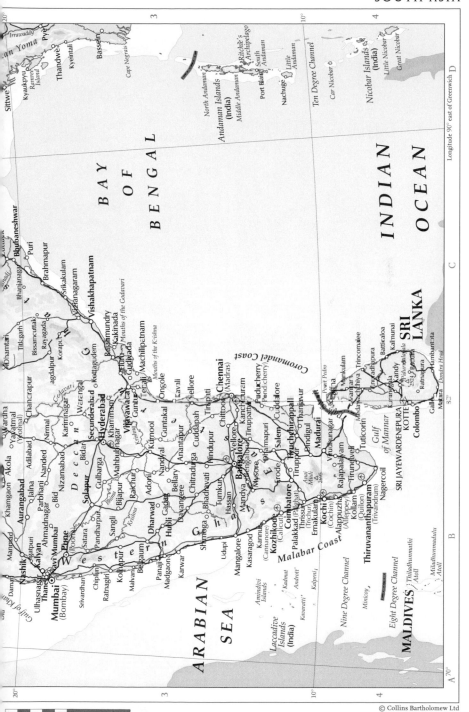

© Collins Bartholomew Ltd

0   250   500 KM

A B

AFGHANISTAN

Garabil Belentligi
TURKMENISTAN
Andkhvoy
Sheberghan
Mazar-e Sharif
Khānābād
Khānābād
Kholm
Fayzābād
Qullai Karl Marks
Barā Gonbad
Pasu
Morghāb
Maymanah
Sar-e Pol
Āybak
Baghlān
Taloqān
Ishkoshim
Tirich Mir
Battura Glacier
Mazar
Serhetabat
Qal'eh-ye Now
Merghab
Dowshī
Pol-e Khomrī
Bāzārak
Mastuj
Chitral
Gilgit
Chilas
Kondu
Qogir Feng (Godwin Austen)
8126
Skardu
Sar-e Pol
Safīd Kūh
(Paropamisus)
Bāmiān
Chārīkār
Nūrestan
Drosh
Dir
Nanga Parbat
8126
Astor
Skardu
Khāplu
Kargil
Kūh-e Bāba
Shāh Fōladi
Maīdān Shahr
Sikaram 4761
Mehtar Lam
Jalālābād
Mardan
Dargai
Mongora
Abbottābād
Sopur
Sutak
Ladākh Ra.
Zanskar
Kūh-e Māzow
Hazare-e Qeyāzi
4112
Ghaznī
Sharan
Orgūn
Khowst
Gardēz
Peshāwar
Nowshera
Wah
Hripur
Baramula
Srinagar
Anantnag
Kishtwar
HIMACHAL
Kūh-e Chehel
Abdalan
Delārām
Chaghcharān
Harī Rūd
Tarīn Kowt
Kowt
Thal
Daud Khel
Talagang
Jhelum
Gujrat
Jammu
Udhampur
Chamba
Kyelang
Kandahār
Qalāt
Bannu
Marwat
Mianwali
Khushāb
Dhera
Wazirābād
Sialkot
Kathua
Sundarnagar
Mandi
Gershk
Lashkar Gāh
Dera Ismail Khan
Takht-i-Sulaiman 3374
Bhakkar
Chiniot
Gujrānwāla
Batāla
PRADESH
Sialkot
Amritsar
Hoshiarpur
Simla
Helmand
Tank
Sargodha
Jhang
Shorkot
LAHORE
Firozpur
Jalandhar
Ludhiana
Chandigarh
Muslimbagh
Zhob
Layyah
Fatehabad
Patiala
Ambala
Saharanpur
Dasht-e Arbū Shomālī
Pishin
Quetta
Mach
Loralai
Taunsa
Ahmadpur Sial
Khanewal
MULTAN
Burewala
Sahiwal
Abohar
Bathinda
Karnal
Rohtak
Meerut

PAKISTAN

Amir Chah
Chagai
Dalbandin
Hāmūn-i Lora
Nushki
Kalāt
Sibi
Lahri
Dera Ghazi Khan
Jāmpur
Rajanpur
Muzaffargarh
Lodhran
Bahawalnagar
Hanumangarh
Lisa
Nohar
Hisar
Bhiwani
Gurgaon
DELHI
Ghaziabad
Moradabad
Rohri
NEW DELHI
Nok Kundi
Yakmach
Hamun-i Māshkel
Siahan Range
Khārān
Nāgha Kalāt
Shahdad Kot
Shikarpur
Kandiaro
Jacobābād
Sukkur
Khairpur
Ghotki
Sadiqabad
Khanpur
Rahimyar Khan
Anupgarh
Pugal
Sardarshahr
Ratangarh
Bikaner
Churu
Rajgarh
Jhunjhunun
Sikar
Alwar
Mathura
Agra
Qila Ladgasht
Washuk
Karodi
Kharan
Khuzdar
Larkāna
Dadu
Nawabshah
Jaisalmer
Pokaran
Phalodi
Nokha
Nāgaur
Sujangarh
Sambhar
JAIPUR
Bharatpur

RAJASTHAN

Zāmarod
Panjgur
Diz
Central Brāhui Range
Pab Range
Wadh
Bela
Diwana
Sakrand
Khipro
Shiv
Balotra
Barmer
Jodhpur
Merta
Ajmer
Sawai Madhopur
Tonk
Devli
Bundi
Shivpuri
Gwalior
Tump
Turbat
Hoshab
Bāghi
Bazdar
Goshanak
Uthal
Tando Adam
Mirpur Khas
Khokhropar
Nagar Parkar
Sirohi
Jalore
Pali
Deogarh
Bhilwara
Chittaurgarh
Kota
Jhalawar
Baran
Lalitpur
Suntsar
Dasht
Makran
Coast Range
Sonmiani Bay
Hyderabad
Thano
Bula Khan
Mithi
Abu Road
Guru Sikhar
Palanpur
Sidhpur
Udaipur
Neemuch
Mandsaur
Garoth
Guna
Bina-Eta
Gwadar
Pasni
Ormara
Sonmiani
Thatta
Sujawal
Badin
Naukot
Sīrohi
Radhanpur
Himatnagar
Dungarpur
Banswara
Jaora
Agar
Ujjain
Biaora
Vidisha

KARACHI

Mouths of the Indus

Tropic of Cancer

Lakhpat
Bhuj
Gandhidham
Mahesana
Gandhinagar
Ratlam
Dewas
BHOPAL
Rann of Kachchh
Radhanpur
Viramgam
Dahod
Dhar
Mhow
Harda
Itarsi
Rapur
Kandla
Gulf of Kachchh
Surendranagar
Dhandhuka
AHMADABAD
Godhra
Nadiad
Indore
MADHYA P.
Okha
Dwarka
Jamnagar
Rajkot
Gondal
Upleta
Dhasa
Bhavnagar
Alirajpur
Rajpur
Narmada
Khargon
Khandwa
Chhindwara

GUJARAT

Porbandar
Junagadh
Keshod
Visavadar
Amreli
Mahuva
Bharuch
Nandurbar
Satpura Range
Burhanpur
Achalpur

**ARABIAN
SEA**

Veraval
Diu
Gulf of Khambhat
Surat
Vyara
Jalgaon
Bhusawal
Akola
Wardha
Hingan

Administrative areas not named on the map:
INDIA
1. DADRA AND NAGAR HAVELI (B2)
2. DAMAN AND DIU (B2)

Dahanu
Silvassa
Daman
Nashik
Manmad
Aurangabad
Pusad
Jalna
Parbhani
Nanded

MAHARASHTRA

Ulhasnagar
Kalyan
Thane
Navi
Narayangaon
Ahmadnagar
Mumbai (Bombay)
Mumbai

Longitude 70° east of Greenwich

1 2 3

5000 16404
3000 9843
2000 6562
1000 3281
500 1640
200 656
0 0
Land below sea level
200 656
4000 13124
6000 19686
M FT

Albers Equal Area Conic Projection

74

1:12 000 000

MILES 0   100   200   300

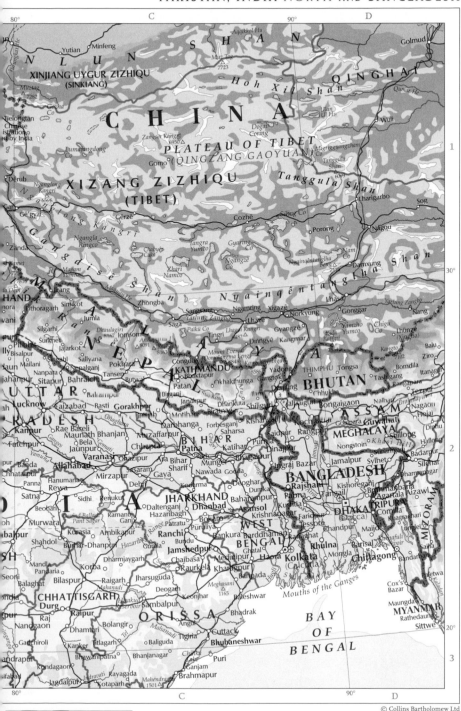

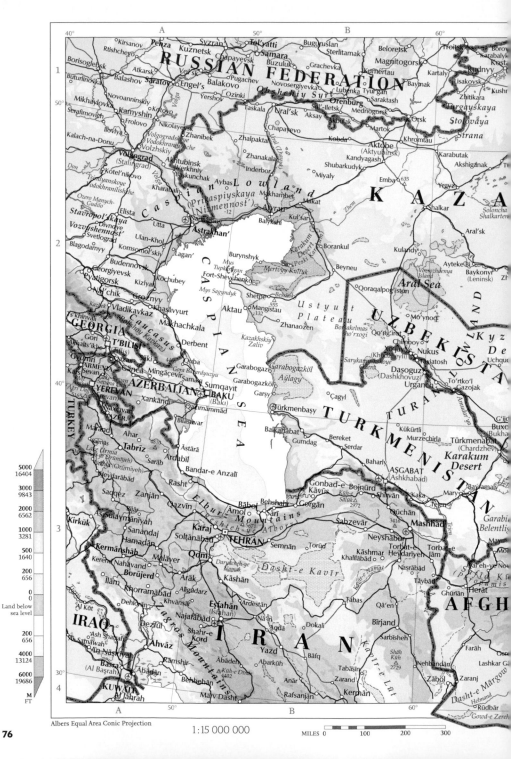

Albers Equal Area Conic Projection

1:15 000 000

MILES 0   100   200   300

Map labels (top to bottom, left to right):

RUSSIAN FEDERATION

KAZA

KAZA

UZBEKISTA

TURKMENISTAN

TURKEY

GEORGIA

ARMENIA

AZERBAIJAN

IRAQ

IRAN

KUWAIT

AFGH

Scale / Elevation legend:

5000 / 16404
3000 / 9843
2000 / 6562
1000 / 3281
500 / 1640
200 / 656
0
Land below sea level
200 / 656
4000 / 13124
6000 / 19686
M
FT

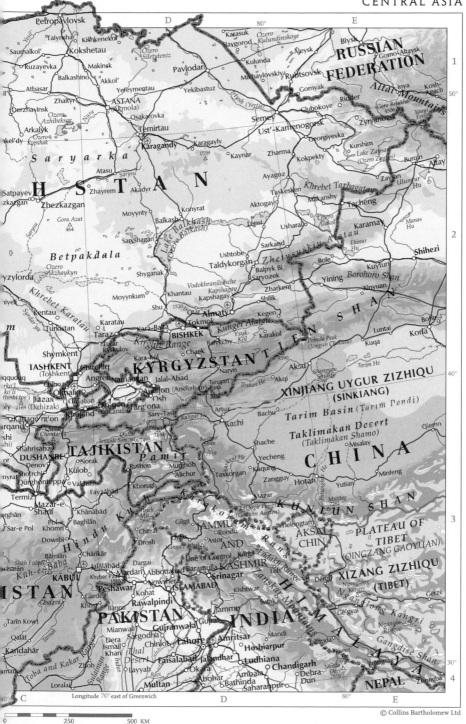

Petropavlovsk
Taiynsha
Saumalkol'
Ruzayevka
Makinsk
Kishkenekol
Karasuk
Ozero
Slavgorod Kulundinskoye
Biysk
Gorno-Altaysk
RUSSIAN
FEDERATION
Kokshetau
Balkashino
Akkol'
Pavlodar
Mikhaylovskiy
Rubtsovsk
Aleysk
Gornyak
Inya
Kosh-Agach
Atbasar
Zhaltyr
Yereymentau
Yekibastuz
Georgiyevka
Gluboskoye
Ridder
Zyryanovsk
Gora Belukha
4506
Youyi
Feng
Derzhavinsk
ASTANA
(Akmola)
Osakarovka
Semey
Ust'-Kamenogorsk
Kurshim
Lake Zaysan
(Ozero Zaysan)
Burqin
Arkalyk
Atasu
Zharma
Kokpekty
Zaysan
Ulungur
Altay
Saryozek

Longitude 70° east of Greenwich

0   250   500 KM

© Collins Bartholomew Ltd

77

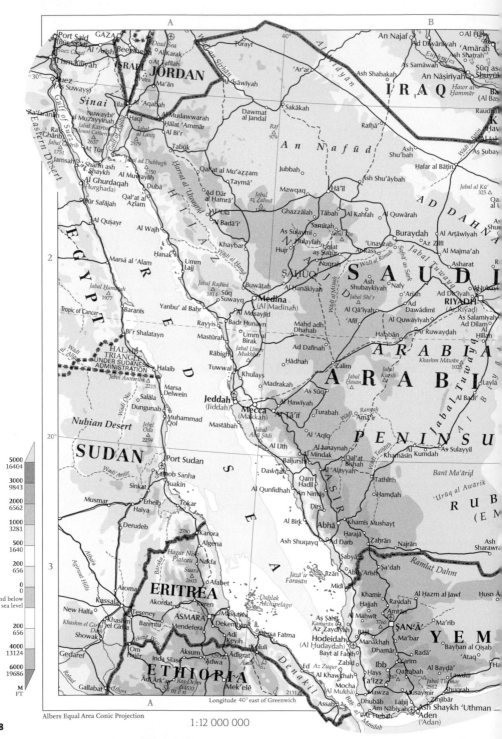

Port Said GAZA Al Ḥayy
(Būr Sa'īd) Dead Sea Al Karak Ṭurayl 40° An Najaf Ad Dīwānīyah Amārah
Suez Canal Al 'Arīsh Beersheba Aş Samāwah Ash Shaṭrah
Ismā'īlīyah Beersheba Al Karak 'Ar'ar Aş Samāwah Sūq as
Suez **Suways** JORDAN 'Isāwīyah An Nāşirīyah Shuyū
(As Suways) **ISRAEL** Ma'ān Petra Sakākah Ash Shabakah **I R A Q** Ḥawr al
Sinai Ilat Al 'Aqabah Dawmat Ḥammār
Za'farānah Al Mudawwarah al Jandal (Al Baş
Ra's Nuwaybi' Ḥaql Ḥālat 'Ammār Raf Rafḥā' Raudhah
Gharib Jabal Al Bi'r 979 Ash Shu'bah Ḥafar al Bāţin
Jabal Katrīnā 2637 Jabal al Lawz Tabūk **A n N a f ū d** Aş Ṣubay Al Ḥal
1751 2579 Jubbah Ash Shu'bah
Jamsah Sharm ash At Tūr Jabal ad Dubbagh Qal'at al Mu'aẓẓam Ḥā'il Jabal al Kū 325 Qa
Shaykh Al Muwaylih Mawqaq **A D D A H N**
Al Ghurdaqah Dubā Taymā' Jabal Al Quwārah Ash Shu
(Hurghada) Qal'at al Al Dār Ghazzālah Tābah Buraydah Al Arṭāwīyah
Azlam al Hamra' As Sulaymī Samīrā' Al Kahfah Az Zilfi
Al Quṣayr Qal'at al Mu'aẓẓam As Sulaymī Al Rass 'Unayzah Al Majma'ah
Al Wajh Al 'Ulā 'Uqlat Jabal Ṭiq Ash Asharat
Al Badā'i' Khaybar Hujr Nuqrah as Ṣuqūr 'Unayzah Az Zilfi Ad Dilam
H I J Ā Z Ḥanak **Ṣ A Ḥ Ū Q** Al Ḥanākīyah Nafy 'Ariah Ad Dir'īyah
Marṣā al 'Alam Umm Wādī al Ḥamḍ Buwāṭ **RIYADH**
Lajj Jabal Shi'r (Al Riyāḍ)
Jabal Ḥamāṭah Jabal Radwā Medina Mahd adh Al Qā'īyah As Salāmīyah
1977 1814 **N J D** Al Qā'īyah Ad Dilam
Baranīs Yanbu' al Baḥr (Al Madīnah) Dhahab 'Afīf Al Quwayḥīyah Ar Ruwaydah Hillah
R Sūq Suwayq As Suwayq Al Qā'īyah
Tropic of Cancer Rayyis Badr Ḥunayn Umm al Ḥalabān
Bi'r Shalatayn Mastūrah Birak Ad Dafīnah **A R A B I A**
Ḥ A L A Ī B Rābigh Jabal Umm Zalim Khashm Mātwān
TRIANGLE Mukhrah Ḥādhah 1025 Jabal Tuwayq
UNDER SUDANESE Khulays Jabal Layla Al Badi'
ADMINISTRATION Tuwwal Ḥasan **A R A B I**
Jebel Asoteriba Halaib **A R** Madrakah As Sūq
2215 Khulays Jabal Al Badi'
Salāla Marsa As Sūq Kursh Turabah
Dungunab Delwein **E** 'Al Hawīyah Wādī
Muhammad Jeddah Mecca **A R A B** Turabah Ramah
Nubian Desert Qol (Jiddah) (Makkah) Aṭ Ṭā'if 'Arnā'ir Kumdah
Jebel **D** Al 'Aqīq Al 'Aqīq **P E N I N S U**
Oda Mastābah Jabal Ramah
2259 Al Līth 'Al Junaynah Al Ḥamāsīn As Sulayyil
20° **SUDAN** Port Sudan Al Junaynah Kumdah
S Dawqah Qal'at Banī Ma'ārid
Wādī 'Amūr Baljurshī Bishah Tathlīth
Kanob Sanha Qam **A S** Tathlīth **R U B**
Sinkat Ḥadlī Aḥmī 'Asīr Aḥ Aḥmī (E
Suakin Al Qunfidhah An Nimāş Ḥamdah
Musmar Erheib **E** Dirs
Haiya Al Birk Khamis Mushayṭ **'Urūq al Awārik**
Derudeb 2780 Abhā Zahrān Najrān Ash
Karora Ash Shuqayq Ad Darb Sharawra
Algena 'Harajā
A Sabyā Sa'dah **Ramlat Dahm**
Hagar Nish Nakfa Jīzān Abū 'Arīsh
New Halfa Plateau Suara Jazā'ir Al Ḥazm al Jawf Husn Ā
Kassala 2603 Afabet Farasan Midī Khamir Raydah
Akordat Keren Dahlak Hajjah 'Amrān
Teseney Archipelago Aş Ṣāfir 3760
ASMARA Massawa Al Maḥwīt Ma'rib
Khashm el Girba Barentu Adi Az Zaydīyah Bājil **ṢAN'Ā'** **Y E M**
Showak Mendefera Keyih Manākhah Ma'bar Bayḥān al Qiṣab
Gedaref Om Adi Grat Koluli Hodeidah Dhamār Radā'
Hajer Inda Silasē Ādīgrat Marsa Fatma (Al Ḥudaydah) Bayt al Faqīh Ibb Yarīm Al Bayḍā'
E R I T R E A Āksum Zabid Qaṭabah 2512 Lawdar
Ādwa Denakil Az Zuqur Ḥays Ta'izz Jabal Thamar Shugrah
E T H I O P I A Ādī Ark'ay Al Khawkhah 867 Aḥsayn Zinjibār
Ras Dejen Mochal Dhubāb Am Nabīyah Shaykh 'Uthman
Mek'elē 4533 2131 Al Mukhā Mawza Lahij **Aden**
m e n Assab Bāb al Laḥij ('Adan)
Mandab Aṭ Ṭurbah

Albers Equal Area Conic Projection

1:12 000 000

Land below
sea level

M	FT
5000	16404
3000	9843
2000	6562
1000	3281
500	1640
200	656
0	0
200	656
4000	13124
6000	19686

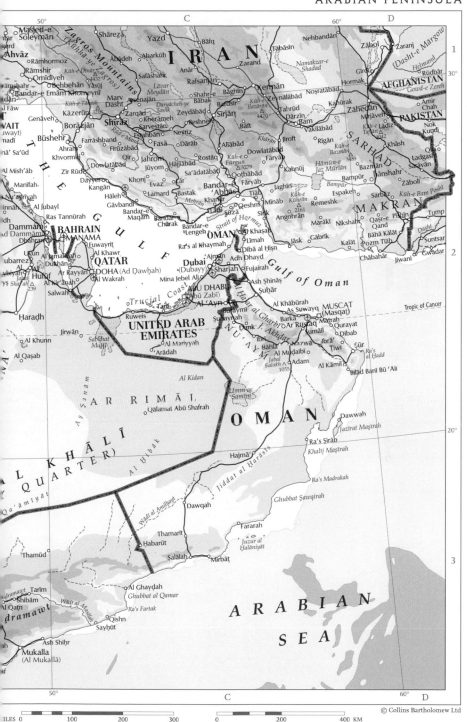

Târgu Mureş
Miercurea-Ciuc
Bacău
CHIŞINĂU
Tighina
Tiraspol
Berezivka
Mykolaïv
Tokmak
Mariupol'
Taganrog

Sebeş
Sighişoara
Vaslui
MOLDOVA
Kherson
Nova
Berdyans'k
Ros
na-D
Tecuci
Kakhovka
Gulf of Taganrog
na-D

Lugoj
Deva
Sibiu
Sfântu
Gheorghe
Comrat
Artsyz
Odessa
Melitopol'
Staropinskaya
Yeysk

Caransebeş
Făgăraş
Focşani
Bârlad
Bolhrad
(Odesa)
Novooleksiyivka
Primors'k
Pavlov

ROMANIA
Piteşti
Braşov
Galaţi
Bilhorod-
Dnistrovs'kyy
Skadovs'k
Dzhankoy
Timashevsk

Reşiţa
Drobeta
Turnu
Severin
Rosiori
de Vede
Ploieşti
Dunărea
Brăila
Izmail
Karkinits'ka Zatoka
Kerms'kyy
Crimea
Pivostriv
Nyzhn'ohirs'kyy
Kerch
Slavyar
Novorossiysk

Zăbreni
Craiova
Slatina
BUCHAREST
(Bucureşti)
Chornomors'ke
Feodosiya
Krymsk
Kuban'
Krasno

Calafat
Caracal
Orabia
Ruse
Călăraşi
Yevpatoriya
Simferopol'
Roman-
Kosh
Sudak
Khadyzhensk

Montana
Vratsa
Pleven
Dobrich
Razgrad
Sevastopol'
Novorossiysk
Tuapse

Pernik
SOFIA
Lovech
Shumen
Mangalia
BLACK SEA
Sc

Kyustendil
Kazanlŭk
Stara Zagora
Varna
Burgas

Blagoevgrad
Smolyan
Plovdiv
Dimitrovgrad
Kavarna

Sandanski
Khaskovo
Cide
İnebolu
İnce Burun

Serres
Drama
Xanthi
Komotini
Saray
Edirne
Kırklareli
Zonguldak
Bartın
Boyabat
Sinop

Kavala
Thasos
Çorlu
İstanbul
Kadıköy
Ereğli
Karabük
Kastamonu
Vezirköprü
Samsun
Trab

Thessaloniki
Gökçeada
Silivri
Körfez
Düzce
Gerede
Tosya
Merzifon
Terme
Ordu
Giresu

Polygyros
Limnos
Gallipoli
Keşan
Sea of Marmara
Bandırma
Adapazarı
(Sakarya)
Bolu
Mudurnu
Çankırı
Kızılırmak
Amasya
Sebinkarahisar
Kelkit

GREECE
Volos
Aegean Sea
Lesbos
Çanakkale
Bursa
Uludağ
Bilecik
Beypazarı
Kalecik
Çorum
Turhal
Tokat
Sivas
Zara
Suşehri

Euboea
Lemnos
Edremit
Ezine
Susurluk
İnegöl
Eskişehir
ANKARA
Polatlı
Sungurlu
Yıldızeli
Diyrigi

Chalkida
Chios
Ayvalık
Soma
Demirci
Simav
Kütahya
Kırıkkale
Yozgat
Akdağmadeni
Kangal

ATHENS
Piraeus
Mytilini
Pergama
Akhisar
Banaz
Afyon
Emirdağ
Yunak
Kaman
Kırşehir
Boğazlıyan
Kayseri
Erciyes
Dağı
Pınarbaşı
Elazığ

Andros
Chios
İzmir
Manisa
Gediz
Sivrihisar
Cihanbeyli
Nevşehir
Diyarbakır

Etmoupoli
Tinos
Kuşadası
Aydın
Nazilli
Dinar
Çivril
Sandıklı
Akşehir
Lake
Tuz
(Tuz Gölü)
Aksaray
Niğde
Yahyalı
Elbistan
Malatya
Sivas

Syros
Samos
Naxos
Yatağan
Büyük Menderes
İsparta
Burdur
Beyşehir
Gölü
Hasan Dağı
3268
Bor
Karapınar
Adıyaman
Sivas

Milos
Paros
Ios
Milas
Bodrum
Muğla
Dalaman
Korkuteli
Konya
Beyşehir
Ereğli
Kahramanmaraş
Euphrates
(Fyrat)
Bilecik

Santorini
(Thira)
Kritiko Pelagos
Marmaris
Fethiye
Elmalı
Antalya
Manavgat
Karaman
Taurus Mountains
(Toros Dağları)
Adana
Osmaniye
Gaziantep

Chania
Irakleio
Agios
Nikolaos
Rhodes
(Rodos)
Kaş
3073
Megisti
Antalya
Körfezi
Anamur
Ermenek
Erdemli
Tarsus
Mersin
(İçel)
İskenderun (Alexandretta)
Aleppo

Rethymno
Sitea
Ierapetra
Lindos
Cape Apostolos
Andreas
Kyrenia
(Keryneia)
Aiglaousa
Silifke
Antakya
(Antioch)
Idlib
Ar Raqqah
Al

CRETE
(KRITI)
Karpathos
(Scarpanto)
NICOSIA
(Lefkosia)
Famagusta
Latakia
Ma'arrat an
Nu'mān
Hamāh

Cape Arnauti
Polis
Evrychou
Larnaca
Tartūs
Bāniyās
SYRIA

MEDITERRANEAN SEA
Paphos
CYPRUS
Limassol
(Lemesos)
Tripoli
(Trâblous)
Homs
Tadmur

LEBANON
BEIRUT
(Beyrouth)
Al Qaryatayn

Sidon
An Nabk

Tyre
Zahlé
Sab' Ābār

Al Qunaytirah
DAMASCUS
(Dimashq)

Al Bardī
Haifa
(Hefa)
Sea of
Galilee
(L. Kinneret)
As Suwaydā'

Umm
Sa'ad
Marsá
Matrūḥ
Alexandria
(Al Iskandariya)
Kafr ash
Shaykh
Balṭīm
Dumyāṭ
ISRAEL
Tel Aviv-Yafo
Nazareth
Irbid
Al Mafraq
Syrian
Desert
(Bādiyat ash

LIBYA
Libyan Plateau
(Ad Diffah)
Al 'Āmirīyah
Al Manṣūrah
Port Said
(Bûr Sa'îd)
Al 'Arīsh
Rehovot
JERUSALEM
GAZA
WEST
BANK
AMMAN
Turayf

Al Hammām
Banhā
Suez
Canal
Al Ismā'īlīyah
Beersheba
Dead Sea
Al Karak
JORDAN
Al 'Īsāwīyah

Qārah
Qattara
Depression
Damanhūr
Shubrā al Khaymah
Az Zaqāzīq
Suez
(As Suways)
At Tafilah
Ma'ān
Wādī an
Sirhān

Siwa
Wahat
Siwa
(Siwa Oasis)
Al Jīzah
Banī Suwayf
GIZA
Pyramids of Giza
CAIRO
Al Qāhirah
Petra
SAUDI

EGYPT
Al Fayyūm
Memphis
Ẓa'farānah
Sinai
Ēlāt
Al Aqabah
Al Mudawwarah
Dawmat al J

Al Bawīṭī
Banī Mazār
Maghāghah
Za'farānah
Gulf of Suez
Nuwaybi' al
Muzayyinah
Jabal Katrīna
Mount Catherine
2637
Ḥaql
Hālat 'Ammār
Al Bi'r

5000
16404

3000
9843

2000
6562

1000
3281

500
1640

200
656

0
0

Land below
sea level

200
656

4000
13124

6000
19686

M
FT

Albers Equal Area Conic Projection

1:12 000 000

MILES 0   100   200   300

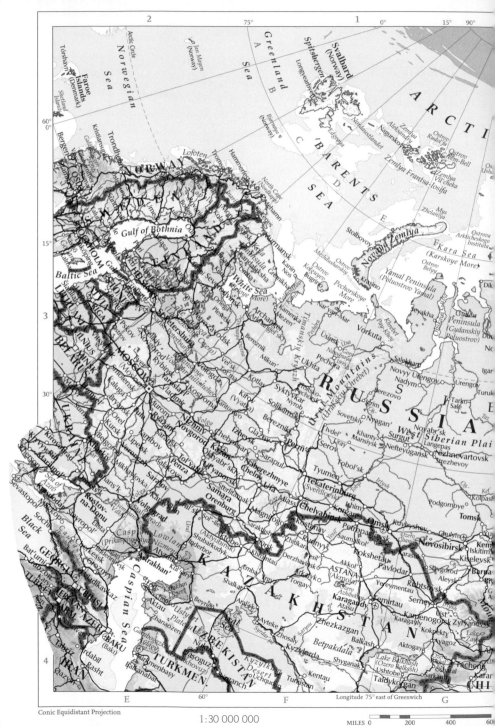

Arctic Circle

Jan Mayen
(Norway)

Norwegian
Sea

Greenland
Sea

A R C T I

Svalbard
(Norway)

Spitsbergen
(Norway)

Longyearbyen

Zemlya
Aleksandry

Nagurskoye

Ostrov
Graham-Bell

Ostrov
Rudolfa

Faroe
Islands
(Denmark)

Tórshavn

Streymoy
Island

Bjørnøya
(Norway)

B A R E N T S

S E A

Zemlya Frantsa-Iosifa

Ostrov
Vil'cheka

Mys
Uslie

Bergen

60°
0°

Kristiansund

Trondheim

Lofoten

Hammerfest

North Cape
(Norway)

Nordkinn
Vardø

Murmansk

Stolbovoy

Novaya

Zemlya

F Kara Sea
(Karskoye More)

Ostrov
Belyy

Ostrov
Arktcheskogo
Instituta

Dik

N O R W A Y

Gulf of Bothnia

Umeå

Kem' Peninsula
(Kola)

Kandalaksha

Kanin
Nos

White Sea
(Beloye More)

Mezhdusharskiy

Pechorskoye
More

Krasino

Yamal Peninsula
(Poluostrov Yamal)

Seyakha

Gydan
Peninsula
(Gydanskiy
Poluostrov)

Nc

15°

STOCKHOLM

Baltic Sea

S W E D E N

F I N L A N D

Kem'

Onega

Severodvinsk

Arkhangel'sk

Mezen'

Nar'yan-Mar

Vorkuta

Khalmer-
Yu

Pay-Khoy

Kharp

LITHUANIA

VILNIUS

RIGA

LATVIA

St Petersburg

Novgorod

Pskov

Ostrov

Olonets

Vologda

Kotlas

Syktyvkar

Timanskiy Kryazh

Usinsk

Pechora

Ukhta

Novyy Urengoy

Nadym

Ostrov
Uselie

Salekhard

Urengoy

Tarko-
Sale

Turuk

3

BELARUS

MINSK

MOSCOW

Smolensk

Kaluga

Tula

Ryazan'

Vladimir

Yaroslavl'

Rybinsk

Tver'

Cherepovets

Kirov
(Vyatka)

Glazov

Nyrob

Izhevsk

Berezniki

Solikamsk

Sovetskiy

Nyagan'

Ural Mountains
(Ural'skiy Khrebet)

Khanty-
Mansiysk

Serov

R U S S I A

West Siberian Plain

Langepas

Nizhnevartovsk

Strezhevoy

S

30°

UKRAINE

Kiev

Bryansk

Orel

Kursk

Belgorod

Tambov

Lipetsk

Voronezh

Penza

Saratov

Balakovo

Syzran'

Tol'yatti

Samara

Ul'yanovsk

Cheboksary

Kazan'

Naberezhnyye
Chelny

Ufa

Sarapul

Perm'

Yekaterinburg
(Sverdlovsk)

Nizhniy
Novgorod

Kostroma

Ivanovo

Volga

Serov

Tyumen'

Tobol'sk

Irtysh

Tara

Podgornoye

Kolpas

Ket'

Kolpashevo

Tomsk

Sea of
Azov

Rostov-
na-Donu

Elista

Krasnodar

Stavropol'

MOLDOVA

Odesa

Sevastopol'

Black Sea

Sochi

GEORGIA

Bat'umi

TBILISI

ARMENIA

YEREVAN

AZER.

BAKU
(Baku)

IRAN

Caspian
Sea

Astrakhan'

Caspian
Lowland

Volgograd

Kamyshin

Ural

Orenburg

Sterlitamak

Magnitogorsk

Miass

Chelyabinsk

Kostanay

Omsk

Petropavlovsk

Omsk

Novosibirsk

Barnaul

Kiselyovsk

Ozero
Chany

Kulundy

Kuybyshev

Slavgorod

Aleysk

Kamen'-na-
Obi

Rubtsovsk

Ust'-

Semey

Zyryanovsk

Ust'-
Kamenogorsk

Lake
Zaysan

Tacheng

CHI

Caspian
Sea

Krasnovodsk

TURKMEN.

Qazvin

Rasht

Tabriz

Ardabil

K A Z A K H S T A N

Atyrau

Aktyubinsk (Aktobe)

Shalkar

Aral'sk

Aral Sea

Kzyl-Orda

Turkistan

Kentau

UZBEKISTAN

ASTANA
(Akmola)

Karagandy

Temirtau

Zhezkazgan

Balkhash

Lake Balkhash
(Ozero Balkash)

Taldykorgan

Ayagoz

Aktogay

Ushtobe

Pavlodar

Semey

Kyzylkum
Desert

Dashhowuz

Nukus

Urganch

4

TURKMEN.

Balkanabat

Turkmenbasy

Conic Equidistant Projection

1:30 000 000

Longitude 75° east of Greenwich

MILES 0   200   400   60

90° 165° 180° 75°

OCEAN

Ostrov Shmidta

Ostrov Komsomolets
Ostrov Oktyabr'skoy
Revolyutsii
Ostrov Bol'shevik

Zemlya

Prolive Vil'kitskogo
Chelyuskina

H Peninsula
Taymyr (Poluostrov Taymyr)
Ozero
Taymyr

Khatanga

Kheta

Uyar

Tembenchi

Tura

E R
Central Siberian
Plateau

Chernyshevskiy
Yerbogachen

Podkamennaya Tunguska

govo
Severo-
Yeniseyskiy

Bor'chany
Angara

ngoyarsk Kansk

Zaozernyy
Nizhneudinsk

Tulun

Abakan

Vostochnyy
Sayan

Kyzyl

aangom

Hyargas
Nuur

Tosontsengel

Har
Nuur

Gichgeniyn Nuruu

Altay

MONGOLIA

New Siberia Islands
(Novosibirskiye Ostrova)

Laptev Sea
(More Laptevykh)

Ostrov Kotel'nyy

Ostrov Bel'kovskiy

Ostrov Malyy
Lyakhovskiy

Yanskiy
Zaliv

Ostrova
De-Longa
Ostrov
Bennetta
Ostrov Zhokhova

Ostrov Novaya Sibir'

East Siberian Sea
(Vostochno-Sibirskoye More)

Anabarskiy
Zaliv

Ust'-
Olenek

Olenek

Anabar

Popigay

Olenek

Saskylakh

Udachnyy

Markha

Vilyuy

Zhigansk

Verkhoyanskiy Khr.

Aykhal

Ust'-Kut

Ust'-
Ilimsk

Kirensk

Bodaybo

Kitoy

Bratsk

Severobaykal'sk

Kumukan

Mogocha

Mama

Vitim

Sretensk

Borzya

Olovyannaya

Chita

Darasun

Mangut

Hailar

Manzhouli

Ulan-Ude

Ulan

Irkutsk

Usinoozersk

Sühbaatar

Darhan

Dzuunmod

ULAN BATOR
(Ulaanbaatar)

Tsetserleg

Mandalgovi

Arvayheer

Bayanhongor

GOBI

105°

H

Saynshand

Choybalsan

Baruun-Urt

Uliastai

Khabarovsk

N

M

L

Ostrov
Ayon

Mys
Shelagskiy

Prolive Longa

Mys
Medvezh'i

Chokurdakh

Nizhneyansk

Deputatskiy

Batagay

Srednekolymsk

Zyryanka

Niznekolymsk

Kolyma

Anyuyskiy

Omolon

Malyy Anyuy

Wrangel Island
(Ostrov Vrangelya)

Ostrov Shmidta

Mys
Billings

Pevek

Chaunskaya Guba

Bilibino

Palyavaam

Chukchi
Sea

Mys Shmidta

Egvekinot

Zaliv
Kresta

Anadyr'

Mys Navarin

Bering Strait

St. Lawrence
Island

Uelen

Kolyuchinskaya Guba

Provideniya

U.S.A.

180°

60°

Yano-Indigirskaya
Nizmennost'

Deputatskiy

Kular

Verkhoyansk

Batagay

Adycha

El'gorskiy

Khrebet Cherskogo

Nera

Omsukchan

Kolymskoye Nagor'ye

Evensk

Severo-Evensk

Gizhiga

Penzhinskaya Guba

Karaginskiy
Zaliv

Ostrov
Karaginskiy

Palana

Srednyy Khrebet

Klyuchi

Kamchatka

Nikol'skoye

Komandorskiye
Ostrova

Y

Oleněk

Yessey

Kovrov

Tyung

Tura

Anabar

Vilyuy

Verkhoyansk

Zhigansk

Batagay

Lazo

Adycha

Namtsy

Yakutsk

Pokrovsk

Mokhsogollokh

Olekminsk

Aldan

Olekma

Aldan

Serymungi

Neryungri

Stanovoy Nagor'ye

Stanovoy Khrebet

Tynda

Zeya

Mayskiy

Zeya

Khani

Khrebet Dzhugdzhur

Chul'man

Ayan

Khrebet Tukuringra

Svobodnyy

Vodokhranilishche

Shilka

Yablonovyy Khrebet

Nerchinsk

Onon

Aginskoye

Priargunsk

Hulun Buir

Hailar

Manzhouli

Xiao Hinggan Ling

Heilong Jiang

Hegang

Yichun

Komsomol'sk-
na-Amure

Vanino

Tatarskiy Proliv

Amguema

Ust'-Maya

Oymyakon

Magadan

Ola

Okhotsk

Ayan

Mys Marii

Nikolayevsk-na-Amure

Okha

Nogliki

Sea of Okhotsk

Kamchatka
Peninsula

Petropavlovsk-
Kamchatskiy

Ozernoy

Oktyabr'skiy

Severo-
Kuril'sk

Kuril Islands

150°

3

Komsomol'sk-
na-Amure

Birobidzhan

Blagoveshchensk

Da Hinggan Ling

Gulian

Mangui

Nenjiang

Qiqihar

Daqing

Bei'an

Suihua

Harbin

Jiamusi

Mudanjiang

Suifenhe

Jixi

Shuangyashan

Ussuriysk

Vladivostok

Nakhodka

Zaliv Petra
Velikogo

Sikhote-Alin'

Dal'nerechensk

Spassk-Dal'niy

Bikin

45°

135°

4

CHINA

Qitaihe

Jilin

Changchun

Siping

Dunhua

Tonghua

Fushun

Shenyang

Benxi

Dandong

N. KOREA

Hyesan

Kimch'aek

Hamhung

Wonsan

PYONGYANG

SEOUL

S. KOREA

Xilinhot

Tongliao

Fuxin

Chifeng

Chengde

120°

J

Ozero
Khanka

Chaoyang

Xinmin

Anshan

200
656

4000
13124

6000
19686

M
FT

5000
16404

3000
9843

2000
6562

1000
3281

500
1640

200
656

0
0

Land below
sea level

ATLANTIC OCEAN

Denmark Strait

Arctic Circle

Greenland (Denmark)

Bjørnøya (Nor.)

Jan Mayen (Nor.)

ICELAND
Reykjavik

NORWEGIAN SEA

Trondheim

Faroe Islands (Den) Tórshavn

Shetland Islands

Orkney Islands

SCOTLAND

Glasgow
Edinburgh

N. Belfast
IRELAND

IRELAND
Dublin

Manchester

WALES
ENGLAND
Cardiff

London

Birmingham

Bergen

Oslo

Gothenburg

Stockholm

NORWAY

SWEDEN

Gulf of Both

Tromsø

Gotland

Baltic Sea

NORTH SEA

UNITED KINGDOM

NETHERLANDS
Amsterdam

The Hague

Brussels

BELGIUM

DENMARK
Copenhagen

Malmö

Hamburg

Hannover

Berlin

RUS.

POLAN

Poznan
Warsaw

Łódź

K.   KOSOVO
LIE.   LIECHTENSTEIN
MACE. MACEDONIA (F.Y.R.O.M.)
MONT. MONTENEGRO

English Channel

Channel Is. (U.K.)

Paris

Seine

Essen

GERMANY

Frankfurt

Luxembourg
LUXEMBOURG

Rhine

Prague

CZECH REPUBLIC

Katow

SLOVAK

Bay of Biscay

Cape Finisterre

Bordeaux

Loire

Munich

Danube

Vienna

Bratislava

Buda

HUNGARY

FRANCE

Lyon

Rhône

SWITZERLAND

Bern

Mont Blanc

LIE.

AUSTRIA

SLOVENIA

Ljubljana

Zagreb

CROATIA

Belgrad

Oporto

Bilbao

Pyrenees

PORTUGAL

SPAIN

Madrid

Andorra la Vella
ANDORRA

Marseille

MONACO

Milan

Turin

ITALY

SAN MARINO

BOSNIA- HERZ.

Sarajevo

MONT.

Podgorica

K.

Prist

Skopj

Lisbon

Cabo de São Vicente

Valencia

Barcelona

Corsica

VATICAN CITY

Rome

Adriatic Sea

Tirana

ALBANIA

Seville

Gibraltar (U.K.)

Balearic Islands

Sardinia

Naples

Palermo

Sicily

Ionian Sea

MEDITERRANEAN SEA

MALTA
Valletta

MOROCCO

ALGERIA

TUNISIA

Chamberlin Trimetric Projection

1 : 25 000 000

MILES 0   250   500

84

BARENTS
SEA

Novaya
Zemlya

Ostrov
Kolguyev

app

Murmansk

Vorkuta

White Sea

Archangel

RUSSIAN FEDERATION

Ob'

Syktyvkar

LAND

Lake
Onega

Perm'

Lake
Ladoga

elsinki

St Petersburg

50°

llinn

Nizhniy
Novgorod

Kazan'

NIA

Yaroslavl'

Volga

IA

Samara

Orenburg

A

Moscow

Ryazan'

Ural Mountains

Minsk

Voronezh

Saratov

KAZAKHSTAN

BELARUS

Homyel'

Volgograd

Kiev

Kharkiv

Don

Aral Sea

Dnipropetrovs'k

Donets'k

Volga

UZBEKISTAN

Chişinău

Rostov
na-Donu

Astrakhan

MOLDOVA

Sea
of Azov

Odessa

Krasnodar

Caspian Sea

MANIA

Grozny

Bucharest

C a u c a s u s

TURKMENISTAN

Black Sea

GEORGIA

ofia

AZERBAIJAN

LGARIA

ARMENIA

AZER.

Istanbul

I R A N

ssaloniki

T U R K E Y

egean
Sea

CE

Athens

Crete

Euphrates

CYPRUS

SYRIA

LEBANON

I R A Q

Tigris

0   250   500   750 KM

© Collins Bartholomew Ltd

85

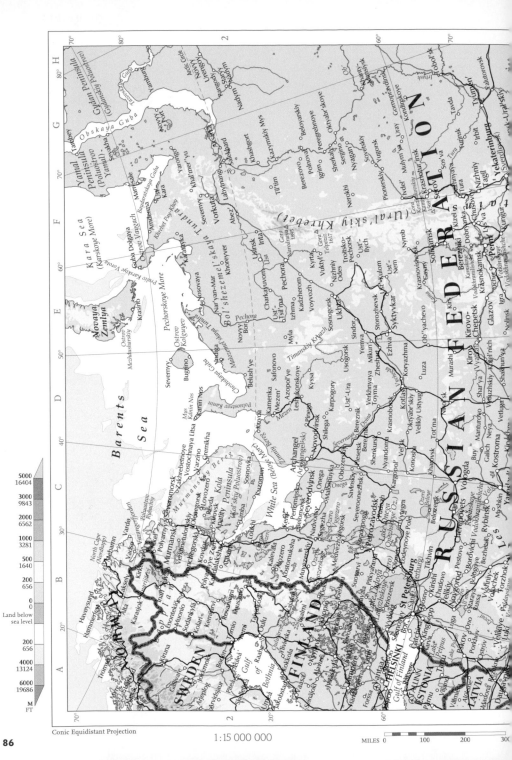

Conic Equidistant Projection

1:15 000 000

MILES 0   100   200   300

5000
16404
3000
9843
2000
6562
1000
3281
500
1640
200
656
0
0
Land below
sea level
200
656
4000
13124
6000
19686
M
FT

Barents
Sea

Kara Sea
(Karskoye More)

Novaya
Zemlya

White Sea (Beloye More)

RUSSIAN FEDERATION

FINLAND

SWEDEN

NORWAY

ESTONIA

LATVIA

St Petersburg

HELSINKI

Murmansk

Arkhangel

Yekaterinburg

Gulf of Finland

Gulf of Bothnia

(Ural'skiy Khrebet)

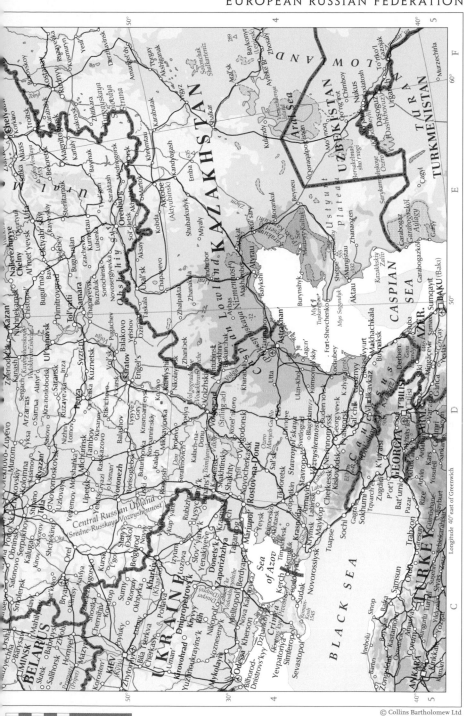

Conic Equidistant Projection

1:6 000 000

Longitude 25° east of Greenwich

MILES 0   50   100   150

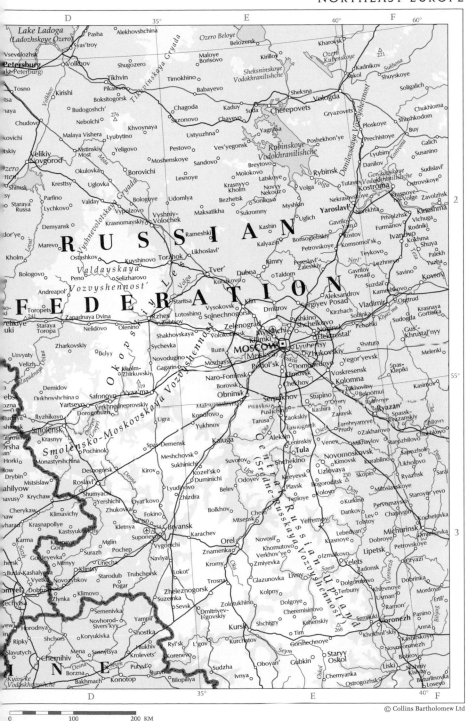

Lake Ladoga
(Ladozhskoye Ozero)
Pasha
Alekhovshchina
Ozero Beloye
Kharovsk
Vsevolozhsk
Syas'troy
Maloye
Borisovo
Belozersk
Ozero
Kubenskoye
221
Petersburg
(kt-Peterburg)
Volkhov
Tikhvin
Shugozero
Timokhino
Kirillov
Kadnikov
Sukhona
Shuyskoye
Soligalich
tsa
Tosno
Kirishi
Pikalevo
Babayevo
Sheksna
Sokol
Vologda
Gryazovets
Chukhloma
Shishkodom
Chudovo
Budogosch'
Nebolchi
276
Chagoda
Kaduy
Suda
Cherepovets
Ploskoye
Prechistoye
Buy
Galich
kovichi
Khvoynaya
Sazonovo
Chayevo
Sheksninskoye
Vodokhranilishche
Posin
Lyubim
Kostroma
Susanino
tskiy
Malaya Vishera
Lyubytino
Ustyuzhna
Yagnitsa
Poshekhon'ye
Danilov
Sudislavl
Velikiy
Novgorod
Mstinskiy
Most
Yeligovo
Pestovo
Ves'yegonsk
Rybinskoye
Vodokhranilishche
Rybinsk
Tutayev
Gorkovskoye
Vodokhranilishche
zero
men
Okulovka
Borovichi
Moshenskoye
Sandovo
Breytovo
Molokovo
Volga
Danilov
Nekras
Nerekhta
Privolzhsk
Kineshma
Vichuga
shinsk
sy
Krestsy
Uglovka
Bologoye
Udomlya
Bezhetsk
Krasnyy
Kholm
Novyy
Nekouz
Sonkovo
Myshkin
Yaroslavl'
Volge
Zavolzhsk
Rodniki
Staraya
Russa
Parfino
Valday
Vypolzovo
Vyshniy-
Volochek
Maksatikha
Sukromny
Uglich
Gavrilov
Yam
Rostov
Furmanov
Ivanovo
Kokhma
Shuya
dor'ye
Lychkovo
Krasnomayskiy
Rameshki
Kashin
Borisoglebsk
Petrovskoye
Komsomol'sk
Teykovo
Palekh
Yuzha
Demyansk
Ostashkov
Kuvshinovo
Torzhok
Likhoslavl'
Kalyazin
Pereslavl'-
Zalesskiy
Nerl'
Lezhnevo
Savino
Kovrov
Kholm
Marevo
Peno
Selizharovo
Tver'
Kimry
Dubna
Taldom
Gavrilov
Posad
Suzdal'
Kameshko
Krasnaya
Gorbitka
Bologovo
Andreapol'
Zapadnaya Dvina
Staritsa
Vysokovsk
Klin
Dmitrov
Sergiyev Posad
Kirzhach
Sobinka
Vladimir
Orgtrud
Gus'-
Khrustal'nyy
elikiye
uki
Toropets
241
Nelidovo
Olenino
Rzhev
Lotoshino
Solnechnogorsk
Volokolamsk
Zelenograd
Pushkino
Shchelkovo
Noginsk
Elektrostal'
Petushki
Sudogda
Melenki
Zharkovskiy
Belyy
Sychevka
Ruza
Mytishchi
MOSCOW
(Moskva)
Lyubertsy
Zhukovskiy
Shatura
Yegor'yevsk
Usvyaty
Velizh
Demidov
Dukhovshchina
Safonovo
Vyaz'ma
Gagarin
Novodugino
Mozhaysk
Podol'sk
Domodedovo
Chekhov
Klimovsk
Voskresensk
Kolomna
Spas-
Klepiki
Kasimov
Yartsevo
Verkhnedneprovsk
Dorogobuzh
Naro-Fominsk
Borovsk
Obninsk
Serpukhov
Stupino
Kashira
Lukhovitsy
Beloomut
Tuma
Rudnya
shevsk
Ryzhikovo
Smolensk
Krasnyy
Pochinok
Spas-Demensk
Ugra
Yukhnov
Kaluga
Maloyaroslavets
Puslichina
Tarusa
Krasnogorsk
Zaoksk
Aleksin
Leninskiy
Venev
Torkhryanyye
Mikhaylov
Prudy
Zakharovo
Starozhilovo
Ryazan'
Spask-
Ryazanskiy
brshch
an'
Horki
Monastyrshchina
Desnogorsk
Kirov
Sukhinichi
Kozel'sk
Duminichi
Odoyev
Shchekino
Tula
Novomoskovsk
Uzlovaya
Kimovsk
Bogoroditsk
Skopin
Korablino
Ukholovo
Ryazhsk
Sapozhok
Sarai
tow
Drybin
Mstsislaw
Roslavl'
Shumyachi
Lyudinovo
Zhizdra
Belev
Plavsk
Teploye
Don
Kurkino
Dankov
Lev
Tolstoy
Pervomayskiy
Staroze'yevo
Kochetovka
navusy
Krychaw
Yershichi
Zhukovka
Fokino
Bolkhov
Mtsensk
Chern
Yefremov
Krasnoye
Dobroye
Michurinsk
Pervomayk
Cherykaw
Klimavichy
Krasnapollye
Kletnya
Bryansk
Karachev
Novosil'
Verkhov'ye
Izmalkovo
Lipetsk
Gryazi
Karma
Kastsyukovichy
Krasnaya
Gora
Mglin
Pochep
Vygonichi
Navlya
Orel
Znamenka
Kromy
Khomutovo
Zmiyevka
Krasnoye
Yelets
Zadonsk
Usman'
Mordovo
Ertil'
hersk
Buda-Kashalyova
Mirnyy
Surazh
Unecha
Lokot'
Oka
Glazunovka
Kolpny
Dolgorukovo
Terbuny
Dobrinka
myel
Dobrush
Vyetka
Novozybkov
Starodub
Trubchevsk
Trosna
Zheleznogorsk
Suzemka
Zolotukhino
Dolgoye
Cheremisinovo
Semiluki
Khokhol'skiy
Voronezh
Anna
echlytsa
Zlynka
Klimovo
Sevsk
Dmitriyev-
L'govskiy
Kursk
Shchigry
Kashenskiy
268
Novovoronezh
Borovo
Semenivka
Novhorod-
Sivers'kyy
Yampil'
Shostka
Ryl'sk
L'gov
Kurchatov
Tim
Gorshechnoye
Kasporskoye
newl
Horodnya
Koryukivka
Hlukhiv
Koreniv
Oboyan'
Gubkin
Staryy
Oskol
Liski
Nizhniy
Kisl'
Ripky
Shchors
Mena
Sosnytsya
Krolevets'
Oskol
Seym
Chernyanka
Ostrogozhsk
Butturlinovka
Loseyo
Slavutych
Chernihiv
Desna
Putyvl'
Sudzha
Ivnya
Staryy
Oskol
Kyivs'ke
Vodoskhovyshche
Bakhmach
Konotop
Bilopillya

Conic Equidistant Projection

1:6 000 000

Longitude 25° east of Greenwich

MILES 0   50   100   150

Elevation scale (left margin):

M	FT
5000	16404
3000	9843
2000	6562
1000	3281
500	1640
200	656
0	0
Land below sea level	
200	656
4000	13124
6000	19686
M	FT

Grid references (top): A   25°   B   30°

Map labels:

POLAND · BELARUS · UKRAINE · SLOVAKIA · HUNGARY · ROMANIA · SERBIA · BULGARIA · MOLDOVA

CARPATHIAN MOUNTAINS · Transylvania Alps (Carpatii Meridionali) · Podishul Moldovei · Pripet Marshes · Nizina Mazowiecka

WARSAW (Warszawa) · KIEV (Kyiv) · CHIŞINĂU (Kishinev) · BUCHAREST (Bucureşti)

Cities include: Białystok, Lublin, Radom, Brest, Pinsk, Mazyr, Lviv (L'vov), Ternopil', Chernivtsi, Iaşi, Cluj-Napoca, Oradea, Debrecen, Constanţa, Galaţi, Brăila, Craiova, Sibiu, Braşov, Ploieşti, Piteşti

0   100   200 KM

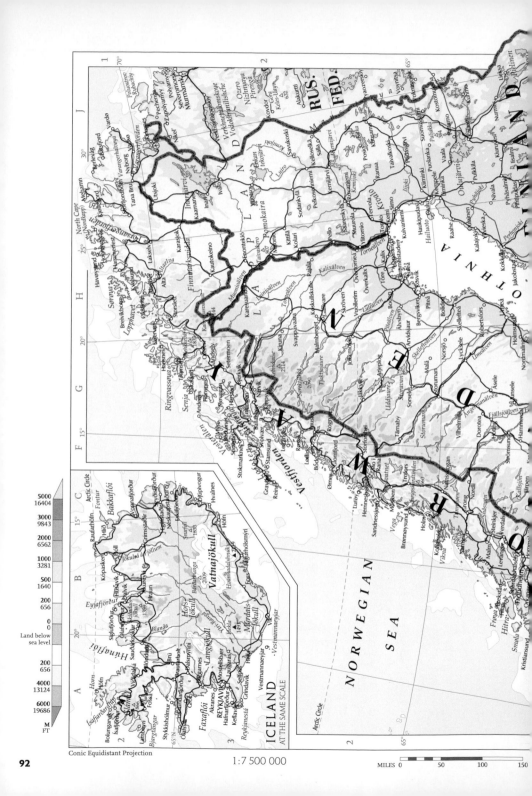

Conic Equidistant Projection

1:7 500 000

MILES 0   50   100   150

ICELAND
AT THE SAME SCALE

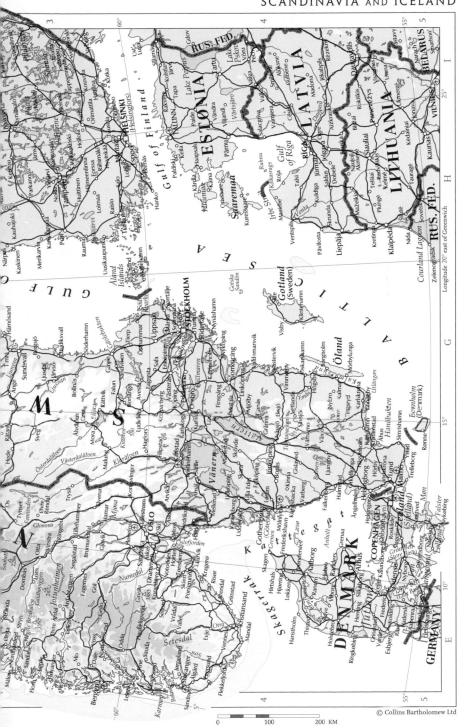

© Collins Bartholomew Ltd

0   100   200 KM

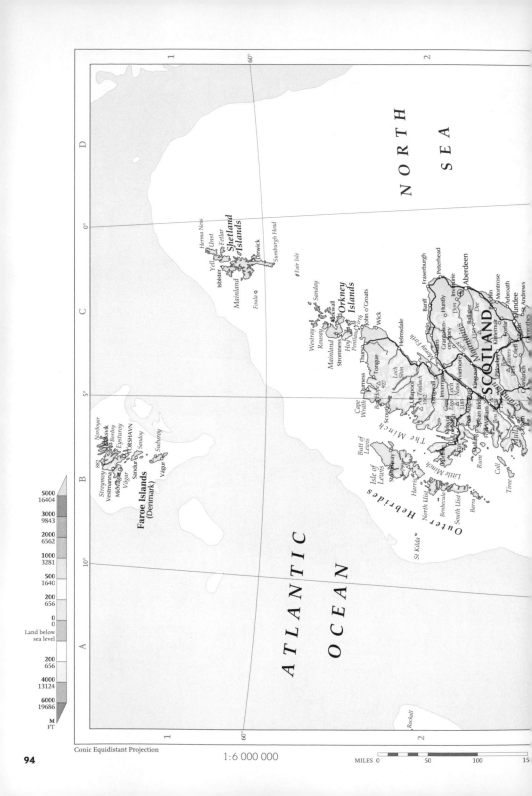

60°

D

0°

C

5°

B

10°

A

N O R T H

S E A

Herma Ness

Yell · Unst
Fetlar
**Shetland
Islands**
Isbister
Mainland · Lerwick
Foula ◦

Sanday
Westray
Rousay
**Orkney
Islands**
Mainland Kirkwall
Stromness
Hoy
Pentland Firth
Thurso
Durness Tongue
Scourie
Cape
Wrath
Ullapool
Loch
Shin

Sumburgh Head

◦ Fair Isle

John o'Groats
Wick

Helmsdale

Fraserburgh
Peterhead

Banff
Huntly
Aberdeen
Keith
Elgin
Nairn Dufftown
Moray Firth
Inverness
Loch
Ness
Grantown-
on-Spey
Ballater
Dee
Kintore
Montrose
Arbroath
Forfar
Dundee
Crieff
St Andrews
Ben Nevis
1343

SCOTLAND

The Minch

Butt of
Lewis
Stornoway
**Isle of
Lewis**
Harris

O u t e r   H e b r i d e s

North Uist
Benbecula
South Uist
Barra

St Kilda ◦

Little Minch

Rum
Eigg
Coll
Tiree
Mull

Fort William
Ben Nevis
1343

Spean Bridge
Oban

Callander

Loch
Lomond

Glasgow
Ayr

A T L A N T I C

O C E A N

**Faroe Islands
(Denmark)**

Norðoyar
882
Streymoy
Eysturoy
Vestmanna
Miðvágur **TÓRSHAVN**
Vágar
Sandur
Suðuroy

Vágur

Borðoy
Klaksvík
Sandoy

Rockall

60°

5000
16404
3000
9843
2000
6562
1000
3281
500
1640
200
656
0
0
Land below
sea level
200
656
4000
13124
6000
19686
M
FT

Conic Equidistant Projection

1:6 000 000

MILES 0   50   100   15

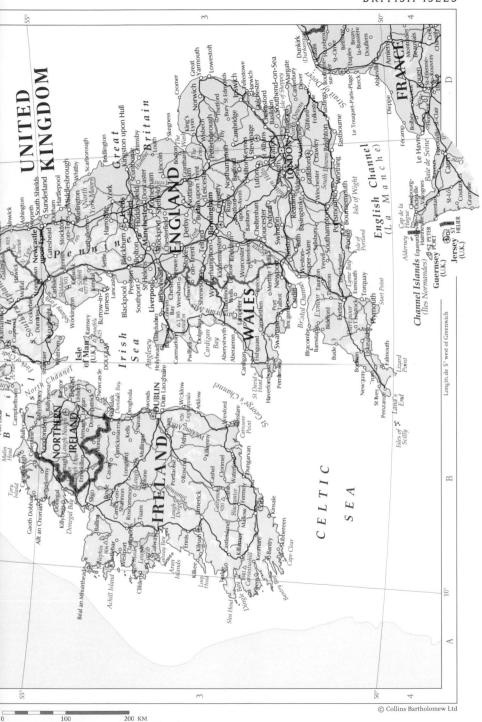

© Collins Bartholomew Ltd

0   100   200 KM

SCOTLAND

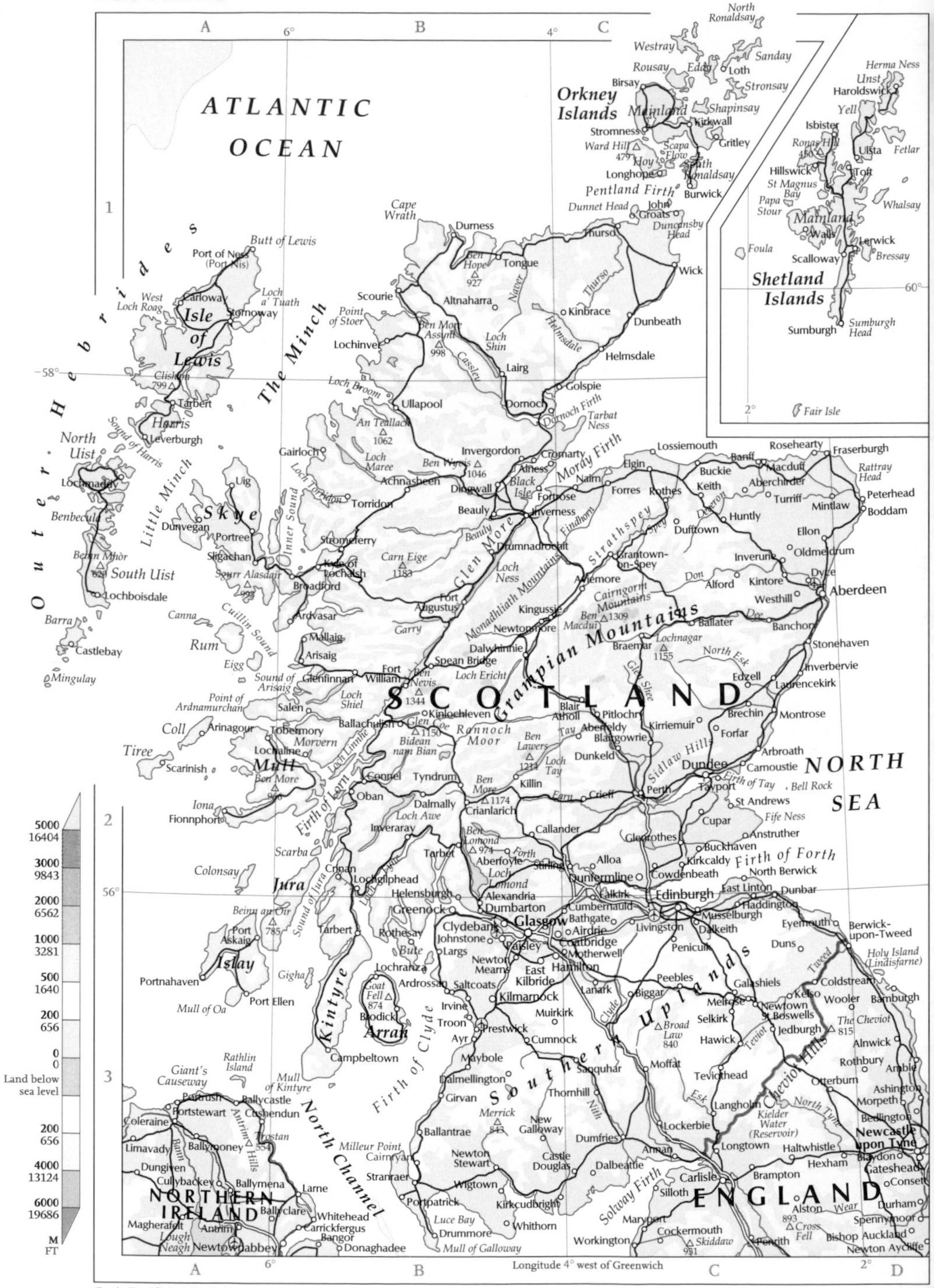

ATLANTIC
OCEAN

Orkney Islands

North Ronaldsay
Westray
Rousay   Eday   Sanday
Birsay   Shapinsay   Loth   Stronsay
Mainland
Stromness   Kirkwall
Ward Hill   Scapa   Gritley
479 △   Flow   St
Hoy   Longhope   Ronaldsay   Burwick
Pentland Firth
Dunnet Head   John   o'Groats   Duncansby Head

Shetland Islands

Herma Ness
Unst
Haroldswick
Yell
Isbister
Ronas Hill   Ulsta   Fetlar
450 △   Toft
Hillswick
St Magnus   Papa   Whalsay
Bay   Stour
Mainland
Foula   Lerwick
Wadis   Bressay
Scalloway
Sumburgh
Sumburgh Head
2°   Fair Isle   60°

Butt of Lewis
(Port of Ness)
Port of Ness
(Port Nis)
West   Stornoway
Loch Roag   Carloway
Isle
of
Lewis
Clisham   Tarbert
799 △
Harris
Leverburgh
North
Uist
Lochmaddy
Benbecula
Beinn Mhòr
South Uist
Lochboisdale
Barra
Castlebay
Mingulay

Cape
Wrath
Durness
Loch   Ben
a' Tuath   Hope   Tongue
927   Thurso
Scourie   Altnaharra
Point   Ben More   Naver   Kinbrace
of Stoer   Assynt   Loch   Helmsdale
998   Shin   Laird   Dunbeath
Lochinver   Helmsdale
Ullapool   Lairg   Golspie
Loch Broom   Dornoch   Tarbat
Dornoch Firth   Ness
The Minch
An Teallach   Bonar Bridge
1062
Gairloch   Invergordon   Cromarty
Loch   Beinn Dearg   Dingwall   Lossiemouth   Rosehearty
Maree   1046   Black   Nairn   Elgin   Banff   Macduff   Fraserburgh
Achnasheen   Isle   Forres   Rothes   Buckie   Aberchirder   Rattray
Torridon   Beauly   Findhorn   Keith   Turriff   Head
Strome Ferry   Inverness   Dufftown   Huntly   Mintlaw   Peterhead
Carn Eige   Drumnadrochit   Grantown-   Ellon   Boddam
1183   Loch   on-Spey   Inverurie   Oldmeldrum
Kyle of   Ness   Strathspey   Alford   Kintore   Dyce
Lochalsh   Fort   Aviemore   Cairngorm   Don   Westhill   Aberdeen
Broadford   Augustus   Mountains   Ballater   Banchory   Stonehaven
Canna   Monadhliath Mountains   Ben   Braemar   Inverbervie
Ardvasar   Kingussie   Macdui   1309   Lochnagar   Laurencekirk
Rum   Garry   Newtonmore   1155   North Esk   Edzell   Montrose
Eigg   Spean Bridge   Blair   Brechin
Glenfinnan   Ben Nevis   Loch Ericht   Atholl   Pitlochry   Kirriemuir   Arbroath
Point of   Fort   1344   Forfar
Ardnamurchan   William   Glen Coe   Rannoch   Aberfeldy   Blairgowrie   Carnoustie
Salen   Ballachulish   1150   Moor   Ben   Dunkeld   Dundee
Tobermory   Morvern   Lawrs   Tay   Sidlaw Hills   NORTH
Arinagour   Bidean   1214   Loch   Perth   Tayport   Firth of Tay   Bell Rock   SEA
Coll   nam Bian   Tay   Crieff   St Andrews
Tiree   Connel   Tyndrum   Ben   Killin   Earn   Cupar   Fife Ness
Ben More   Oban   Dalmally   More   Callander   Glenrothes   Anstruther
Iona   966   Loch Awe   1174   Crianlarich   Buckhaven
Fionnphort   Inveraray   Ben   Stirling   Kirkcaldy   North Berwick
Scarba   Lomond   Forth   Alloa   Cowdenbeath   Firth of Forth
Colonsay   974   Aberfoyle   Dunfermline   East Linton   Dunbar
Crinan   Loch   Falkirk   Edinburgh   Eyemouth
Beinn an Oir   Lochgilphead   Lomond   Cumbernauld   Musselburgh   Berwick-
Jura   785   Helensburgh   Dumbarton   Livingston   Dalkeith   Haddington   upon-Tweed
Sound of Jura   Greenock   Glasgow   Bathgate   Penicuik   Duns   Holy Island
Islay   Tarbert   Rothesay   Paisley   Airdrie   Peebles   Galashiels   (Lindisfarne)
Port   Gigha   Bute   Johnstone   Coatbridge   Selkirk   Coldstream
Portnahaven   Askaig   Largs   Motherwell   Lanark   Melrose   Kelso   Wooler
Lochranza   East   Hamilton   St Boswells   Newtown   The Cheviot
Port Ellen   Ardrossan   Kilbride   Biggar   Hawick   Jedburgh   815
Mull of Oa   Goat   Saltcoats   Muirkirk   Galashiels   Cheviot Hills   Alnwick
Fell   Kilmarnock   Teviot   Langholm   Rothbury
874   Brodick   Ayr   Troon   Cumnock   Broad   Otterburn   Amble
Arran   Prestwick   Sanquhar   Law   Ashington
Kintyre   Maybole   Thornhill   840   Teviothead   North Tyne   Morpeth
Campbeltown   Dalmellington   Moffat   Bedlington
Girvan   New   Nith   Lockerbie   Kielder   Newcastle
Giant's   Mull   Ballantrae   Galloway   Esk   Water   upon Tyne
Causeway   of Kintyre   Merrick   Longtown   (Reservoir)   Blaydon
Portrush   843   Castle   Dumfries   Haltwhistle   Gateshead
Portstewart   Coleraine   Newton   Douglas   Carlisle   Hexham   Consett
Coleraine   Ballymoney   Stewart   Dalbeattie   Brampton   Wear
Cushendun   Tristan   Stranraer   Kirkcudbright   Silloth   Alston   Durham
Limavady   Antrim   Milleur Point   Wigtown   893   Spennymoor
Dungiven   Cross   Bishop Auckland
Ballymena   Portpatrick   Whithorn   Maryport   Fell   Newton Aycliffe
NORTHERN   Ballyclare   Luce Bay   Solway Firth   Cockermouth   791
IRELAND   Larne   Workington   Skiddaw
Magherafelt   Whitehead   Mull of Galloway   ENGLAND
Lough   Carrickfergus   Drummore
Neagh   Newtownabbey   Bangor   Donaghadee

ENGLAND

Firth of Clyde
North Channel
Rathlin Island

Longitude 4° west of Greenwich

Conic Equidistant Projection

1:3 000 000

MILES 0   20   40   60

Land below sea level

5000   16404
3000   9843
2000   6562
1000   3281
500   1640
200   656
0   0
200   656
4000   13124
6000   19686
M   FT

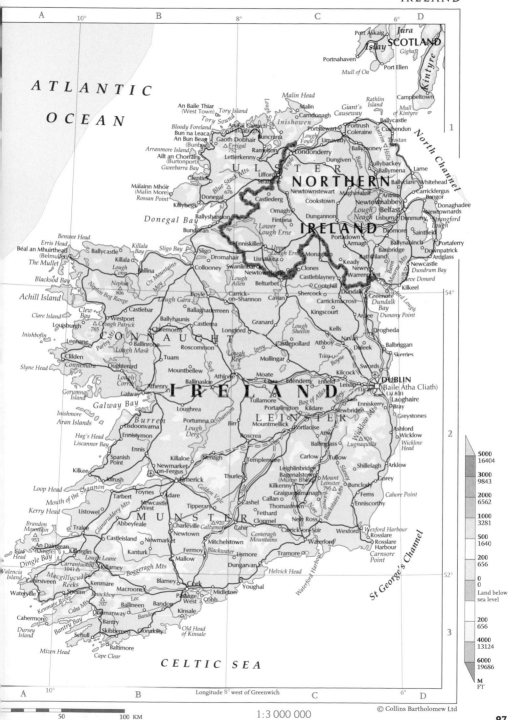

ATLANTIC

OCEAN

Port Askaig
Jura
SCOTLAND
Islay
Gighan
Portnahaven
Port Ellen
Mull of Oa
Kintyre

Malin Head
Port na Blagh
An Baile Thiar
(West Town) *Tory Island*
Tory Sound
Malin
Camdonagh
Giant's
Causeway
Campbeltown
Bloody Foreland
Bun na Leaca
An Bun Beag
(Bunbeg) **Gaoth Dobhair**
Inishowen
Portstewart **Portrush**
Rathlin Island
Mull of Kintyre
Ballycastle
Cushendun
Arranmore Island
Ailt an Chorráin
(Burtonport) **Errigal**
752 **Ramelton**
Lough Foyle
Coleraine
Limavady
Trostan
Gweebarra Bay
Letterkenny
Londonderry
Ballymoney
Ballycastle
Dungiven
Antrim Hills
Cullybackey
Larne
Glenties
Blue Stack Mts
Lifford
Ballymena
Ballyclare
Whitehead
Carrickfergus
Málainn Mhóir
(Malin More) 676
Strabane
NORTHERN
Antrim
Bangor
Donagadee
Rossan Point
Donegal
Castlederg
Newtownstewart
Magherafelt
Newtownabbey
Newtownards
Killybegs
Omagh
Cookstown
IRELAND
Belfast
Strangford Lough
Ballyshannon
Fintona
Dungannon
Lough Neagh
Lisburn
Dunmurry
Saintfield
Donegal Bay
Bundoran
Lower Lough Erne
Portadown
Dromore
Ballynahinch
Portaferry
Benwee Head
Enniskillen
Upper Lough Erne
Armagh
Banbridge
Downpatrick
Erris Head
Ballycastle
Killala Bay
Dromahair
Lisnaskea
Monaghan
Rathfriland
Ardglass
Béal an Mhuirthead
(Belmullet)
Killala
Sligo
Clones
Keady
Newcastle
Dundrum Bay
The Mullet
Ballina
Ox Mountains
Colloney
Swanlinbar
Newtownbutler
Castleblayney
Newry
Slieve Donard
Warrenpoint
Blacksod Bay
Lough Conn
Nephin
806
Lough Allen
Belturbet
Cootehill
Dundalk
Carlingford Lough
Kilkeel
Nephin Beg Range
Boyle
Carrick-on-Shannon
Cavan
Shercock
Dundalk
Achill Island
Lough Gara
Ballaghaderreen
Carrickmacross
Dundalk Bay
Greenore
Clare Island
Clew Bay
Castlebar
Westport
Croagh Patrick 765
Kingscourt
Ardee
Dunany Point
Louisburgh
Ballyhaunis
Castlerea
Longford
Granard
Kells
Drogheda
Inishbofin
Leenane
CONNAUGHT
Claremorris
Roscommon
Lough Sheelin
Castlepollard
Navan
Balbriggan
Clifden
Lough Mask
Ballinrobe
Tuam
Lough Ree
Mullingar
Trim
Duleek
Skerries
Slyne Head
Contemmara
Oughterard
Mountbellew
Lough Corrib
Athlone
Moate
Edenderry
Enfield
Kilcock
Swords
Gorumna Island
Athenry
Ballinasloe
DUBLIN
(Baile Átha Cliath)
Galway Bay
Galway
IRELAND
Tullamore
Bog of Allen
Kildare
Leixlip
Laoghaire
Inishmore
Loughrea
Portumna
Birr
Portarlington
Newbridge
Bray
Aran Islands
Burren
Lough Derg
Mountmellick
Portlaoise
LEINSTER
Enniskerry
Greystones
Hag's Head
Lisdoonvarna
Roscrea
Wicklow Mts
Ashford
Liscannor Bay
Ennistymon
Mountmellick
926
Lugnaquilla
Wicklow
Ennis
Golden Vale
Templemore
Carlow
Tullow
Wicklow Head
Spanish Point
Killaloe
Thurles
Baltinglass
Shillelagh
Arklow
Kilkee
Newmarket-on-Fergus
Leighlinbridge
Gorey
Kilrush
Limerick
Bagenalstown
(Muine Bheag)
Mount Leinster 795
Bunclody
Loop Head
Foynes
Adare
Kilkenny
Graiguenamanagh
Ferns
Cahore Point
Mouth of the Shannon
Tarbert
Newcastle West
Tipperary
Callan
Thomastown
Enniscorthy
Kerry Head
Listowel
Glanaruddery Mts
Cashel
Clonmel
Brandon Mountain 953
Tralee
Abbeyfeale
Charleville
Galtymore 920
Cahir
Fethard
Comeragh Mountains
Carrick-on-Suir
New Ross
Wexford Harbour
Wexford
An Daingean
(Dingle)
Castleisland
Newmarket
Newtown
Mitchelstown
Blackwater
Waterford
Rosslare
Sleá Head
Killorglin
Kanturk
Fermoy
Lismore
Tramore
Rosslare Harbour
Dingle Bay
Lough Leane
Boggeragh Mts
Mallow
Dungarvan
Carnsore Point
Valencia Island
Killarney
Carrantuohill 1041
Macroom
Helvick Head
Macgillycuddy's Reeks
Blarney
Cork
Youghal
Cahirsiveen
Kenmare
Lee
Midleton
Sneem
Macroom
Knockboy 707
Ballineen
Passage West
Cobh
Waterville
Caha Mts
Dunmanway
Bandon
Kinsale
Kenmare River
Bantry
Clonakilty
Cahermore
Bantry Bay
Schull
Skibbereen
Old Head of Kinsale
Dursey Island
Baltimore
Mizen Head
Cape Clear

CELTIC SEA

North Channel

St George's Channel

MUNSTER

Longitude 8° west of Greenwich

5000	**16404**
3000	**9843**
2000	**6562**
1000	**3281**
500	**1640**
200	**656**
0	**0**
Land below sea level	
200	**656**
4000	**13124**
6000	**19686**
M	
FT	

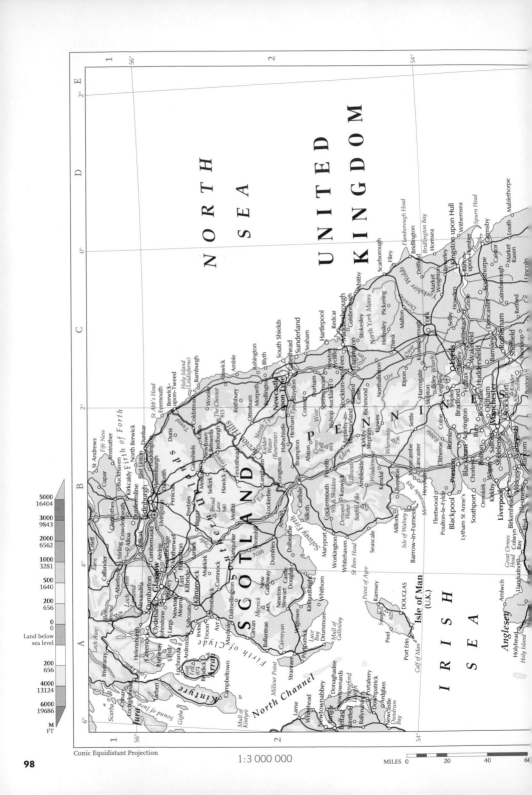

ENGLAND

WALES

Cardigan Bay

Cambrian Mountains

ENGLISH CHANNEL (LA MANCHE)

Strait of Dover

FRANCE

Bristol Channel

Isle of Wight

The Weald

South Downs

Cotswold Hills

Chiltern Hills

The Fens

The Wash

Lyme Bay

Dartmoor

Exmoor

Brecon Beacons

Mendip Hills

Black Mountains

LONDON

Birmingham

© Collins Bartholomew Ltd

0   50   100 KM

99

© Collins Bartholomew Ltd

0   50   100 KM

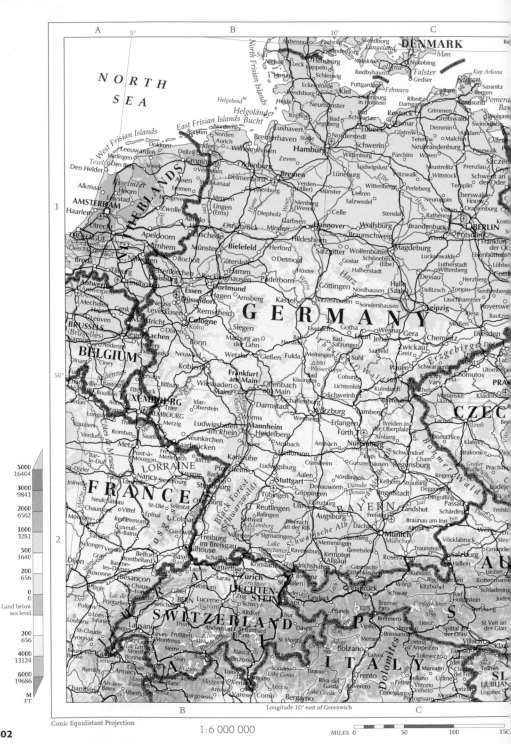

A   5°   B   10°   C

DENMARK

Møn

Langeland

Kap Arkona

Rügen

Sassnitz

Bergen

Stralsund

Pomera

Bay

NORTH

SEA

Helgoland

Helgoländer

Bucht

East Frisian Islands

North Frisian Islands

West Frisian Islands

Borkum

Norderney

West Frisian Islands

Texel

Den Helder

Alkmaar

AMSTERDAM

Haarlem

THE HAGUE

s-Gravenhage

(Den Haag)

Rotterdam

Breda

Antwerp

Antwerpen

Mechelen

Hasselt

Genk

BRUSSELS

(Bruxelles)

Namur

BELGIUM

Dinant

Charleville-Mézières

LUXEMBOURG

LUXEMBOURG

Trier

FRANCE

LORRAINE

Nancy

GERMANY

Hamburg

Bremen

Hannover

BERLIN

Potsdam

Leipzig

Dresden

Frankfurt

am Main

Darmstadt

Mannheim

Heidelberg

Nuremberg

CZEC

PRA

SWITZERLAND

BERN

Zürich

LIECHTEN-

STEIN

VADUZ

ITALY

AU

SL

LJUBLJAN

Longitude 10° east of Greenwich

B   C

5000
16404

3000
9843

2000
6562

1000
3281

500
1640

200
656

0
0

Land below
sea level

200
656

4000
13124

6000
19686

M
FT

Conic Equidistant Projection

1:6 000 000

MILES 0   50   100   150

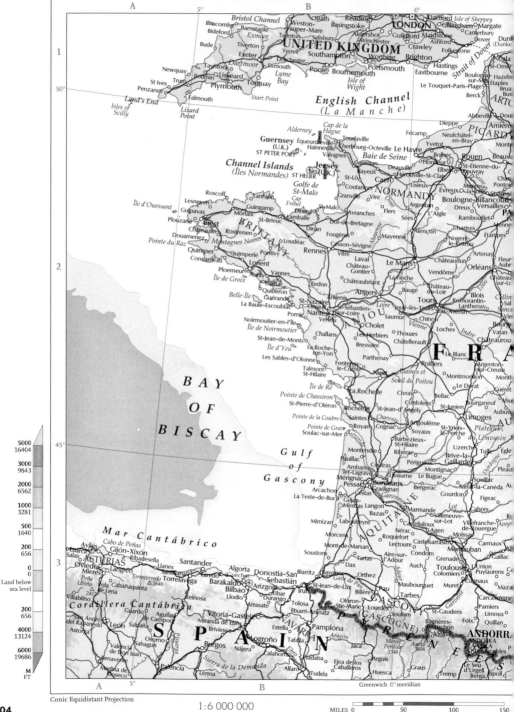

Conic Equidistant Projection

1:6 000 000

MILES 0   50   100   150

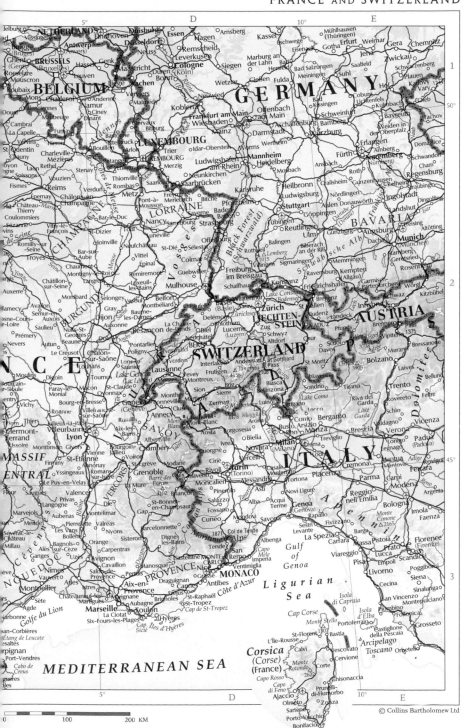

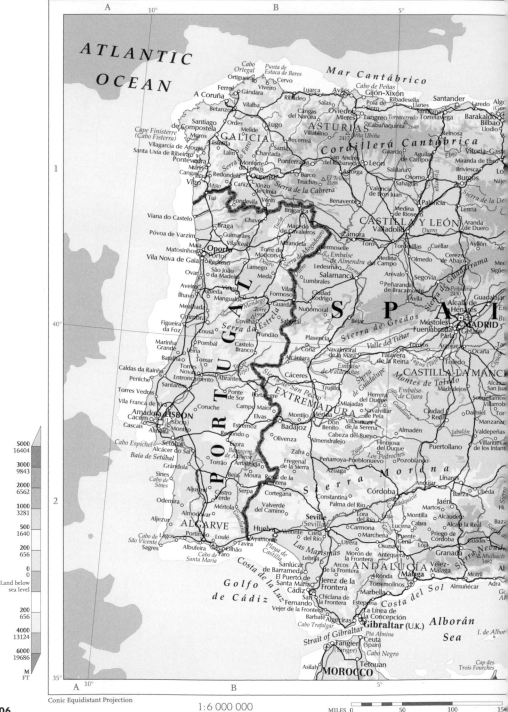

ATLANTIC

OCEAN

Mar Cantábrico

Cabo
Ortegal
Punta de
Estaca de Bares
Ortiguelra
Cervo
Luarca
Avilés
Cabo de Peñas
Gijón–Xixón
Santander
Ferrol
Viveiro
Ribadeo
Salas
Pola de
Ribadesella
Laredo
Algc
A Coruña
Gándara
Villablino
Oviedo
Llanes
Santoña
Barakaldo
Betanzos
Vilalba
Ribadeo
Cangas
del Narcea
Mieres
Torrecerredo
Ribadesella
Torrelavega
Bilbao
Santiago
de Compostela
Ordes
Melide
Lugo
ASTURIAS
Peña Ubiña
Cabañaquinta 2648
Reinosa
Vitoria–Gast
Llodio
Muros
Estrada
Becerreá
2417
Cordillera Cantábrica
Miranda de Ebro
GALICIA
Lalín
Chantada
San Andrés
del Rabanedo
Guardo
Aguilar
de Campoo
Briviesca
Ná
Vilagarcia de Arousa
Sarria
Ponferrada
León
Ebro
Na
Santa Uxía de Ribeira
Pontevedra
Monforte
de Lemos
Barco de
Valdeorras
Astorga
Saldaña
Osorno
Burgos
Cangas
Redondela
Caniza
Xinzo
de Limia
Verín
Sierra de la Cabrera
Benavente
Valencia
de Don Juan
Medina
de Rioseco
Palencia
Lerma
Vigo
Miño
Tui
Ribadavia
Ourense
El Teleno
Trujas
2188
Zamora
Valladolid
Tordesillas
Cuéllar
Aranda
de Duero
Ayllón
Viana do Castelo
Braga
Chaves
Bragança
Macedo
de Cavaleiros
CASTILLA Y LEÓN
Duero
Cere
de
Abajo
Sigüe

MOROCCO

Conic Equidistant Projection
1:6 000 000
MILES 0   50   100   15(

Gulf of Gascony

FRANCE

GASCOGNE

PYRENEES

ANDORRA

CATALUÑA

Barcelona

Zaragoza

Costa Dorada

Costa del Azahar

VALENCIA

Golfo de Valencia

Valencia

Costa Blanca

Alicante

Murcia

Cartagena

Majorca
(Mallorca)

Palma de Mallorca

Minorca
(Menorca)

Ibiza
(Eivissa)

Formentera

Illa de Cabrera

BALEARIC ISLANDS
(ISLAS BALEARES)
(Spain)

MEDITERRANEAN SEA

ALGERIA

ALGIERS
(Alger)

Oran

LANGUEDOC

Marseille

0   100   200 KM

Conic Equidistant Projection

1:6 000 000

MILES 0   50   100   150

Longitude 10° east of Greenwich

0   100   200 KM

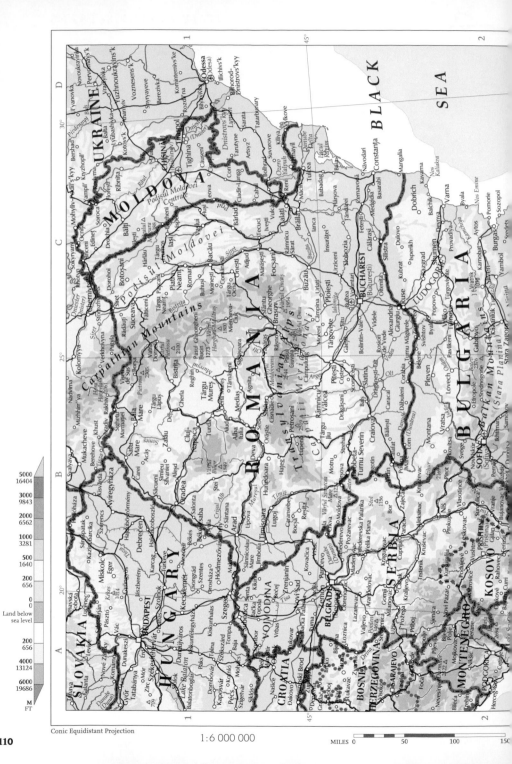

Conic Equidistant Projection

1:6 000 000

MILES 0   50   100   150

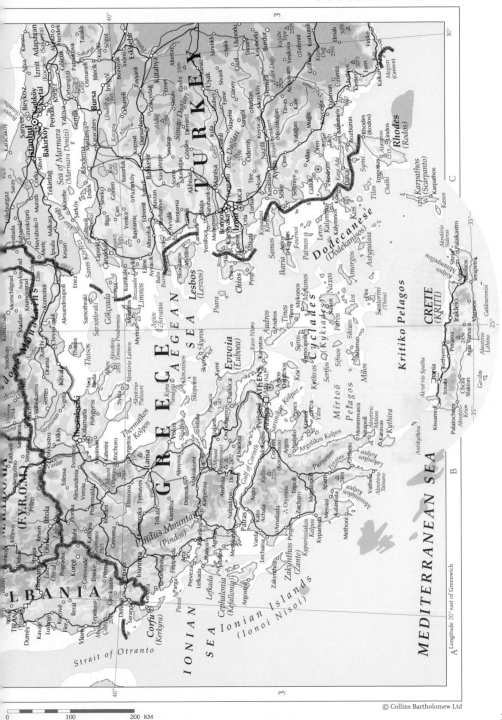

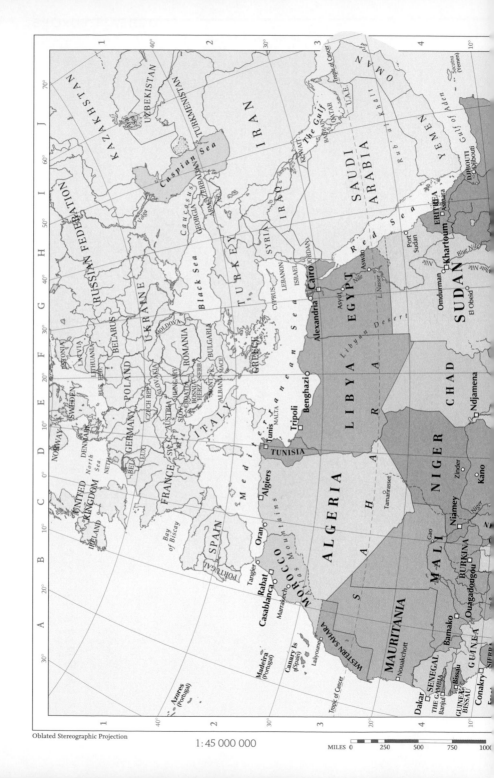

Oblated Stereographic Projection

1:45 000 000

MILES 0   250   500   750   1000

SEYCHELLES
Coëtivy
Victoria□ Mahé
MAURITIUS
Port Louis□
Réunion
(France)
St-Denis□
Tropic of Capricorn

MADAGASCAR

Antananarivo
Fianarantsoa
Tanjona
Vohimena

Agalega
Islands
(Mauritius)

Farquhar
Group
(Seychelles)

Tanjona
Bobaomby

INDIAN

OCEAN

Mogadishu

COMOROS
Moroni□
Mayotte
(France)
Aldabra Islands
(Seychelles)

Mahajanga

Nampula

Mombasa
Zanzibar
Dar es Salaam

Mozambique Channel

KENYA
Nairobi
Lake
Turkana

Dodoma
Kilimanjaro
5892

TANZANIA
Tabora
Lake
Victoria

MOZAMBIQUE

Beira

UGANDA
Kampala
Kisangani

RWANDA
Kigali□
BURUNDI
Bujumbura

MALAWI
Lilongwe
Blantyre
Lake
Nyasa

Zambezi

Maputo
Mbabane
SWAZILAND

DEMOCRATIC
REPUBLIC
OF THE CONGO

Lubumbashi

Kalemie

Lake
Tanganyika

ZAMBIA
Ndola
Lusaka

ZIMBABWE
Harare
Bulawayo

Limpopo

Pretoria
(Tshwane)

Durban

Bangui

Mbandaka

Kananga

Livingstone

Zambezi

Johannesburg

REPUBLIC
OF
SOUTH AFRICA

Port Elizabeth

Congo

ANGOLA
Huambo

Francistown

BOTSWANA
Gaborone

Kalahari
Desert

Orange

LESOTHO
Maseru

Kinshasa

CONGO
Brazzaville
CABINDA
(Angola)

Luanda

NAMIBIA
Windhoek

Namib

Namibe

Cape Town
Cape of Good Hope
Cape
Agulhas

CAMEROON
Douala
Yaoundé

GABON
Libreville
Port Gentil

Gulf of Guinea
Abidjan
Accra

Malabo
Bioko
EQUAT.
GUINEA
SÃO TOMÉ
AND PRÍNCIPE
São Tomé
São Tomé

ATLANTIC

OCEAN

St Helena
(U.K.)

Ascension
Island
(U.K.)

Tropic of Capricorn

Greenwich 0° meridian

Equator

Tristan da Cunha
(U.K.)

500   1000   1500 KM

© Collins Bartholomew Ltd

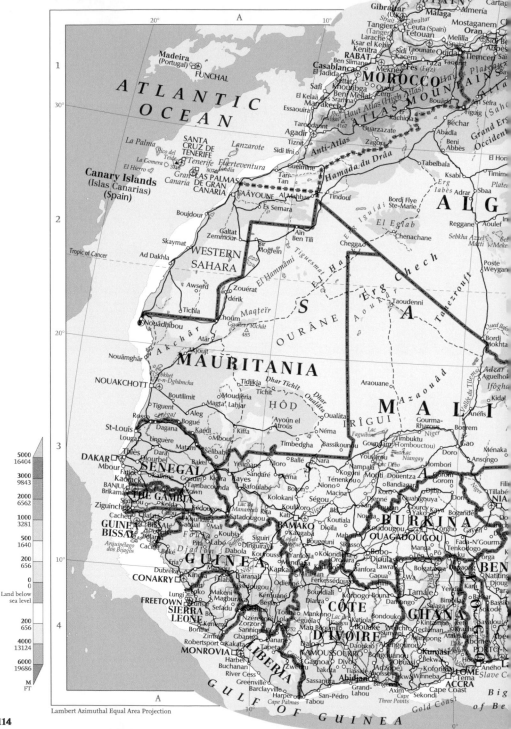

Lambert Azimuthal Equal Area Projection

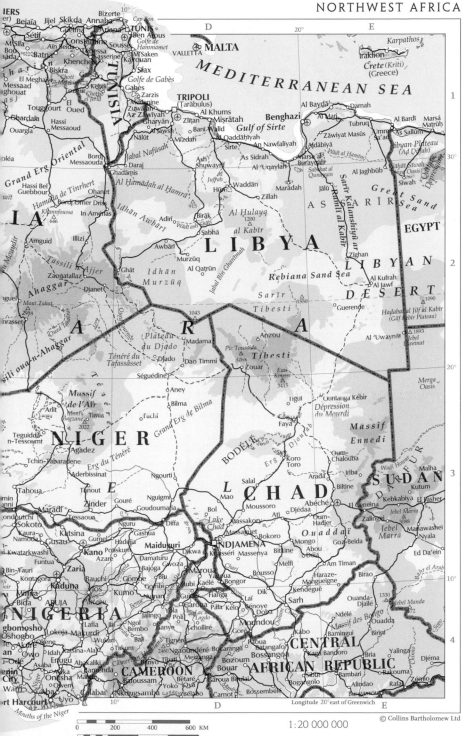

1:20 000 000

© Collins Bartholomew Ltd

Longitude 20° east of Greenwich

0   200   400   600 KM
0   100   200   300   400 MILES

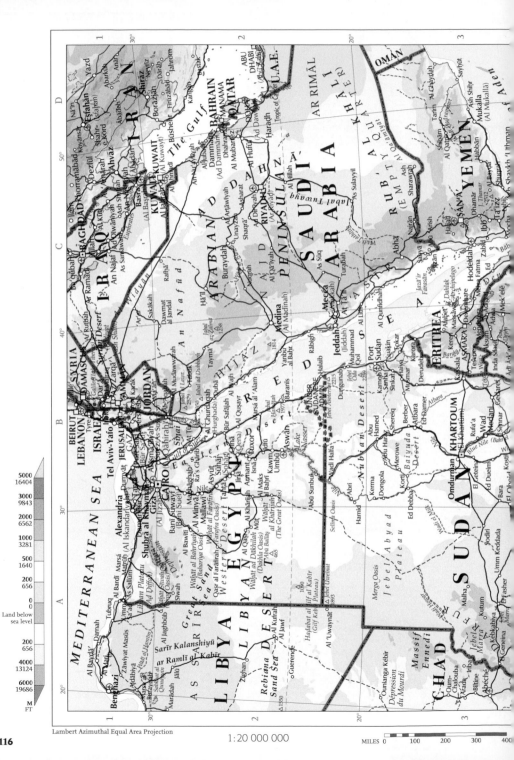

MEDITERRANEAN SEA

IRAN

IRAQ

SYRIA

JORDAN

ISRAEL

LEBANON

BEIRUT

DAMASCUS

AMMAN

JERUSALEM

Tel Aviv-Yafo

CAIRO

Alexandria

EGYPT

LIBYA

LIBYAN DESERT

Western Desert

Great Sand Sea

Benghazi

CHAD

SUDAN

KHARTOUM

Omdurman

ERITREA

ASMARA

YEMEN

SAN'A'

SAUDI ARABIA

ARABIAN PENINSULA

RUB' AL KHALI (EMPTY QUARTER)

AR RIMAL

OMAN

U.A.E.

ABU DHABI

QATAR

BAHRAIN

MANAMA

KUWAIT

BAGHDAD

The Gulf

Mecca (Makkah)

Medina (Al Madīnah)

Jeddah (Jiddah)

RIYADH

NAJD

AD DAHNĀ'

AN NAFŪD

Nubian Desert

Jebel Abyad Plateau

Massif Ennedi

Lake Nasser

Land below
sea level

M	FT
5000	16404
3000	9843
2000	6562
1000	3281
500	1640
200	656
0	0
200	656
4000	13124
6000	19686

116

Lambert Azimuthal Equal Area Projection

1 : 20 000 000

MILES   0   100   200   300   400

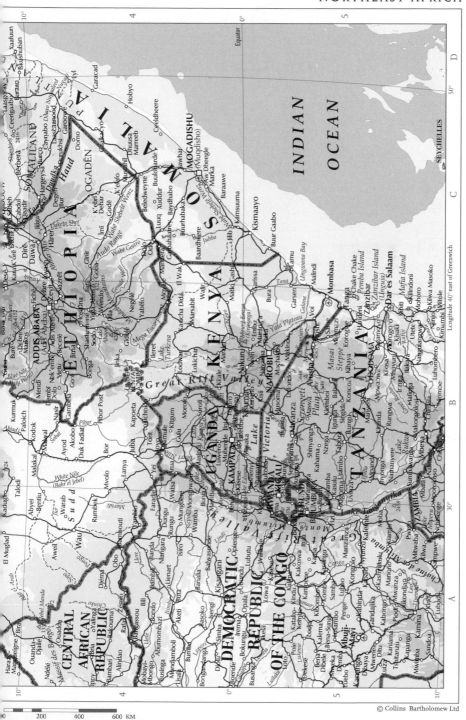

INDIAN OCEAN

SEYCHELLES

Equator

Longitude 40° east of Greenwich

200   400   600 KM

© Collins Bartholomew Ltd

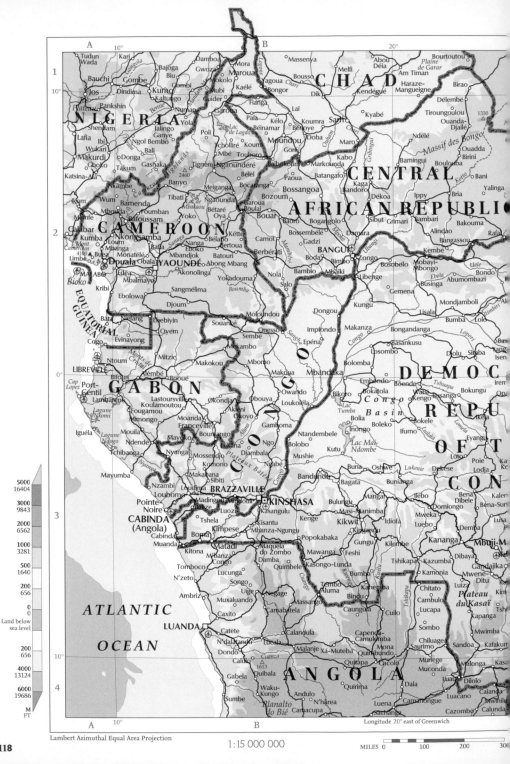

Lambert Azimuthal Equal Area Projection

1:15 000 000

Longitude 20° east of Greenwich

MILES 0   100   200   300

Elevation scale (left margin):

M	FT
5000	16404
3000	9843
2000	6562
1000	3281
500	1640
200	656
0	0
Land below sea level	
200	656
4000	13124
6000	19686
M	FT

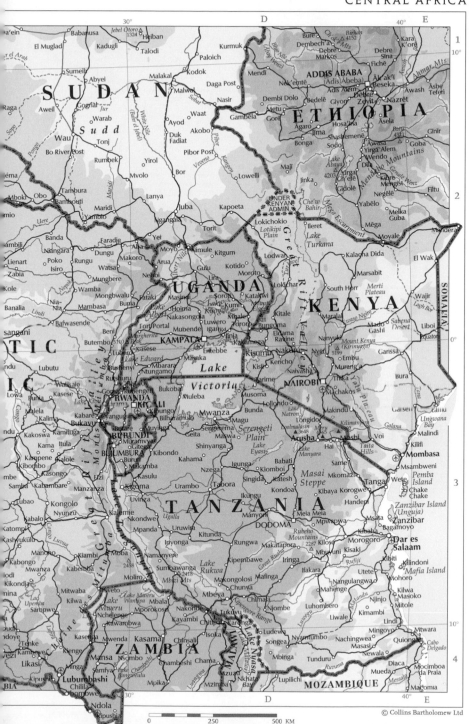

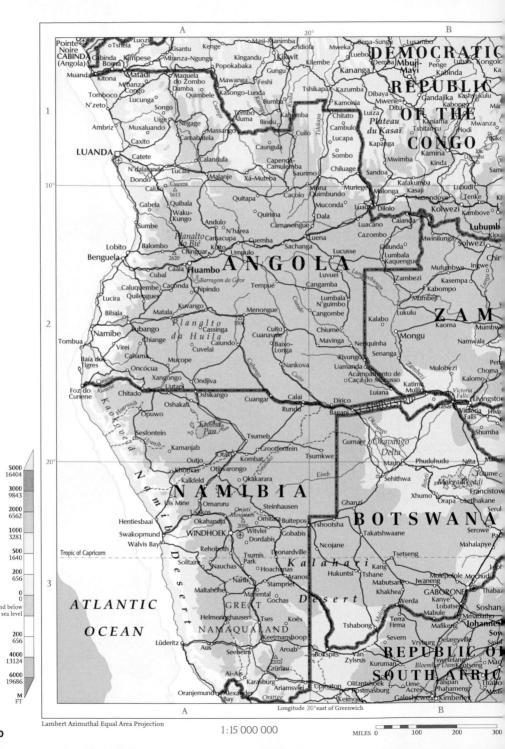

A B

Pointe-
Noire
CABINDA
(Angola) Cabinda
Boma Luozi Tshela Kisantu Kenge Masi-Manimba Idiofa Mweka Bena-Sungu Lusambo
20° DEMOCRATIC
Muanda Kitona Matadi Kimpese Mbanza-Ngungu Kingandu Kikwit Kilembe Luebo Demba Mbuji- Penge Lubao Kongolc
Mbanza Maquela Popokabaka Gungu Mayi Kabinda Ka
Congo do Zombo Damba Kasongo-Lunda Tshikapa Kazumba Dibaya Gandajika Kashyukulu
Tomboco Lucunga Quimbele Bumba Kamonia Mwene- Kabongo Mar
N'zeto Uige Negage Massango Tembo Bindu Kahemba Ditu Tshitanzu Piodi
Ambriz Muxaluando Camabatela Aluma Chitato Cambulo Luiza Kaniama Mwanza
Caxito Lucapa Kapanga Kamina Kinda
LUANDA Catete Calandula Capenda- Sombo Plateau Mwinba REPUBLIC OF THE
N'dalatando Lucala Xá-Muteba Camulemba Saurimo Chiluage du Kasaï Kafakumba Lubudi
Dondo Malanje Mona Sandoa CONGO
10° Caculo Quimbundo Muconda Muriege Malonga Kasaji
Gabela Quibala Quitapa Cacolo Muconda Dala Luau Dilolo Nasondoye Tenke Ki
Waku- Quirima Luacano Cazombo Caianda Kolwezi Kambove
Sumbe Kungo Andulo Camanongue Luena Caxinga Mwinilunga Solwezi Lubumb
Lobito Balombo N'harea Cuemba Sachanga Lucusse Lumbala Kasempa Ingw
Benguela Huambo Chinguar Umpulo Kaquengue Mufumbwe Kabompo Chir
Caála Catabola Luvuei Zambezi ZAM
Cubal Barragem do Gove Cangamba Mumbeji
Caluquembe Caconda Tempué Lukulu
Quilengues Chipindo Lumbala Kaoma Mumbwe
Lucira Kuvango Menongue N'guimbo Kalabo Namwala
Bibala Matala Planalto Cuito Cangombe Lukulu
Namibe Lubango da Huíla Cassinga Cuanavale Chiume Mongu Choma
Tombua Virei Chiange Caiundo Baixo- Mavinga Nenquinha Mulobezi Pem
Baía dos Cahama Mucope Cuvelai Longa Rivungo Senanga Kalomo
Tigres Oncócua Nankova Uamanda Acampamento de Katima Livingst
Foz do Chitado Xangongo Ondjiva Cuito Caça do Mucusso Mulilo Victoria
Cunene Uutapi Oshikango Cuangar Calai Luiana Bukalo Kasane Falls
Oshakati Rundu Dirico Victoria Hwa
Opuwo Bagani Falls
Sesfontein Etosha Tsumeb Gumare Okavango Shumba
Pan Delta
Kamanjab Grootfontein Phuduhudu Nata Maller
Outjo Otavi Tsumkwe Maun Tsume
Khorixas Kombat Sehithwa Makgadikgadi Francistow
20° Otjiwarongo Eiseb Xhumo Orapa Jetlhakane Serul
Kalkfeld Okakarara Ghanzi Phudutlhakane
Uis Mine NAMIBIA BOTSWANA Serowe
Omaruru Onjati Steinhausen Tshootsha Takatshwaane Serowe Pal
Hentiesbaai Usakos Mountain Omitara Buitepos Mahalapye
Okahandja 2050 Mabutsane Jwaneng Molepolole Mochudi
Swakopmund WINDHOEK Witvlei Gobabis Ncojane Tsetseng Kanye Lobatse Soshan
Walvis Bay Dordabis Tshane Khakhea GABORONE Thaba
Rehoboth Leonardville Kang Hukuntsi Werda Mabule Mmabatho Sow
Tropic of Capricorn Tsumis Kalahari Mafikeng Johannes
Solitaire Park Hoachanas Mabule Delareyville Sas
Nauchas Aranos Desert Severn Vryburg
Narib Stampriet Werda Lime Phahameng Masil
Maltahöhe Mariental Gochas Tshabong Terra Acres Klasin
Gochas Firma Kuruman Thabo
ATLANTIC Helmeringhausen GREAT Bokspits Van Vryburg Delareyville
Tses Koës Zylsrus Bloem Dam Kootsong Mag
OCEAN Lüderitz NAMAQUALAND Keetmanshoop REPUBLIC OF
Aus Aroab Upington Olifantshoek SOUTH AFRIC
Seeheim Kenhardt Postmasburg Kimberle
Ai-Ais 2002 Grünau Galeshewe
Karasburg Keimoos
Oranjemund Alexander Ariamsvlei Orange
Bay

1:15 000 000

MILES 0 100 200 300

Lambert Azimuthal Equal Area Projection

Longitude 20° east of Greenwich

5000
16404
3000
9843
2000
6562
1000
3281
500
1640
200
656
0
0
Land below
sea level
200
656
4000
13124
6000
19686
M
FT

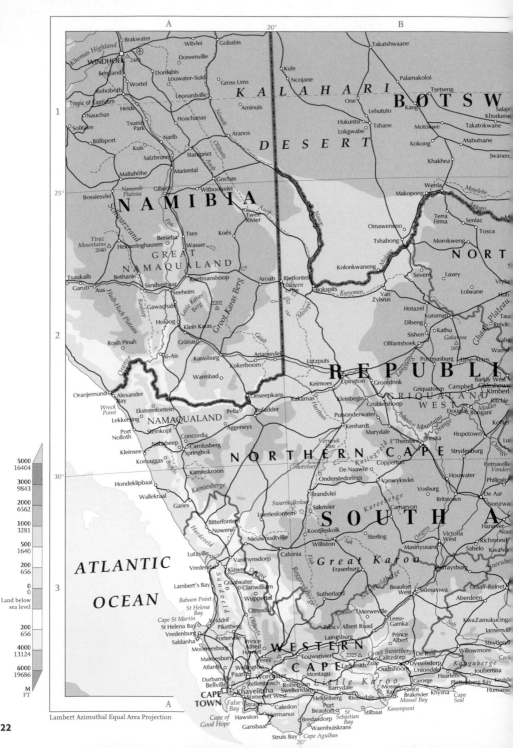

A | 20° | B

Khomas Highland
Brakwater
WINDHOEK 2480
Witvlei
Gobabis
Takatshwaane

Bergland
Doreenville
Kule
Palamakaloi

Rehoboth
Wortel
Dordabis
Louwater-Suid
Gross Ums
Ncojane
Tsetseng
Salajy
Khudume

Tropic of Capricorn
Leonardville
One
Lehututu
Kang
BOTSW

Nauchas
Heide
Aminuis
Hukuntsi
Tshane
Motokwe
Takatokwane

Solitare
Tsumis Park
Hoachanas
Lokgwabe
Kokong
Mabutsane

Büllsport
Narib
Aranos
DESERT
Khakhea
Jwanen

Kuis
Salzbrunn
Stampriet
Werda
Makopong
Moselebe

Maltahöhe
Mariental
Gochas
Omaweneno
Terra Firma
Senlac
Tosca

Nananib Plateau
Gibeon
Witbooisvlei
Tshabong
Morokweng
NORT

Bossiesvlei
N A M I B I A
Twee Rivier
Kolonkwaneng
Laxey
Vryb

Tiraz Mountains 2040
Helmeringhausen
Tses
Koës
Severn
Lolwane
Hun

Tsaukaib
GREAT NAMAQUALAND
Aroab
Rietfontein
Bokspits
Van Zylsrus
Hotazel
Kuruman
Tau
Reivilo

Garub
Bethanie
Sandverhaar
Keetmanshoop
Dibeng
Kathu
Valsp

Aus
Seeheim
2202
Holoog
Klein Karas
Sishen
Olifantshoek
Warrel

Rosh Pinah
Grünau
Ai-Ais
Karasburg
Kokerboom
Ariamsvlei
Lutzputs
Keimoes
Upington
Groottrink
Postmasburg
Lime-Acres

Warmbad
Onseepkans
Kakamas
Kleinbegin
Groblershoop
Griquatown
Campbell
Kimberl

Oranjemund
Alexander Bay
Eksteenfontein
Pella
Pofadder
Pusonderwater
Kenhardt
Marydale
E'Thembini
Prieska
Hopetown

Port Nolloth
Steinkopf
Concordia
Aggeneys
Verneuk Pan
Copperton
Strydenburg

Kleinsee
Nababeep
Carolusberg
Springbok
NORTHERN CAPE
De Naawte
Vanwyksvlei
Houwater
Philipsto

Komaggas
Kamieskroon
Grootvloer
Onderstedorings
Vosburg
Britstown
De Aar

Hondeklipbaai
Garies
Brandvlei
Sakrivier
Camarvon

ATLANTIC
Wallekraal
Loeriesfontein
SOUTHA

OCEAN
Bitterfontein
Nuwerus
Kootjieskolk
Sterling
Victoria West
Richmond

Lutzville
Vredendal
Nieuwoudtville
Williston
Masinyusane
Sabelo
KwaNor

Lambert's Bay
Klawer
Vanrhynsdorp
Calvinia
Great Karoo
Murraysburg

Baboon Point
Graafwater
Clanwilliam
Fraserburg
Beaufort West
Sidesaviwa
Graaff-Reinet

St Helena Bay
Wupperal
Sutherland
Aberdeen

Saldanha
Citrusdal
WESTERN
Prince Albert Road
KwaZamukucinga

Vredenburg
Porterville
Laingsburg
Prince Albert
Jansenvill

Malmesbury
Moorreesburg
CAPE
De Rust
Willowmore
Steytlerv

Atlantis
Wellington
Worcester
Montagu
Oudtshoorn
Uniondale
Joubertina

Durbanville
Paarl
Little Karoo
George
Knysna

CAPE TOWN
Stellenbosch
Swellendam
Barrydale
Mossel Bay
Humans

Cape of Good Hope
Hermanus
Bredasdorp
Cape Agulhas

Lambert Azimuthal Equal Area Projection

122

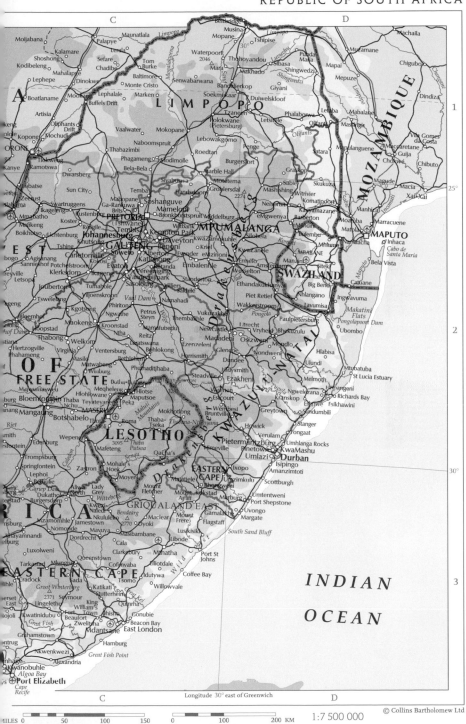

INDIAN

OCEAN

Longitude 30° east of Greenwich

1:7 500 000

© Collins Bartholomew Ltd

MILES 0   50   100   150   0   100   200 KM

ICELAND

Arctic Circle

Denmark Strait

Ammassalik

Kong Frederik VI Kyst

Kong Christian IX Land

Labrador Sea

NEWFOUNDLAND AND LABRADOR

St John's
C. Race

St Pierre
St Pierre and Miquelon (Fr.)
Strait
PRINCE EDWARD ISLAND
Gulf of St Lawrence
Anticosti I.
NEW BRUNSWICK
NOVA SCOTIA

Greenland (Denmark)

Kong Christian IX Land

Kong Frederik VIII Land

Nuuk

Davis Strait

Labrador

Québec
Montréal

Baffin Bay

Baffin Island

Hudson Strait

QUÉBEC

Iqaluit

Kuujjuaq

Labrador City

Ottawa

Foxe Basin

Quttinirpaaq

Queen Elizabeth Islands

Ellesmere Island

Devon Island

Dundas

Achuguttuaaq

Bylot Island

Prince of Wales

Southampton I.

Coats I.

Mansel I.

Repulse Bay

Hudson Bay

Belcher Is.

Churchill

Otisaabi

ONTARIO

Lake Superior

Thunder Bay

MICH

N U N A V U T

Victoria Island

Banks Island

Sachs Harbour

Amundsen Gulf

Great Bear Lake

Great Slave Lake

Yellowknife

MANITOBA

Lake Winnipeg

Winnipeg

MINNESOTA

St Pau

ARCTIC OCEAN

Beaufort Sea

Mackenzie

Inuvik

NORTHWEST TERRITORIES

Lake Athabasca

SASKATCHEWAN

Peace

Saskatoon

Regina

Edmonton

Calgary

NORTH DAKOTA

Bismarck

SOUTH DAKOTA

Rapid City

Barrow

U.S.A.
ALASKA

YUKON

Whitehorse

ALBERTA

BRITISH COLUMBIA

Prince George

ROCKY MOUNTAIN

Billings

Helena
MONTANA

WYOMING

Fairbanks

Mt McKinley
6194

Anchorage

Juneau

Kamloops

Vancouver

Seattle

WASHINGTON

IDAHO

Boise

RUS
FED.

Arctic Circle

Nome

Bering Str.

St Lawrence I.

Gulf of Alaska

Alexander Archipelago

Vancouver Island

Victoria

Olympia

Portland

Salem

OREGON

Columbia

Salt Lake Cit

St Matthew Island

Kodiak I.

Queen Charlotte Islands

Reno

Carson C

CALI

Nunivak I.

Pribilof Islands

Aleutian Range

Sacramento

San Francisco

Bering Sea

Aleutian Islands

Bi-Polar Oblique Projection

1:40 000 000

MILES 0   200   400   600   800

Lambert Azimuthal Equal Area Projection

1:25 000 000

MILES 0   250   50

I 45° J 75° 30° K 15°

2

60°

Kong Christian X
Land

ICELAND
Keflavik
REYKJAVÍK
Vestmannaeyjar
Vik

Denmark
Strait
Horn
Ísafjörður
Kangerlussuaq

Greenland
(Kalaallit Nunaat)
(Denmark)

Lauge Koch Kyst 2000
2500
3000
1500
1000

Knud Rasmussen
Land

Thule
(Qaanaaq)
Dundas
Innaanganeq
Qimusseriarsuaq

Baffin
Bay

Nuussuaq
Kangersuatsiaq
Upernavik
Sigguup Nunaa
Nuttaal
Uummannaq
Qeqertarsuaq
Ilulissat
Qasigiannguit
Qeqertarsuaq
Aasiaat
Uummannaq

Kong Frederik VI Kyst

Ammassalik
Qilak
Kulusuk

Uummannarsuaq
Kangerlua

Cape
Liverpool
Bylot Island
Pond Inlet
Cape
Christian
Clyde River

Baffin
Island

Barnes
Ice cap
Cape
Henry Kater
Home Bay

Maniitsoq
Sisimiut
Kangerlussuaq
Napasoq
NUUK
(Godthåb)
Paamiut

Qeqertarsuatsiaat
Ivittuut
Nanortalik
Qassimiut
Qaqortoq
Qagartoq

Cape Farewell
(Nunap Isua)

ATLANTIC
OCEAN

3

30°

43°

Prince
Charles
Maria
Nettilling
Lake

Cumberland
Peninsula
Pangnirtung
Cape Mercy
Cumberland Sound

Iqaluit
(Frobisher Bay)
Frobisher Bay
Meta Incognita
Peninsula
Resolution
Island
Loks Land

Labrador
Sea

U
T

Melville
Peninsula
Hall
Beach

Repulse Bay
Foxe
Basin
Foxe
Peninsula

Foxe
Channel

Southampton
Island
Coral
Harbour

Hudson
Strait

Cape Dorset

Kimmirut

Cape Chidley

Akpatok
Island
Killiniq

NEWFOUNDLAND AND LABRADOR

Evans Strait
field Inlet

Coats
Island

Ivujivik
Salluit

NUNAVIK

Kangiqsujuaq

Ungava
Bay

D
A

Mansel
Island

Puvirnituq

Kangirsuk

Ungava
Peninsula
d'Ungava

Kangiqsualujjuaq
George

Nain

Cape
Harrison

45°

Inukjuak

Rivière
aux Mélèzes

Kuujjuaq

HUDSON

BAY

King George Islands
Sanikiluaq

Belcher
Islands

Lac à l'Eau
Claire

Schefferville

Hopedale

Labrador

Churchill
Falls

Happy Valley-
Goose Bay
Red

Strait of Belle Isle
St Anthony

Fort
Severn

Cape Henrietta
Maria

Grande Rivière de la Baleine

Réservoir
Robert-Bourassa
La Grande 3

Réservoir
de Caniapiscau

Petit
Mécatina

St-Augustin

Belle
Island

Winisk (abandoned)

James
Bay

Chisasibi
Wemindji

Eastmain

Lac Bienville

Réservoir
Opinaca

Lac Pletipi

Port aux
Choix

Newfoundland

Gander
Grand Falls-
Windsor
Grand Bank

St John's

Cape Race

out Lake
Ekwan

Attawapiskat
Akimiski
Island
Fort George
Fort Albany
Moosonee

Broadback
Rupert
Eastmain

Waskaganish
Fort Rupert

Réservoir
Gouin

Lac Mistassini

Réservoir
Pipmuacan

Réservoir
Manicouagan
Manicouagan
Baie-
Comeau

Sept-
Îles
Port-
Cartier
Havre-
St-Pierre

Anticosti

Gaspé

Corner
Brook

Port-aux-Basques
Cabot Strait

ARIO

Attawapiskat
Missinaibi

Chibougamau

Broadback

Lac
Casgrain

Jonquière
Chicoutimi

La Tuque

Rimouski

Mont-Joli

St Lawrence

Îles de la
Madeleine

Gulf of
St Lawrence

St Pierre and
Miquelon (France)

ay

Nakina
Hearst
Kapuskasing

Lake
Abitibi
La Sarre

Amos

Val-d'Or

Réservoir
Cabonga

Mont-
Laurier

Québec
Trois-Rivières
Sherbrooke

PR. EDWARD I.
Charlottetown

Sydney
Cape Breton
Island

Sable
Island

4

out Lake

Beardmore
Geraldton
Homepayne
Timmins

Kirkland
Lake
Rouyn-Noranda

Lake
Nipigon

Marathon
Michipicoten
River
Chapleau
Sault Sainte
Marie

North
Bay

New Liskeard

Pembroke
Huntsville

OTTAWA

Kingston

MAINE

Bangor

Augusta

NEW
BRUNSWICK

Fredericton
Saint John

Truro
Dartmouth
Halifax
Bridgewater
Liverpool
Yarmouth

NOVA SCOTIA

Cape
Sable

Lake
Nipigon
kan

Nipigon
Thunder
Bay

Hancock

Isle
Royale

Sudbury

Georgian
Bay

Lake
Huron

Orillia
Peterborough

Toronto

Lake Ontario

Portland

Concord
Lowell

CONN.
Boston
MASS.

60°

MICH-
CONSIN
Wausau

Green Bay
Oshkosh

Cadillac

Grand
Rapids

Lansing
Flint

Saginaw
Bay

Owen
Sound

Hamilton
Buffalo
Syracuse
Utica

NEW YORK

Hartford
R.I.
Providence

Long Island

Cape Cod

ilwaukee
Sheboygan

Ann Arbor
Detroit

PENNSYLVANIA

Warren
Scranton

Newark

New York

Trenton

90° G 75° H 60° I

© Collins Bartholomew Ltd

0 250 500 750 KM

Lambert Azimuthal Equal Area Projection

1:12 000 000

MILES 0   100   200   300

Longitude 120° west of Greenwich

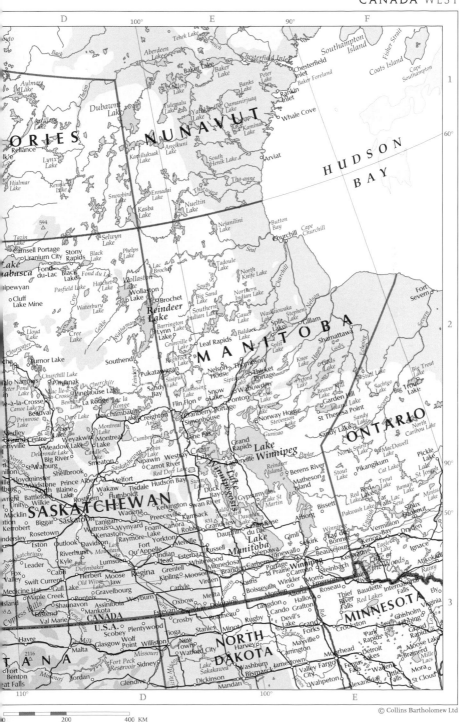

Suto

Back

Aylmer Lake

Artillery Lake

Hjalmar Lake

Kennie Lake

Lynx Lake

Reliance

ORIES

NUNAVUT

Tehek Lake

Aberdeen Lake

Mallery Lake

Baker Lake

Baker Lake

Banks Lake

Yathkyed Lake

Tulemalu Lake

Qamanirjuaq Lake

Angikuni Lake

Kaminak Lake

Kamilukuak Lake

South Henik Lake

Tha-anne

Ennadai Lake

Snowbird Lake

Thlewiaza

Nueltin Lake

Kasba Lake

Dubawnt Lake

Chesterfield Inlet

Southampton Island

Chesterfield Inlet

Peter Lake

Rankin Inlet

Baker Foreland

Coats Island

Cape Southampton

Whale Cove

Arviat

Fisher Strait

HUDSON

BAY

60°

Tazin

594

Selwyn Lake

Camsell Portage Uranium City

Stony Rapids

Black Lake

Phelps Lake

Nejanilini Lake

Button Bay

Churchill

Cape Churchill

Seal

Cluff Lake Mine

Fond-du-Lac

Black Lake

Fond du Lac

Pasfield Lake

Hatchet Lake

Wollaston Lake

Brochet

Big Sand Lake

Tadoule Lake

North Knife Lake

South

Churchill

Lac Brochet

Northern Indian Lake

Fort Severn

abasca

Lake

ipewyan

Cree Lake

Waterbury Lake

Cree Lake

Reindeer Lake

Southern Indian Lake

Gauer Lake

Waskaiowaka Lake

Gillam

Stephens Lake

Nelson

MANITOBA

2

Lloyd Lake

Turnor Lake

Clearwater

Southend

Barrington Lake

Lynn Lake

Granville Lake

Leaf Rapids

Baldock Lake

Split Lake

Thompson

Shamattawa

Gods

Echoing

Severn

ONTARIO

alo Narrows

Churchill Lake

Patuanak

Pukatawagan

Highrock Lake

Knee Lake

Gods Lake

Big Trout Lake

Big Trout Lake

Peter Pond Lake

h-Churchill

Besnard

Pinehouse Lake

Sandy Bay

Sisipuk Lake

Nelson House

Sipiwesk Lake

Thicket Portage

Oxford Lake

Beaver Hill Lake

Garden Hill

Island Lake

Sachigo Lake

Big Trout Lake

le-a-la-Crosse

Beauval

Canoe Lake

La Ronge

Lac la Ronge

Dore Lake

Kississing Lake

Snow Lake

Wabowden

St Theresa Point

Sandy Lake

Primrose Lake

Green Lake

Montreal Lake

Trout Lake

Flin Flon

Cranberry Portage

Wabowden

Norway House

Stevenson

Gunisao

North Spirit Lake

MacDowell

Pickle Lake

90°

Medley

Grand Centre

nnyville

Weyakwin

Meadow Lake

Delaronde Lake

Montreal Lake

Candle Lake

La Ronge

Creighton

Amisk Lake

Simonhouse

Cumberland House

The Pas

Grand Rapids

Lake Winnipeg

Leven ville

Poplar

Berens River

Reindeer Island

Matheson Island

Stout Lake

Pikangikum

Cat Lake

Sandy Lake

Red Lake

Trout Lake

Eat Lake

Keeper

Red Lake

Falls

Seul

Sioux Lookout

MacDowell

St Joseph

50°

Big River

Smeaton

Saskatchewan

Nipawin

Carrot River

Robin

Red Deer Lake

Swan Lake

Little Grand Rapids

Bissett

Pakwash Lake

Falls

Ignace

Walburg

Lloydminster

Shellbrook

Prince Albert

Wakaw

Tisdale

Hudson Bay

Swan River

Duck Bay

Lake St Martin

Arborg

Vermillion Bay

Dryden

MANITOBA

Maidstone

North Blaine

Battleford

Rosthern

Humboldt

Kelvington

Kamsack

Waldy

Dauphin

Gimli

Selkirk

Kenora

Eagle Lake

Atikokan

burg

Unity

Wilkie

SASKATCHEWAN

Saskatoon

Tanigan

Wadena

Preeceville

Grandview

831 △

Dauphin Lake

Lake Manitoba

Rose

Winnipeg

Lac du Bonnet

Rainy

Macklin

Biggar

Watrous

Wynyard

Foam Lake

Cahora

Russell

Dauphin

Neepawa

Beausejour

Kerrobert

Rosetown

Raymore

Melville

Esterhazy

Whitewood

Minnedosa

Carberry

Portage la Prairie

Carman

Winnipeg

Steinbach

Keewatin

Lake of the Woods

3

ndersley

Eston

Outlook

Davidson

Fort Qu'Appelle

Yorkton

Brandon

Souris

Winkler

Morris

Emerson

Roseau River

Baudette

International Falls

Ely

Kyle

Riverhurst

Cabri

Lumsden

Indian Head

Grenfell

Kipling

Virden

Boissevain

Melita

Winkler

Altona

Hallock

Thief River Falls

Red Lakes

Virginia

saskatchewan

Fox Valley

Herbert

Diefenbaker

Regina

Moosomin

Carlyle

Souris

Deloraine

Langdon

Cando

Devil's Lake

Crookston

Park Rapids

Chisholm

Hibbing

Medicine Hat

Gull Lake

Swift Current

Old Wives Lake

Moose Jaw

Gravelbourg

Weyburn

Oxbow

Melita

Rugby

Minot

Grafton

Red Lake

MINNESOTA

Grand Rapids

Moose Lake

Maple Creek

Shaunavon

Assiniboia

Ponteix

Mankota

Val Marie

Eastend

CANADA

U.S.A.

Estevan

Bonneau

Sheyenne

Grand Forks

Mayville

Fergus Falls

Detroit Lakes

Brainerd

Mille

Mora

St Cloud

Havre

Scobey

Plentywood

Crosby

Stanley

Minot

Harvey

Carrington

Moorhead

Wadena

Little

ANA

2116 △

Malta

Glasgow

Wolf Point

Williston

New Town

Watford City

Missouri

NORTH DAKOTA

Washburn

Jamestown

Valley City

Wahpeton

Alexandria

Fort Benton

Milk

Jordan

Fort Peck Reservoir

Sidney

Glendive

Lake Sakakawea

Dickinson

Bismarck

Mandan

0   200   400 KM

HUDSON

BAY

NUNAVUT

MANITOBA

ONTARIO

QUÉ

JAMES
Bay

Puvirnituq

Gilmour
Island

Lac
Payne

Ottawa
Islands

Lac
Tasiat

Lasia

Lac
Le Roy

Inukjuak

Lac
Bacqueville

Lac
Chavigny

Lac
Minto

North
Knife Lake

Cape
Churchill

Churchill

Stephens
Lake

Nelson

Churchill

Hopewell Islands

Lac
L'EAU
Claire

Nastapoca Islands

Lac
Guillaume-Delisle

Gillam

Knee
Lake

Hayes

Shamattawa

Gods

Oxford
Lake

Sleeper
Islands

North Belcher
Islands

King-George
Islands

Lac
Bienville

Gods
Lake

Stull Lake

Fort
Severn

Belcher
Islands

Sanikiluaq

Grande Rivière de la Baleine

Sachigo
Lake

North Spirit
Lake

Severn

Flaherty Island

Cape Henrietta
Maria

Long Island

Kuujjuarapik
(Poste-de-la-Baleine)

Lac Burton

Réservoir
La Grande 4

Stout
Lake

Sandy
Lake

Big Trout Lake

Big Trout
Lake

Winisk
(abandoned)

Kasabonika
Lake

Winisk

Ekwan

Réservoir
Robert-Bourassa

Réservoir
La Grande 3

MacDowell
Lake

Webequie

Winisk
Lake

Chisasibi
(Fort George)

Pikangikum

North
Caribou Lake

Nibinamik

Attawapiskat
Lake

Attawapiskat

Akimiski
Island

North
Twin
Island

Radisson

Réservoir
Opinaca

Red
Lake

Red
Lake

Cat Lake

Trout
Lake

Pickle Lake

St Joseph
Lake

Missisa
Lake

Kapiskau

Fort Albany

South
Twin
Island

Wemindji

Pakwash Lake

Far
Falls

Bamaji
Lake

Whitewater
Lake

Ogoki
Reservoir

Albany

Charlton
Island

Eastmain

Eastmain

English

Vermilion

Lac
Seul

Miniss
Lake

Savant
Lake

Ogoki

Ogoki

Pledger
Lake

Moosonee

Waskaganish
(Fort Rupert)

Rupert

Lac Evans

Mistissi

Kenora

Sioux
Lookout

Sturgeon Lake

Caribou Lake

Albany

Missinaibi

Moose
Factory

Rupert Rivière de Rupert

Lac
Comencho

Lac
Opataca

Mistissi

Lake of
the Woods

Eagle
Lake

Dryden

Vermilion
Bay

Armstrong

Nakina

Kesagami
Lake

Moose

Nottaway Rivière Nottaway

Broadback

Fort
Frances

Ignace

Lac des
Mille Lacs

Lake
Nipigon

Longlac

Hearst

Fraserdale

Rivière d'Harricana

Lac au Goéland

Chibougam

Rainy Lake

Atikokan

Thunder
Bay

Beardmore

Nipigon

Terrace
Bay

Kapuskasing

Otter Rapids

Smooth Rock Falls

Cochrane

Lac Matagami

Matagami

Lac Waswanipi

Lebel-sur-
Quévillon

Lac au
Opataca

Mistissi

CANADA

U.S.A.

Pigeon
River

Royale

Kahinakagami
Lake

Hornepayne

Manitouwadge

Missinaibi
Lake

Groundhog

Iroquois
Falls

Timmins

Night Hawk
Lake

La Sarre

Rouyn-
Noranda

Amos

Réservoir
Gouin

St-Michel-
des-

Lac St

Rober

Métabet

Grand
Marais

Pigeon

Copper
Harbor

Michipicoten
Island

Michipicoten
River

Wawa

Chapleau

Foleyet

Kirkland
Lake

Englehart

Val-d'Or

Malartic

Senneterre

Réservoir
Parent

Ashland

Gogebic
Range

Keweenaw
Peninsula

Batchawana
Mountain

Lac
Abitibi

New Liskeard

Temagami

Lac
Simard

Réservoir
Dozois

Parent
Cabonga

La Tuque

Hancock

Houghton

MICHIGAN

Sault Sainte
Marie

Sault Sainte
Marie

Blind
River

Sudbury

Sturgeon
Falls

North
Bay

Lac
Kipawa

St-Michel-
des-Saints

Ishpeming

Marquette

Newberry

St Ignace
Island

Espanola

Mattawa

Petawawa

Mont-
Laurier

Shawinigan

Trois-
Rivières

Crystal
Falls

Iron
Mountain

Escanaba

St Ignace

Manitoulin
Island

Wikwemikong

Nipissing

Deep River

Pembroke

Maniwaki

Mont-Tremblant

Joliette

Park
Falls

Menominee

Marinette

Cheboygan

Little Current

South
Baymouth

Georgian Bay

Huntsville

Barry's
Bay

Amprior

Carleton Place

Hull

Montréal

Valleyfield

Merrill

Rhinelander

Petoskey

Alpena

Bruce
Peninsula

Owen
Sound

Bracebridge

Gravenhurst

Parry
Sound

South River

Rideau

Ottawa

Smiths
Falls

Cornwall

Ste-Adèle

Wausau

Shawano

Green
Bay

Manistique

Traverse
City

Gaylord

Midland

Kawartha Lakes

Carleton
Place

Brockville

Ogdensburg

Massena

Plattsburgh

St-Jean

Wisconsin
Rapids

Oshkosh

Appleton

Sturgeon Bay

Ludington

Cadillac

Standish

Kincardine

Orillia

Lindsay

Peterborough

Belleville

Watertown

Burlington

VERM

Portage

Fond
du Lac

Sheboygan

Big Rapids

Mount Pleasant

Saginaw

Bay
City

Port
Elgin

Goderich

Guelph

Oswego

Rome

Lowville

Madison

Milwaukee

Grand
Rapids

Lansing

Flint

Huron

Stratford

Kitchener

Toronto

Oshawa

Rochester

Oneida

Utica

WISCONSIN

Waukesha

Racine

Owosso

Pontiac

London

Brantford

Hamilton

Scarborough

Syracuse

Auburn

Finger Lakes

Adirondack

NEW

Marcy

Rutland

Rockford

Kenosha

Battle Creek

Livonia

St Catharines

Lake Ontario

Oswego

Ithaca

Cortland

Glens Falls

Aurora

Elgin

Waukegan

Kalamazoo

Jackson

Ann
Arbor

Detroit

Windsor

St
Thomas

Buffalo

Batavia

Corning

Norwich

Oneonta

Albany

Troy

Pittsfield

Chicago

South Bend

Michigan
City

Jackson

Adrian

Sylvania

Toledo

Lake Erie

Erie

Dunkirk

Jamestown

Olean

Elmira

Binghamton

York

Schenectady

Joliet

Ottawa

Plymouth

Wayne

Maumee

Lorain

Cleveland

Warren

Bradford

Sayre

Springfield

ILLINOIS

Pontiac

INDIANA

Watseka

OHIO

Lambert Azimuthal Equal Area Projection

1:12 000 000

MILES 0   100   200   300

Longitude 80° west of Greenwich

5000
16404

3000
9843

2000
6562

1000
3281

500
1640

200
656

0
0

Land below
sea level

200
656

4000
13124

6000
19686

M
FT

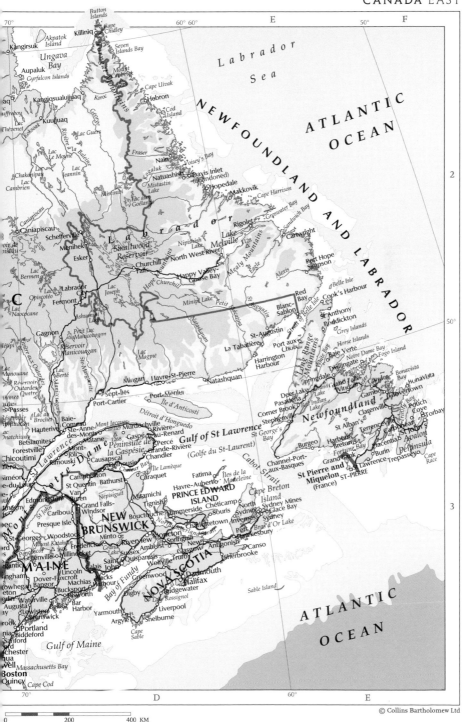

70°   60° 60°   E   50°   F

Button
Islands
Cape
Chidley
Killinia
Akpatok
Island
Kangirsuk
Aupaluk
Gyrfalcon Islands
Seven
Islands Bay
Ungava
Bay
Mont
d'Iberville
Cape Uivak
Kangiqsualujjuaq
Kuujjuaq
Koroc
Hebron
Cod
Island
Lac Guers
Lac
Thévenet
Lac
Koksoak
Lac
Le Moyne
Fraser
Nain
Voisey's Bay
Lac
Jeannin
Davis Inlet
(abandoned)
Natuashish
Chakonipau
Lac
Cambrien
Mistanibi
Lake
Hopedale
Lac à
Goéland
Mistinibi
Makkovik
Cape Harrison

Caniapiscau

Labrador
Sea

NEWFOUNDLAND AND LABRADOR

ATLANTIC
OCEAN

Caniapiscau
Scheffervillle
Menihek
Esker
Smallwood
Reservoir
Nipishish
Lake
North West River
Lake
Melville
Rigolet
Churchill
Falls
Happy Valley-
Goose Bay
Meally Mountains
Eagle
Alexis
Cartwright
Port Hope
Simpson

Lac
Bermen
Lac
Opiscotéo
Labrador
City
Fermont
Lac
Joseph
Hope
Churchill
Belle Isle
Cook's Harbour
Minipi Lake
Petit
Mécatina
Red
Bay
St Anthony
Roddickton
Lac
Naococane
Ashuanipi
Lake
Blanc-
Sablon
St-Augustin
Grey Islands
Gagnon
Petit Lac
Manicouagan
Lac
Magpie
La Tabatière
Port aux
Choix
Baie
Verte
Horse Islands
Fogo Island
Reservoir
Manicouagan
Harrington
Harbour
Springdale
Notre Dame Bay
Lac
Berté
Minghan
Havre-St-Pierre
Natashquan
Twillingate

C

Lac
Brochet
Baie-
Comeau
Ste-Anne-
des-Monts
Mont Jacques
Cartier
Murdochville
Rivière-
au-Renard
Port-Menier
Île d'Anticosti
Détroit d'Honguedo
Deer Lake
Pasadena
Indian
Lake
Gambo
Bonavista
Bay
Bonavista
Hauterive
Sept-Îles
Port-Cartier
Gulf of St Lawrence
Corner Brook
Gander
Clarenville
Glovertown
Trinity
Bay
Catalina
Betsiamites
Matane
Gaspé
Percé
(Golfe du St-Laurent)
Stephenville
St Alban's
Terra Nova
Conception Bay
Pouch
Cove
Forestville
Mont-
Joli
Rimouski
Péninsule de
la Gaspésie
Grande-Rivière
Chandler
St George's
Burgeo
Harbour
Breton
Avalon
Peninsula
St John's
Torbay
Chicoutimi
Causapscal
Cabot Strait
Placentia Bay
Grand
Bank
Fortune Bay
Fortune
Chapeau
du-Loup
Chaleur Bay
La Lamèque
Channel-Port-
aux-Basques
St Pierre and
Miquelon
(France) ST-PIERRE
Burin
Trepassey
Cape
Race
Campbellton
St Quentin Bathurst
Caraquet
Fatima
Îles de la
Madeleine
Havre-Aubert
Edmundston
Van
Buren
Grand Falls-
Windsor
Nepisiguit
Miramichi
PRINCE EDWARD
ISLAND
Cape Breton
Island
Sydney Mines
Glace Bay
Caribou
Presque Isle
Woodstock
Bouctouche
Chéticamp
North
Sydney
Sydney
Summerside
Souris
Charlottetown
Inverness
Port
Hawkesbury
Bras d'Or Lake

MAINE

NEW
BRUNSWICK
Minto
Grand
Lake
Riverview
Moncton
Springhill
Amherst
New
Glasgow
Antigonish
Canso
Mount Katahdin
Fredericton
Sussex
Oromocto
Parrsboro
Truro
Sherbrooke
Dover-Foxcroft
Lincoln
Saint
John
Quispamsis
NOVA SCOTIA
Bangor
Machias
Buckspoit
St Stephen
St George
Wolfville
Dartmouth
Ellsworth
Calais
Digby
Greenwood
Halifax
Bar
Harbor
Kentville
Bridgewater
Waterville
Belfast
Yarmouth
Liverpool
Augusta
Lewiston
Cape
Sable
Argyle
Shelburne
Lac
Rossignol
Sable Island

ATLANTIC

OCEAN

Portland
Biddeford
Sanford
Gulf of Maine

Bay of Fundy

Massachusetts Bay
Boston
Quincy
Cape Cod

70°   D   60°   E

50° 130° A 120° B 110° C 10

Port
Hardy
Gold River
Campbell River
Vancouver Island
Nanaimo
Victoria
Cape Flattery
Powell River
Kamloops
Mount Waddington 4019
100 Mile House
Jasper
Hedley
Edmonton
Vegreville
Noydminster
Meadow Lake
BRITISH COLUMBIA
Wetaskiwin
Red Deer
Prince Albert
Nipawin
SASKATCHEWAN
Winnipegosis
CANADA
MANITOBA
Wainwright
Medicine Hat
Humboldt
Canora
Yorkton
Melville
Regina
Weyburn
Estevan
Virden
Brandon

PACIFIC
OCEAN

WASHINGTON
Seattle
Tacoma
Olympia
Mount Rainier 4392
Mount St. Helens
Yakima
Spokane
Portland
Salem
Oregon City
Eugene
Albany
Coos Bay
OREGON
Bend
Pendleton
La Grande
Burns
MONTANA
Great Falls
Helena
Butte
Bozeman
Billings
Miles City
Glendive
N. DAKO
Bismarck
Dickinson
Bowman
Mobridge
Lake Oahe
S. DAKO
Pierre

IDAHO
Boise
Idaho Falls
Twin Falls
Pocatello
WYOMING
Casper
Lander
Sheridan
Gillette
Buffalo
Rapid City
Black Hills
Chadron
Scottsbluff
NEBRAS
Ogallala

Crescent City
Eureka
Redding
Red Bluff
Ukiah
Santa Rosa
Sacramento
San Francisco
Oakland
San Jose
Modesto
Fresno
Salinas
Monterey Bay
CALIFORNIA
NEVADA
Winnemucca
Reno
Sparks
Carson City
Elko
Wendover
Great Salt Lake
Salt Lake City
Provo
UTAH
Logan
Green River
Evanston
Rock Springs
Laramie
Cheyenne
Greeley
Boulder
Denver
Aurora
COLORADO
North Platte
McCook

Bakersfield
Santa Maria
Santa Barbara
Point Conception
Oxnard
Los Angeles
Long Beach
Pasadena
Riverside
Oceanside
San Diego
Tijuana
Ensenada
UNITED STAT
Las Vegas
Henderson
Barstow
St George
Cedar City
Richfield
Grand Junction
Moab
Durango
Colorado Springs
Pueblo
Alamosa
Trinidad
Ulysses
Dodge City
Liberal

ARIZONA
Phoenix
Mesa
Glendale
Tucson
Nogales
Flagstaff
Kingman
Lake Havasu City
Prescott
Winslow
Tuba City
Kayenta
Gallup
Santa Fe
Albuquerque
Los Alamos
Taos
Las Vegas
Clayton
Dumas
Amarillo
NEW MEXICO
Socorro
Alamogordo
Roswell
Lubbock
Wichita
Post
Snyder
Clovis
Portales
Vernon

Mexicali
Yuma
San Felipe
Puerto Peñasco
Casa Grande
Nogales
Douglas
Agua Prieta
Ciudad Juárez
El Paso
Las Cruces
Deming
Van Horn
Fort Stockton
Odessa
Midland
San Angelo
Edwards Plateau
TEX

Lázaro Cárdenas
Cabo San Quintín
Guadalupe (Mexico)
Bahía Sebastián Vizcaíno
Isla Cedros
Punta Eugenia
Bahía Tortugas
Santa Rosalía
Rosarito
Caborca
Magdalena
Benjamín Hill
Hermosillo
Nuevo Casas Grandes
Madera
Moctezuma
Chihuahua
Ciudad Delicias
Hidalgo del Parral
Ciudad Camargo
Jiménez
Bolsón de Mapimí
Gómez Palacio
Torreón
Sabinas
Piedras Negras
Ciudad Acuña
Nuevo Laredo
Monclova
Sabinas Hidalgo
Matamoros
Saltillo
Mont

Santa Margarita
Villa Insurgentes
La Paz
San José del Cabo
MEXICO
Ciudad Obregón
Navojoa
Los Mochis
Guamúchil
Culiacán
Guasave
Guaymas
Cuauhtémoc
Durango
Mazatlán
Río Grande
Matehuala
Monterrey
Costa Rica

Tropic of Cancer

Longitude 110° west of Greenwich

5000 16404
3000 9843
2000 6562
1000 3281
500 1640
200 656
0 0
Land below sea level
200 656
4000 13124
6000 19686
M FT

Lambert Azimuthal Equal Area Projection

1:20 000 000

MILES 0   100   200   300   400

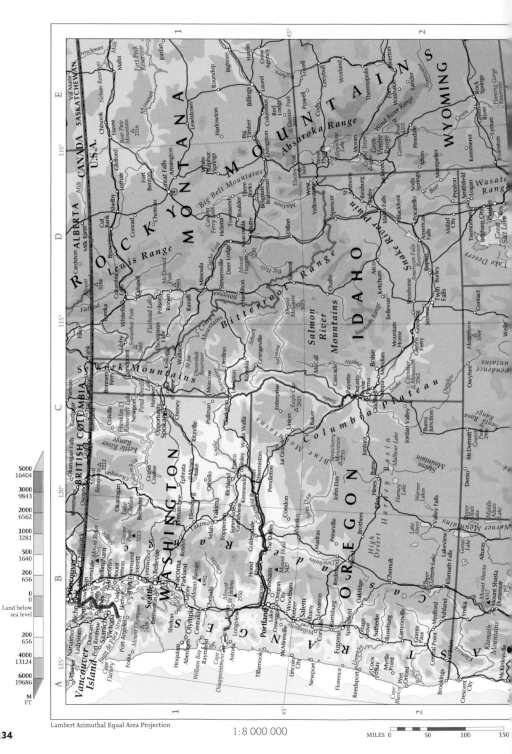

Lambert Azimuthal Equal Area Projection

1 : 8 000 000

MILES 0   50   100   150

Land below sea level scale:

M	FT
5000	16404
3000	9843
2000	6562
1000	3281
500	1640
200	656
0	0
Land below sea level	
200	656
4000	13124
6000	19686

CANADA
U.S.A.
ALBERTA
SASKATCHEWAN
BRITISH COLUMBIA

ROCKY MOUNTAINS
MONTANA
WYOMING
IDAHO
WASHINGTON
OREGON

Lewis Range
Kirk Mountains
Bitterroot Range
Salmon River Mountains
Big Belt Mountains
Absaroka Range
Wind River Range
Wasatch Range
Snake River Plain
Columbia Plateau
Blue Mountains
Cascade Range
Warner Mountains
Klamath Mountains
Kettle River Range

Vancouver Island
Juan de Fuca Strait

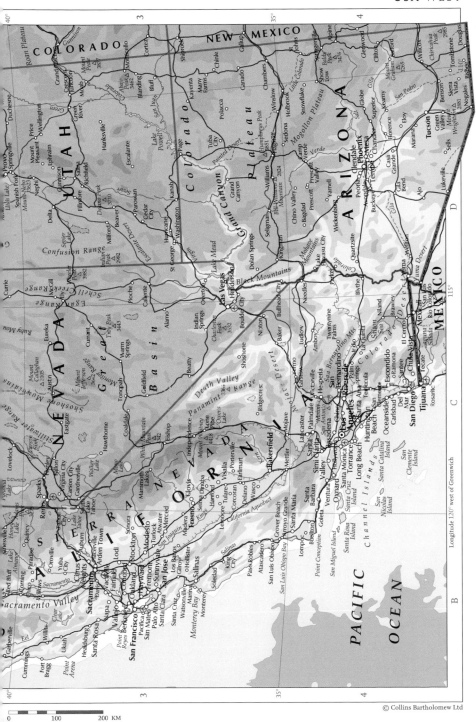

135

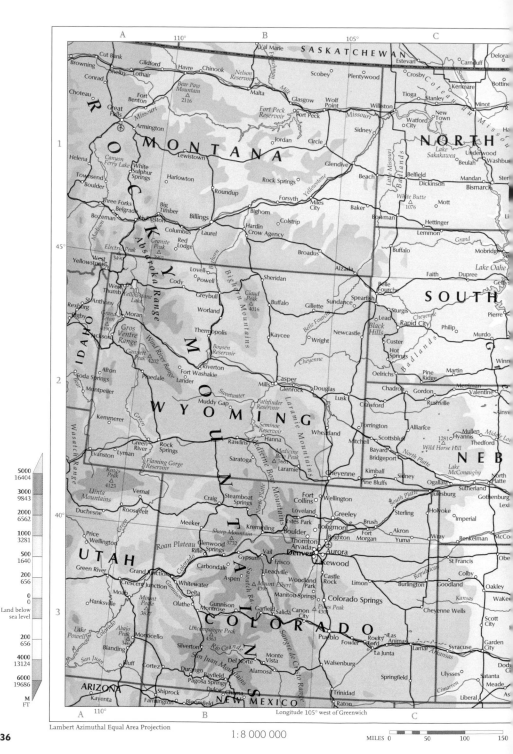

MILES 0   50   100   150

1 : 8 000 000

Lambert Azimuthal Equal Area Projection

Longitude 105° west of Greenwich

136

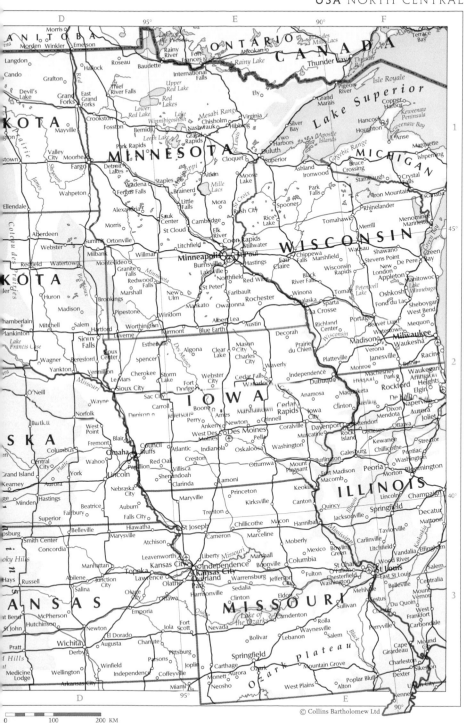

MINNESOTA

Ely
Virginia
Chisholm
Duluth
Superior
Cloquet

Two Harbors
Silver Bay
Copper Harbor
Grand Marais

Pigeon River

Thunder Bay
Nipigon

St Ignace Island
Terrace Bay
Marathon

Kabinakagami Lake

ONTARIO

Wawa

Nighthawk Lake
Timmins
Foleyet
Iroquois F

Missinaibi Lake

Lake Superior

Isle Royale

Michipicoten
Michipicoten Bay
Michipicoten River
Michipicoten Island

Chapleau
Sultan

Ramsey Lake
Onaping Lake

Tema

Wanapitei Lake

Hancock
Houghton
Keweenaw Peninsula
Keweenaw Bay

Bashawana Mountain
653

L'Anse

Sault Sainte Marie
Sault Sainte Marie

Thessalon
Elliot Lake
Blind River
Espanola
Sudbury

Stu

M I C H I G A N

Marquette
Ishpeming
Newberry

St Joseph I.
North Channel
Little Current
Wikwemikong

Georgian Bay

WISCONSIN

Hastings
Eau Claire
Chippewa Falls
Marshfield

Ironwood
Park Falls
Gogebic Range
Bruce Crossing
Ashland

Stambaugh
Crystal Falls
Iron Mountain
Tomahawk
Rhinelander
Merrill

Manistique
Escanaba
Menominee
Marinette

St Ignace
Drummond Island

Manitoulin Island

Beaver Island

Cheboygan
Petoskey
Charlevoix

Rogers City

Tobermory
Bruce Peninsula

Owen Sound

Rice Lake
Spooner

Chippewa
Falls
Wausau
Shawano

Door Peninsula
Green Bay
Sturgeon Bay

Manitou Islands
Traverse City
Frankfort

Gaylord
Au Sable
Grayling
Oscoda
Tawas City

Alpena

Collingwood

Port Elgin
Kincardine

Lake Huron

Winona
Sparta
Onalaska
La Crosse

Stevens Point
New London
Appleton
Oshkosh

De Pere
Manitowoc

Manistee
Ludington
Shelby

Cadillac
Big Rapids
Mount Pleasant

Standish
Saginaw Bay

Harbor Beach

Goderich

Orange

Gu

Black River Falls
Wisconsin Rapids

Fond du Lac
Sheboygan

Midland
Bay City

Kitchener
Stratford
Woodstock

Decorah
Richland Center

Portage
Beaver Dam

West Bend
Mequon
Glendale

Muskegon
Grand Haven
Holland

Grand Rapids
Wyoming

Owosso
East Lansing

Saginaw
Lapeer
Flint

Port Huron
Sarnia

London
Brant
Sim

Prairie du Chien
Platteville

Madison
Waukesha
Verona
Janesville
Monroe

Milwaukee
Racine
Kenosha

Lansing

Jackson

Pontiac
Sterling Heights

Lake St Clair

St Thomas

IOWA
Independence
Dubuque
Anamosa

Freeport
Beloit
Belvidere

Arlington Heights
Evanston
Waukegan

Kalamazoo
Benton Harbor
Three Rivers
Battle Creek

Ann Arbor
Brighton
Livonia

Detroit
Windsor
Chatham

Cedar Rapids
Clinton

Rockford
Dixon
De Kalb

Elgin
Wheaton
Chicago

Niles
Sturgis

Adrian
Monroe

Pelee Island

Lake Erie

Iowa City
Davenport

Sterling
Rock Island

Mendota
Ottawa
Aurora
Joliet
Oak Lawn

South Bend
Elkhart
Angola

Toledo
Sandusky
Sylvania
Perrysburg

Norwalk
Euclid
Cleveland
Warren
Youngstown

Painesville
Sha

Muscatine
Washington
Mount Pleasant
Burlington
Fort Madison

Geneseo
Kewanee
Galesburg
Chillicothe
Streator
Kankakee

Merrillville
Plymouth
Warsaw
Fort Wayne
Defiance

Bowling Green
Fremont
Findlay
Tiffin
Ashland

Wooster
Massillon
Akron
Canton
Alliance

Keokuk
Macomb
Peoria
Washington
Pontiac
Watseka

Huntington
Van Wert
Lima
Mansfield

East Liverpool
Steubenville
Washing

ILLINOIS
Canton
Morton
Bloomington
Lincoln

Logansport
Peru
Marion

Kokomo

Bellefontaine
Mount Vernon
Sidney

Delaware

Philadelphia
New

Cambridge
Zanesville
Moundsvill

Bowling Green
St Charles
Springfield

Lafayette
Danville
Champaign
Decatur

Crawfordsville
Noblesville
Muncie
Anderson

Springfield

Columbus
Lancaster

Washington Court House
Newark

Vienna
Marietta
Fairm

O'Fallon
Carlinville
Taylorville
Litchfield

Terre Haute
Greencastle
Indianapolis
Lawrence
Richmond

Middletown
Dayton
Kettering
Hamilton

Wilmington
Hillsboro

Chillicothe
Athens
Point Pleasant

Parkersburg

Morga

WES
Virgi

St Louis
East St Louis
Belleville
Vandalia
Centralia

Bloomington
Greensburg
Columbus

INDIANA

Charleston
Sullivan
Mattoon

Effingham

Olney

Bedford
Seymour

Covington
Cincinnati
Reading

Portsmouth
Ironton

Ohio

Weston
Sutton

Festus
Chester
Du Quoin
West Frankfort

Washington
Vincennes

Princeton
Jasper

New Albany
Frankfort
Georgetown

Madison
Maysville

Ashland
St Albans

Huntington
Morehead

Charleston

Oak Hill

MISSOURI
Perryville
Cape Girardeau
Charleston

Carbondale
Harrisburg
Marion

Evansville
Henderson
Radcliff
Elizabethtown

Ridge Park
Louisville
Lexington
Winchester
Richmond

Danville

London
Somerset

Morgan
Salyersville
Pikeville
Hazard

Madison
Beckley

Williamson
Welch

Summersville

Poplar Bluff
Dexter
Sikeston

Mayfield
Owensboro
Madisonville

Hopkinsville
Russellville

Munfordville
Campbellsville
Columbia

Williamsburg
Middlesboro

Norton

Abingdon
Bluefield
Lewisbu

Blacksburg

Kennett
Paragould

Murray
Clarksville
Union City
Paris

TENNESSEE

Paducah

Bowling Green
Glasgow

Dale Hollow Lake

Springfield
Gallatin

Kingsport
Bristol

Holston
Mount Rogers

Marion
Wytheville

Blue R

Lambert Azimuthal Equal Area Projection

Longitude 85° west of Greenwich

1 : 8 000 000

MILES 0   50   100   150

5000
16404
3000
9843
2000
6562
1000
3281
500
1640
200
656
0
0
Land below
sea level
200
656
4000
13124
6000
19686
M
FT

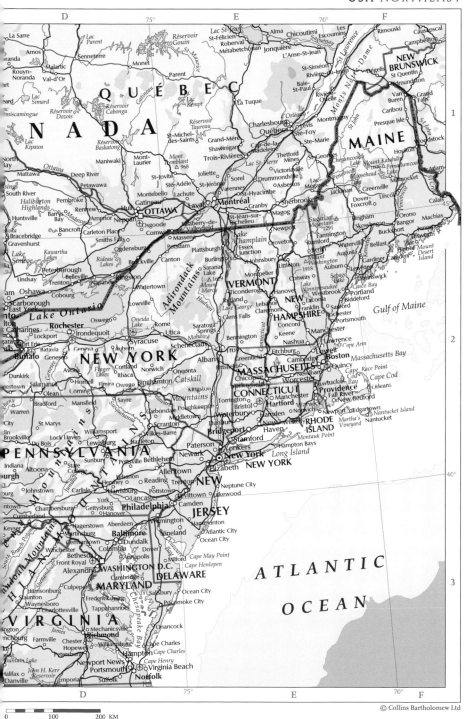

La Sarre
Amos
Rouyn-
Noranda   Malartic
Val-d'Or
Lac
Simard
Réservoir
Dozois
Lac
miscamingue
North
Bay
Mattawa
South River
Haliburton
Highlands
Huntsville
Bracebridge
Gravenhurst
Lake
Simcoe
Oshawa
Scarborough
East York
to
ton
Catharines
ara
ort Erie
ls
Buffalo
Dunkirk
nestown   Salamanca
Olean
Bradford
Warren
City
St Marys
Brookville
Du Bois
Indiana
Altoona
urgh
Johnstown

Lac
Parent
Senneterre
Monet

Réservoir
Gouin
St-Maurice
St-Michel-
des-Saints
Mont
Tremblant △ 968
Maniwaki
Mont-
Laurier
Deep River
Petawawa
Pembroke
Renfrew
Barry's
Bay
Bancroft
Carleton Place
Smiths Falls
Kawartha
Lakes
Peterborough
Belleville
Lindsay
Trenton
Cobourg
Rideau
Lakes
Kingston
Napanee
Lowville
Oswego
Geneva
Avoca
Ithaca
Hornell
Elmira   Owego
Mansfield
Williamsport
Lock Haven
Lewisburg
Hazleton

Réservoir
Cabonga
Gatineau
Ottawa
Hull
OTTAWA
Nepean
Osgoode
Cornwall
Massena
Potsdam
Ogdensburg
Canton
Watertown
Rome
Utica
Syracuse
Oneida
Lake
Rochester
Irondequoit
Cortland
Oneonta
Norwich
Binghamton
Sayre
Scranton
Wilkes-Barre
Carbondale
Middletown
Pottsville   Bethlehem

QUÉBEC
NADA

Réservoir
Baskatong
Réservoir
Taureau
Grand-Mère
Shawinigan
Trois-Rivières
Joliette
St-Jovite
St-Adèle   St-Jérôme
Montebello
Lachute
Laval   Montréal
Valleyfield
Châteauguay-
de-
Beauharnois
Lake
Champlain
Saranac
Lake
Saratoga
Springs

La Tuque

Lac St-Jean
St-Félicien
Roberval
Métabetchouan   Jonquière

Alma   Chicoutimi
Les
Escoumins
L'Anse-St-Jean
Rivière-du-Loup
St-Siméon
Baie-
St-Paul
Rivière-
Ouelle

Rimouski
Causapscal
Campbellton

NEW
BRUNSWICK

St Quentin

charged content continues...

Île
d'Orléans
Charlesbourg
Québec
Ste-Foy
Lévis

St-Michel-
de-
Madeleine
Victoriaville   Thetford
Mines
Drummondville
Asbestos
Sorel
St-Hyacinthe
Varennes
Granby
Magog
Newport
Groveton
Essex
Junction
Burlington
St Johnsbury
Montpelier
Ticonderoga
Rutland
Lebanon
Glens Falls
Claremont

Montmagny
Lac
Mégantic
Jackman
Greenville
Dover-
Foxcroft
Lincoln
Calais

St-Georges
Chesuncook
Lake

Mount Katahdin
1606 △

Moosehead
Lake

Millinocket

Old
Town
Orono
Bangor

Lincoln

Van
Buren
Caribou
Presque Isle

Houlton

Woodstock

Grand
Falls

St Adam

Machias

MAINE

Kingfield
Skowhegan
Waterville
Augusta

Sugarloaf △
Mountain
1291
Farmington

Bethel

Mount
Washington
1918
Conway

Lake
Winnipesaukee
Laconia
Franklin

NEW
HAMPSHIRE
Concord

Keene

Bennington

NEW YORK
Schenectady
Albany
Troy

Saratoga
Springs
Glens Falls

Hanover

Auburn

Sebago
Lake
Westbrook
Rochester
Sanford

Lewiston
Auburn

Gardiner
Casco Bay
Portland
Biddeford

Belfast
Bucksport

Bar
Harbor
Mount
Desert
Island

Gulf of Maine

Camden

Ellsworth

Manchester
Nashua
Lowell
Lawrence

Portsmouth

Greenfield
Fitchburg

Schenectady

Cambridge
Worcester
Boston
Quincy

Cape Ann

Massachusetts Bay

Northampton
Chicopee
Springfield

MASSACHUSETTS

Torrington
Bristol
Waterbury
Hamden
Bridgeport
Stamford

CONNECTICUT
Hartford
Manchester
Norwich

New
Haven
Weatherly
New London

Providence
Fall River
New Bedford

Newport

RHODE
ISLAND

Race Point
Cape
Cod Bay

Provincetown
Orleans

Cape Cod

Nantucket Island
Martha's
Vineyard   Nantucket

Catskill
Kingston
Poughkeepsie
Middletown

Paterson
Newark

Yonkers
NEW YORK
Elizabeth

NEW
YORK

Long Island

Hampton Bays
Montauk Point

Allentown

NEW
JERSEY

PENNSYLVANIA

State
College
Lebanon
Reading
Harrisburg
Hershey
Lancaster

Trenton

Neptune City
Lakewood

40°

Carlisle
Chambersburg
Gettysburg
Hanover

York

Levittown

Camden
Philadelphia

Wilmington
Vineland

Hammonton

Atlantic City
Ocean City

ntown
Cumberland
Keyser

Hagerstown
Martinsburg
Germantown

Aberdeen
Dundalk
Columbia

Baltimore

Dover
Milford

DELAWARE

Cape May Point
Cape Henlopen

Winchester
Bethesda
Front Royal
Alexandria   WASHINGTON D.C.
Columbia
Annapolis
Cambridge
Easton

MARYLAND

ATLANTIC

Harrisonburg
Staunton
Waynesboro
Charlottesville
Culpeper
Fredericksburg
Tappahannock

Salisbury   Ocean City
Pocomoke City

OCEAN

3

VIRGINIA

Mechanicsville
Richmond
Chester
Williamsburg
Hopewell
Petersburg
Hampton
Newport News
Portsmouth   Virginia Beach
Norfolk
Suffolk

Farmville
Charles
John H. Kerr
Reservoir
Halifax
Danville
Emporia

Onancock

Cape Charles
Cape Charles

Cape Henry

Chesapeake Bay

Chesapeake Bay

0   100   200 KM

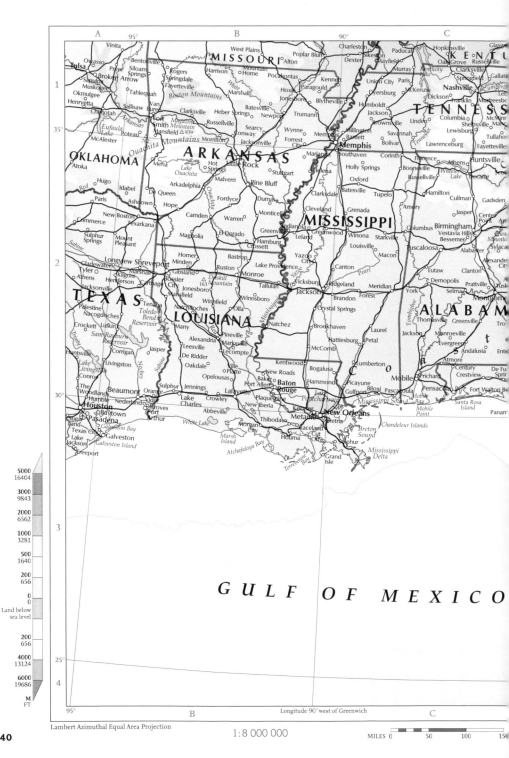

A
B
C

95°
90°

Vinita

West Plains
Poplar Bluff   Charleston   Paducah
Hopkinsville   Glasg

Owasso   Bentonville   **MISSOURI**   Alton   Sikeston   Mayfield   Oak Grove   Russellville
Tulsa   Prior   Siloam   Rogers   Harrison   Mountain   Pocahontas   Kennett   Union City   Paris   Murray   Kentucky   Clarksville   Springfield   Gallat
Sapulpa   Broken   Springs   Springdale   Home   Hoxie   Paragould   Dyersburg   McKenzie   Dickson   **Nashville**   Franklin   Le
Muskogee   Arrow   Fayetteville   White   Jonesboro   Blytheville   Humboldt   Columbia   McMinn
Okmulgee   Tahlequah   Marshall   Searcy   Trumann   Brownsville   Linden   Shelbyville   Lewisburg   **TENNESS**
Henryetta   Sallisaw   Van   Clarksville   Heber Springs   Batesville   Jackson   Memphis   Bartlett   Savannah   Lawrenceburg   Tullaho   Fayettevi
Checotah   Buren   Fort   Magazine   Russellville   Newport   Wynne   Millington   Bolivar   Athens   Huntsville
OKLAHOMA   Eufaula   Smith   Mountain   Conway   Jacksonville   Forrest   Helena   Corinth   Florence   Decatur
Lake   Poteau   Mansfield △ 839   Morrilton   City   Marianna   Southaven   Booneville   Russellville   Wheele   Lake
Atoka   Mena   **ARKANSAS**   Little Rock   Stuttgart   Holly Springs   Oxford   Hamilton   Cullman   Gadsden
Hugo   Idabel   Hot   Malvern   Helena   Batesville   Tupelo   Amory   Jasper   Cente
Springs   Arkadelphia   Pine Bluff   Clarksdale   Grenada   Birmingham   Point   Am
Paris   Ashdown   De Queen   Fordyce   Dumas   Cleveland   **MISSISSIPPI**   Columbus   Vestavia Hills   Sylaca
Commerce   New Boston   Hope   Camden   Warren   Monticello   Indianola   Greenwood   Winona   Starkville   Tuscaloosa   Bessemer   Alabaster   Alexande
Texarkana   El Dorado   Greenville   Leland   Louisville   Macon   York   Eutaw   Clanton
Sulphur   Magnolia   Hamburg   Yazoo   Canton   Demopolis   Prattville   Montg
Springs   Mount   Homer   Crossett   City   Meridian   Selma
Pleasant   Minden   Bastrop   Ruston   Lake Providence   Vicksburg   Ridgeland   Forest   Brandon   York   **ALABAM**
Gladewater   Longview   Shreveport   Monroe   Tallulah   Jackson   Thomasville   Greenville
Tyler   Marshall   Bossier   Driskill   Crystal Springs   Montgomery   a
Athens   Kilgore   Henderson   City   Mountain   Winnsboro   Brookhaven   Laurel   Jackson   Monroeville
TEXAS   Carthage   Jonesboro   △ 163   Olla   Petal   Evergreen   Enterp
Palestine   Tenaha   Mansfield   Winnfield   Natchez   McComb   Andalusia
Nacogdoches   Toledo   Natchitoches   Alexandria   Pineville   Kentwood   Atmore   Ente
Crockett   Bend   Many   Marksville   Bogalusa   Century   De Fu
Corrigan   Reservoir   Leesville   Lecompte   Picayune   Crestview   Spri
Jasper   De Ridder   Ville   Hammond   Lumberton   Mobile   Prichard   Pensacola   Fort Walton
Lufkin   Oakdale   Platte   New Roads   Port Allen   **Baton**   Biloxi   Pascagoula
Livingston   Opelousas   Rouge   Gulfpor   Mobile   Santa Rosa   Panam
Conroe   Sulphur   Jennings   Lafayette   Plaquemine   New Orleans   Bay   Island
The   Crowley   Baker   Hammond   Gretna   Point
Woodlands   Beaumont   Lake   New Iberia   Thibodaux   Metairie   **New Orleans**   Breton   Chandeleur Islands
Baytown   Humble   Orange   Charles   Abbeville   Morgan   Raceland   Sound
Houston   Nederland   Vidor   City   Cut   Port
Pasadena   Port   White Lake   Houma   Off   Sulphur   **Mississippi**
Texas City   Arthur   Marsh   Grand   **Delta**
Galveston   Island   Isle
Lake   Atchafalaya Bay   Terrebonne Bay
Jackson   Galveston Island
Freeport

35°

30°

25°

1

2

3

4

GULF OF MEXICO

95°
Longitude 90° west of Greenwich

1 : 8 000 000

MILES 0   50   100   150

5000
16404
3000
9843
2000
6562
1000
3281
500
1640
200
656
0
0
Land below
sea level
200
656
4000
13124
6000
19686
M
FT

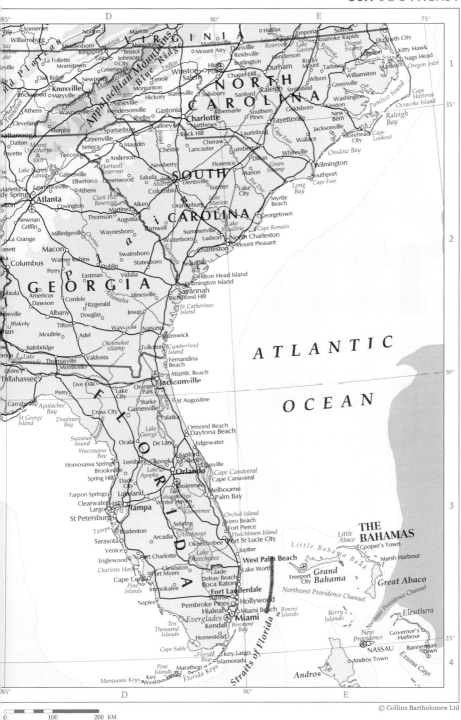

0   100   200 KM

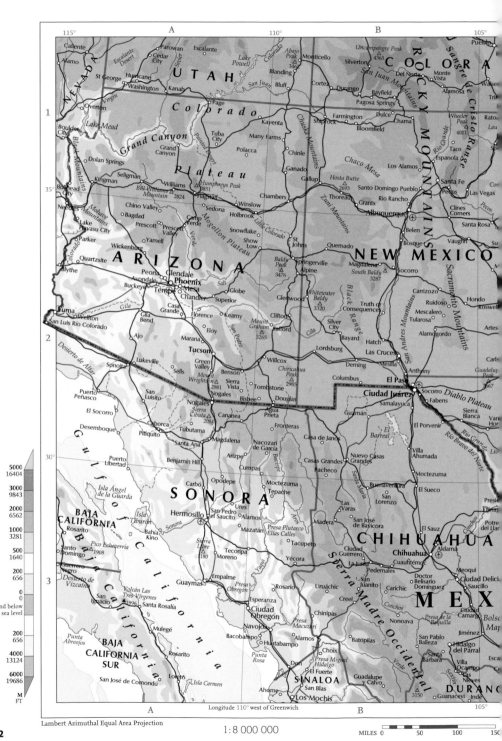

Lambert Azimuthal Equal Area Projection

Longitude 110° west of Greenwich

1:8 000 000

MILES 0   50   100   150

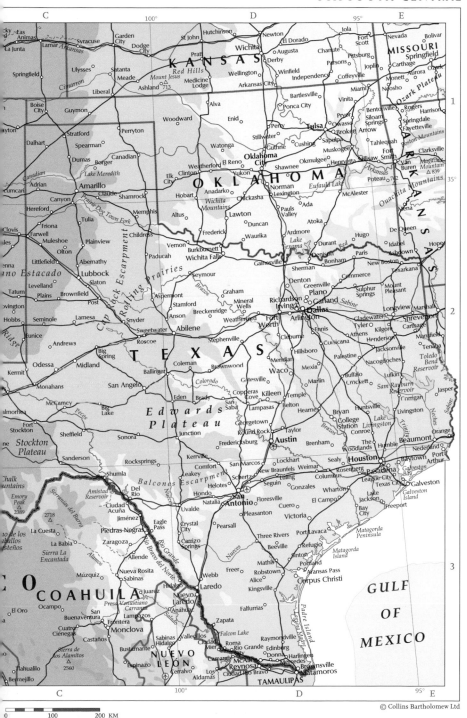

Lambert Azimuthal Equal Area Projection

1:12 000 000

MILES 0   100   200   300

Longitude 110° west of Greenwich

Elevation scale (left margin):

M	FT
5000	16404
3000	9843
2000	6562
1000	3281
500	1640
200	656
0	0
Land below sea level	
200	656
4000	13124
6000	19686
M	FT

Map labels:

A   110°   B

El Centro, Brawley, Yuma, Florence, Superior, Clifton, Truth or Consequences, Ruidoso, Roswell, Levelland, Brownfield

Tecate, Tijuana, Mexicali, Gila Bend, Casa Grande, Kearny, Lordsburg, Tularosa, Alamogordo, Lovington, Artesia, Hobbs, Seminole

Rosarito, Ensenada, Tucson, Green Valley, Willcox, Benson, Bisbee, Douglas, Deming, Las Cruces, Columbus, El Paso, Carlsbad, Eunice, Andrews, Midland

San Vicente, San Felipe, Nogales, Sierra Vista, Agua Prieta, Ciudad Juárez, Fabens, Van Horn, Fort Stockton, Big

ARIZONA   NEW MEXICO   UNITE

Hermosillo, Guaymas, Ciudad Obregón, Los Mochis, Culiacán, La Paz, Mazatlán, Durango, Chihuahua, Ciudad Delicias, Torreón, Gómez Palacio, Matamoros

Tropic of Cancer

20°

30°

Islas Revillagigedo (Mexico)

Isla Clarión, Isla Socorro, Isla San Benedicto

PACIFIC

OCEAN

Guadalajara, Tepic, Colima, Manzanillo, Lázaro Cárdenas, MEXICO

Gulf of California

Baja California

Sierra Madre

2

3

GULF

OF

MEXICO

Tropic of Cancer

Bahía
de Campeche

Gulf of
Tehuantepec

© Collins Bartholomew Ltd

0   200   400 KM

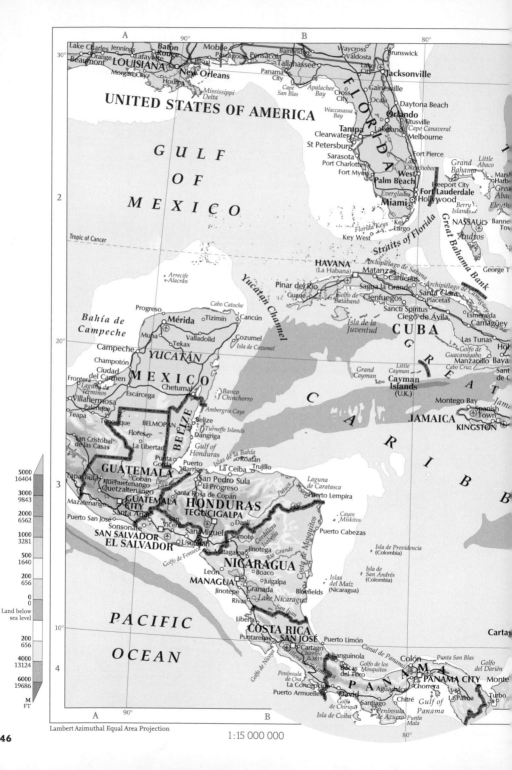

Lambert Azimuthal Equal Area Projection

1:15 000 000

70°   D   60°   E

30°

A T L A N T I C

O C E A N

Tropic of Cancer

2

B A H A M A S

land

San Salvador

Rum Cay

ng Island

Crooked Island

Mayaguana

Acklins Island

Great Inagua

Caicos Islands

Turks and Caicos Islands (U.K.)

GRAND TURK (Cockburn Town)

Turks Islands

W E S T   I N D I E S

20°

acoa

antanamo

Port-de-Paix

H i s p a n i o l a

L E E W A R D   I S L A N D S

Puerto Plata

Cap-Haïtien

Santiago

Gonaïves Hinche

St-Marc

Pico Duarte 3175

DOMINICAN REPUBLIC

Puerto Rico (U.S.A.)

SAN JUAN

Virgin Is (U.K.)

Anguilla (U.K.)

THE VALLEY

Jérémie

HAÏTI

San Juan

Barahona

La Romana

SANTO DOMINGO

Mayagüez

Ponce

St Croix

Virgin Is (U.S.A.)

St-Maarten (Neth.)

St-Martin (Fr.)

St-Barthélemy (Fr.)

ANTIGUA AND BARBUDA

ST JOHN'S

oteaux

Les Cayes

Jacmel

PORT-AU-PRINCE

Isla Beata

Cabo Beata

BASSETERRE

ST KITTS AND NEVIS

BRADES

Plymouth (abandoned)

Montserrat (U.K.)

Guadeloupe (Fr.)

BASSE-TERRE

Marie-Galante

Antigua

Windward

A N T I L L E S

DOMINICA

ROSEAU

Martinique (Fr.)

FORT-DE-FRANCE

St Lucia Channel

3

C A R I B B E A N   S E A

L e s s e r   A n t i l l e s

ST LUCIA

CASTRIES

St Vincent Passage

BARBADOS

BRIDGETOWN

ST VINCENT AND THE GRENADINES

KINGSTOWN

ST GEORGE'S

GRENADA

W I N D W A R D   I S L A N D S

nnel

Aruba (Neth.)

ORANJESTAD

Netherlands Antilles

Curaçao

Bonaire

Isla Blanquilla

Scarborough

Tobago

Punta Gallinas

Península de la Guajira

WILLEMSTAD

Santa Marta

Ríohacha

Maicao

Punto Fijo

Golfo de Venezuela

Coro

Península de Paraguaná

Islas Los Roques

Isla La Tortuga

La Asunción

Isla de Margarita

PORT OF SPAIN

TRINIDAD AND TOBAGO

Barranquilla

Ciénaga

Campo Mara

Maracaibo

Cabimas

San Felipe

Maiquetía

Valencia

CARACAS

Cumaná

Carúpano

Güiria

Gulf of Paria

San Fernando

Trinidad

10°

Sabanalarga

Valledupar

Rosario

Maracay

Los Teques

Petare

Barcelona

Maturín

Plato

San Carlos

Machiques

Lake Maracaibo

Lagunillas

El Tocuyo

San Carlos

Barquisimeto

Acarigua

San Juan de los Morros

Valle de la Pascua

Zaraza

Anaco

Guanipa

Orinoco Delta

Tucupita

Mabaruma

elejo

El Banco

San Carlos del Zulia

Trujillo

Mérida

Guanare

Barinas

Calabozo

El Tigre

angue

El Banco

Valera

Libertad

Baúl

Orinoco

Ciudad Guayana

GUYANA

COLOMBIA

Cúcuta

Pico Bolívar 5007

San Cristóbal

Cordillera de Mérida

Ciudad Bolivia

Apure

Puerto Miranda

San Fernando de Apure

Ciudad Bolívar

Embalse de Guri

VENEZUELA

4

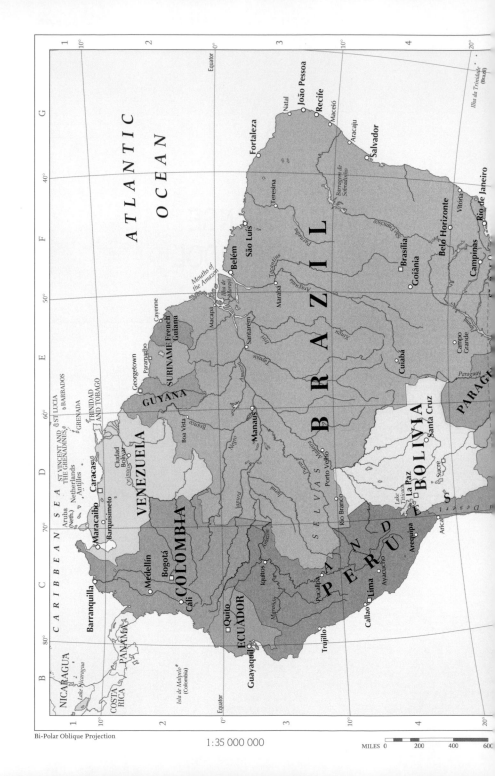

Bi-Polar Oblique Projection

1:35 000 000

MILES 0   200   400   600

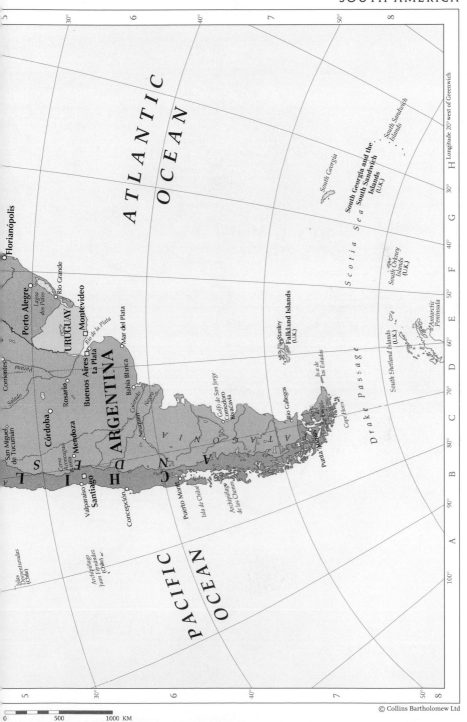

ATLANTIC OCEAN

PACIFIC OCEAN

Florianópolis

Porto Alegre

Lagoa dos Patos

Rio Grande

URUGUAY

Montevideo

Río de la Plata

La Plata

Buenos Aires

Mar del Plata

Corrientes

Paraná

Salado

Rosario

Córdoba

Bahía Blanca

ARGENTINA

Colorado

Neuquén

Negro

San Miguel de Tucumán

Cerro Aconcagua

Mendoza

Valparaíso

Santiago

CHILE

ANDES

PATAGONIA

Golfo de San Jorge

Comodoro Rivadavia

Concepción

Puerto Montt

Isla de Chiloé

Archipiélago de los Chonos

Río Gallegos

Punta Arenas

Cape Horn

Isla de los Estados

Drake Passage

Scotia Sea

South Georgia

South Georgia and the South Sandwich Islands (U.K.)

South Sandwich Islands

Falkland Islands (U.K.)

Stanley

South Orkney Islands (U.K.)

South Shetland Islands (U.K.)

Antarctic Peninsula

Islas Desventuradas (Chile)

Archipiélago Juan Fernández (Chile)

H Longitude 20° west of Greenwich

0   500   1000 KM

© Collins Bartholomew Ltd

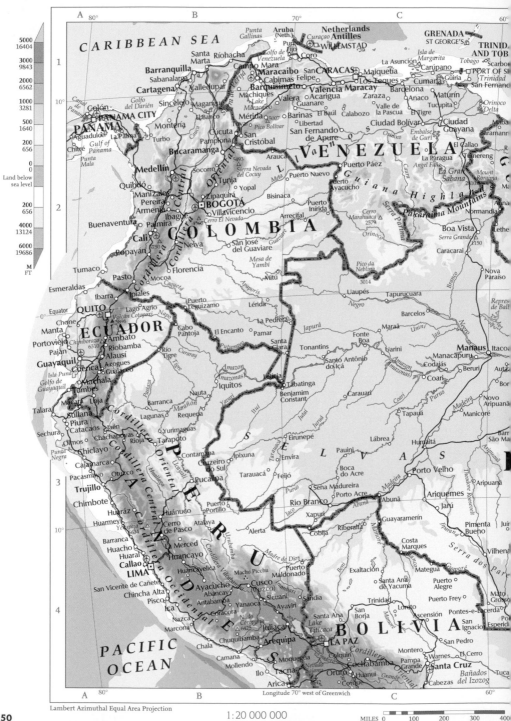

Lambert Azimuthal Equal Area Projection

1 : 20 000 000

MILES 0   100   200   300   400

D   50°   E   40°   F

1

10°

ATLANTIC

OCEAN

RGETOWN
adise
New Amsterdam
Totness   PARAMARIBO
Albina   St-Laurent-du-Maroni
le Brokopondo   Sinnamary
Professor van   Kourou   CAYENNE
Blommestein Mar   Guisanbourg
URINAME   **French**   Cayenne
Juliana Top   **Guiana**   Oiapoque
1230   Inini
Pontoetoe

2

Lourenço   Calçoene
Ilha de   Amapá   *Maracá*
Serra Tumucumaque

Mouths of the
Amazon

Macapá
Porto Santana   *Ilha*
Arere   *Paru*   Mazagão   Chaves   *Caviana*
Trombetas   *Cabo*
Serra   Almeirim   *Ilha de*   Salinópolis
Parauaquara   *Marajó*   Bragança
riximiná,   Óbidos   Breves   **Belém**   Viseu
Juruti   Monte   Portel   Capité   Gurupu
ra   Parintins   Alegre   Cametá   Acará   Castanhal
untuba   Santarém   Pinheiro
Ilha de São Marcos
Altamira   Tucuruí   Vizeu   **São Luís**
Itaituba   *Represa de*   Santa   Itapicuru   Parnaíba   Camocim
Itaituba   *Tucuruí*   Luzia   Mirim   Luzilândia   Caucaia   **Fortaleza**
areacanga   Jacunda   Pedreiras   Codó   Tianguá   Sobral   Cascavel
Araras   *Serra dos Carajás*   Pres. Dutra   Caxias   Piripiri   Campo Maior   Cininde   Aracati
Marabá   Imperatriz   Timon   **Teresina**   Crateús   Quixadá   Macau
Manuelzinho   São   Grajaú   Barra do   Buriti Bravo   Taua   Boa   Mossoró
Félix   Tocantinópolis   Corda   Palmirais   Viagem   **Natal**
Porto Franco   Araguaína   Floriano   Iguatu   Icó   Piranhas
Carolina   Balsas   Jerumenha   Mamanguape
RAZIL   Conceição   Uruçuí   Oeiras   Picos   Crato   **João**
do Araguaia   Canto do Buriti   Juazeiro   Campina   **Pessoa**
Santa Maria   Paulistana   do Norte   Grande   Olinda
Serra   das Barreiras   Pedro   Caracol   São Raimundo   Salgueiro   Jaboatão dos   **Recife**
do Cachimbo   Afonso   Nonato   Floresta   Guararapes
Peixoto de   Palmas   Gilbués   Nova   Petrolina   Caruaru   Cabo de Santo
Azevedo   Remanso   São   Garanhuns   Agostinho
Porto Nacional   Corrente   *Barragem de*   Juazeiro   Paulo   Rio Largo
dos   Dianópolis   *Sobradinho*   Senhor do Bonfim   Afonso   **Maceió**
nos   Natividade   Monte Santo   Arapiraca
Porto Artur   Gurupi   Barreiras   Irecê   Lagarto
Ilha do   Ibotirama   Jacobina   Aracaju
Diamantino   *Bananal*   Xique   Estância
Rosário Oeste   São   Santana   Bom Jesus   Itaberaba   Serrinha
arta do Bugres   Félix   Cavalcante   da Lapa   Feira   Camaçari
Cuiabá   Porangatu   Correntina   de Santana   **Salvador**
Rondonópolis   Uruaçu   Posse   Brumado   Jequié   Santo Antônio de Jesus
eres   Barra do   Niquelândia   Januária   Guanambi   Itabuna   Ipiaú
Garças   **BRASÍLIA**   Formosa   Vitória da   Ubaitaba   Ilhéus
Itiquira   Goiás   Anápolis   Arinos   Espinosa   Conquista   Itapetinga   Una
Iporá   Trindade   Luziânia   Janaúba   Salinas
el   Coxim   Alto   Vianópolis   Montes   Almenara   Porto Seguro
Garças   **Goiânia**   Unaí   Claros
rumbá   Itumbiara   Paraúna   Paracatu   Jequitaí   Teófilo   Alcobaça
Rio Verde de Mato Grosso   Itumbiara   Araguari   Patos   Otoni
Jataí   Rio Verde   Uberlândia   de Minas

Equator 0°

3

10°

4

D   50°   E   40°   F

© Collins Bartholomew Ltd

0   200   400   600 KM

151

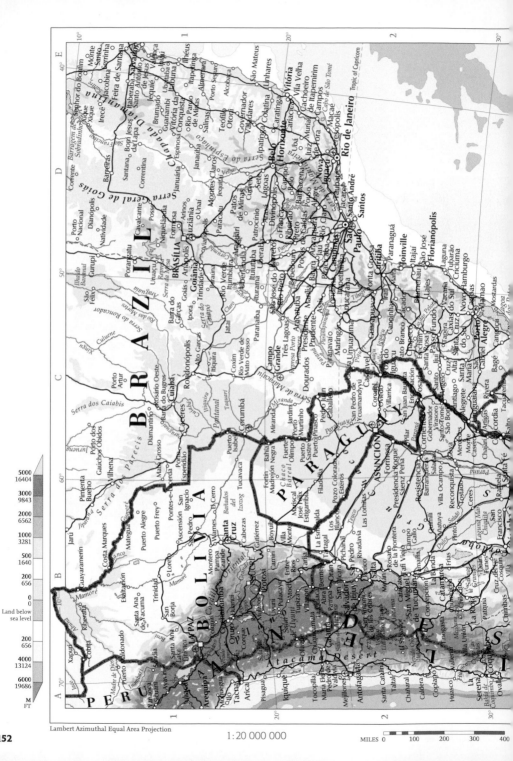

Lambert Azimuthal Equal Area Projection

1 : 20 000 000

MILES 0   100   200   300   400

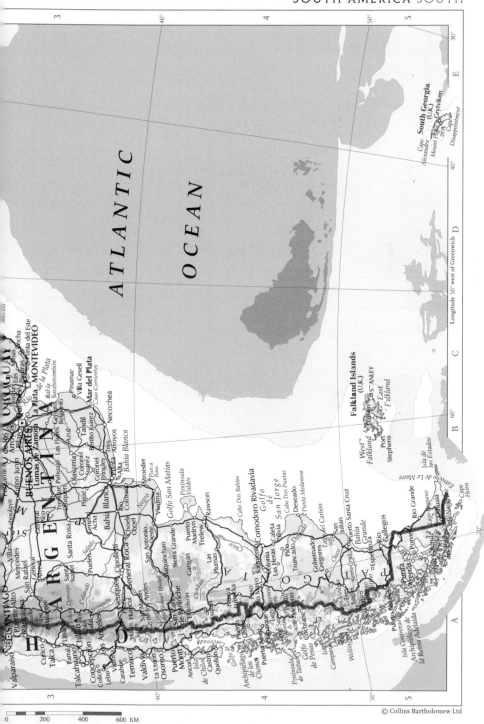

ATLANTIC

OCEAN

South Georgia
(U.K.)
Grytviken
Cape
Alexandra
Mount Paget
2934
Cape
Disappointment

Longitude 50° west of Greenwich

URUGUAY
MONTEVIDEO
Mar del Plata
Necochea

Falkland Islands
(U.K.)
STANLEY
East
Falkland
West
Falkland
Port
Stephens

BUENOS AIRES
Lomas de Zamora
La Plata

Bahía Blanca

ARGENTINA

Comodoro Rivadavia
Golfo
de
San Jorge

Puerto Santa Cruz

Río Gallegos

Punta
Arenas

Cape
Horn

SANTIAGO

Valparaíso

Talca

Concepción
Los Ángeles

Temuco
Valdivia
Osorno
Puerto
Montt

© Collins Bartholomew Ltd

0   200   400   600 KM

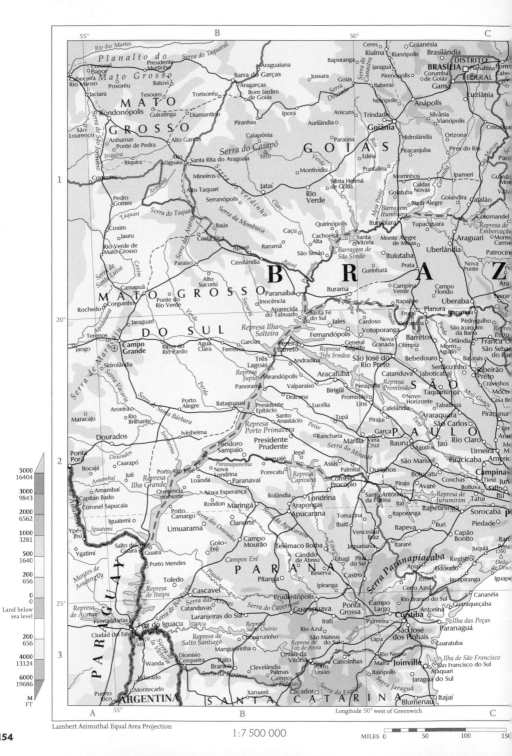

Lambert Azimuthal Equal Area Projection

1:7 500 000

MILES 0   50   100   150

154

Map labels

A   B   C

55°   50°

Rio das Mortes

Planalto do
Mato Grosso

Serra do Taquaral

Ceres   Goianésia
Rialma   Rianópolis   Brasilândia

Coronel
Murtinho   President

Jaraguá   da
Camastra   Corumbá
de Goiás

BRASÍLIA   Planaltina
DISTRITO
FEDERAL

Cabeceira   Coronel
Rio Manso   Ponce   Poxoréu   Batovi

Barra do Garças   Araguaiana

Itapuranga

Pirenópolis   Gama
Luziânia

Jaciara   President

Rondonópolis

Tesouro   Torixoréu

Aragarças
Bom Jardim
de Goiás

Jussara   Goiás

Itaberaí

Neropólis

MATO   Guiratinga   Diamantino
GROSSO

Piranhas   Iporá

Anicuns

Trindade   **GOIÂNIA**

Anápolis

Silvânia   Vianópolis
Cristalina

São
Lourenço   Anhumas
Ponte de Pedra

Caiapônia   Paraúna

Aurilândia

Hidrolândia   Orizona

Itaguru   Itiquira

Alto Garças   Serra do Caiapó
1010

Edéia   Piracanjuba   Pires do Rio

Guarda
Mor
Vaza

Pedro
Gomes   Corrente

Alto   Santa Rita do Araguaia
Araguaia

Mineiros   Jataí

Rio
Verde   Montividiu

Santa Helena
de Goiás

Pontalina   Morrinhos   Caldas
Novas   Goiandira   Catalão

Coromandel
Monte
Carme
Patrocínio

Coxim   Jauru

Baús   Serra da Mombuca   Caçu

Quirinópolis   Cachoeira
Alta

Santa
Vitória   Monte Alegre
de Minas   Tupaciguara   Araguari

Represa de
Emborcação

Rio Verde de
Mato Grosso   Costa Rica   Aporé   Itarumã

São Simão

Barragem de
São Simão   Ituiutaba   Prata

Uberlândia

Nova
Ponte

Paraíso   Cassilândia

Itaguru   Gurinhatã

Campina
Verde   Campo
Florido

Uberaba

MATO GROSSO   Alto
Sucuriú   Paranaíba
Inocência   Iturama

Itapajipe   Planura

Ará

Rochedo   Corguinho   Ponte do
Rio Verde   Aparecida
do Tabuado   Santa Fé
do Sul   Jales   Cardoso   Votuporanga

Frutal   Pedregulho
Colômbia   Barretos   São Joaquim
da Barra   Franca   São Sebastião

Campo
Grande   Jaraguari   **DO SUL**   Represa Ilha
Solteira   Fernandópolis   Nova
Granada   General
Salgado   Olímpia   Orlândia   São
Sebastião

Jango   Terenos   Ribas do
Rio Pardo   Água
Clara   Garcias   Pereira
Barreto   Andradina   Morro
Agudo   Batatais

Ribeirão
Preto   Moço

Sidrolândia   Três
Lagoas   Mirandópolis   Represa
Três Irmãos   São José do
Rio Preto   Catanduva   Jaboticabal   Sertãozinho

Cravinhos
Casa Br

Maracaju   Aroeira   Rio
Brilhante   Porto
Alegre   Panorama   Valparaíso   Birigüi   Penápolis   Represa
Promissão   Novo
Horizonte   **SÃO**   Taquaritinga   Tabatinga   Moji-Guaçu

Dracena   Lucélia   Promissão   Cafelândia   Araraquara   Piração

Dourados   Ivinheima   Bataguassu   President
Epitácio   Santo
Anastácio   Tupã   Pirajuí   Lins   São Carlos   Iler

Ponta
Porã   Caarapó   Represa
Porto Primavera   Teodoro
Sampaio   Rancharia   Marília   Vera   Garça   **PAULO**   Rio Claro   Limeira   M

Bocajá   Juti   Porto São José   Presidente
Prudente   Iepê   Assis   Palmital   Ourinhos   Bauru   São Manuel   Agudos   Piracicaba   Americ

Amambaí   Nova   Loanda   Paranavaí   Porecatu   Represa
Capivara   Cornélio
Procópio   Santo Antônio
da Platina   Piraju   Avaré   Botucatu   Conchas   **Campinas**
Tietê   Jun
Boituva   Salto   Itu

Capitán Bado   Iguatemi   Represa
Ilha Grande   Londrina   Rolândia   Itaí   Itaporanga   Itapetininga   Tatuí   Sorocaba

Coronel Sapucaia   Queréncia
do Norte   Nova Esperança   Apucarana   Tomazina   Ibaiti   Vençeslau
Bráz   Itapeva   Buri   Capão
Bonito   Juquiá   Piedade

Ypê   Porto
Camargo   Rondon   Maringá   Cianorte   Campo
Mourão   Jaguariaíva   Itararé   Itan
1350

Jhú   Salto del   Guaíra   Goio-   Telêmaco Borba   Piraí
do Sul   **PARANÁ**   Apiaí   Jacupiranga   Iguape

Ygatimí   Umuarama   Erê   Cândido
de Abreu   Reserva   Tibagi   Serra Paranapiacaba   Eldorado   Dedo
de Deu

Porto Mendes   Pitanga   Castro   Cerro Azul   Registro

Toledo   Campos Eré   Prudentópolis   Ipiranga   Campo
Largo   Antonina   Cananéia

Cascavel   Laranjeiras do Sul   Guarapuava   Ponta
Grossa   Rio Branco do Sul   Guaraqueçaba

Represa
de Itaipu   Catanduvas   Guarapuava   Palmeira   Curitiba   Ilha das Peças

Hernandarias   Irati   São José
dos Pinhais   Paranaguá

Foz do Iguaçu   Iguaçu   Rio Azul   Lapa   Guaratuba

Ciudad del Este   Iguaçu
Falls   Represa
Salto Osório   Chopinzinho   São Mateus
do Sul   Rio Negro   Ilha de São Francisco

Dionísio   Represa de
Foz de Areia   União da
Vitória   Mafra   Joinville   São Francisco do Sul

Wanda   Cerqueira   Clevelândia   Canoinhas   Araquari

Eldorado   Rato
Branco   Palmas   Porto
União   Itaiópolis   Jaraguá do Sul

Puerto   Xanxerê   Caçador   Serra do Espigão   Serra Jaraguá

ARGENTINA   **SANTA CATARINA**   Blumenau   Itajaí

55°   Longitude 50° west of Greenwich

PARAGUAY

B R A Z

G O I A S

Elevation scale

5000
16404

3000
9843

2000
6562

1000
3281

500
1640

200
656

0
0

Land below
sea level

200
656

4000
13124

6000
19686

M
FT

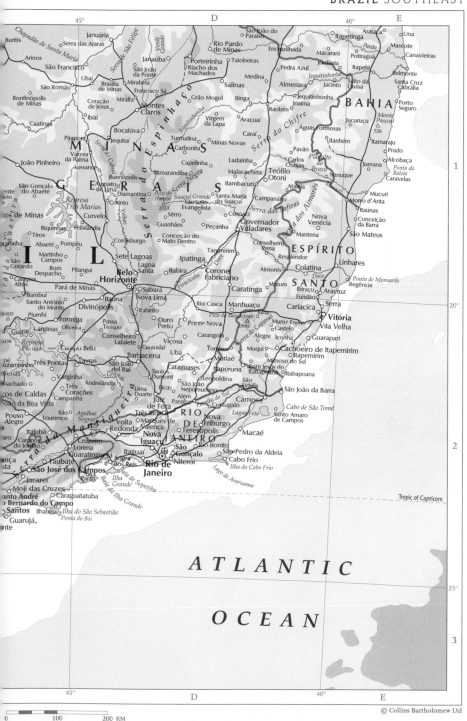

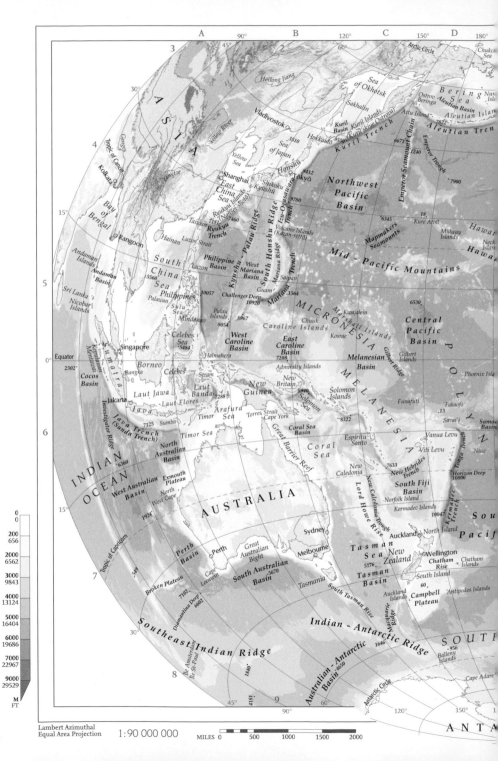

Lambert Azimuthal
Equal Area Projection
1 : 90 000 000

MILES 0   500   1000   1500   2000

156

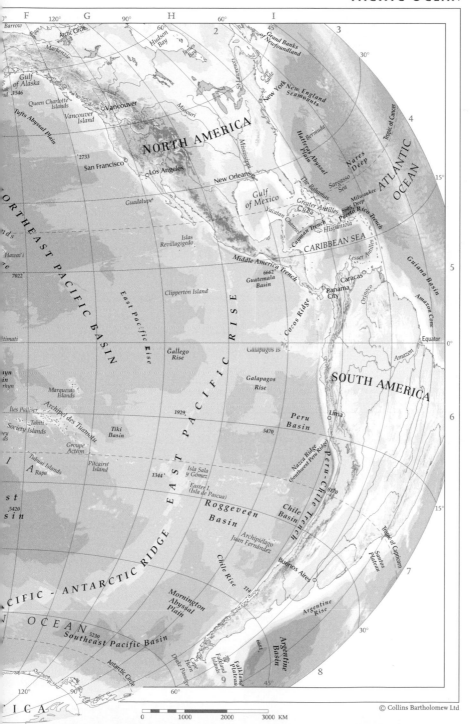

F 120° G 90° H 60° I

Barrow

Arctic Circle

Mackenzie

Hudson Bay

James Bay

Gulf of Alaska

Queen Charlotte Islands

Vancouver Island

Vancouver

Tufts Abyssal Plain

2733

San Francisco

Los Angeles

Guadalupe

NORTH AMERICA

Missouri

Mississippi

New Orleans

Gulf of Mexico

Yucatan Channel

Islas Revillagigedo

Middle America Trench

Guatemala Basin 6662

Clipperton Island

East Pacific Basin

NORTHEAST PACIFIC BASIN

Hawai'i

7022

Gallego Rise

Galapagos is

timati

yn in rhyn

Marquesas Islands

Iles Palliser

Society Islands

Tahiti

Groupe Actéon

Tiki Basin

Tubuai Islands

Rapa

Pitcairn Island

1929

1344

Isla Sala y Gómez

5420 sin

A

st

I

Easter I. (Isla de Pascua)

Roggeveen Basin

2 H 60° I 45°

Grand Banks of Newfoundland

Cape Sable

New York New England Seamounts

Bermuda

Hatteras Abyssal Plain

Nares Deep

Sargasso Sea

The Bahamas

Greater Antilles

Cuba

Cayman Trench

Hispaniola

Milwaukee 8605 Deep

Puerto Rico Trench

Lesser Antilles

CARIBBEAN SEA

Caracas

Panama City

Orinoco

Galapagos Rise

Peru Basin

5470

Nazca Ridge (Southwest Peru Ridge)

Lima

SOUTH AMERICA

Amazon

Equator

Peru - Chile Trench

8170

Chile Basin

Archipiélago Juan Fernández

Chile Rise 118

ATLANTIC OCEAN

Guiana Basin

Amazon Cone

3

Tropic of Cancer

4

15°

5

0°

6

PACIFIC - ANTARCTIC RIDGE

EAST PACIFIC RISE

Cocos Ridge

Middle America Trench

Cape Horn

Drake Passage

Mornington Abyssal Plain

Buenos Aires

Argentine Rise

Falkland Islands

Falkland Plateau

Santos Plateau

Argentine Basin

Tropic of Capricorn

15°

7

30°

OCEAN 5230

Southeast Pacific Basin

Antarctic Circle

120° 90°

ICA

60°

9 8 45°

0 1000 2000 3000 KM

ATLANTIC OCEAN

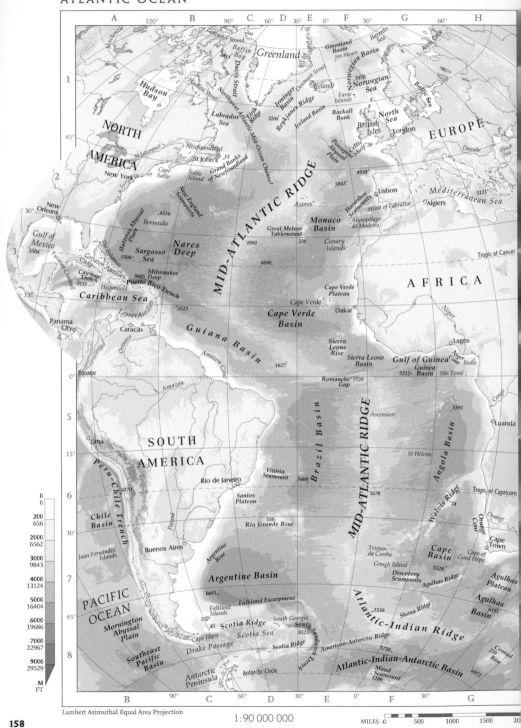

Lambert Azimuthal Equal Area Projection

1:90 000 000

MILES 0   500   1000   1500   20

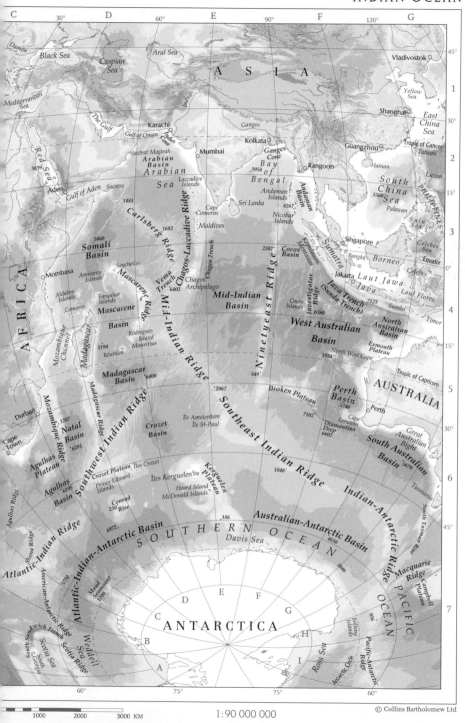

1 : 90 000 000

© Collins Bartholomew Ltd

1000   2000   3000 KM

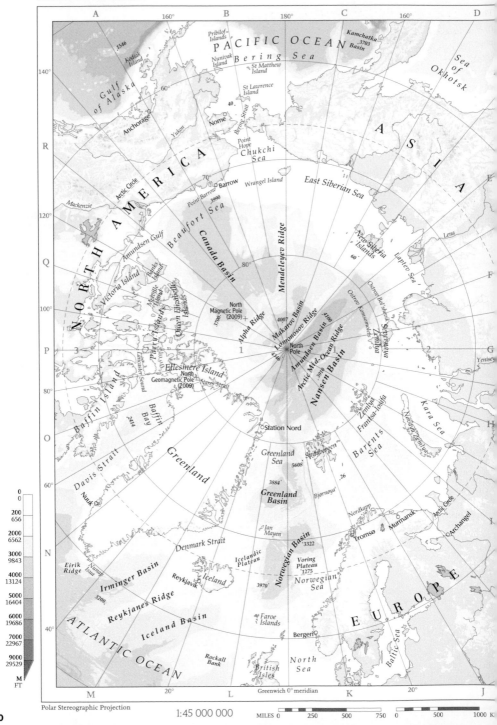

A 160° B 180° C 160° D

PACIFIC OCEAN

Pribilof Islands
Kamchatka Basin 3703
Nunivak Island
Kodiak Island 1546
Bering Sea
St Matthew Island
Sea of Okhotsk
Gulf of Alaska
St Lawrence Island
Anchorage
Nome
40
Bering Strait
ASIA

Yukon
Point Hope
Chukchi Sea

NORTH AMERICA
Arctic Circle
70°
Point Barrow Barrow
Wrangel Island
East Siberian Sea

Mackenzie
120°
3990
Beaufort Sea
Mendeleyev Ridge
New Siberia Islands
Lena
Laptev Sea

Amundsen Gulf
Banks Island
Canada Basin
60
Ostrov Bol'shevik
Ostrov Komsomolets
Severnaya Zemlya

Victoria Island
80°
Makarov Basin
Lomonosov Ridge
4007
Kara Sea

North Magnetic Pole (2009)
3700
Alpha Ridge
1
North Pole
Amundsen Basin
Novaya Zemlya

Queen Elizabeth Islands
Arctic Mid-Ocean Ridge
Nansen Basin
Franz Josef Land
Zemlya

Prince of Wales Island
Ellesmere Island
North Geomagnetic Pole (2009)
Nares Strait
Station Nord
Spitsbergen
Barents Sea

Lancaster Sound
Baffin Bay
2414
Greenland Sea
5608

Baffin Island
Davis Strait
Greenland
3884
Greenland Basin
Bjørnøya
.26
Nordkapp
Murmansk
Arctic Circle
Archangel

Jan Mayen
Nuuk
Denmark Strait
Icelandic Plateau
Norwegian Basin
3322
Voring Plateau 1275
Tromsø

Eirik Ridge
Nanup Sida
Irminger Basin
Iceland
Reykjavik
3970
Norwegian Sea
EUROPE

3208
Reykjanes Ridge
Faroe Islands
Bergen
Baltic Sea

ATLANTIC OCEAN
Iceland Basin
Rockall Bank
British Isles
North Sea
Greenwich 0° meridian

Scale bar (left side):
0 / 0
200 / 656
2000 / 6562
3000 / 9843
4000 / 13124
5000 / 16404
6000 / 19686
7000 / 22967
9000 / 29529
M / FT

Polar Stereographic Projection

1:45 000 000

MILES 0   250   500   750

0   500   1000 K

INTRODUCTION TO THE INDEX

The index includes all names shown on the maps in the Atlas of the World. Names are referenced by page number and by a grid reference. The grid reference correlates to the alphanumeric values which appear within each map frame. Each entry also includes the country or geographical area in which the feature is located. Entries relating to names appearing on insets are indicated by a small box symbol: □, followed by a grid reference if the inset has its own alphanumeric values.

Name forms are as they appear on the maps, with additional alternative names or name forms included as cross-references which refer the user to the entry for the map form of the name. Names beginning with Mc or Mac are alphabetized exactly as they appear. The terms Saint, Sainte, Sankt, etc, are abbreviated to St, Ste, St, etc, but alphabetized as if in the full form.

Names of physical features beginning with generic geographical terms are permuted – the descriptive term is placed after the main part of the name. For example, Lake Superior is indexed as Superior, Lake; Mount Everest as Everest, Mount. This policy is applied to all languages.

Entries, other than those for towns and cities, include a descriptor indicating the type of geographical feature. Descriptors are not included where the type of feature is implicit in the name itself.

Administrative divisions are included to differentiate entries of the same name and feature type within the one country. In such cases, duplicate names are alphabetized in order of administrative division. Additional qualifiers are also included for names within selected geographical areas.

INDEX ABBREVIATIONS

admin. div.	administrative division	g.	gulf	Port.	Portugal
Afgh.	Afghanistan	Ger.	Germany	prov.	province
Alg.	Algeria	Guat.	Guatemala	pt	point
Arg.	Argentina	hd	headland	r.	river
Austr.	Australia	Hond.	Honduras	r. mouth	river mouth
aut. comm.	autonomous community	i.	island	reg.	region
aut. reg.	autonomous region	imp. l.	impermanent lake	resr	reservoir
aut. rep.	autonomous republic	Indon.	Indonesia	rf	reef
Azer.	Azerbaijan	is.	islands	Rus. Fed.	Russian Federation
b.	bay	isth.	isthmus	S.	South
B.I.O.T.	British Indian Ocean Territory	Kazakh.	Kazakhstan	salt l.	salt lake
		Kyrg.	Kyrgyzstan	sea chan.	sea channel
Bangl.	Bangladesh	l.	lake	special admin. reg.	special administrative region
Bol.	Bolivia	lag.	lagoon		
Bos.-Herz.	Bosnia Herzegovina	Lith.	Lithuania	str.	strait
Bulg.	Bulgaria	Lux.	Luxembourg	Switz.	Switzerland
c.	cape	Madag.	Madagascar	Tajik.	Tajikistan
Can.	Canada	Maur.	Mauritania	Tanz.	Tanzania
C.A.R.	Central African Republic	Mex.	Mexico	terr.	territory
Col.	Colombia	Moz.	Mozambique	Thai.	Thailand
Czech Rep.	Czech Republic	mt.	mountain	Trin. and Tob.	Trinidad and Tobago
Dem. Rep.	Democratic	mts	mountains	Turkm.	Turkmenistan
Congo	Republic of the Congo	mun.	municipality	U.A.E.	United Arab Emirates
depr.	depression	N.	North	U.K.	United Kingdom
des.	desert	Neth.	Netherlands	Ukr.	Ukraine
Dom. Rep.	Dominican Republic	Neth. Antilles	Netherlands Antilles	union terr.	union territory
Equat.	Equatorial Guinea	Nic.	Nicaragua	Uru.	Uruguay
Guinea		N.Z.	New Zealand	U.S.A.	United States of America
esc.	escarpment	Pak.	Pakistan	Uzbek.	Uzbekistan
est.	estuary	Para.	Paraguay	val.	valley
Eth.	Ethiopia	pen.	peninsula	Venez.	Venezuela
Fin.	Finland	Phil.	Philippines	vol.	volcano
for.	forest	plat.	plateau	vol. crater	volcanic crater
Fr. Guiana	French Guiana	P.N.G.	Papua New Guinea		
Fr. Polynesia	French Polynesia	Pol.	Poland		

1

128 B2 100 Mile House Can.

A

93 E4 Aabenraa Denmark
100 C2 Aachen Ger.
93 E4 Aalborg Denmark
102 C2 Aalen Ger.
100 B2 Aalst Belgium
93 I3 Äänekoski Fin.
105 D2 Aarau Switz.
100 B2 Aarschot Belgium
70 A2 Aba China
119 D2 Aba Dem. Rep. Congo
115 C4 Aba Nigeria
81 C2 Ābādān Iran
81 D2 Ābādān Iran
81 D3 Ābādeh Ţashk Iran
114 B1 Abadla Alg.
155 C1 Abaeté Brazil
   Abagnar Qi China see Xilinhot
135 E3 Abajo Peak U.S.A.
115 C4 Abakaliki Nigeria
83 H3 Abakan Rus. Fed.
150 B4 Abancay Peru
81 D2 Abarkūh Iran
66 D2 Abashiri Japan
66 D2 Abashiri-wan b. Japan
59 D3 Abau P.N.G.
   Abaya, Lake l. Eth. see Lake Abaya
   Ābay Wenz r. Eth. see Blue Nile
83 H3 Abaza Rus. Fed.
108 A2 Abbasanta Italy
104 C1 Abbeville France
141 C2 Abbeville AL U.S.A.
140 B3 Abbeville LA U.S.A.
97 B2 Abbeyfeale Ireland
55 R2 Abbot Ice Shelf Antarctica
74 B1 Abbottabad Pak.
115 E3 Abéché Chad
114 B4 Abengourou Côte d'Ivoire
114 C4 Abeokuta Nigeria
99 A3 Aberaeron U.K.
96 C2 Aberchirder U.K.
   Abercorn Zambia see Mbala
99 B4 Aberdare U.K.
99 A3 Aberdaron U.K.
53 D2 Aberdeen Austr.
122 B3 Aberdeen S. Africa
96 C2 Aberdeen U.K.
139 D3 Aberdeen MD U.S.A.
137 D1 Aberdeen SD U.S.A.
134 B1 Aberdeen WA U.S.A.
129 E1 Aberdeen Lake Can.
96 C2 Aberfeldy U.K.
96 B2 Aberfoyle U.K.
99 B4 Abergavenny U.K.
   Abergwaun U.K. see Fishguard
   Aberhonddu U.K. see Brecon
143 C2 Abernathy U.S.A.
134 B2 Abert, Lake U.S.A.
   Abertawe U.K. see Swansea
   Aberteifi U.K. see Cardigan
99 B4 Abertillery U.K.
99 A3 Aberystwyth U.K.
86 F2 Abez' Rus. Fed.
78 B3 Abhā Saudi Arabia
81 C2 Abhar Iran
   Abiad, Bahr el r. Sudan/Uganda see White Nile
114 B4 Abidjan Côte d'Ivoire
137 D3 Abilene KS U.S.A.
143 D2 Abilene TX U.S.A.
99 C4 Abingdon U.K.
138 C3 Abingdon U.S.A.
91 D3 Abinsk Rus. Fed.
130 B3 Abitibi, Lake Can.
   Åbo Fin. see Turku
74 B1 Abohar India
114 B4 Aboisso Côte d'Ivoire
114 C4 Abomey Benin
60 A1 Abongabong, Gunung mt. Indon.
118 B2 Abong Mbang Cameroon
64 A3 Aborlan Phil.
115 D3 Abou Déia Chad
106 B2 Abrantes Port.
152 B3 Abra Pampa Arg.

142 A3 Abreojos, Punta pt Mex.
116 B2 'Abri Sudan
136 A2 Absaroka Range mts U.S.A.
81 C1 Abşeron Yarımadası pen. Azer.
78 B3 Abū 'Arīsh Saudi Arabia
116 A2 Abū Ballāş h. Egypt
79 C2 Abu Dhabi U.A.E.
116 B3 Abu Hamed Sudan
116 B3 Abu Haraz Sudan
115 C4 Abuja Nigeria
81 C2 Abū Kamāl Syria
118 C2 Abumombazi Dem. Rep. Congo
152 B1 Abunã r. Bol./Brazil
150 C3 Abunã Brazil
74 B2 Abu Road India
78 B2 Abū Şādi, Jabal h. Saudi Arabia
116 B2 Abū Sunbul Egypt
116 A3 Abu Zabad Sudan
   Abū Zabī U.A.E. see Abu Dhabi
117 A4 Abyei Sudan
145 B2 Acambaro Mex.
120 B2 Acampamento de Caça do Mucusso Angola
106 B1 A Cañiza Spain
144 B2 Acaponeta Mex.
145 C3 Acapulco Mex.
151 E3 Acará Brazil
154 A1 Acaray, Represa de resr Para.
150 C2 Acarigua Venez.
110 B1 Acâş Romania
145 C3 Acatlán Mex.
145 C3 Acayucán Mex.
114 B4 Accra Ghana
98 B3 Accrington U.K.
74 B2 Achalpur India
97 A2 Achill Island Ireland
101 D1 Achim Ger.
96 B2 Achnasheen U.K.
91 D2 Achuyevo Rus. Fed.
111 C3 Acıpayam Turkey
109 C3 Acireale Italy
147 C2 Acklins Island Bahamas
153 A3 Aconcagua, Cerro mt. Arg.
106 B1 A Coruña Spain
108 A2 Acqui Terme Italy
103 D2 Ács Hungary
49 N6 Actéon, Groupe is Fr. Polynesia
145 C2 Actopán Mex.
143 D2 Ada U.S.A.
   Adabazar Turkey see Adapazarı
79 C2 Adam Oman
111 B3 Adamas Greece
135 B3 Adams Peak U.S.A.
   'Adan Yemen see Aden
80 B2 Adana Turkey
111 D2 Adapazarı Turkey
97 B2 Adare Ireland
55 M2 Adare, Cape Antarctica
108 A1 Adda r. Italy
78 B2 Ad Dafinah Saudi Arabia
78 B2 Ad Dahnā' des. Saudi Arabia
79 B2 Ad Dahnā' des. Saudi Arabia
114 A2 Ad Dakhla Western Sahara
   Ad Dammām Saudi Arabia see Dammam
78 A2 Ad Dār al Ḩamrā' Saudi Arabia
78 B3 Ad Darb Saudi Arabia
78 B2 Ad Dawādimī Saudi Arabia
   Ad Dawḩah Qatar see Doha
   Aḑ Ḑiffah plat. Egypt/Libya see Libyan Plateau
78 B2 Ad Dilam Saudi Arabia
78 B2 Ad Dir'īyah Saudi Arabia
117 B4 Addis Ababa Eth.
81 C2 Ad Dīwānīyah Iraq
141 D2 Adel U.S.A.
52 A2 Adelaide Austr.
55 A3 Adelaide Island Antarctica
50 C1 Adelaide River Austr.
101 D2 Adelebsen Ger.
55 K2 Adélie Land reg. Antarctica
78 B3 Aden Yemen
116 C3 Aden, Gulf of Somalia/Yemen
100 C2 Adenau Ger.
115 C3 Aderbissinat Niger
79 C2 Adh Dhayd U.A.E.
59 C3 Adi i. Indon.
78 B3 Āḏī Ārk'ay Eth.
108 B1 Adige r. Italy
116 B3 Ādīgrat Eth.

78 A3 Adi Keyih Eritrea
74 B3 Adilabad India
115 D2 Adīrī Libya
139 E2 Adirondack Mountains U.S.A.
   Ādīs Ābeba Eth. see Addis Ababa
117 B4 Ādīs Alem Eth.
80 B2 Adıyaman Turkey
110 C1 Adjud Romania
50 B1 Admiralty Gulf Austr.
128 A2 Admiralty Island U.S.A.
59 D3 Admiralty Islands P.N.G.
73 B3 Adoni India
104 B3 Adour r. France
106 C2 Adra Spain
114 B2 Adrar Alg.
138 C2 Adrian MI U.S.A.
143 C1 Adrian TX U.S.A.
108 B2 Adriatic Sea Europe
   Adua Eth. see Ādwa
116 B3 Ādwa Eth.
83 K2 Adycha r. Rus. Fed.
91 D3 Adygeysk Rus. Fed.
114 B4 Adzopé Côte d'Ivoire
111 B3 Aegean Sea Greece/Turkey
101 D1 Aerzen Ger.
106 B1 A Estrada Spain
116 B3 Afabet Eritrea
   Affreville Alg. see Khemis Miliana
76 C3 Afghanistan country Asia
78 B2 'Afīf Saudi Arabia
136 A2 Afton U.S.A.
80 B2 Afyon Turkey
115 C3 Agadez Niger
114 B1 Agadir Morocco
113 I7 Agalega Islands Mauritius
   Agana Guam see Hagåtña
74 B2 Agar India
119 D2 Āgaro Eth.
75 D2 Agartala India
81 C2 Ağdam Azer.
105 C3 Agde France
   Agedabia Libya see Ajdābiyā
104 C3 Agen France
122 A2 Aggeneys S. Africa
111 C3 Agia Varvara Greece
111 B3 Agios Dimitrios Greece
111 C3 Agios Efstratios i. Greece
111 C3 Agios Kirykos Greece
111 C3 Agios Nikolaos Greece
78 A3 Agirwat Hills Sudan
123 C2 Agisanang S. Africa
110 B1 Agnita Romania
74 B2 Agra India
81 C2 Ağrı Turkey
   Ağrı Dağı mt. Turkey see Ararat, Mount
108 B3 Agrigento Italy
111 B3 Agrinio Greece
109 B2 Agropoli Italy
87 E3 Agryz Rus. Fed.
144 B2 Agua Brava, Laguna lag. Mex.
154 B3 Água Clara Brazil
145 C3 Aguada Brazil
146 B4 Aguadulce Panama
144 B2 Aguanaval r. Mex.
144 B1 Agua Prieta Mex.
144 B2 Aguascalientes Mex.
155 D1 Águas Formosas Brazil
154 C2 Agudos Brazil
106 B1 Agueda Port.
114 B3 Aguelhok Mali
106 C1 Aguilar de Campoo Spain
107 C2 Águilas Spain
144 B3 Aguililla Mex.
122 B3 Agulhas, Cape S. Africa
158 F7 Agulhas Basin Southern Ocean
155 D2 Agulhas Negras mt. Brazil
158 F7 Agulhas Plateau Southern Ocean
158 F7 Agulhas Ridge S. Atlantic Ocean
111 C2 Ağva Turkey
115 C2 Ahaggar plat. Alg.
115 C2 Ahaggar, Tassili oua-n- plat. Alg.
81 C2 Ahar Iran
100 C1 Ahaus Ger.
111 C3 Ahlat Turkey
100 C2 Ahlen Ger.
74 B2 Ahmadabad India
74 B3 Ahmadnagar India
74 B2 Ahmadpur East Pak.
74 B1 Ahmadpur Sial Pak.

117 C4 Ahmar Eth.
   Ahmedabad India see Ahmadabad
   Ahmednagar India see Ahmadnagar
144 B2 Ahome Mex.
81 D3 Ahram Iran
101 E1 Ahrensburg Ger.
104 C2 Ahun France
93 F4 Åhus Sweden
81 C2 Ahvāz Iran
   Ahvenanmaa is Fin. see Åland Islands
122 A2 Ai-Ais Namibia
80 B2 Aigialousa Cyprus
111 B3 Aigio Greece
   Aihui China see Heihe
   Aijal India see Aizawl
141 D2 Aiken U.S.A.
97 B1 Ailt an Chorráin Ireland
155 D1 Aimorés Brazil
155 D1 Aimorés, Serra dos hills Brazil
105 D2 Ain r. France
107 E2 Aïn Azel Alg.
115 C1 Aïn Beïda Alg.
114 B2 'Aïn Ben Tili Maur.
107 D2 Aïn Defla Alg.
114 B1 Aïn Sefra Alg.
136 D2 Ainsworth U.S.A.
   Aintab Turkey see Gaziantep
107 D2 Aïn Taya Alg.
107 D2 Aïn Tédélès Alg.
107 C2 Aïn Temouchent Alg.
115 C3 Aïr, Massif de l' mts Niger
60 A1 Airbangis Indon.
128 C2 Airdrie Can.
96 C3 Airdrie U.K.
104 B3 Aire-sur-l'Adour France
101 E3 Aisch r. Ger.
128 A1 Aishihik Lake Can.
100 A3 Aisne r. France
59 D3 Aitape P.N.G.
137 E1 Aitkin U.S.A.
110 B1 Aiud Romania
105 D3 Aix-en-Provence France
   Aix-la-Chapelle Ger. see Aachen
105 D2 Aix-les-Bains France
75 D2 Aizawl India
88 C2 Aizkraukle Latvia
88 B2 Aizpute Latvia
67 C3 Aizu-Wakamatsu Japan
105 D3 Ajaccio France
   Ajayameru India see Ajmer
115 E1 Ajdābiyā Libya
115 C2 Ajjer, Tassili n' plat. Alg.
79 C2 'Ajman U.A.E.
74 B2 Ajmer India
   Ajmer-Merwara India see Ajmer
142 A2 Ajo U.S.A.
77 D2 Akadyr Kazakh.
119 D2 Āk'ak'ī Beseka Eth.
   Akamagaseki Japan see Shimonoseki
54 B2 Akaroa N.Z.
87 E3 Akbulak Rus. Fed.
80 B2 Akçakale Turkey
114 A3 Akchâr reg. Maur.
111 C3 Akdağ mt. Turkey
80 B2 Akdağmadeni Turkey
88 A2 Åkersberga Sweden
118 C2 Aketi Dem. Rep. Congo
81 C1 Akhalk'alak'i Georgia
81 C1 Akhalts'ikhe Georgia
79 C2 Akhḑar, Jabal mts Oman
111 C3 Akhisar Turkey
87 D4 Akhtubinsk Rus. Fed.
118 B3 Akiéni Gabon
130 B2 Akimiski Island Can.
66 D3 Akita Japan
114 A3 Akjoujt Maur.
   Akkerman Ukr. see Bilhorod-Dnistrovs'kyy
77 D1 Akkol' Kazakh.
   Ak-Mechet Kazakh. see Kyzylorda
88 B2 Akmenrags pt Latvia
   Akmola Kazakh. see Astana
67 B4 Akō Japan
117 B4 Akobo Sudan
74 B2 Akola India
118 B2 Akonolinga Cameroon
78 A3 Akordat Eritrea
131 D1 Akpatok Island Can.

77 D2 Akqi China
92 □A3 Akranes Iceland
36 C2 Akron CO U.S.A.
38 C2 Akron OH U.S.A.
75 B1 Aksai Chin terr. Asia
86 F2 Aksarka Rus. Fed.
80 B2 Aksaray Turkey
76 B1 Aksay Kazakh.
91 D2 Aksay Rus. Fed.
80 B2 Akşehir Turkey
76 C2 Akshiganak Kazakh.
77 E2 Aksu China
16 B3 Āksum Eth.
76 B2 Aktau Kazakh.
76 B1 Aktobe Kazakh.
77 D2 Aktogay Kazakh.
88 C3 Aktsyabrski Belarus
Aktyubinsk Kazakh. see
Aktobe
67 B4 Akune Japan
15 C4 Akure Nigeria
92 □B2 Akureyri Iceland
Akyab Myanmar see Sittwe
11 D2 Akyazı Turkey
77 C2 Akzhaykyn, Ozero salt l.
Kazakh.
40 C2 Alabama r. U.S.A.
40 C2 Alabama state U.S.A.
40 C2 Alabaster U.S.A.
11 C3 Alaçatı Turkey
45 D2 Alacrán, Arrecife rf Mex.
81 C1 Alagir Rus. Fed.
51 F4 Alagoinhas Brazil
107 C1 Alagón Spain
77 E2 Al Aḥmadī Kuwait
77 C2 Alakol', Ozero salt l. Kazakh.
92 J2 Alakurtti Rus. Fed.
81 C2 Al 'Alayyah Saudi Arabia
81 C2 Al 'Amādīyah Iraq
81 C2 Al 'Amārah Iraq
80 A2 Al 'Āmirīyah Egypt
43 C3 Alamitos, Sierra de los mt. Mex.
35 C3 Alamo U.S.A.
42 B2 Alamogordo U.S.A.
44 B2 Alamos Mex.
44 B2 Alamos Mex.
44 B2 Alamos r. Mex.
36 B3 Alamosa U.S.A.
Åland is Fin. see
Åland Islands
93 G3 Åland Islands is Fin.
80 B2 Alanya Turkey
80 B3 Alappuzha India
80 B3 Al 'Aqabah Jordan
76 B2 Al 'Aqīq Saudi Arabia
07 C2 Alarcón, Embalse de resr Spain
80 B2 Al 'Arīsh Egypt
78 B2 Al Arṭāwīyah Saudi Arabia
61 C2 Alas Indon.
11 C3 Alaşehir Turkey
28 A2 Alaska state U.S.A.
24 D4 Alaska, Gulf of U.S.A.
26 C2 Alaska Peninsula U.S.A.
26 C2 Alaska Range mts U.S.A.
81 C2 Alāt Azer.
87 D3 Alatyr' Rus. Fed.
50 B3 Alausí Ecuador
93 H3 Alavus Fin.
52 B2 Alawoona Austr.
79 C2 Al 'Ayn U.A.E.
08 A2 Alba Italy
07 C1 Albacete Spain
78 A2 Al Bada'i' Saudi Arabia
78 B2 Al Badī' Saudi Arabia
10 B1 Alba Iulia Romania
05 C2 Albania country Europe
50 A3 Albany Austr.
30 B2 Albany r. Can.
41 D2 Albany GA U.S.A.
39 E2 Albany NY U.S.A.
34 B2 Albany OR U.S.A.
15 E1 Al Bardī Libya
Al Başrah Iraq see Basra
51 D1 Albatross Bay Austr.
16 A2 Al Bawītī Egypt
15 E1 Al Baydā' Libya
78 B3 Al Baydā' Yemen
41 D1 Albemarle U.S.A.
41 E1 Albemarle Sound sea chan. U.S.A.
08 A2 Albenga Italy
51 B2 Alberga watercourse Austr.
19 D2 Albert, Lake Dem. Rep. Congo/Uganda
28 C2 Alberta prov. Can.
00 B2 Albert Kanaal canal Belgium

137 E2 Albert Lea U.S.A.
117 B4 Albert Nile r. Sudan/Uganda
123 C2 Alberton S. Africa
Albertville Dem. Rep. Congo see Kalemie
105 D2 Albertville France
104 C3 Albi France
151 D2 Albina Suriname
78 A2 Al Bi'r Saudi Arabia
78 B3 Al Birk Saudi Arabia
78 B2 Al Biyāḍh reg. Saudi Arabia
106 C2 Alborán, Isla de i. Spain
106 C2 Alborán Sea Europe
Alborz, Reshteh-ye mts Iran see Elburz Mountains
93 E4 Albox Spain
106 B2 Albufeira Port.
106 B2 Albufeira Port.
142 B1 Albuquerque U.S.A.
79 C2 Al Buraymī Oman
53 C3 Albury Austr.
106 B2 Alcácer do Sal Port.
106 C1 Alcalá de Henares Spain
106 C2 Alcalá la Real Spain
108 B3 Alcamo Italy
107 C1 Alcañiz Spain
106 B2 Alcántara Spain
107 C2 Alcantarilla Spain
106 C2 Alcaraz Spain
106 C2 Alcaraz, Sierra de mts Spain
106 C2 Alcázar de San Juan Spain
Alcazarquivir Morocco see Ksar el Kebir
91 D2 Alchevs'k Ukr.
155 E1 Alcobaça Brazil
107 C2 Alcoy-Alcoi Spain
107 D2 Alcúdia Spain
113 H6 Aldabra Islands Seychelles
142 B3 Aldama Mex.
145 C2 Aldama Mex.
83 J3 Aldan Rus. Fed.
83 J2 Aldan r. Rus. Fed.
99 D3 Aldeburgh U.K.
95 C4 Alderney i. Channel Is
99 C4 Aldershot U.K.
114 A3 Aleg Maur.
155 D2 Alegre Brazil
152 C2 Alegrete Brazil
89 D1 Alekhovshchina Rus. Fed.
89 E2 Aleksandrov Rus. Fed.
Aleksandrovsk Ukr. see Zaporizhzhya
83 K3 Aleksandrovsk-Sakhalinskiy Rus. Fed.
87 E1 Aleksandry, Zemlya i. Rus. Fed.
Alekseyevka Kazakh. see Akkol'
91 D1 Alekseyevka Rus. Fed.
91 D1 Alekseyevka Rus. Fed.
89 E3 Aleksin Rus. Fed.
109 D2 Aleksinac Serbia
118 B3 Alèmbé Gabon
155 D2 Além Paraíba Brazil
93 F3 Ålen Norway
104 C2 Alençon France
80 B2 Aleppo Syria
150 B4 Alerta Peru
128 B2 Alert Bay Can.
105 C3 Alès France
110 B1 Aleşd Romania
Aleshki Ukr. see Tsyurupyns'k
108 A2 Alessandria Italy
Alessio Albania see Lezhë
93 E3 Ålesund Norway
156 D2 Aleutian Basin Bering Sea
124 A4 Aleutian Islands U.S.A.
83 L3 Alevina, Mys i. Rus. Fed.
128 A2 Alexander Archipelago is U.S.A.
122 A2 Alexander Bay S. Africa
140 C2 Alexander City U.S.A.
55 A2 Alexander Island Antarctica
53 C3 Alexandra Austr.
54 A3 Alexandra N.Z.
153 E5 Alexandra, Cape S. Georgia
Alexandra Land i. Rus. Fed. see Aleksandry, Zemlya
111 B2 Alexandreia Greece
116 A1 Alexandria Egypt
110 C2 Alexandria Romania
123 C3 Alexandria S. Africa
96 B3 Alexandria U.K.
140 B2 Alexandria LA U.S.A.
137 D1 Alexandria MN U.S.A.
139 D3 Alexandria VA U.S.A.

52 A3 Alexandrina, Lake Austr.
111 C2 Alexandroupoli Greece
131 E2 Alexis r. Can.
128 B2 Alexis Creek Can.
77 E1 Aleysk Rus. Fed.
107 C1 Alfaro Spain
81 C3 Al Fāw Iraq
80 B3 Al Fayyūm Egypt
101 D2 Alfeld (Leine) Ger.
155 C2 Alfenas Brazil
96 C2 Alford U.K.
Al Fujayrah U.A.E. see Fujairah
Al Furāt r. Iraq/Syria see Euphrates
93 E4 Ålgård Norway
106 B2 Algarve reg. Port.
106 B2 Algeciras Spain
107 C2 Algemesí Spain
78 A3 Algena Eritrea
Alger Alg. see Algiers
114 C2 Algeria country Africa
79 C3 Al Ghaydah Yemen
108 A2 Alghero Italy
116 B2 Al Ghurdaqah Egypt
79 B2 Al Ghwaybiyah Saudi Arabia
115 C1 Algiers Alg.
123 C3 Algoa Bay S. Africa
137 E2 Algona U.S.A.
106 C1 Algorta Spain
Algueirao Moz. see Hacufera
81 C2 Al Ḥadīthah Iraq
79 C2 Al Ḥajar al Gharbī mts Oman
107 C2 Alhama de Murcia Spain
80 A2 Al Ḥammām Egypt
78 B2 Al Ḥanākīyah Saudi Arabia
81 C2 Al Ḥasakah Syria
78 B2 Al Ḥawīyah Saudi Arabia
81 C2 Al Ḥayy Iraq
78 B3 Al Ḥazm al Jawf Yemen
79 C3 Al Ḥibāk des. Saudi Arabia
Al Ḥillah Iraq see Hillah
78 B2 Al Ḥillah Saudi Arabia
79 B2 Al Ḥinnāh Saudi Arabia
Al Ḥudaydah Yemen see Hodeidah
79 B2 Al Ḥufūf Saudi Arabia
115 D2 Al Ḥulayq al Kabīr hills Libya
79 C2 'Alīābād Iran
111 C3 Aliağa Turkey
111 B2 Aliakmonas r. Greece
81 C2 Āli Bayramlı Azer.
107 C2 Alicante Spain
143 D3 Alice U.S.A.
109 C3 Alice, Punta pt Italy
51 C2 Alice Springs Austr.
77 D3 Alichur Tajik.
74 B2 Aligarh India
81 C2 Alīgūdarz Iran
69 E1 Alihe China
118 B3 Alima r. Congo
118 C2 Alindao C.A.R.
111 C3 Aliova r. Turkey
74 B2 Alirajpur India
117 C3 Ali Sabieh Djibouti
78 A1 Al 'Īsāwiyah Saudi Arabia
Al Iskandarīyah Egypt see Alexandria
116 B1 Al Ismā'īlīyah Egypt see
123 C3 Aliwal North S. Africa
115 E2 Al Jaghbūb Libya
79 C2 Al Jahrah Kuwait
79 C2 Al Jamalīyah Qatar
115 E2 Al Jawf Libya
115 D1 Al Jawsh Libya
106 B2 Aljezur Port.
Al Jīzah Egypt see Giza
79 B2 Al Jubayl Saudi Arabia
78 B2 Al Jubaylah Saudi Arabia
115 D2 Al Jufrah Libya
78 B2 Al Junaynah Saudi Arabia
106 B2 Aljustrel Port.
78 B2 Al Kahfah Saudi Arabia
79 C2 Al Kāmil Oman
80 B2 Al Karak Jordan
81 C2 Al Kāẓimīyah Iraq
79 C2 Al Khābūrah Oman
116 B2 Al Khārijah Egypt
79 C2 Al Khasab Oman
78 B3 Al Khawkhah Yemen
79 C2 Al Khawr Qatar
115 D1 Al Khums Libya
79 C2 Al Khunn well Saudi Arabia
79 C2 Al Kidan well Saudi Arabia

79 C2 Al Kir'ānah Qatar
100 B1 Alkmaar Neth.
115 E2 Al Kufrah Libya
81 C2 Al Kūt Iraq
Al Kuwayt Kuwait see Kuwait
Al Lādhiqīyah Syria see Latakia
75 C2 Allahabad India
83 K2 Allakh-Yun' Rus. Fed.
78 A2 'Allāqī, Wādī al watercourse Egypt
139 D2 Allegheny r. U.S.A.
139 C3 Allegheny Mountains U.S.A.
97 B1 Allen, Lough l. Ireland
145 B2 Allende Mex.
145 B2 Allende Mex.
139 D2 Allentown U.S.A.
Alleppey India see Alappuzha
101 D1 Aller r. Ger.
136 C2 Alliance NE U.S.A.
138 C2 Alliance OH U.S.A.
78 B2 Al Lith Saudi Arabia
96 C2 Alloa U.K.
131 C3 Alma Can.
Alma-Ata Kazakh. see Almaty
106 B2 Almada Port.
106 C2 Almadén Spain
Al Madīnah Saudi Arabia see Medina
80 B2 Al Mafraq Jordan
114 B2 Al Mahbas Western Sahara
78 B3 Al Maḥwīt Yemen
78 B2 Al Majma'ah Saudi Arabia
116 B2 Al Maks al Baḥrī Egypt
Al Manāmah Bahrain see Manama
135 B2 Almanor, Lake U.S.A.
107 C2 Almansa Spain
80 B2 Al Manṣūrah Egypt
79 C2 Al Mariyyah U.A.E.
115 E1 Al Marj Libya
77 D2 Almaty Kazakh.
Al Mawṣil Iraq see Mosul
81 C2 Al Mayādīn Syria
106 C1 Almazán Spain
151 D3 Almeirim Brazil
100 C1 Almelo Neth.
155 D1 Almenara Brazil
106 B1 Almendra, Embalse de resr Spain
106 B2 Almendralejo Spain
106 C2 Almería Spain
106 C2 Almería, Golfo de b. Spain
87 E3 Al'met'yevsk Rus. Fed.
78 B2 Al Mindak Saudi Arabia
116 B2 Al Minyā Egypt
79 B2 Al Mish'āb Saudi Arabia
106 B2 Almodôvar Port.
106 B2 Almonte Spain
75 B2 Almora India
79 B2 Al Mubarrez Saudi Arabia
79 C2 Al Muḍaibī Oman
80 B3 Al Mudawwarah Jordan
Al Mukallā Yemen see Mukalla
Al Mukhā Yemen see Mocha
106 C2 Almuñécar Spain
81 C2 Al Muqdādīyah Iraq
78 A2 Al Musayjid Saudi Arabia
78 A2 Al Muwayliḥ Saudi Arabia
111 B3 Almyros Greece
96 B2 Alness U.K.
98 C2 Alnwick U.K.
62 A1 Along India
111 B3 Alonnisos i. Greece
59 C3 Alor i. Indon.
59 C3 Alor, Kepulauan is Indon.
Alor Setar Malaysia see Alor Star
60 B1 Alor Star Malaysia
Alost Belgium see Aalst
59 E3 Alotau P.N.G.
86 C2 Alozero Rus. Fed.
138 C1 Alpena U.S.A.
160 Q1 Alpha Ridge Arctic Ocean
100 B1 Alphen aan den Rijn Neth.
142 B2 Alpine AZ U.S.A.
143 C2 Alpine TX U.S.A.
84 E4 Alps mts Europe
79 B3 Al Qa'āmīyāt reg. Saudi Arabia
115 D1 Al Qaddāḥīyah Libya
Al Qāhirah Egypt see Cairo
78 B2 Al Qā'īyah Saudi Arabia
81 C2 Al Qāmishlī Syria

06	C2	Antequera Spain
42	B2	Anthony U.S.A.
14	B2	Anti-Atlas *mts* Morocco
151	F4	Antibes France
05	D3	Anticosti, Île d' *i.* Can.
31	D3	Antigonish Can.
47	D3	Antigua *i.* Antigua
47	D3	Antigua and Barbuda *country* West Indies
45	C2	Antiguo-Morelos Mex.
11	B3	Antikythira *i.* Greece
		Antioch Turkey *see* Antakya
49	I8	Antipodes Islands N.Z.
		An t-Ob U.K. *see* Leverburgh
52	A2	Antofagasta Chile
54	C3	Antonina Brazil
		António Enes Moz. *see* Angoche
97	C1	Antrim U.K.
97	C1	Antrim Hills U.K.
21	□D2	Antsalova Madag.
		Antseranana Madag. *see* Antsirañana
21	□D2	Antsirabe Madag.
21	□D2	Antsohihy Madag.
00	B2	Antwerp Belgium
		Antwerpen Belgium *see* Antwerp
		An Uaimh Ireland *see* Navan
74	B2	Anupgarh India
73	C4	Anuradhapura Sri Lanka
		Anvers Belgium *see* Antwerp
51	C3	Anxious Bay Austr.
70	B2	Anyang China
65	B2	Anyang S. Korea
108	B2	Anzio Italy
67	C4	Aoga-shima *i.* Japan
66	D2	Aomori Japan
54	B2	Aoraki N.Z.
08	A1	Aosta Italy
14	B2	Aoukâr *reg.* Mali/Maur.
14	C2	Aoulef Alg.
15	D2	Aozou Chad
41	D1	Apalachee Bay U.S.A.
50	C3	Apaporis *r.* Col.
54	B2	Aparecida do Tabuado Brazil
64	B2	Aparri Phil.
86	C2	Apatity Rus. Fed.
44	B3	Apatzingán Mex.
00	B1	Apeldoorn Neth.
00	C1	Apen Ger.
08	A2	Apennines *mts* Italy
49	J5	Apia Samoa
54	C2	Apiaí Brazil
64	B3	Apo, Mount *vol.* Phil.
01	E2	Apolda Ger.
52	B3	Apollo Bay Austr.
41	D3	Apopka U.S.A.
41	D3	Apopka, Lake U.S.A.
54	B1	Aporé Brazil
54	B1	Aporé *r.* Brazil
38	A1	Apostle Islands U.S.A.
80	B2	Apostolos Andreas, Cape Cyprus
91	C2	Apostolove Ukr.
133	F3	Appalachian Mountains U.S.A.
		Appennino *mts* Italy *see* Apennines
53	D2	Appin Austr.
00	C1	Appingedam Neth.
98	B2	Appleby-in-Westmorland U.K.
138	B2	Appleton U.S.A.
08	B2	Aprilia Italy
62	A1	Aprunyi India
91	D3	Apsheronsk Rus. Fed.
		Apsheronskaya Rus. Fed. *see* Apsheronsk
154	B2	Apucarana Brazil
154	B2	Apucarana, Serra da *hills* Brazil
64	A3	Apurahuan Phil.
147	D4	Apure *r.* Venez.
78	A2	Aqaba, Gulf of Asia
81	D2	'Aqdā Iran
75	C1	Aqqikkol Hu *salt l.* China
154	A1	Aquidauana *r.* Brazil
104	B3	Aquitaine *reg.* France
75	C2	Ara India
117	A4	Arab, Bahr el *watercourse* Sudan
		Arabian Gulf *g.* Asia *see* The Gulf
78	B2	Arabian Peninsula Asia
56	B4	Arabian Sea Indian Ocean
151	F4	Aracaju Brazil
154	A2	Aracanguy, Montes de *hills* Para.
151	F3	Aracati Brazil
154	B2	Araçatuba Brazil
155	D1	Aracruz Brazil
155	D1	Araçuaí Brazil
110	B1	Arad Romania
115	E3	Arada Chad
79	C2	'Arādah U.A.E.
156	C6	Arafura Sea Austr./Indon.
154	B1	Aragarças Brazil
107	C1	Aragón *aut. comm.* Spain
107	C1	Aragón *r.* Spain
151	E3	Araguaia *r.* Brazil
154	B1	Araguaiana Brazil
151	E3	Araguaína Brazil
154	C1	Araguari Brazil
67	C3	Arai Japan
81	C2	Arak Alg.
81	C2	Arāk Iran
62	A1	Arakan Yoma *mts* Myanmar
81	C1	Arak's *r.* Armenia
76	C2	Aral Sea *salt l.* Kazakh./Uzbek.
76	C2	Aral'sk Kazakh.
		Aral'skoye More *salt l.* Kazakh./Uzbek. *see* Aral Sea
106	C1	Aranda de Duero Spain
109	D2	Arandelovac Serbia
97	B2	Aran Islands Ireland
106	C1	Aranjuez Spain
122	A1	Aranos Namibia
143	D3	Aransas Pass U.S.A.
67	B4	Arao Japan
114	B3	Araouane Mali
151	F3	Arapiraca Brazil
154	B2	Arapongas Brazil
154	C3	Araquari Brazil
78	B1	'Ar'ar Saudi Arabia
154	C2	Araraquara Brazil
151	D3	Araras Brazil
154	C2	Araras Brazil
154	B3	Araras, Serra das *hills* Brazil
154	B3	Araras, Serra das *mts* Brazil
81	C2	Ararat Armenia
52	B3	Ararat Austr.
81	C2	Ararat, Mount Turkey
155	D2	Araruama, Lago de *lag.* Brazil
155	E1	Arataca Brazil
		Aratürük China *see* Yiwu
150	B2	Arauca Col.
154	C1	Araxá Brazil
81	C2	Arbīl Iraq
129	E2	Arborg Can.
96	C2	Arbroath U.K.
74	A2	Arbū Shomālī, Dasht-e *des.* Afgh.
104	B3	Arcachon France
141	D3	Arcadia U.S.A.
134	B2	Arcata U.S.A.
145	B3	Arcelia Mex.
86	D2	Archangel Rus. Fed.
51	D1	Archer *r.* Austr.
49	M5	Archipel des Tuamotu *is* Fr. Polynesia
149	B6	Archipiélago Juan Fernández S. Pacific Ocean
134	D2	Arco U.S.A.
106	B2	Arcos de la Frontera Spain
127	G2	Arctic Bay Can.
		Arctic Institute Islands *is* Rus. Fed. *see* Arkticheskogo Instituta, Ostrova
160	J1	Arctic Mid-Ocean Ridge Arctic Ocean
160		Arctic Ocean
126	D2	Arctic Red *r.* Can.
81	C2	Ardabīl Iran
81	C1	Ardahan Turkey
93	E3	Årdalstangen Norway
97	C2	Ardee Ireland
100	B3	Ardennes *plat.* Belgium
135	B3	Arden Town U.S.A.
81	D2	Ardestān Iran
97	D1	Ardglass U.K.
53	C2	Ardlethan Austr.
143	D2	Ardmore U.S.A.
96	A2	Ardnamurchan, Point of U.K.
52	A2	Ardrossan Austr.
96	B3	Ardrossan U.K.
96	A2	Ardvasar U.K.
135	B3	Arena, Point U.S.A.
93	E4	Arendal Norway
100	B2	Arendonk Belgium
101	E1	Arendsee (Altmark) Ger.
150	B4	Arequipa Peru
151	D3	Arere Brazil
106	C1	Arévalo Spain
108	B2	Arezzo Italy
108	B2	Argenta Italy
104	B2	Argentan France
153	C2	Argentina *country* S. America
158	D7	Argentine Basin S. Atlantic Ocean
157	I8	Argentine Rise S. Atlantic Ocean
153	A5	Argentino, Lago *l.* Arg.
104	C2	Argenton-sur-Creuse France
110	C2	Argeş *r.* Romania
74	A1	Arghandāb, Rūd-e *r.* Afgh.
111	B3	Argolikos Kolpos *b.* Greece
111	B3	Argos Greece
111	B3	Argostoli Greece
107	C1	Arguis Spain
69	E1	Argun' *r.* China/Rus. Fed.
131	D3	Argyle Can.
50	B1	Argyle, Lake Austr.
		Argyrokastron Albania *see* Gjirokastër
		Ar Horqin Qi China *see* Tianshan
93	F4	Århus Denmark
122	A2	Ariamsvlei Namibia
152	A1	Arica Chile
96	A2	Arinagour U.K.
155	C1	Arinos Brazil
150	D4	Aripuanã Brazil
150	C3	Aripuanã *r.* Brazil
150	C3	Ariquemes Brazil
154	B1	Ariranhá *r.* Brazil
96	B2	Arisaig U.K.
96	B2	Arisaig, Sound of *sea chan.* U.K.
104	B3	Arizgoiti Spain
142	A2	Arizona *state* U.S.A.
144	A1	Arizpe Mex.
78	B2	'Arjah Saudi Arabia
61	C2	Arjasa Indon.
92	G2	Arjeplog Sweden
140	B2	Arkadelphia U.S.A.
77	C1	Arkalyk Kazakh.
140	B2	Arkansas *r.* U.S.A.
140	B1	Arkansas *state* U.S.A.
137	D3	Arkansas City U.S.A.
		Arkhangel'sk Rus. Fed. *see* Archangel
97	C2	Arklow Ireland
102	C1	Arkona, Kap *c.* Ger.
82	G1	Arkticheskogo Instituta, Ostrova *is* Rus. Fed.
105	C3	Arles France
143	D2	Arlington U.S.A.
138	B2	Arlington Heights U.S.A.
115	C3	Arlit Niger
100	B3	Arlon Belgium
97	C1	Armagh U.K.
116	B2	Armant Egypt
87	D4	Armavir Rus. Fed.
81	C1	Armenia *country* Asia
150	B2	Armenia Col.
		Armenopolis Romania *see* Gherla
144	B3	Armeria Mex.
53	D2	Armidale Austr.
134	D1	Armington U.S.A.
130	B2	Armstrong Can.
91	C2	Armyans'k Ukr.
		Armyanskaya S.S.R. *country* Asia *see* Armenia
		Arnaoutis, Cape *c.* Cyprus *see* Arnauti, Cape
130	D2	Arnaud *r.* Can.
80	B2	Arnauti, Cape Cyprus
100	B2	Arnhem Neth.
51	C1	Arnhem, Cape Austr.
51	C1	Arnhem Bay Austr.
51	C1	Arnhem Land *reg.* Austr.
108	B2	Arno *r.* Italy
52	A2	Arno Bay Austr.
130	C3	Arnprior Can.
101	D2	Arnsberg Ger.
101	E2	Arnstadt Ger.
122	A2	Aroab Namibia
154	B2	Aroeira Brazil
101	D2	Arolsen Ger.
78	A3	Aroma Sudan
108	A1	Arona Italy
144	B2	Aros *r.* Mex.
		Arquipélago dos Açores *aut. reg.* Port. *see* Azores
		Arrah India *see* Ara
81	C2	Ar Ramādī Iraq
96	B3	Arran *i.* U.K.
97	B1	Arranmore Island Ireland
80	B2	Ar Raqqah Syria
105	C1	Arras France
106	C1	Arrasate Spain
78	B2	Ar Rass Saudi Arabia
79	C2	Ar Rayyān Qatar
150	C2	Arrecifal Col.
145	C3	Arriagá Mex.
79	C2	Ar Rimāl *reg.* Saudi Arabia
		Ar Riyāḍ Saudi Arabia *see* Riyadh
54	A2	Arrowtown N.Z.
135	B3	Arroyo Grande U.S.A.
145	C2	Arroyo Seco Mex.
79	C2	Ar Rustāq Oman
78	B2	Ar Ruṭbah Iraq
78	B2	Ar Ruwaydah Saudi Arabia
81	D3	Arsanjan Iran
66	B2	Arsen'yev Rus. Fed.
117	C3	Arta Djibouti
111	B3	Arta Greece
144	B3	Arteaga Mex.
66	B2	Artem Rus. Fed.
91	D2	Artemivs'k Ukr.
104	C2	Artenay France
142	C2	Artesia U.S.A.
51	E2	Arthur Point Austr.
54	B2	Arthur's Pass N.Z.
152	C3	Artigas Uru.
129	D1	Artillery Lake Can.
123	C1	Artisia Botswana
104	C1	Artois *reg.* France
90	B2	Artsyz Ukr.
		Artur de Paiva Angola *see* Kuvango
77	D3	Artux China
81	C1	Artvin Turkey
59	C3	Aru, Kepulauan *is* Indon.
119	D2	Arua Uganda
147	D3	Aruba *terr.* West Indies
75	C2	Arun *r.* Nepal
119	D3	Arusha Tanz.
136	B3	Arvada U.S.A.
68	C1	Arvayheer Mongolia
129	E1	Arviat Can.
92	G2	Arvidsjaur Sweden
93	F4	Arvika Sweden
108	A2	Arzachena Italy
87	D3	Arzamas Rus. Fed.
107	C2	Arzew Alg.
100	C2	Arzfeld Ger.
		Arzila Morocco *see* Asilah
103	C2	Aš Czech Rep.
115	C4	Asaba Nigeria
74	B1	Asadābād Afgh.
66	D2	Asahi-dake *vol.* Japan
66	D2	Asahikawa Japan
78	B3	Asale *l.* Eth.
75	C2	Asansol India
117	C3	Asayita Eth.
130	C3	Asbestos Can.
122	B2	Asbestos Mountains S. Africa
119	E2	Asbe Teferi Eth.
109	C2	Ascea Italy
152	B1	Ascensión Bol.
113	B6	Ascension *i.* S. Atlantic Ocean
145	D3	Ascensión, Bahía de la *b.* Mex.
101	D3	Aschaffenburg Ger.
100	C2	Ascheberg Ger.
101	E2	Aschersleben Ger.
108	B2	Ascoli Piceno Italy
119	D2	Asela Eth.
92	G3	Åsele Sweden
111	B2	Asenovgrad Bulg.
76	B3	Aşgabat Turkm.
78	B2	Asharat Saudi Arabia
50	A2	Ashburton *watercourse* Austr.
54	B2	Ashburton N.Z.
140	B2	Ashdown U.S.A.
141	D1	Asheville U.S.A.
53	D1	Ashford Austr.
97	C2	Ashford Ireland
99	D3	Ashford U.K.
66	D2	Ashibetsu Japan
98	C2	Ashington U.K.
67	B4	Ashizuri-misaki *pt* Japan
		Ashkhabad Turkm. *see* Aşgabat

136 D3 Ashland *KS* U.S.A.
138 C3 Ashland *KY* U.S.A.
138 C2 Ashland *OH* U.S.A.
134 B2 Ashland *OR* U.S.A.
138 A1 Ashland *WI* U.S.A.
53 C1 Ashley Austr.
88 C3 Ashmyany Belarus
66 D2 Ashoro Japan
81 C2 Ash Shabakah Iraq
78 B3 Ash Sharawrah Saudi Arabia
Ash Shāriqah U.A.E. *see* Sharjah
81 C2 Ash Sharqāt Iraq
81 C2 Ash Shaṭrah Iraq
78 B3 Ash Shaykh 'Uthman Yemen
79 B3 Ash Shiḥr Yemen
79 C2 Ash Shināṣ Oman
78 B2 Ash Shu'aybah Saudi Arabia
78 B2 Ash Shu'bah Saudi Arabia
78 B2 Ash Shubaykīyah Saudi Arabia
78 B2 Ash Shumlūl Saudi Arabia
78 B3 Ash Shuqayq Saudi Arabia
115 D2 Ash Shuwayrif Libya
138 C2 Ashtabula U.S.A.
131 D2 Ashuanipi Lake Can.
75 B3 Asifabad India
106 B2 Asilah Morocco
108 A2 Asinara, Golfo dell' *b.* Italy
108 A2 Asinara, Isola *i.* Italy
82 G3 Asino Rus. Fed.
88 C3 Asipovichy Belarus
78 B2 'Asīr *reg.* Saudi Arabia
93 F4 Asker Norway
93 F4 Askim Norway
68 C1 Askiz Rus. Fed.
116 B3 Asmara Eritrea
93 F4 Åsnes *l.* Sweden
116 B2 Asoteriba, Jebel *mt.* Sudan
103 D2 Aspang-Markt Austria
136 B3 Aspen U.S.A.
143 C2 Aspermont U.S.A.
54 A2 Aspiring, Mount N.Z.
116 C3 Assab Eritrea
78 B3 Aş Şahīf Yemen
Aş Şaḥrā' al Gharbīyah *des.* Egypt *see* Western Desert
Aş Şaḥrā' ash Sharqīyah *des.* Egypt *see* Eastern Desert
78 B2 As Salamiyah Saudi Arabia
75 D2 Assam *state* India
81 C2 As Samāwah Iraq
79 C2 Aş Şanām *reg.* Saudi Arabia
115 E2 As Sarīr *reg.* Libya
108 A3 Assemini Italy
100 C1 Assen Neth.
100 B2 Assesse Belgium
115 D1 As Sidrah Libya
129 D3 Assiniboia Can.
128 C2 Assiniboine, Mount Can.
154 B2 Assis Brazil
78 B2 Aş Şubayḥīyah Kuwait
81 C2 As Sulaymānīyah Iraq
78 B2 As Sulaymī Saudi Arabia
78 B2 As Sulayyil Saudi Arabia
78 B2 As Sūq Saudi Arabia
80 B2 As Suwaydā' Syria
79 C2 As Suwayq Oman
As Suways Egypt *see* Suez
111 B3 Astakos Greece
77 D1 Astana Kazakh.
81 C2 Āstārā Iran
100 B2 Asten Neth.
108 A2 Asti Italy
74 B1 Astor Pak.
106 B1 Astorga Spain
134 B1 Astoria U.S.A.
Astrabad Iran *see* Gorgān
87 D4 Astrakhan' Rus. Fed.
Astrakhan' Bazar Azer. *see* Cälilabad
88 C3 Astravyets Belarus
Astrida Rwanda *see* Butare
106 B1 Asturias *aut. comm.* Spain
111 C3 Astypalaia *i.* Greece
152 C2 Asunción Para.
116 B2 Aswān Egypt
116 B2 Asyūṭ Egypt
Atacama, Desierto de *des.* Chile *see* Atacama Desert
152 B2 Atacama, Puna de *plat.* Arg.
152 B2 Atacama, Salar de *salt flat* Chile
152 B3 Atacama Desert *des.* Chile
114 C4 Atakpamé Togo

111 B3 Atalanti Greece
150 B4 Atalaya Peru
155 D1 Ataléia Brazil
77 C3 Atamyrat Turkm.
78 B3 'Ataq Yemen
114 A2 Atâr Maur.
135 B3 Atascadero U.S.A.
77 D2 Atasu Kazakh.
111 C3 Atavyros *mt.* Greece
116 B3 Atbara Sudan
116 B3 Atbara *r.* Sudan
77 C1 Atbasar Kazakh.
140 B3 Atchafalaya Bay U.S.A.
137 D3 Atchison U.S.A.
Ateransk Kazakh. *see* Atyrau
108 B2 Aterno *r.* Italy
108 B2 Atessa Italy
100 A2 Ath Belgium
128 C2 Athabasca Can.
129 C2 Athabasca *r.* Can.
129 D2 Athabasca, Lake Can.
97 C2 Athboy Ireland
97 B2 Athenry Ireland
111 B3 Athens Greece
140 C2 Athens *AL* U.S.A.
141 D2 Athens *GA* U.S.A.
138 C3 Athens *OH* U.S.A.
141 D1 Athens *TN* U.S.A.
143 D2 Athens *TX* U.S.A.
51 D1 Atherton Austr.
Athina Greece *see* Athens
97 C2 Athlone Ireland
111 B2 Athos *mt.* Greece
97 C2 Athy Ireland
115 D3 Ati Chad
130 A3 Atikokan Can.
87 D3 Atkarsk Rus. Fed.
141 D2 Atlanta U.S.A.
137 D2 Atlantic U.S.A.
141 D2 Atlantic Beach U.S.A.
139 E3 Atlantic City U.S.A.
55 D3 Atlantic-Indian-Antarctic Basin S. Atlantic Ocean
158 E8 Atlantic-Indian Ridge Southern Ocean
158 Atlantic Ocean
122 A3 Atlantis S. Africa
114 B1 Atlas Mountains Africa
114 C1 Atlas Saharien *mts* Alg.
128 A2 Atlin Can.
128 A2 Atlin Lake Can.
140 C2 Atmore U.S.A.
152 B2 Atocha Bol.
143 D2 Atoka U.S.A.
75 C2 Atrai *r.* India
80 B2 Aṭ Ṭafīlah Jordan
78 B2 Aṭ Ṭā'if Saudi Arabia
63 B2 Attapu Laos
130 B2 Attawapiskat Can.
130 B2 Attawapiskat *r.* Can.
130 B2 Attawapiskat Lake Can.
100 C2 Attendorn Ger.
100 A3 Attichy France
116 B2 Aṭ Ṭūr Egypt
78 B3 At Turbah Yemen
135 B3 Atwater U.S.A.
76 B2 Atyrau Kazakh.
105 D3 Aubagne France
105 C3 Aubenas France
126 D2 Aubry Lake Can.
140 C2 Auburn *AL* U.S.A.
135 B3 Auburn *CA* U.S.A.
138 B3 Auburn *IN* U.S.A.
139 E2 Auburn *ME* U.S.A.
137 D2 Auburn *NE* U.S.A.
139 D2 Auburn *NY* U.S.A.
104 C2 Aubusson France
104 C3 Auch France
54 B1 Auckland N.Z.
48 H9 Auckland Islands N.Z.
117 C4 Audo *mts* Eth.
101 F2 Aue Ger.
101 F2 Auerbach Ger.
102 C2 Augsburg Ger.
50 A3 Augusta Austr.
109 C3 Augusta Italy
141 D2 Augusta *GA* U.S.A.
137 D3 Augusta *KS* U.S.A.
139 F2 Augusta *ME* U.S.A.
155 D1 Augusto de Lima Brazil
103 E1 Augustów Pol.
50 A2 Augustus, Mount Austr.
100 A2 Aulnoye-Aymeries France
Aumale Alg. *see* Sour el Ghozlane
62 A2 Aunglan Myanmar
122 B2 Auob *watercourse* Namibia/S. Africa

131 D2 Aupaluk Can.
74 B3 Aurangabad India
104 B2 Auray France
100 C1 Aurich Ger.
154 B1 Aurilândia Brazil
104 C3 Aurillac France
136 C3 Aurora *CO* U.S.A.
138 B2 Aurora *IL* U.S.A.
137 E3 Aurora *MO* U.S.A.
137 D2 Aurora *NE* U.S.A.
122 A2 Aus Namibia
138 C2 Au Sable *r.* U.S.A.
137 E2 Austin *MN* U.S.A.
135 C3 Austin *NV* U.S.A.
143 D2 Austin *TX* U.S.A.
Australes, Îles *is* Fr. Polynesia *see* Tubuai Islands
50 B2 Australia *country* Oceania
55 K4 Australian-Antarctic Basin *sea feature* Southern Ocean
53 C3 Australian Capital Territory *admin. div.* Austr.
102 C2 Austria *country* Europe
150 D3 Autazes Brazil
144 B3 Autlán Mex.
105 C2 Autun France
104 C3 Auvergne *reg.* France
105 C2 Auvergne, Monts d' *mts* France
105 C2 Auxerre France
105 D2 Auxonne France
105 C2 Avallon France
131 E3 Avalon Peninsula Can.
154 C2 Avaré Brazil
91 D2 Avdiyivka Ukr.
106 B1 Aveiro Port.
109 B2 Avellino Italy
108 B2 Aversa Italy
100 A2 Avesnes-sur-Helpe France
93 G3 Avesta Sweden
104 C3 Aveyron *r.* France
108 B2 Avezzano Italy
96 C2 Aviemore U.K.
109 C2 Avigliano Italy
105 C3 Avignon France
106 C1 Ávila Spain
106 B1 Avilés Spain
52 B3 Avoca Austr.
139 D2 Avoca U.S.A.
109 C3 Avola Italy
99 B3 Avon *r. England* U.K.
99 B4 Avon *r. England* U.K.
99 C4 Avon *r. England* U.K.
142 A2 Avondale U.S.A.
104 B2 Avranches France
54 B1 Awanui N.Z.
78 B3 Awārik, 'Urūq al *des.* Saudi Arabia
119 C3 Āwasa Eth.
117 C4 Āwash Eth.
117 C3 Āwash *r.* Eth.
115 D2 Awbārī Libya
115 D2 Awbārī, Idhān *des.* Libya
117 C4 Aw Dheegle Somalia
96 B2 Awe, Loch *l.* U.K.
117 A4 Aweil Sudan
115 C4 Awka Nigeria
114 A2 Awserd Western Sahara
126 F1 Axel Heiberg Island Can.
114 B4 Axim Ghana
150 B4 Ayacucho Peru
77 E2 Ayagoz Kazakh.
Ayaguz Kazakh. *see* Ayagoz
68 B2 Ayakkum Hu *salt l.* China
106 B2 Ayamonte Spain
83 K3 Ayan Rus. Fed.
150 B4 Ayaviri Peru
74 A1 Aybak Afgh.
76 A2 Aybas Kazakh.
91 D2 Aydar *r.* Ukr.
77 C2 Aydarko'l ko'li *l.* Uzbek.
111 C3 Aydın Turkey
Ayers Rock *h.* Austr. *see* Uluru
83 I2 Aykhal Rus. Fed.
99 C4 Aylesbury U.K.
106 C1 Ayllón Spain
129 D1 Aylmer Lake Can.
117 B4 Ayod Sudan
83 M2 Ayon, Ostrov *i.* Rus. Fed.
114 B3 'Ayoûn el 'Atroûs Maur.
51 D1 Ayr Austr.
96 B3 Ayr U.K.
98 A2 Ayre, Point of Isle of Man
76 C2 Ayteke Bi Kazakh.
110 C2 Aytos Bulg.
145 C3 Ayutla Mex.

63 B2 Ayutthaya Thai.
111 C3 Ayvacık Turkey
111 C3 Ayvalık Turkey
91 C3 Ayya, Mys *pt* Ukr.
114 B3 Azaouâd *reg.* Mali
114 C3 Azaouagh, Vallée de *watercourse* Mali/Niger
115 D3 Azare Nigeria
77 C2 Azat, Gory *h.* Kazakh.
Azbine *mts* Niger *see* Aïr, Massif de l'
81 C1 Azerbaijan *country* Asia
Azerbaydzhanskaya S.S.R *country* Asia *see* Azerbaijan
77 C1 Azhibeksor, Ozero *salt l.* Kazakh.
150 B3 Azogues Ecuador
86 D2 Azopol'ye Rus. Fed.
112 A2 Azores *aut. reg.* Port.
91 D2 Azov Rus. Fed.
91 D2 Azov, Sea of Rus. Fed./Uk
Azraq, Bahr el *r.* Sudan *see* Blue Nile
106 B2 Azuaga Spain
146 B4 Azuero, Península de *pen.* Panama
153 C3 Azul Arg.
108 A3 Azzaba Alg.
Az Ẓahrān Saudi Arabia *se* Dhahran
80 B2 Az Zaqāziq Egypt
80 B2 Az Zarqā' Jordan
115 D1 Az Zāwiyah Libya
78 B3 Az Zaydīyah Yemen
114 C2 Azzel Matti, Sebkha *salt p* Alg.
78 B2 Az Zilfī Saudi Arabia
78 B3 Az Zuqur *i.* Yemen

B

63 B2 Ba, Sông *r.* Vietnam
117 C4 Baardheere Somalia
77 C3 Bābā, Kūh-e *mts* Afgh.
111 C3 Baba Burnu *pt* Turkey
110 C2 Babadag Romania
111 C2 Babaeski Turkey
116 C3 Bāb al Mandab *str.* Africa/Asia
61 C2 Babana Indon.
119 C1 Babanusa Sudan
59 C3 Babar *i.* Indon.
119 D3 Babati Tanz.
89 E2 Babayevo Rus. Fed.
59 C2 Babeldaob *i.* Palau
128 B2 Babine *r.* Can.
128 B2 Babine Lake Can.
59 C3 Babo Indon.
81 D2 Bābol Iran
122 A3 Baboon Point S. Africa
88 C3 Babruysk Belarus
Babu China *see* Hezhou
64 B2 Babuyan *i.* Phil.
64 B2 Babuyan Channel Phil.
64 B2 Babuyan Islands Phil.
151 E3 Bacabal Brazil
145 D3 Bacalar Mex.
59 C3 Bacan *i.* Indon.
110 C1 Bacău Romania
52 B3 Bacchus Marsh Austr.
Băc Giang Vietnam
77 D3 Bachu China
129 E1 Back *r.* Can.
109 C1 Bačka Palanka Serbia
109 C1 Bačka Topola Serbia
52 A3 Backstairs Passage Austr.
63 B3 Băc Liêu Vietnam
142 B3 Bacobampo Mex.
64 B2 Bacolod Phil.
130 C2 Bacqueville, Lac *l.* Can.
Bada China *see* Xilin
70 A1 Badain Jaran Shamo *des.* China
106 B2 Badajoz Spain
Badaojiang China *see* Baishan
75 D2 Badarpur India
101 E2 Bad Berka Ger.
101 E1 Bad Berleburg Ger.
101 E1 Bad Bevensen Ger.
101 D3 Bad Dürkheim Ger.
100 C2 Bad Ems Ger.
103 D2 Baden Austria
101 D3 Baden-Baden Ger.
101 E1 Bad Harzburg Ger.
101 E2 Bad Hersfeld Ger.

02	C2	Bad Hofgastein Austria
01	D2	Bad Homburg vor der Höhe Ger.
01	D1	Bad Iburg Ger.
74	A2	Badin Pak.
		Bādiyat ash Shām des. Asia see Syrian Desert
01	E2	Bad Kissingen Ger.
00	C3	Bad Kreuznach Ger.
36	C1	Badlands reg. ND U.S.A.
36	C2	Badlands reg. SD U.S.A.
01	E2	Bad Lauterberg im Harz Ger.
01	D2	Bad Lippspringe Ger.
01	D3	Bad Mergentheim Ger.
01	D2	Bad Nauheim Ger.
00	C2	Bad Neuenahr-Ahrweiler Ger.
01	E2	Bad Neustadt an der Saale Ger.
01	E1	Bad Oldesloe Ger.
01	D2	Bad Pyrmont Ger.
78	A2	Badr Ḥunayn Saudi Arabia
01	D1	Bad Salzuflen Ger.
01	D2	Bad Salzungen Ger.
02	C1	Bad Schwartau Ger.
01	E1	Bad Segeberg Ger.
00	C3	Bad Sobernheim Ger.
73	C4	Badulla Sri Lanka
01	D1	Bad Zwischenahn Ger.
06	C2	Baeza Spain
14	A3	Bafatá Guinea-Bissau
27	H2	Baffin Bay sea Can./Greenland
27	H2	Baffin Island Can.
18	B2	Bafia Cameroon
14	A3	Bafing r. Africa
14	A3	Bafoulabé Mali
18	B2	Bafoussam Cameroon
81	D2	Bāfq Iran
80	B1	Bafra Turkey
79	C2	Bāft Iran
19	C2	Bafwasende Dem. Rep. Congo
19	D3	Bagamoyo Tanz.
		Bagan Datoh Malaysia see Bagan Datuk
60	B1	Bagan Datuk Malaysia
64	B3	Baganga Phil.
20	B2	Bagani Namibia
60	B1	Bagansiapiapi Indon.
18	B3	Bagata Dem. Rep. Congo
91	E2	Bagayevskiy Rus. Fed.
42	A2	Bagdad U.S.A.
52	C3	Bagé Brazil
97	C2	Bagenalstown Ireland
81	C2	Baghdād Iraq
79	C1	Bāghīn Iran
77	C3	Baghlān Afgh.
04	C3	Bagnères-de-Luchon France
05	C3	Bagnols-sur-Cèze France
		Bago Myanmar see Pegu
88	B3	Bagrationovsk Rus. Fed.
		Bagrax China see Bohu
64	B2	Baguio Phil.
15	C3	Bagzane, Monts mts Niger
79	D2	Bāhā Kālāt Iran
46	C2	Bahamas, The country West Indies
75	C2	Baharampur India
		Bahariya Oasis oasis Egypt see Wāḥāt al Baḥrīyah
76	B3	Baharly Turkm.
60	B1	Bahau Malaysia
74	B2	Bahawalnagar Pak.
74	B2	Bahawalpur Pak.
		Bahia Brazil see Salvador
55	E1	Bahia state Brazil
46	B3	Bahía, Islas de la is Hond.
53	B3	Bahía Blanca Arg.
44	A2	Bahía Kino Mex.
52	C2	Bahía Negra Para.
44	A2	Bahía Tortugas Mex.
17	B3	Bahir Dar Eth.
79	C2	Bahlā Oman
75	C2	Bahraich India
79	C2	Bahrain country Asia
89	D3	Bahushewsk Belarus
10	C2	Baia Romania
20	A2	Baía dos Tigres Angola
10	B1	Baia Mare Romania
18	B2	Baïbokoum Chad
69	C2	Baicheng China
		Baidoa Somalia see Baydhabo
		Baie-aux-Feuilles Can. see Tasiujaq
31	D3	Baie-Comeau Can.
		Baie-du-Poste Can. see Mistissini
31	C3	Baie-St-Paul Can.
31	E3	Baie Verte Can.
62	A1	Baihanchang China
65	B1	Baihe China
69	D1	Baikal, Lake Rus. Fed.
		Baile Átha Cliath Ireland see Dublin
10	B2	Băileşti Romania
68	C2	Baima China
41	D2	Bainbridge U.S.A.
		Baingoin China see Porong
48	I3	Bairiki Kiribati
		Bairin Youqi China see Daban
53	C3	Bairnsdale Austr.
65	B1	Baishan Jilin China
65	B1	Baishan Jilin China
65	B1	Baitou Shan mt. China/N. Korea
20	A2	Baixo-Longa Angola
70	A2	Baiyin China
16	B3	Baiyuda Desert Sudan
03	D2	Baja Hungary
44	A1	Baja California pen. Mex.
61	D2	Bajawa Indon.
78	B3	Bājil Yemen
15	D3	Bajoga Nigeria
09	D2	Bajram Curri Albania
14	A3	Bakel Senegal
35	C3	Baker CA U.S.A.
40	B2	Baker LA U.S.A.
36	C1	Baker MT U.S.A.
34	C2	Baker OR U.S.A.
34	B1	Baker, Mount vol. U.S.A.
29	E1	Baker Foreland hd Can.
49	J3	Baker Island terr. N. Pacific Ocean
29	E1	Baker Lake Can.
29	E1	Baker Lake l. Can.
35	C3	Bakersfield U.S.A.
		Bakharden Turkm. see Baharly
91	C3	Bakhchysaray Ukr.
91	C1	Bakhmach Ukr.
		Bakhmut Ukr. see Artemivs'k
		Bākhtarān Iran see Kermānshāh
		Baku Azer. see Baku
11	C2	Bakırköy Turkey
92	DC2	Bakkaflói b. Iceland
18	C2	Bakouma C.A.R.
81	C1	Baku Azer.
99	B3	Bala U.K.
64	A3	Balabac Phil.
64	A3	Balabac i. Phil.
61	C1	Balabac Strait Malaysia/Phil.
75	C2	Balaghat India
60	C2	Balaiberkuak Indon.
52	A2	Balaklava Austr.
91	C3	Balaklava Ukr.
91	D2	Balakliya Ukr.
87	D3	Balakovo Rus. Fed.
45	C3	Balancán Mex.
11	C3	Balan Dağı h. Turkey
64	B2	Balanga Phil.
87	D3	Balashov Rus. Fed.
03	D2	Balatonboglár Hungary
50	D3	Balbina, Represa de resr Brazil
97	C2	Balbriggan Ireland
52	A2	Balcanoona Austr.
10	C2	Balchik Bulg.
54	A3	Balclutha N.Z.
43	E2	Balcones Escarpment U.S.A.
29	E2	Baldock Lake Can.
38	D2	Baldwin U.S.A.
29	D2	Baldy Mountain h. Can.
42	B2	Baldy Peak U.S.A.
		Baleares, Islas is Spain see Balearic Islands
07	D2	Balearic Islands is Spain
55	E1	Baleia, Ponta da pt Brazil
30	C2	Baleine, Grande Rivière de la r. Can.
31	D2	Baleine, Rivière à la r. Can.
75	C2	Baleshwar India
08	A2	Balestrieri, Punta mt. Italy
59	B3	Balige Indon.
79	B3	Balȟaf Yemen
61	C2	Bali i. Indon.
15	D4	Bali Nigeria
60	A1	Balige Indon.
75	C2	Baliguda India
11	C3	Balıkesir Turkey
61	C2	Balikpapan Indon.
64	A3	Balimbing Phil.
59	D3	Balimo P.N.G.
02	B2	Balingen Ger.
64	B2	Balintang Channel Phil.
		Bali Sea sea Indon. see Laut Bali
78	B3	Baljurshī Saudi Arabia
76	B3	Bālkanabat Turkm.
10	B2	Balkan Mountains Bulg./Serbia
77	D2	Balkash Kazakh.
77	C1	Balkashino Kazakh.
77	D2	Balkhash, Lake Kazakh.
		Balkhash, Ozero l. Kazakh. see Balkhash, Lake
		Balla Balla Zimbabwe see Mbalabala
96	B2	Ballachulish U.K.
50	B3	Balladonia Austr.
97	B2	Ballaghaderreen Ireland
92	G2	Ballangen Norway
96	B3	Ballantrae U.K.
52	B3	Ballarat Austr.
50	B2	Ballard, Lake imp. l. Austr.
96	C2	Ballater U.K.
14	B3	Ballé Mali
55	M3	Balleny Islands Antarctica
53	D1	Ballina Austr.
97	B1	Ballina Ireland
97	B2	Ballinasloe Ireland
97	B3	Ballineen Ireland
43	D2	Ballinger U.S.A.
97	B2	Ballinrobe Ireland
97	B1	Ballycastle Ireland
97	C1	Ballycastle U.K.
97	C1	Ballyclare U.K.
97	B2	Ballyhaunis Ireland
97	C1	Ballymena U.K.
97	C1	Ballymoney U.K.
97	D1	Ballynahinch U.K.
97	B1	Ballyshannon Ireland
95	R7	Ballyvoy U.K.
52	B3	Balmoral Austr.
43	C2	Balmorhea U.S.A.
20	A2	Balombo Angola
51	D2	Balonne r. Austr.
74	B2	Balotra India
77	D2	Balpyk Bi Kazakh.
75	C2	Balrampur India
52	B2	Balranald Austr.
10	B2	Balş Romania
51	E3	Balsas Brazil
45	C3	Balsas Mex.
45	B3	Balsas r. Mex.
90	B2	Balta Ukr.
90	B2	Bălţi Moldova
93	G4	Baltic Sea g. Europe
80	B2	Baltīm Egypt
97	B3	Baltimore S. Africa
39	D3	Baltimore U.S.A.
97	C2	Baltinglass Ireland
88	A3	Baltiysk Rus. Fed.
75	C2	Balu India
52	A2	Balumbah Austr.
88	C2	Balvi Latvia
77	D2	Balykchy Kyrg.
87	E4	Balykshi Kazakh.
79	C2	Bam Iran
71	A3	Bama China
51	D1	Bamaga Austr.
30	A2	Bamaji Lake Can.
14	B3	Bamako Mali
18	C2	Bambari C.A.R.
01	E3	Bamberg Ger.
19	C2	Bambili Dem. Rep. Congo
18	B2	Bambouti C.A.R.
19	C2	Bambouti C.A.R.
55	C2	Bambuí Brazil
98	C2	Bamburgh U.K.
18	B2	Bamenda Cameroon
77	C3	Bāmiān Afgh.
18	C2	Bamingui C.A.R.
79	D2	Bampūr Iran
79	D2	Bampūr watercourse Iran
19	C2	Banalia Dem. Rep. Congo
71	A3	Bana China
51	D4	Bananal, Ilha do i. Brazil
74	B2	Banas r. India
11	C3	Banaz Turkey
62	B2	Ban Ban Laos
97	C1	Banbridge U.K.
99	C3	Banbury U.K.
96	C2	Banchory U.K.
30	C3	Bancroft Can.
		Bancroft Zambia see Chililabombwe
19	C2	Banda Dem. Rep. Congo
75	C2	Banda India
59	C3	Banda, Kepulauan is Indon.
59	C3	Banda, Laut Indon.
60	A1	Banda Aceh Indon.
		Bandar India see Machilipatnam
		Bandar Abbas Iran see Bandar-e ‘Abbās
75	D2	Bandarban Bangl.
79	C2	Bandar-e ‘Abbās Iran
81	C2	Bandar-e Anzalī Iran
79	C2	Bandar-e Anzalī Iran
81	C2	Bandar-e Emām Khomeynī Iran
79	C2	Bandar-e Lengeh Iran
79	C2	Bandar-e Maqām Iran
		Bandar-e Pahlavī Iran see Bandar-e Anzalī
		Bandar-e Shāhpūr Iran see Bandar-e Emām Khomeynī
60	B2	Bandar Lampung Indon.
61	C1	Bandar Seri Begawan Brunei
		Banda Sea sea Indon. see Banda, Laut
55	D2	Bandeiras, Pico de mt. Brazil
23	C1	Bandelierkop S. Africa
44	B2	Banderas, Bahía de b. Mex.
14	B3	Bandiagara Mali
11	C2	Bandırma Turkey
		Bandjarmasin Indon. see Banjarmasin
97	B3	Bandon Ireland
97	B3	Bandon r. Ireland
18	B3	Bandundu Dem. Rep. Congo
60	B2	Bandung Indon.
28	C2	Banff Can.
96	C2	Banff U.K.
14	B3	Banfora Burkina
64	B3	Banga Phil.
73	B3	Bangalore India
18	C2	Bangassou C.A.R.
61	D2	Banggai Indon.
61	D2	Banggai, Kepulauan is Indon.
61	C1	Banggi i. Malaysia
		Banghāzī Libya see Benghazi
60	B2	Bangka i. Indon.
60	B2	Bangka, Selat sea chan. Indon.
61	C2	Bangkalan Indon.
60	B1	Bangkinang Indon.
60	B2	Bangko Indon.
63	B2	Bangkok Thai.
75	C2	Bangladesh country Asia
63	B2	Ba Ngoi Vietnam
97	D1	Bangor Northern Ireland U.K.
98	A3	Bangor Wales U.K.
39	F2	Bangor U.S.A.
63	A3	Bang Saphan Yai Thai.
64	B2	Bangued Phil.
18	B2	Bangui C.A.R.
64	B2	Bangui Phil.
21	B2	Bangweulu, Lake Zambia
80	B2	Banhā Egypt
62	B2	Ban Huai Khon Thai.
18	C2	Bani r. C.A.R.
80	B3	Banī Mazār Egypt
16	B2	Banī Suwayf Egypt
15	D1	Banī Walīd Libya
80	B2	Bāniyās Syria
09	C2	Banja Luka Bos.-Herz.
61	C2	Banjarmasin Indon.
14	A3	Banjul Gambia
63	A3	Ban Khok Kloi Thai.
28	A2	Banks Island B.C. Can.
26	D2	Banks Island N.W.T. Can.
48	H5	Banks Islands Vanuatu
29	E1	Banks Lake Can.
54	B3	Banks Peninsula N.Z.
51	D4	Banks Strait Austr.
75	C2	Bankura India
62	A1	Banmauk Myanmar
62	B2	Ban Mouang Laos
97	C1	Bann r. U.K.
62	B2	Ban Napè Laos
62	A2	Ban Na San Thai.
46	C2	Bannerman Town Bahamas
		Banningville Dem. Rep. Congo see Bandundu

71	A4	Ban Nong Kung Thai.
74	B1	Bannu Pak.
62	B2	Ban Phôn-Hông Laos
103	D2	Banská Bystrica Slovakia
111	B2	Bansko Bulg.
74	B2	Banswara India
62	B2	Ban Taviang Laos
63	A3	Ban Tha Kham Thai.
62	A2	Ban Tha Song Yang Thai.
63	B2	Ban Tôp Laos
97	B3	Bantry Ireland
97	B3	Bantry Bay Ireland
60	A1	Banyak, Pulau-pulau is Indon.
118	B2	Banyo Cameroon
107	D1	Banyoles Spain
61	C2	Banyuwangi Indon.
		Banzville Dem. Rep. Congo see Mobayi-Mbongo
		Bao'an China see Shenzhen
70	B1	Baochang China
70	B2	Baoding China
70	A2	Baoji China
62	B1	Bao Lac Vietnam
63	B2	Bao Lôc Vietnam
66	B1	Baoqing China
118	B2	Baoro C.A.R.
62	A1	Baoshan China
70	B1	Baotou China
74	B2	Bap India
81	C2	Ba'qūbah Iraq
109	C2	Bar Montenegro
90	B2	Bar Ukr.
116	B3	Bara Sudan
117	C4	Baraawe Somalia
147	C2	Baracoa Cuba
53	C2	Baradine Austr.
147	C3	Barahona Dom. Rep.
116	B3	Baraka watercourse Eritrea/Sudan
106	C1	Barakaldo Spain
61	C1	Baram r. Malaysia
150	D2	Baramanni Guyana
74	B1	Baramulla India
89	D3	Baran' Belarus
74	B2	Baran India
88	C3	Baranavichy Belarus
116	B2	Baranis Egypt
90	B1	Baranivka Ukr.
128	A2	Baranof Island U.S.A.
		Baranovicze Belarus see Baranavichy
59	C3	Barat Daya, Kepulauan is Indon.
155	D2	Barbacena Brazil
147	E3	Barbados country West Indies
107	C1	Barbastro Spain
106	B2	Barbate Spain
123	D2	Barberton S. Africa
104	B2	Barbezieux-St-Hilaire France
51	D2	Barcaldine Austr.
107	D1	Barcelona Spain
150	C1	Barcelona Venez.
105	D3	Barcelonnette France
150	C3	Barcelos Brazil
114	B4	Barclayville Liberia
51	D2	Barcoo watercourse Austr. see Cooper Creek
		Barcoo Creek watercourse Austr. see Cooper Creek
103	D2	Barcs Hungary
92	□B3	Bárðarbunga mt. Iceland
75	C2	Barddhaman India
103	E2	Bardejov Slovakia
		Bardera Somalia see Baardheere
79	C2	Bardsīr Iran
75	B2	Bareilly India
82	D1	Barents Sea Arctic Ocean
78	A3	Barentu Eritrea
72	C1	Barga China
108	B2	Barga Italy
75	C2	Barh India
52	B3	Barharwa India
139	F2	Bar Harbor U.S.A.
109	C2	Bari Italy
107	E2	Barika Alg.
74	B1	Barī Kowt Afgh.
150	B2	Barinas Venez.
75	C2	Baripada India
75	D2	Barisal Bangl.
60	B2	Barisan, Pegunungan mts Indon.
61	C2	Barito r. Indon.
79	C2	Barkā Oman
88	C2	Barkava Latvia
74	A2	Barkhan Pak.
51	C1	Barkly Tableland reg. Austr.
122	B2	Barkly West S. Africa
68	C2	Barkol China
110	C1	Bârlad Romania
105	D2	Bar-le-Duc France
50	A2	Barlee, Lake imp. l. Austr.
109	C2	Barletta Italy
53	C2	Barmedman Austr.
		Barmen-Elberfeld Ger. see Wuppertal
74	B2	Barmer India
52	B2	Barmera Austr.
99	A3	Barmouth U.K.
101	D1	Barmstedt Ger.
98	C2	Barnard Castle U.K.
53	B2	Barnato Austr.
82	G3	Barnaul Rus. Fed.
127	H2	Barnes Icecap Can.
100	B1	Barneveld Neth.
98	C3	Barnsley U.K.
99	A4	Barnstaple U.K.
99	A4	Barnstaple Bay U.K.
		Bideford Bay b. U.K. see Barnstaple Bay
141	D2	Barnwell U.S.A.
		Baroda India see Vadodara
150	C1	Barquisimeto Venez.
96	A2	Barra i. U.K.
53	D2	Barraba Austr.
151	D4	Barra do Bugres Brazil
151	E3	Barra do Corda Brazil
154	B1	Barra do Garças Brazil
150	D3	Barra do São Manuel Brazil
		Barraigh i. U.K. see Barra
150	B4	Barranca Lima Peru
150	B3	Barranca Loreto Peru
152	C2	Barranqueras Arg.
150	B1	Barranquilla Col.
105	D3	Barre des Écrins mt. France
151	E4	Barreiras Brazil
63	A2	Barren Island India
154	C2	Barretos Brazil
128	C2	Barrhead Can.
130	C3	Barrie Can.
128	B2	Barrière Can.
52	B2	Barrier Range hills Austr.
53	D2	Barrington, Mount Austr.
129	D2	Barrington Lake Can.
53	C1	Barringun Austr.
97	C2	Barrow r. Ireland
126	B2	Barrow U.S.A.
126	B2	Barrow, Point U.S.A.
51	C2	Barrow Creek Austr.
98	B2	Barrow-in-Furness U.K.
50	A2	Barrow Island Austr.
126	C2	Barrow Strait Can.
99	B4	Barry U.K.
122	B3	Barrydale S. Africa
130	C3	Barrys Bay Can.
74	B2	Barsalpur India
101	D1	Barsinghausen Ger.
135	C4	Barstow U.S.A.
105	C2	Bar-sur-Aube France
102	C1	Barth Ger.
80	B1	Bartın Turkey
51	D1	Bartle Frere, Mount Austr.
143	D1	Bartlesville U.S.A.
137	D2	Bartlett NE U.S.A.
140	C1	Bartlett TN U.S.A.
98	C3	Barton-upon-Humber U.K.
103	E1	Bartoszyce Pol.
61	C2	Barung i. Indon.
69	D1	Baruun-Urt Mongolia
91	D2	Barvinkove Ukr.
53	C2	Barwon r. Austr.
88	C3	Barysaw Belarus
118	B2	Basankusu Dem. Rep. Congo
110	C2	Basarabi Romania
64	B1	Basco Phil.
105	D2	Basel Switz.
71	C3	Bashi Channel Phil./Taiwan
91	C2	Bashtanka Ukr.
64	B3	Basilan i. Phil.
99	D4	Basildon U.K.
99	C4	Basingstoke U.K.
81	C2	Başkale Turkey
130	C3	Baskatong, Réservoir resr Can.
		Basle Switz. see Basel
118	C2	Basoko Dem. Rep. Congo
81	C2	Basra Iraq
128	C2	Bassano Can.
114	C4	Bassar Togo
63	A2	Bassein Myanmar
147	D3	Basse-Terre Guadeloupe
147	D3	Basseterre St Kitts and Nevis
114	B3	Bassikounou Maur.
114	C4	Bassila Benin
51	D3	Bass Strait Austr.
79	C2	Bastak Iran
101	E2	Bastheim Ger.
75	C2	Basti India
105	D3	Bastia France
100	B2	Bastogne Belgium
140	B2	Bastrop U.S.A.
		Basuo China see Dongfang
		Basutoland country Africa see Lesotho
118	C2	Bata Equat. Guinea
146	B2	Batabanó, Golfo de b. Cuba
83	J2	Batagay Rus. Fed.
154	B2	Bataguassu Brazil
74	B1	Batala India
106	B2	Batalha Port.
64	B1	Batan i. Phil.
118	B2	Batangafo C.A.R.
64	B2	Batangas Phil.
60	B2	Batanghari r. Indon.
64	B1	Batan Islands Phil.
154	C2	Batatais Brazil
139	D2	Batavia U.S.A.
91	D2	Bataysk Rus. Fed.
130	B3	Batchawana Mountain h. Can.
50	C1	Batchelor Austr.
63	B2	Bătdâmbâng Cambodia
118	B3	Batéké, Plateaux Congo
53	D3	Batemans Bay Austr.
140	B1	Batesville AR U.S.A.
140	C2	Batesville MS U.S.A.
89	D2	Batetskiy Rus. Fed.
99	B4	Bath U.K.
96	C3	Bathgate U.K.
74	B1	Bathinda India
53	C2	Bathurst Austr.
131	D3	Bathurst Can.
		Bathurst Gambia see Banjul
126	E2	Bathurst Inlet inlet Can.
126	E2	Bathurst Inlet (abandoned) Can.
50	C1	Bathurst Island Austr.
126	F1	Bathurst Island Can.
78	B1	Bāţin, Wādī al watercourse Asia
53	C3	Batlow Austr.
81	C2	Batman Turkey
115	C1	Batna Alg.
140	B2	Baton Rouge U.S.A.
144	B2	Batopilas Mex.
118	B2	Batouri Cameroon
154	B1	Batovi Brazil
		Batrâ' tourist site Jordan see Petra
92	I1	Båtsfjord Norway
73	C4	Batticaloa Sri Lanka
109	B2	Battipaglia Italy
128	D2	Battle r. Can.
138	B2	Battle Creek U.S.A.
135	C2	Battle Mountain U.S.A.
74	B1	Battura Glacier Pak.
117	B4	Batu mt. Eth.
60	A2	Batu, Pulau-pulau is Indon.
61	D2	Batuata i. Indon.
61	D2	Batudaka i. Indon.
64	B3	Batulaki Phil.
		Batum Georgia see Bat'umi
81	C1	Bat'umi Georgia
60	B1	Batu Pahat Malaysia
61	D2	Baubau Indon.
115	C3	Bauchi Nigeria
137	E1	Baudette U.S.A.
		Baudouinville Dem. Rep. Congo see Moba
104	B2	Baugé France
105	D2	Baume-les-Dames France
154	C2	Bauru Brazil
154	B1	Baús Brazil
88	B2	Bauska Latvia
102	C1	Bautzen Ger.
144	B2	Bavispe r. Mex.
87	E3	Bawdwin Myanmar
62	A1	Bawean i. Indon.
114	B3	Bawku Ghana
		Baxian China see Banan
146	C2	Bayamo Cuba
		Bayan Gol China see Dengkou
68	C2	Bayanhongor Mongolia
70	A2	Bayan Hot China
70	A1	Bayan Kuang China
69	D1	Bayan-Uul Mongolia
136	C2	Bayard NE U.S.A.
142	B2	Bayard NM U.S.A.
64	B3	Bayawan Phil.
81	C1	Bayburt Turkey
138	C2	Bay City MI U.S.A.
143	D3	Bay City TX U.S.A.
86	F2	Baydaratskaya Guba Rus. Fed.
117	C4	Baydhabo Somalia
102	C2	Bayern reg. Ger.
104	B2	Bayeux France
136	B3	Bayfield U.S.A.
78	B3	Bayḥān al Qişab Yemen
		Bay Islands is Hond. see Bahía, Islas de la
81	C2	Bayjī Iraq
		Baykal, Ozero l. Rus. Fed. see Baikal, Lake
		Baykal Range mts Rus. Fed. see Baykal'skiy Khrebet
83	I3	Baykal'skiy Khrebet mts Rus. Fed.
76	C2	Baykonyr Kazakh.
87	E3	Baymak Rus. Fed.
64	B2	Bayombong Phil.
104	B3	Bayonne France
76	C3	Bayramaly Turkm.
111	C3	Bayramiç Turkey
101	E3	Bayreuth Ger.
78	B3	Bayt al Faqīh Yemen
143	D3	Baytown U.S.A.
106	C2	Baza Spain
74	B1	Bāzā'ī Gonbad Afgh.
106	C2	Baza, Sierra de mts Spain
74	A1	Bāzārak Afgh.
76	A2	Bazardyuzyu, Gora mt. Azer./Rus. Fed.
104	B3	Bazas France
74	A2	Bazdar Pak.
70	A2	Bazhong China
79	D2	Bazmān Iran
79	D2	Bazmān, Kūh-e mt. Iran
		Bé, Nossi i. Madag. see Bé, Nosy
121	□D2	Bé, Nosy i. Madag.
136	C1	Beach U.S.A.
52	B3	Beachport Austr.
99	D4	Beachy Head hd U.K.
123	C3	Beacon Bay S. Africa
50	B1	Beagle Gulf Austr.
121	□D2	Bealanana Madag.
97	B1	Béal an Mhuirthead Ireland
121	□D3	Beampingaratra mts Madag.
134	D2	Bear r. U.S.A.
130	B3	Beardmore Can.
		Bear Island i. Arctic Ocean see Bjørnøya
134	E1	Bear Paw Mountain U.S.A.
147	C3	Beata, Cabo c. Dom. Rep.
147	C3	Beata, Isla i. Dom. Rep.
137	D2	Beatrice U.S.A.
135	C3	Beatty U.S.A.
53	D1	Beaudesert Austr.
52	B3	Beaufort Austr.
61	C1	Beaufort Malaysia
141	D2	Beaufort U.S.A.
126	C2	Beaufort Sea Can./U.S.A.
122	B3	Beaufort West S. Africa
96	B2	Beauly U.K.
96	B2	Beauly r. U.K.
100	B2	Beaumont Belgium
54	A3	Beaumont N.Z.
143	E2	Beaumont U.S.A.
105	C2	Beaune France
100	B2	Beauraing Belgium
129	E2	Beauséjour Can.
104	C2	Beauvais France
129	D2	Beauval Can.
135	D3	Beaver r. Can.
128	A1	Beaver Creek Can.
138	B2	Beaver Dam U.S.A.
129	E2	Beaver Hill Lake Can.
138	B1	Beaver Island U.S.A.
128	C2	Beaverlodge Can.
74	B2	Beawar India
154	C2	Bebedouro Brazil
101	D2	Bebra Ger.
99	D3	Beccles U.K.
109	D1	Bečej Serbia
106	B1	Becerreá Spain
114	B1	Béchar Alg.
		Bechuanaland country Africa see Botswana
136	C2	Beckley U.S.A.
117	B4	Bedelē Eth.
99	C3	Bedford U.K.
138	B3	Bedford U.S.A.
98	C2	Bedlington U.K.

63 B2 Bolavén, Phouphiang plat.
Laos
04 C2 Bolbec France
77 E2 Bole China
18 B3 Boleko Dem. Rep. Congo
14 B3 Bolgatanga Ghana
90 B2 Bolhrad Ukr.
66 B1 Boli China
18 B3 Bolia Dem. Rep. Congo
92 H3 Boliden Sweden
10 C2 Bolintin-Vale Romania
37 E3 Bolivar MO U.S.A.
40 C1 Bolivar TN U.S.A.
50 B2 Bolívar, Pico mt. Venez.
52 B1 Bolivia country S. America
89 E3 Bolkhov Rus. Fed.
05 C3 Bollène France
93 G3 Bollnäs Sweden
53 C1 Bollon Austr.
01 E2 Bollstedt Ger.
93 F4 Bolmen l. Sweden
18 B3 Bolobo Dem. Rep. Congo
08 B2 Bologna Italy
89 D2 Bologovo Rus. Fed.
89 D2 Bologoye Rus. Fed.
23 C2 Bolokanang S. Africa
18 B2 Bolomba Dem. Rep. Congo
08 B2 Bolsena, Lago di l. Italy
83 H1 Bol'shevik, Ostrov i.
Rus. Fed.
86 E2 Bol'shezemel'skaya Tundra
lowland Rus. Fed.
66 B2 Bol'shoy Kamen' Rus. Fed.
Asia/Europe see Caucasus
83 K2 Bol'shoy Lyakhovskiy,
Ostrov i. Rus. Fed.
Bol'shoy Tokmak Kyrg. see
Tokmok
Bol'shoy Tokmak Ukr. see
Tokmak
44 B2 Bolsón de Mapimí des.
Mex.
00 B1 Bolsward Neth.
98 B3 Bolton U.K.
80 D1 Bolu Turkey
59 E3 Bolubulu P.N.G.
92 □A2 Bolungarvík Iceland
82 C1 Bolzano Italy
18 B3 Boma Dem. Rep. Congo
53 D2 Bomaderry Austr.
53 C3 Bombala Austr.
Bombay India see Mumbai
55 C1 Bom Despacho Brazil
75 D2 Bomdila India
54 B1 Bom Jardim de Goiás Brazil
1 C4 Bom Jesus da Lapa Brazil
55 D2 Bom Jesus do Itabapoana
Brazil
15 D1 Bon, Cap c. Tunisia
17 D3 Bonaire i. Neth. Antilles
34 C1 Bonaparte, Mount U.S.A.
50 B1 Bonaparte Archipelago is
Austr.
31 E3 Bonavista Can.
31 E3 Bonavista Bay Can.
18 C2 Bondo Dem. Rep. Congo
14 B4 Bondoukou Côte d'Ivoire
Bône Alg. see Annaba
61 D2 Bonerate, Kepulauan is
Indon.
55 C1 Bonfinópolis de Minas
Brazil
17 B4 Bonga Eth.
75 D2 Bongaigaon India
18 C2 Bongandanga
Dem. Rep. Congo
22 B2 Bongani S. Africa
18 C2 Bongo, Massif des mts
C.A.R.
21 □D2 Bongolava mts Madag.
15 D3 Bongor Chad
14 B4 Bongouanou Côte d'Ivoire
53 B2 Bông Sơn Vietnam
43 D2 Bonham U.S.A.
08 A2 Bonifacio France
Bonifacio, Strait of
France/Italy
59 F3 Bonin Islands is Japan
00 C2 Bonn Ger.
34 C1 Bonners Ferry U.S.A.
05 D2 Bonneville France
00 A3 Bonnie Rock Austr.
29 C2 Bonnyville Can.
22 B2 Bonorva Italy
53 D1 Bonshaw Austr.
61 C1 Bontang Indon.
14 A4 Bonthe Sierra Leone

64 B2 Bontoc Phil.
61 C2 Bontosunggu Indon.
123 C3 Bontrug S. Africa
53 C1 Boolba Austr.
52 B2 Booligal Austr.
53 C1 Boomi Austr.
53 D1 Boonah Austr.
137 E2 Boone U.S.A.
140 C2 Booneville U.S.A.
137 E3 Boonville U.S.A.
52 B2 Booroorban Austr.
53 C2 Boorowa Austr.
117 C3 Boosaaso Somalia
126 G2 Boothia, Gulf of Can.
126 F2 Boothia Peninsula Can.
118 B3 Booué Gabon
100 C2 Boppard Ger.
144 B2 Boquilla, Presa de la resr
Mex.
109 D2 Bor Serbia
117 B4 Bor Sudan
80 B2 Bor Turkey
119 E2 Bor, Lagh watercourse
Kenya/Somalia
121 □E2 Boraha, Nosy i. Madag.
76 B2 Borankul Kazakh.
93 F4 Borås Sweden
79 C2 Borāzjān Iran
150 D3 Borba Brazil
104 B3 Bordeaux France
126 E1 Borden Island Can.
127 G2 Borden Peninsula Can.
52 B3 Bordertown Austr.
107 D2 Bordj Bou Arréridj Alg.
107 D2 Bordj Bounaama Alg.
114 B2 Bordj Flye Ste-Marie Alg.
115 C1 Bordj Messaouda Alg.
114 C2 Bordj Mokhtar Alg.
Bordj Omer Driss Alg. see
Bordj Omer Driss
Bordj Omer Driss Alg. see
94 B1 Borðoy i. Faroe Is
Borgå Fin. see Porvoo
92 □A3 Borgarnes Iceland
143 C1 Borger U.S.A.
93 G4 Borgholm Sweden
100 B2 Borgloon Belgium
108 A1 Borgosesia Italy
87 D3 Borisoglebsk Rus. Fed.
89 E2 Borisoglebskiy Rus. Fed.
91 D1 Borisovka Rus. Fed.
119 C2 Bo River Post Sudan
100 C2 Borken Ger.
92 G2 Borkenes Norway
100 C1 Borkum Ger.
100 C1 Borkum i. Ger.
93 G3 Borlange Sweden
101 F2 Borna Ger.
100 C1 Borne Neth.
61 C1 Borneo i. Asia
93 F4 Bornholm i. Denmark
111 C3 Bornova Turkey
90 B1 Borodyanka Ukr.
77 E2 Borohoro Shan mts China
114 B3 Boron Mali
89 D2 Borovichi Rus. Fed.
89 E2 Borovsk Rus. Fed.
76 C1 Borovskoy Kazakh.
51 C1 Borroloola Austr.
110 B1 Borşa Romania
76 B2 Borsakelmas sho'rxogi
salt marsh Uzbek.
90 B2 Borshchiv Ukr.
69 D1 Borshchovochnyy Khrebet
mts Rus. Fed.
101 E1 Börßum Ger.
Bortala China see Bole
81 C2 Borüjerd Iran
90 A2 Boryslav Ukr.
90 C1 Boryspil' Ukr.
91 C1 Borzna Ukr.
69 D1 Borzya Rus. Fed.
109 C1 Bosanska Dubica
Bos.-Herz.
109 C1 Bosanska Gradiška
Bos.-Herz.
109 C1 Bosanska Krupa Bos.-Herz.
109 C1 Bosanski Novi Bos.-Herz.
109 C2 Bosansko Grahovo
Bos.-Herz.
71 A3 Bose China
123 C2 Boshof S. Africa
109 C2 Bosnia-Herzegovina country
Europe
118 B2 Bosobolo Dem. Rep. Congo
111 C2 Bosporus str. Turkey
142 B2 Bosque U.S.A.
118 B2 Bossangoa C.A.R.

118 B2 Bossembélé C.A.R.
140 B2 Bossier City U.S.A.
122 A2 Bossiesvlei Namibia
68 B2 Bosten Hu l. China
99 C3 Boston U.K.
139 E2 Boston U.S.A.
140 B1 Boston Mountains U.S.A.
53 D2 Botany Bay Austr.
120 B3 Boteti r. Botswana
110 B2 Botev mt. Bulg.
80 A1 Botevgrad Bulg.
92 G3 Bothnia, Gulf of Fin./
Sweden
110 C1 Botoşani Romania
70 B2 Botou China
123 C2 Botshabelo S. Africa
120 B3 Botswana country Africa
109 C3 Botte Donato, Monte mt.
Italy
136 C1 Bottineau U.S.A.
100 C2 Bottrop Ger.
154 C2 Botucatu Brazil
114 B4 Bouaké Côte d'Ivoire
118 B2 Bouar C.A.R.
114 B1 Bouârfa Morocco
131 D3 Bouctouche Can.
107 E2 Bougaa Alg.
48 G4 Bougainville Island P.N.G.
Bougie Alg. see Bejaïa
114 B3 Bougouni Mali
100 B3 Bouillon Belgium
107 D2 Bouira Alg.
114 A2 Boujdour Western Sahara
50 B3 Boulder Austr.
136 B2 Boulder CO U.S.A.
134 D1 Boulder MT U.S.A.
135 D3 Boulder City U.S.A.
Boulhaut Morocco see
Ben Slimane
51 C2 Boulia Austr.
104 C2 Boulogne-Billancourt
France
104 C1 Boulogne-sur-Mer France
118 B3 Boulouba C.A.R.
118 B3 Boumango Gabon
118 B2 Boumba r. Cameroon
107 D2 Boumerdes Alg.
114 B4 Bouna Côte d'Ivoire
114 B4 Boundiali Côte d'Ivoire
134 D2 Bountiful U.S.A.
49 I8 Bounty Islands N.Z.
114 D3 Bourem Mali
104 C2 Bourganeuf France
105 D2 Bourg-en-Bresse France
104 C2 Bourges France
Bourgogne reg. France see
Burgundy
105 D2 Bourgoin-Jallieu France
53 C2 Bourke Austr.
99 C3 Bourne U.K.
99 C4 Bournemouth U.K.
118 C1 Bourtoutou Chad
115 C1 Bou Saâda Alg.
115 D3 Bousso Chad
100 A2 Boussu Belgium
114 A3 Boutilimit Maur.
128 C3 Bow r. Can.
Bowa China see Muli
51 D2 Bowen Austr.
53 C3 Bowen, Mount Austr.
129 C3 Bow Island Can.
138 B3 Bowling Green KY U.S.A.
137 E3 Bowling Green MO U.S.A.
138 C2 Bowling Green OH U.S.A.
136 C1 Bowman U.S.A.
53 D2 Bowral Austr.
101 D3 Boxberg Ger.
100 B2 Boxtel Neth.
80 B1 Boyabat Turkey
Boyang China see Poyang
97 B2 Boyle Ireland
97 C2 Boyne r. Ireland
136 B2 Boysen Reservoir U.S.A.
152 B2 Boyuibe Bol.
111 C3 Bozburun Turkey
111 C3 Bozcaada i. Turkey
111 D3 Bozdağ mt. Turkey
111 C3 Boz Dağları mts Turkey
111 C3 Bozdoğan Turkey
134 D1 Bozeman U.S.A.
118 B2 Bozoum C.A.R.
111 D3 Bozüyük Turkey
109 C2 Brač i. Croatia
130 C3 Bracebridge Can.
93 G3 Bräcke Sweden
99 C4 Bracknell U.K.
109 C2 Bradano r. Italy
141 D3 Bradenton U.S.A.

147 B3 Brades Montserrat
98 C3 Bradford U.K.
139 D2 Bradford U.S.A.
143 D2 Brady U.S.A.
96 C2 Braemar U.K.
106 B1 Braga Port.
151 E3 Bragança Brazil
106 B1 Bragança Port.
155 C2 Bragança Paulista Brazil
89 D3 Brahin Belarus
75 D2 Brahmanbaria Bangl.
75 C3 Brahmapur India
62 A1 Brahmaputra r. China/India
53 C3 Braidwood Austr.
110 C1 Brăila Romania
137 E1 Brainerd U.S.A.
99 D4 Braintree U.K.
100 B2 Braives Belgium
101 D1 Brake (Unterweser) Ger.
122 A1 Brakwater Namibia
98 B2 Brampton U.K.
101 D1 Bramsche Ger.
150 C3 Branco r. Brazil
101 F1 Brandenburg Ger.
129 E3 Brandon Can.
140 C2 Brandon U.S.A.
97 A2 Brandon Mountain h.
Ireland
122 B3 Brandvlei S. Africa
103 D1 Braniewo Pol.
130 B3 Brantford Can.
53 D2 Branxton Austr.
131 D3 Bras D'Or Lake Can.
155 D1 Brasil, Planalto do plat.
Brazil
154 C1 Brasilândia Brazil
154 C1 Brasília Brazil
155 D1 Brasília de Minas Brazil
88 C2 Braslaw Belarus
110 C1 Braşov Romania
103 D2 Bratislava Slovakia
83 H3 Bratsk Rus. Fed.
102 C2 Braunau am Inn Austria
101 E1 Braunschweig Ger.
92 □A2 Brautarholt Iceland
Bravo del Norte, Río r.
Mex./U.S.A. see Rio Grande
135 C4 Brawley U.S.A.
97 C2 Bray Ireland
150 D2 Brazil country S. America
158 E6 Brazil Basin
S. Atlantic Ocean
143 D3 Brazos r. U.S.A.
118 B3 Brazzaville Congo
109 C2 Brčko Bos.-Herz.
96 C2 Brechin U.K.
100 B2 Brecht Belgium
143 D2 Breckenridge U.S.A.
103 D2 Břeclav Czech Rep.
99 B4 Brecon U.K.
99 B4 Brecon Beacons reg. U.K.
100 B2 Breda Neth.
122 B3 Bredasdorp S. Africa
102 B2 Bregenz Austria
92 H1 Breivikbotn Norway
92 E3 Brekstad Norway
101 D1 Bremen Ger.
101 D1 Bremerhaven Ger.
Bremersdorp Swaziland see
Manzini
134 B1 Bremerton U.S.A.
143 D2 Brenham U.S.A.
105 E2 Brennero Italy
102 C2 Brenner Pass Austria/Italy
99 D4 Brentwood U.K.
108 B1 Brescia Italy
100 A2 Breskens Neth.
105 E2 Bressanone Italy
96 □ Bressay i. U.K.
104 B2 Bressuire France
88 B3 Brest Belarus
104 B2 Brest France
Brest-Litovsk Belarus see
Brest
Bretagne reg. France see
Brittany
140 C3 Breton Sound b. U.S.A.
151 D3 Breves Brazil
53 C1 Brewarrina Austr.
134 C1 Brewster U.S.A.
89 E2 Breytovo Rus. Fed.
Brezhnev Rus. Fed. see
Naberezhnyye Chelny
109 D2 Brezovo Polje plain
Croatia
118 C2 Bria C.A.R.
105 D3 Briançon France

Briceni

90 B2 **Briceni** Moldova
Brichany Moldova *see* **Briceni**
99 B4 **Bridgend** U.K.
139 E2 **Bridgeport** *CT* U.S.A.
136 C2 **Bridgeport** *NE* U.S.A.
147 E3 **Bridgetown** Barbados
131 D3 **Bridgewater** Can.
99 B3 **Bridgnorth** U.K.
99 B4 **Bridgwater** U.K.
99 B4 **Bridgwater Bay** U.K.
98 C2 **Bridlington** U.K.
98 C2 **Bridlington Bay** U.K.
99 B4 **Bridport** U.K.
105 D2 **Brig** Switz.
134 D2 **Brigham City** U.S.A.
53 C3 **Bright** Austr.
54 B3 **Brighton** N.Z.
99 C4 **Brighton** U.K.
136 C3 **Brighton** *CO* U.S.A.
138 C2 **Brighton** *MI* U.S.A.
105 D3 **Brignoles** France
114 A3 **Brikama** Gambia
101 D2 **Brilon** Ger.
109 C2 **Brindisi** Italy
Brinlack Ireland *see* **Bun na Leaca**
53 D1 **Brisbane** Austr.
99 B4 **Bristol** U.K.
139 E2 **Bristol** *CT* U.S.A.
141 D1 **Bristol** *TN* U.S.A.
99 A4 **Bristol Channel** *est.* U.K.
128 B2 **British Columbia** *prov.* Can.
British Guiana *country* S. America *see* **Guyana**
British Honduras *country* Central America *see* **Belize**
56 C6 **British Indian Ocean Territory** *terr.* Indian Ocean
95 B2 **British Isles** *is* Europe
British Solomon Islands *country* S. Pacific Ocean *see* **Solomon Islands**
123 C2 **Brits** S. Africa
122 B3 **Britstown** S. Africa
104 B2 **Brittany** *reg.* France
104 C2 **Brive-la-Gaillarde** France
106 C1 **Briviesca** Spain
99 B4 **Brixham** U.K.
103 D2 **Brno** Czech Rep.
Broach India *see* **Bharuch**
141 D2 **Broad** *r.* U.S.A.
130 C2 **Broadback** *r.* Can.
53 C3 **Broadford** Austr.
96 B2 **Broadford** U.K.
96 C3 **Broad Law** *h.* U.K.
136 B1 **Broadus** U.S.A.
129 D2 **Brochet** Can.
129 D2 **Brochet, Lac** *l.* Can.
131 D3 **Brochet, Lac au** *l.* Can.
101 E1 **Bröckel** Ger.
101 E2 **Brocken** *mt.* Ger.
126 E1 **Brock Island** Can.
130 C3 **Brockville** Can.
127 G2 **Brodeur Peninsula** Can.
96 B3 **Brodick** U.K.
103 D1 **Brodnica** Pol.
90 B1 **Brody** Ukr.
143 D1 **Broken Arrow** U.S.A.
137 D2 **Broken Bow** U.S.A.
52 B2 **Broken Hill** Austr.
Broken Hill Zambia *see* **Kabwe**
159 F6 **Broken Plateau** Indian Ocean
151 D2 **Brokopondo** Suriname
99 B3 **Bromsgrove** U.K.
93 E4 **Brønderslev** Denmark
123 C2 **Bronkhorstspruit** S. Africa
92 F2 **Brønnøysund** Norway
64 A3 **Brooke's Point** Phil.
140 B2 **Brookhaven** U.S.A.
134 B2 **Brookings** *OR* U.S.A.
137 D2 **Brookings** *SD* U.S.A.
128 C2 **Brooks** Can.
126 C2 **Brooks Range** *mts* U.S.A.
141 D3 **Brooksville** U.S.A.
139 D2 **Brookville** U.S.A.
96 B2 **Broom, Loch** *inlet* U.K.
50 B1 **Broome** Austr.
134 B2 **Brothers** U.S.A.
Broughton Island Can. *see* **Qikiqtarjuaq**
90 C1 **Brovary** Ukr.
143 C2 **Brownfield** U.S.A.
128 C3 **Browning** U.S.A.
140 C1 **Brownsville** *TN* U.S.A.
143 D3 **Brownsville** *TX* U.S.A.
143 D2 **Brownwood** U.S.A.
92 □A2 **Brú** Iceland
104 C1 **Bruay-la-Bussière** France
138 B1 **Bruce Crossing** U.S.A.
130 B3 **Bruce Peninsula** Can.
103 D2 **Bruck an der Mur** Austria
Bruges Belgium *see* **Brugge**
100 A2 **Brugge** Belgium
62 A1 **Bruint** India
128 C2 **Brûlé** Can.
151 E4 **Brumado** Brazil
93 F3 **Brumunddal** Norway
61 C1 **Brunei** *country* Asia
Brunei Brunei *see* **Bandar Seri Begawan**
102 C2 **Brunico** Italy
Brünn Czech Rep. *see* **Brno**
101 D1 **Brunsbüttel** Ger.
141 D2 **Brunswick** *GA* U.S.A.
139 F2 **Brunswick** *ME* U.S.A.
53 D1 **Brunswick Heads** Austr.
123 D2 **Bruntville** S. Africa
136 C2 **Brush** U.S.A.
100 B2 **Brussels** Belgium
Bruxelles Belgium *see* **Brussels**
143 D2 **Bryan** U.S.A.
89 D3 **Bryansk** Rus. Fed.
91 D2 **Bryn'kovskaya** Rus. Fed.
91 D2 **Bryukhovetskaya** Rus. Fed.
103 D1 **Brzeg** Pol.
Brześć nad Bugiem Belarus *see* **Brest**
114 A3 **Buba** Guinea-Bissau
111 C3 **Buca** Turkey
80 B2 **Bucak** Turkey
150 B2 **Bucaramanga** Col.
90 B2 **Buchach** Ukr.
53 C3 **Buchan** Austr.
114 A4 **Buchanan** Liberia
110 C2 **Bucharest** Romania
101 D1 **Bucholz in der Nordheide** Ger.
110 C1 **Bucin, Pasul** *pass* Romania
101 D1 **Bückeburg** Ger.
142 A2 **Buckeye** U.S.A.
96 C2 **Buckhaven** U.K.
96 C2 **Buckie** U.K.
99 C3 **Buckingham** U.K.
51 C1 **Buckingham Bay** Austr.
51 D2 **Buckland Tableland** *reg.* Austr.
52 A2 **Buckleboo** Austr.
139 F2 **Bucksport** U.S.A.
103 D2 **Bučovice** Czech Rep.
București Romania *see* **Bucharest**
89 D3 **Buda-Kashalyova** Belarus
103 D2 **Budapest** Hungary
75 B2 **Budaun** India
108 A2 **Budduso** Italy
99 A4 **Bude** U.K.
87 D4 **Budennovsk** Rus. Fed.
Budennoye Rus. Fed. *see* **Krasnogvardeyskoye**
89 D2 **Budogoshch'** Rus. Fed.
108 A2 **Budoni** Italy
Budweis Czech Rep. *see* **České Budějovice**
118 A2 **Buea** Cameroon
135 B4 **Buellton** U.S.A.
150 B2 **Buenaventura** Col.
144 B2 **Buenaventura** Mex.
Buena Vista *i.* N. Mariana Is *see* **Tinian**
106 C1 **Buendía, Embalse de** *resr* Spain
155 D1 **Buenópolis** Brazil
153 C3 **Buenos Aires** Arg.
153 A4 **Buenos Aires, Lago** *l.* Arg./Chile
139 D2 **Buffalo** *NY* U.S.A.
136 C1 **Buffalo** *SD* U.S.A.
143 D2 **Buffalo** *TX* U.S.A.
136 B2 **Buffalo** *WY* U.S.A.
129 D2 **Buffalo Narrows** Can.
121 C3 **Buffalo Range** Zimbabwe
122 A2 **Buffels** *watercourse* S. Africa
123 C1 **Buffels Drift** S. Africa
110 C2 **Buftea** Romania
103 E1 **Bug** *r.* Pol.
61 C2 **Bugel, Tanjung** *pt* Indon.
109 C2 **Bugojno** Bos.-Herz.
86 D2 **Bugrino** Rus. Fed.
64 A3 **Bugsuk** *i.* Phil.
87 E3 **Bugul'ma** Rus. Fed.
87 E3 **Buguruslan** Rus. Fed.
121 C2 **Buhera** Zimbabwe
110 C1 **Buhuși** Romania
99 B3 **Builth Wells** U.K.
69 D1 **Buir Nur** *l.* Mongolia
120 A3 **Buitepos** Namibia
109 D2 **Bujanovac** Serbia
119 C3 **Bujumbura** Burundi
69 D1 **Bukachacha** Rus. Fed.
120 B2 **Bukalo** Namibia
119 C3 **Bukavu** Dem. Rep. Congo
Bukhara Uzbek. *see* **Buxoro**
60 B2 **Bukittinggi** Indon.
119 D3 **Bukoba** Tanz.
103 D1 **Bukowiec** *h.* Pol.
59 C3 **Bula** Indon.
53 D2 **Bulahdelah** Austr.
121 B3 **Bulawayo** Zimbabwe
111 C3 **Buldan** Turkey
123 D2 **Bulembu** Swaziland
68 C1 **Bulgan** Mongolia
110 C2 **Bulgaria** *country* Europe
54 B2 **Buller** *r.* N.Z.
142 A1 **Bullhead City** U.S.A.
52 B1 **Bulloo** *watercourse* Austr.
52 B1 **Bulloo Downs** Austr.
122 A1 **Büllsport** Namibia
61 D2 **Bulukumba** Indon.
118 B3 **Bulungu** Dem. Rep. Congo
118 B3 **Bumba** Dem. Rep. Congo
118 C2 **Bumba** Dem. Rep. Congo
62 A1 **Bumhkang** Myanmar
118 B3 **Buna** Dem. Rep. Congo
Bun Beg Ireland *see* **An Bun Beag**
50 A3 **Bunbury** Austr.
97 C2 **Bunclody** Ireland
97 C1 **Buncrana** Ireland
119 D3 **Bunda** Tanz.
51 E2 **Bundaberg** Austr.
53 C1 **Bundaleer** Austr.
53 D2 **Bundarra** Austr.
74 B2 **Bundi** India
97 B1 **Bundoran** Ireland
75 C2 **Bundu** India
53 C3 **Bungendore** Austr.
119 D2 **Bungoma** Kenya
67 B4 **Bungo-suidō** *sea chan.* Japan
119 D2 **Bunia** Dem. Rep. Congo
118 C3 **Bunianga** Dem. Rep. Congo
97 B1 **Bun na Leaca** Ireland
63 B2 **Buôn Ma Thuôt** Vietnam
119 D3 **Bura** Kenya
117 C3 **Buraan** Somalia
Burang China *see* **Jirang**
78 B2 **Buraydah** Saudi Arabia
101 D2 **Burbach** Ger.
117 C4 **Burco** Somalia
100 B1 **Burdaard** Neth.
111 D3 **Burdur** Turkey
Burdwan India *see* **Barddhaman**
117 B3 **Burē** Eth.
99 D3 **Bure** *r.* U.K.
69 E1 **Bureinskiy Khrebet** *mts* Rus. Fed.
101 D2 **Büren** Ger.
74 B1 **Burewala** Pak.
Bureya Range *mts* Rus. Fed. *see* **Bureinskiy Khrebet**
110 C2 **Burgas** Bulg.
101 E1 **Burg bei Magdeburg** Ger.
101 E1 **Burgdorf** *Niedersachsen* Ger.
101 E1 **Burgdorf** *Niedersachsen* Ger.
131 E3 **Burgeo** Can.
123 C3 **Burgersdorp** S. Africa
123 D1 **Burgersfort** S. Africa
100 A2 **Burgh-Haamstede** Neth.
101 F3 **Burglengenfeld** Ger.
145 C2 **Burgos** Mex.
106 C1 **Burgos** Spain
105 C2 **Burgundy** *reg.* France
111 C3 **Burhaniye** Turkey
74 B2 **Burhanpur** India
75 C2 **Burhar-Dhanpuri** India
101 D1 **Burhave (Butjadingen)** Ger.
154 C2 **Buri** Brazil
60 B2 **Buriai** Indon.
64 B2 **Burias** *i.* Phil.
131 E3 **Burin** Can.
63 B2 **Buriram** Thai.
154 C1 **Buriti Alegre** Brazil
151 E3 **Buriti Bravo** Brazil
155 C1 **Buritis** Brazil
107 C2 **Burjassot** Spain
143 D2 **Burkburnett** U.S.A.
51 C1 **Burketown** Austr.
114 B3 **Burkina** *country* Africa
134 D2 **Burley** U.S.A.
136 C3 **Burlington** *CO* U.S.A.
137 E2 **Burlington** *IA* U.S.A.
141 E1 **Burlington** *NC* U.S.A.
139 E2 **Burlington** *VT* U.S.A.
134 B2 **Burney** U.S.A.
51 D4 **Burnie** Austr.
98 B3 **Burnley** U.K.
134 C2 **Burns** U.S.A.
134 C2 **Burns Junction** U.S.A.
128 B2 **Burns Lake** Can.
137 E2 **Burnsville** U.S.A.
101 F1 **Burow** Ger.
77 E2 **Burqin** China
52 A2 **Burra** Austr.
109 D2 **Burrel** Albania
97 B2 **Burren** *reg.* Ireland
53 C2 **Burrendong, Lake** Austr.
53 C2 **Burren Junction** Austr.
107 C2 **Burriana** Spain
53 C2 **Burrinjuck Reservoir** Aus
144 B2 **Burro, Serranías del** *mts* Mex.
111 C2 **Bursa** Turkey
116 B2 **Būr Safājah** Egypt
Bûr Sa'îd Egypt *see* **Port Said**
Būr Sa'īd Egypt *see* **Port Said**
Būr Sudan Sudan *see* **Port Sudan**
130 C2 **Burton, Lac** *l.* Can.
Burtonport Ireland *see* **Ailt an Chorráin**
99 C3 **Burton upon Trent** U.K.
52 B2 **Burtundy** Austr.
59 C3 **Buru** *i.* Indon.
119 C3 **Burundi** *country* Africa
119 C3 **Bururi** Burundi
96 C1 **Burwick** U.K.
98 B3 **Bury** U.K.
91 C1 **Buryn'** Ukr.
76 B2 **Burynshyk** Kazakh.
99 D3 **Bury St Edmunds** U.K.
118 C3 **Busanga** Dem. Rep. Cong
81 D3 **Būshehr** Iran
119 D3 **Bushenyi** Uganda
Bushire Iran *see* **Būshehr**
118 C2 **Businga** Dem. Rep. Cong
50 A3 **Busselton** Austr.
143 C3 **Bustamante** Mex.
108 A1 **Busto Arsizio** Italy
64 A2 **Busuanga** Phil.
118 C2 **Buta** Dem. Rep. Congo
119 C3 **Butare** Rwanda
96 B3 **Bute** *i.* U.K.
119 C2 **Butembo** Dem. Rep. Con
123 C2 **Butha-Buthe** Lesotho
139 D2 **Butler** U.S.A.
61 D2 **Buton** *i.* Indon.
134 D1 **Butte** U.S.A.
60 B1 **Butterworth** Malaysia
96 A1 **Butt of Lewis** *hd* U.K.
129 E2 **Button Bay** Can.
131 D1 **Button Islands** Can.
64 B3 **Butuan** Phil.
89 F3 **Buturlinovka** Rus. Fed.
75 C2 **Butwal** Nepal
101 D2 **Butzbach** Ger.
117 C4 **Buulobarde** Somalia
117 C5 **Buur Gaabo** Somalia
117 C4 **Buurhabaka** Somalia
78 A2 **Buwāṭah** Saudi Arabia
76 C3 **Buxoro** Uzbek.
101 D1 **Buxtehude** Ger.
98 C3 **Buxton** U.K.
89 F2 **Buy** Rus. Fed.
87 D4 **Buynaksk** Rus. Fed.
111 C3 **Büyükmenderes** *r.* Turkey
65 A1 **Buyun Shan** *mt.* China
110 C1 **Buzău** Romania
110 C1 **Buzău** *r.* Romania
121 C2 **Búzi** Moz.
87 E3 **Buzuluk** Rus. Fed.
88 C3 **Byahoml'** Belarus
110 C2 **Byala** *Sliven* Bulg.
110 C2 **Byala** *Varna* Bulg.
88 C3 **Byalynichy** Belarus
88 D3 **Byarezina** *r.* Belarus
88 B3 **Byaroza** Belarus
88 C3 **Byarozawka** Belarus
103 D1 **Bydgoszcz** Pol.
Byelorussia *country* Euro *see* **Belarus**
88 C3 **Byerazino** Belarus
88 C2 **Byeshankovichy** Belarus
89 D3 **Bykhaw** Belarus
127 G2 **Bylot Island** Can.
53 C2 **Byrock** Austr.

143 C2 Cap Rock Escarpment U.S.A.
143 C1 Capulin U.S.A.
150 C3 Caquetá r. Col.
110 B2 Caracal Romania
150 C2 Caracarai Brazil
150 C1 Caracas Venez.
151 E3 Caracol Brazil
155 C2 Caraguatatuba Brazil
153 A3 Carahue Chile
155 D1 Caraí Brazil
151 D3 Carajás, Serra dos hills Brazil
155 D2 Carandaí Brazil
155 D2 Carangola Brazil
110 B1 Caransebeș Romania
131 D3 Caraquet Can.
146 B3 Caratasca, Laguna de lag. Hond.
155 D1 Caratinga Brazil
150 C3 Carauari Brazil
107 C2 Caravaca de la Cruz Spain
155 E1 Caravelas Brazil
129 E3 Carberry Can.
144 A2 Carbó Mex.
107 C2 Carbon, Cap c. Alg.
153 B5 Carbón, Laguna del l. Arg.
108 A3 Carbonara, Capo c. Italy
136 B3 Carbondale CO U.S.A.
138 B3 Carbondale IL U.S.A.
139 D2 Carbondale PA U.S.A.
131 E3 Carbonear Can.
155 D1 Carbonita Brazil
107 C2 Carcaixent Spain
104 C3 Carcassonne France
128 A1 Carcross Can.
146 B2 Cárdenas Cuba
145 C3 Cárdenas Mex.
99 B4 Cardiff U.K.
99 A3 Cardigan U.K.
99 A3 Cardigan Bay U.K.
154 C2 Cardoso Brazil
128 C3 Cardston Can.
110 B1 Carei Romania
104 B2 Carentan France
50 B2 Carey, Lake imp. l. Austr.
155 D2 Cariacica Brazil
146 B3 Caribbean Sea N. Atlantic Ocean
128 B2 Cariboo Mountains Can.
139 F1 Caribou U.S.A.
130 B2 Caribou Lake Can.
128 C2 Caribou Mountains Can.
144 B2 Carichic Mex.
100 B3 Carignan France
53 C2 Carinda Austr.
107 C1 Cariñena Spain
130 C3 Carleton Place Can.
123 C2 Carletonville S. Africa
135 C2 Carlin U.S.A.
97 C1 Carlingford Lough inlet Ireland/U.K.
138 B3 Carlinville U.S.A.
98 B2 Carlisle U.K.
139 D2 Carlisle U.S.A.
155 D1 Carlos Chagas Brazil
97 C2 Carlow Ireland
96 A1 Carloway U.K.
135 C4 Carlsbad CA U.S.A.
142 C2 Carlsbad NM U.S.A.
129 D3 Carlyle Can.
128 A1 Carmacks Can.
129 E3 Carman Can.
99 A4 Carmarthen U.K.
99 A4 Carmarthen Bay U.K.
104 C3 Carmaux France
145 C3 Carmelita Guat.
144 A2 Carmen, Isla i. Mex.
155 C1 Carmo do Paranaíba Brazil
Carmona Angola see Uíge
106 B2 Carmona Spain
104 B2 Carnac France
50 A2 Carnarvon Austr.
122 B3 Carnarvon S. Africa
97 C1 Carndonagh Ireland
129 D3 Carnduff Can.
50 B2 Carnegie, Lake imp. l. Austr.
96 B2 Carn Eige mt. U.K.
55 P2 Carney Island Antarctica
73 D4 Car Nicobar i. India
118 B2 Carnot C.A.R.
52 A2 Carnot, Cape Austr.
96 C2 Carnoustie U.K.
97 C2 Carnsore Point Ireland
151 E3 Carolina Brazil
49 L4 Caroline Island Kiribati
59 D2 Caroline Islands N. Pacific Ocean

122 A2 Carolusberg S. Africa
103 D2 Carpathian Mountains Europe
Carpaţii Meridionali mts Romania see Transylvanian Alps
51 C1 Carpentaria, Gulf of Austr.
105 D3 Carpentras France
108 B2 Carpi Italy
141 D3 Carrabelle U.S.A.
97 B3 Carrantuohill mt. Ireland
108 B2 Carrara Italy
97 D1 Carrickfergus U.K.
97 C2 Carrickmacross Ireland
97 B2 Carrick-on-Shannon Ireland
97 C2 Carrick-on-Suir Ireland
137 D1 Carrington U.S.A.
143 D3 Carrizo Springs U.S.A.
142 B2 Carrizozo U.S.A.
137 E2 Carroll U.S.A.
141 C2 Carrollton U.S.A.
129 D2 Carrot River Can.
135 C3 Carson City U.S.A.
135 C3 Carson Sink l. U.S.A.
Carstensz-top mt. Indon. see Jaya, Puncak
150 B1 Cartagena Col.
107 C2 Cartagena Spain
146 B4 Cartago Costa Rica
54 C2 Carterton N.Z.
137 E3 Carthage MO U.S.A.
143 E2 Carthage TX U.S.A.
131 E2 Cartwright Can.
151 F3 Caruaru Brazil
150 C1 Carúpano Venez.
52 B1 Caryapundy Swamp Austr.
114 B1 Casablanca Morocco
154 C2 Casa Branca Brazil
144 B1 Casa de Janos Mex.
142 A2 Casa Grande U.S.A.
108 A1 Casale Monferrato Italy
109 C2 Casarano Italy
144 B1 Casas Grandes Mex.
134 C2 Cascade U.S.A.
134 B2 Cascade Range mts Can./U.S.A.
106 B2 Cascais Port.
151 F3 Cascavel Brazil
154 B2 Cascavel Brazil
139 F2 Casco Bay U.S.A.
108 B2 Caserta Italy
97 C2 Cashel Ireland
153 B3 Casilda Arg.
53 D1 Casino Austr.
Casnewydd U.K. see Newport
107 C1 Caspe Spain
136 B2 Casper U.S.A.
76 A2 Caspian Lowland Kazakh./Rus. Fed.
81 C1 Caspian Sea Asia/Europe
Cassaigne Alg. see Sidi Ali
154 C2 Cássia Brazil
128 B2 Cassiar Can.
128 A2 Cassiar Mountains Can.
154 B1 Cassilândia Brazil
120 A2 Cassinga Angola
108 B2 Cassino Italy
96 B2 Cassley r. U.K.
151 E3 Castanhal Brazil
152 B3 Castaño r. Arg.
144 B2 Castaños Mex.
105 C3 Casteljaloux France
105 D3 Castellane France
107 C2 Castellón de la Plana Spain
155 D2 Castelo Brazil
106 B2 Castelo Branco Port.
104 C3 Castelsarrasin France
108 B3 Castelvetrano Italy
52 B3 Casterton Austr.
108 B2 Castiglione della Pescaia Italy
106 C2 Castilla-La Mancha aut. comm. Spain
106 B2 Castilla, Playa de coastal area Spain
106 C1 Castilla y León aut. comm. Spain
97 B2 Castlebar Ireland
96 A2 Castlebay U.K.
97 C1 Castleblayney Ireland
97 C1 Castlederg U.K.
96 C3 Castle Douglas U.K.
128 C3 Castlegar Can.
97 B2 Castleisland Ireland
52 B3 Castlemaine Austr.
97 C2 Castlepollard Ireland

97 B2 Castlerea Ireland
53 C2 Castlereagh r. Austr.
136 C3 Castle Rock U.S.A.
128 C2 Castor Can.
104 C3 Castres France
100 B1 Castricum Neth.
147 D3 Castries St Lucia
154 C2 Castro Brazil
153 A4 Castro Chile
106 B2 Castro Verde Port.
109 C3 Castrovillari Italy
150 A3 Catacaos Peru
155 D2 Cataguases Brazil
154 C1 Catalão Brazil
Catalonia aut. comm. Spain see Cataluña
107 D1 Cataluña aut. comm. Spain
152 B2 Catamarca Arg.
64 B2 Catanduanes i. Phil.
154 C2 Catanduva Brazil
154 B3 Catanduvas Brazil
109 C3 Catania Italy
109 C3 Catanzaro Italy
64 B2 Catarman Phil.
107 C2 Catarroja Spain
64 B2 Catbalogan Phil.
145 C3 Catemaco Mex.
120 A1 Catete Angola
Catherine, Mount mt. Egypt see Kātrīnā, Jabal
147 C2 Cat Island Bahamas
130 A2 Cat Lake Can.
145 D2 Catoche, Cabo c. Mex.
139 E2 Catskill Mountains U.S.A.
123 D2 Catuane Moz.
64 B3 Cauayan Phil.
131 D2 Caubvick, Mount Can.
150 B2 Cauca r. Col.
151 F3 Caucaia Brazil
81 C1 Caucasus mts Asia/Europe
100 A2 Caudry France
109 C3 Caulonia Italy
120 A1 Caungula Angola
150 C2 Caura r. Venez.
131 D3 Causapscal Can.
90 B2 Căuşeni Moldova
105 D3 Cavaillon France
151 E4 Cavalcante Brazil
114 B4 Cavally r. Côte d'Ivoire/Liberia
97 C2 Cavan Ireland
154 B3 Cavernoso, Serra do mts Brazil
151 D2 Caviana, Ilha i. Brazil
Cawnpore India see Kanpur
151 E3 Caxias Brazil
152 C2 Caxias do Sul Brazil
120 A1 Caxito Angola
151 D2 Cayenne Fr. Guiana
146 B3 Cayman Islands terr. West Indies
158 C3 Cayman Trench Caribbean Sea
117 C4 Caynabo Somalia
120 B2 Cazombo Angola
Ceará Brazil see Fortaleza
Ceatharlach Ireland see Carlow
144 B2 Ceballos Mex.
64 B2 Cebu Phil.
64 B2 Cebu i. Phil.
108 B2 Cecina Italy
137 F2 Cedar r. U.S.A.
135 D3 Cedar City U.S.A.
137 E2 Cedar Falls U.S.A.
129 D2 Cedar Lake Can.
137 E2 Cedar Rapids U.S.A.
144 A2 Cedros, Isla i. Mex.
51 C3 Ceduna Austr.
117 C4 Ceeldheere Somalia
117 C3 Ceerigaabo Somalia
108 B3 Cefalù Italy
145 B2 Celaya Mex.
61 D2 Celebes i. Indon.
156 C5 Celebes Sea Indon./Phil.
145 C2 Celestún Mex.
101 E1 Celle Ger.
95 B3 Celtic Sea Ireland/U.K.
59 D3 Cenderawasih, Teluk b. Indon.
140 C2 Center Point U.S.A.
150 B1 Central, Cordillera mts Col.
150 B4 Central, Cordillera mts Peru
64 B2 Central, Cordillera mts Phil.
Central African Empire country Africa see Central African Republic

118 C2 Central African Republic country Africa
74 A2 Central Brahui Range mt Pak.
137 D2 Central City U.S.A.
138 B3 Centralia IL U.S.A.
134 B1 Centralia WA U.S.A.
74 A2 Central Makran Range m Pak.
156 D5 Central Pacific Basin Pacific Ocean
134 B2 Central Point U.S.A.
Central Provinces state India see Madhya Pradesh
59 D3 Central Range mts P.N.G.
89 E3 Central Russian Upland hills Rus. Fed.
83 I2 Central Siberian Plateau plat. Rus. Fed.
140 C2 Century U.S.A.
Ceos i. Greece see Kea
111 B3 Cephalonia i. Greece
Ceram i. Indon. see Seram
Ceram Sea sea Indon. see Laut Seram
101 F3 Čerchov mt. Czech Rep.
152 B2 Ceres Arg.
154 C1 Ceres Brazil
122 A3 Ceres S. Africa
105 C3 Céret France
106 C1 Cerezo de Abajo Spain
109 C2 Cerignola Italy
Cerigo i. Greece see Kythira
110 C2 Cernavodă Romania
145 C2 Cerralvo Mex.
144 B2 Cerralvo, Isla i. Mex.
145 B2 Cerritos Mex.
154 C1 Cerro Azul Brazil
145 C2 Cerro Azul Mex.
150 B4 Cerro de Pasco Peru
105 D3 Cervione France
107 C1 Cervo Spain
108 B2 Cesena Italy
108 B2 Cesenatico Italy
88 C2 Cēsis Latvia
102 C2 České Budějovice Czech Rep.
101 F3 Český les mts Czech Rep.
111 C3 Çeşme Turkey
53 D2 Cessnock Austr.
104 B2 Cesson-Sévigné France
104 B3 Cestas France
109 C2 Cetinje Montenegro
109 C3 Cetraro Italy
106 B2 Ceuta N. Africa
105 C3 Cévennes mts France
Ceylon country Asia see Sri Lanka
79 D2 Chābahār Iran
75 C1 Chabyêr Caka salt l. China
150 B3 Chachapoyas Peru
89 D3 Chachersk Belarus
63 B2 Chachoengsao Thai.
152 C2 Chaco Boreal reg. Para.
142 B1 Chaco Mesa plat. U.S.A.
115 D3 Chad country Africa
115 D3 Chad, Lake Africa
68 C1 Chadaasan Mongolia
68 C1 Chadan Rus. Fed.
123 C1 Chadibe Botswana
136 C2 Chadron U.S.A.
Chadyr-Lunga Moldova see Ciadîr-Lunga
77 D2 Chaek Kyrg.
65 B2 Chaeryŏng N. Korea
74 A2 Chagai Pak.
77 C3 Chaghcharān Afgh.
89 E2 Chagoda Rus. Fed.
56 I10 Chagos Archipelago is B.I.O.T.
159 E4 Chagos-Laccadive Ridge Indian Ocean
159 E4 Chagos Trench Indian Ocean
75 C2 Chaibasa India
63 B2 Chainat Thai.
63 A3 Chaiya Thai.
63 B2 Chaiyaphum Thai.
152 C3 Chajarí Arg.
119 D3 Chake Chake Tanz.
131 D2 Chakonipau, Lac Can.
150 B4 Chala Peru
121 C2 Chaláua Moz.
131 D3 Chaleur Bay inlet Can.
74 B2 Chalisgaon India
111 C3 Chalki i. Greece
111 B3 Chalkida Greece
143 C3 Chalk Mountains U.S.A.

65 B2	Chinhae S. Korea
121 C2	Chinhoyi Zimbabwe
	Chini India see Kalpa
	Chining China see Jining
74 B1	Chiniot Pak.
144 B2	Chinipas Mex.
65 B2	Chinju S. Korea
118 C2	Chinko r. C.A.R.
142 B1	Chinle U.S.A.
71 B3	Chinmen Taiwan
	Chinnamp'o N. Korea see Namp'o
67 C3	Chino Japan
135 C4	Chino U.S.A.
104 C2	Chinon France
134 E1	Chinook U.S.A.
142 A2	Chino Valley U.S.A.
77 C2	Chinoz Uzbek.
121 C2	Chinsali Zambia
108 B1	Chioggia Italy
111 C3	Chios Greece
111 C3	Chios i. Greece
121 C2	Chipata Zambia
120 A2	Chipindo Angola
	Chipinga Zimbabwe see Chipinge
121 C3	Chipinge Zimbabwe
73 B3	Chiplun India
99 B4	Chippenham U.K.
138 A2	Chippewa Falls U.S.A.
99 C4	Chipping Norton U.K.
	Chipuriro Zimbabwe see Guruve
145 D3	Chiquimula Guat.
77 C2	Chirchiq Uzbek.
121 C3	Chiredzi Zimbabwe
142 B2	Chiricahua Peak U.S.A.
146 B4	Chiriquí, Golfo de b. Panama
65 B2	Chiri-san mt. S. Korea
146 B4	Chirripó mt. Costa Rica
121 B2	Chirundu Zimbabwe
130 C2	Chisasibi Can.
137 E1	Chisholm U.S.A.
	Chisimaio Somalia see Kismaayo
90 B2	Chişinău Moldova
87 E3	Chistopol' Rus. Fed.
69 D1	Chita Rus. Fed.
120 A2	Chitado Angola
	Chitaldrug India see Chitradurga
121 C2	Chitambo Zambia
120 B1	Chitato Angola
121 C1	Chitipa Malawi
121 C3	Chitobe Moz.
	Chitor India see Chittaurgarh
66 D2	Chitose Japan
73 B3	Chitradurga India
74 B1	Chitral Pak.
146 B4	Chitré Panama
75 D2	Chittagong Bangl.
74 B2	Chittaurgarh India
73 B3	Chittoor India
	Chittorgarh India see Chittaurgarh
121 C2	Chitungwiza Zimbabwe
120 B2	Chiume Angola
121 C2	Chivhu Zimbabwe
70 B2	Chizhou China
	Chkalov Rus. Fed. see Orenburg
114 C1	Chlef Alg.
107 D2	Chlef, Oued r. Alg.
101 F2	Chodov Czech Rep.
153 B3	Choele Choel Arg.
	Chogori Feng mt. China/Pakistan see K2
48 G4	Choiseul i. Solomon Is
144 B2	Choix Mex.
102 C1	Chojna Pol.
103 D1	Chojnice Pol.
117 B3	Ch'ok'ē Eth.
	Chokue Moz. see Chókwé
83 K2	Chokurdakh Rus. Fed.
121 C3	Chókwé Moz.
104 B2	Cholet France
145 C3	Cholula Mex.
120 B2	Choma Zambia
	Chomo China see Yadong
102 C1	Chomutov Czech Rep.
83 I2	Chona r. Rus. Fed.
65 B2	Ch'ŏnan S. Korea
58 A2	Chon Buri Thai.
150 A3	Chone Ecuador
	Chong'an China see Wuyishan
65 B1	Ch'ŏngjin N. Korea
65 B2	Chŏngju N. Korea
65 B2	Ch'ŏngju S. Korea
65 B2	Chŏngp'yŏng N. Korea
70 A3	Chongqing China
70 A2	Chongqing mun. China
65 B2	Chŏngŭp S. Korea
121 B2	Chongwe Zambia
71 A3	Chongzuo China
65 B2	Chŏnju S. Korea
153 A4	Chonos, Archipiélago de los is Chile
154 B3	Chopimzinho Brazil
111 B3	Chora Sfakion Greece
98 B3	Chorley U.K.
91 C2	Chornobay Ukr.
90 C1	Chornobyl' Ukr.
91 C2	Chornomors'ke Ukr.
90 B2	Chortkiv Ukr.
65 B2	Ch'ŏrwŏn S. Korea
65 B1	Ch'osan N. Korea
67 D3	Chōshi Japan
153 A3	Chos Malal Arg.
103 D1	Choszczno Pol.
134 D1	Choteau U.S.A.
114 A2	Choûm Maur.
69 D1	Choybalsan Mongolia
69 D1	Choyr Mongolia
54 B2	Christchurch N.Z.
99 C4	Christchurch U.K.
127 H2	Christian, Cape Can.
123 C2	Christiana S. Africa
	Christianshåb Greenland see Qasigiannguit
54 A2	Christina, Mount N.Z.
58 B3	Christmas Island terr. Indian Ocean
111 B2	Chrysoupoli Greece
	Chu Kazakh. see Shu
	Chubarovka Ukr. see Polohy
153 B4	Chubut r. Arg.
89 F3	Chuchkovo Rus. Fed.
90 B1	Chudniv Ukr.
89 D2	Chudovo Rus. Fed.
126 C2	Chugach Mountains U.S.A.
67 B4	Chūgoku-sanchi mts Japan
	Chuguchak China see Tacheng
66 B2	Chuguyevka Rus. Fed.
91 D2	Chuhuyiv Ukr.
	Chukchi Peninsula pen. Rus. Fed. see Chukotskiy Poluostrov
160 J3	Chukchi Sea sea Rus. Fed./U.S.A.
89 F2	Chukhloma Rus. Fed.
83 N2	Chukotskiy Poluostrov pen. Rus. Fed.
	Chulaktau Kazakh. see Karatau
135 C4	Chula Vista U.S.A.
82 G3	Chulym Rus. Fed.
152 B2	Chumbicha Arg.
83 K3	Chumikan Rus. Fed.
63 A2	Chumphon Thai.
65 B2	Ch'unch'ŏn S. Korea
	Chungking China see Chongqing
	Ch'ungmu S. Korea see T'ongyŏng
71 C3	Chungyang Shanmo mts Taiwan
83 H2	Chunya r. Rus. Fed.
119 D3	Chunya Tanz.
150 B4	Chuquibamba Peru
152 B2	Chuquicamata Chile
105 D2	Chur Switz.
62 A1	Churachandpur India
83 J2	Churapcha Rus. Fed.
129 E2	Churchill Can.
129 E2	Churchill r. Man. Can.
131 D2	Churchill r. Nfld. and Lab. Can.
129 E2	Churchill, Cape Can.
131 D2	Churchill Falls Can.
129 D2	Churchill Lake Can.
74 B2	Churu India
63 B2	Chu Sê Vietnam
142 B1	Chuska Mountains U.S.A.
86 E3	Chusovoy Rus. Fed.
131 C3	Chute-des-Passes Can.
48 G3	Chuuk is Micronesia
62 B1	Chuxiong China
91 C2	Chyhyryn Ukr.
	Chymyshliya Moldova see Cimişlia
	Ciadâr-Lunga Moldova see Ciadîr-Lunga
90 B2	Ciadîr-Lunga Moldova
60 B2	Ciamis Indon.
60 B2	Cianjur Indon.
154 B2	Cianorte Brazil
142 A2	Cibuta, Sierra mt. Mex.
80 B1	Cide Turkey
103 E1	Ciechanów Pol.
146 C2	Ciego de Ávila Cuba
147 C3	Ciénaga Col.
146 B2	Cienfuegos Cuba
107 C2	Cieza Spain
106 C2	Cigüela r. Spain
80 B2	Cihanbeyli Turkey
144 B3	Cihuatlán Mex.
106 C2	Cíjara, Embalse de resr Spain
109 C2	Çikës, Maja e mt. Albania
60 B2	Cilacap Indon.
	Cill Airne Ireland see Killarney
	Cill Chainnigh Ireland see Kilkenny
143 C1	Cimarron r. U.S.A.
90 B2	Cimişlia Moldova
108 B2	Cimone, Monte mt. Italy
	Cîmpina Romania see Câmpina
	Cîmpulung Romania see Câmpulung
60 B2	Cina, Tanjung c. Indon.
138 C3	Cincinnati U.S.A.
	Cinco de Outubro Angola see Xá-Muteba
111 C3	Çine Turkey
100 B2	Ciney Belgium
134 C2	Cinnabar Mountain U.S.A.
145 C3	Cintalapa Mex.
71 B3	Ciping China
	Ciping China see Ciping
153 B3	Cipolletti Arg.
126 C2	Circle AK U.S.A.
136 B1	Circle MT U.S.A.
60 B2	Cirebon Indon.
99 C4	Cirencester U.K.
108 A1	Ciriè Italy
109 C3	Cirò Marina Italy
110 B1	Cisnădie Romania
109 C2	Čitluk Bos.-Herz.
122 A3	Citrusdal S. Africa
135 B3	Citrus Heights U.S.A.
110 C1	Ciucaş, Vârful mt. Romania
145 B2	Ciudad Acuña Mex.
145 B3	Ciudad Bolívar Venez.
150 C2	Ciudad Bolívar Venez.
147 C4	Ciudad Bolivia Venez.
144 B2	Ciudad Camargo Mex.
144 A2	Ciudad Constitución Mex.
145 C3	Ciudad Cuauhtémoc Mex.
145 C3	Ciudad del Carmen Mex.
154 B3	Ciudad del Este Para.
144 B2	Ciudad Delicias Mex.
145 C2	Ciudad de Valles Mex.
150 C2	Ciudad Guayana Venez.
142 B3	Ciudad Guerrero Mex.
144 B3	Ciudad Guzmán Mex.
145 C3	Ciudad Hidalgo Mex.
145 C3	Ciudad Ixtepec Mex.
144 B1	Ciudad Juárez Mex.
145 C2	Ciudad Madero Mex.
145 C2	Ciudad Mante Mex.
145 C2	Ciudad Mier Mex.
144 B2	Ciudad Obregón Mex.
106 C2	Ciudad Real Spain
145 C2	Ciudad Río Bravo Mex.
106 B1	Ciudad Rodrigo Spain
	Ciudad Trujillo Dom. Rep. see Santo Domingo
145 C2	Ciudad Victoria Mex.
107 D1	Ciutadella Spain
108 B1	Cividale del Friuli Italy
108 B2	Civitanova Marche Italy
108 B2	Civitavecchia Italy
104 C2	Civray France
111 C3	Çivril Turkey
70 C2	Cixi China
99 D4	Clacton-on-Sea U.K.
128 C2	Claire, Lake Can.
105 C2	Clamecy France
140 C2	Clanton U.S.A.
122 A3	Clanwilliam S. Africa
97 C2	Clara Ireland
52 A2	Clare r. Ireland
138 C2	Clare U.S.A.
97 A2	Clare Island Ireland
139 E2	Claremont U.S.A.
97 B2	Claremorris Ireland
54 B2	Clarence N.Z.
55 B3	Clarence Island Antarctica
131 E3	Clarenville Can.
128 C2	Claresholm Can.
137 D2	Clarinda U.S.A.
144 A3	Clarión, Isla i. Mex.
123 C3	Clarkebury S. Africa
141 D2	Clark Hill Reservoir U.S.A.
138 C3	Clarksburg U.S.A.
140 B2	Clarksdale U.S.A.
134 C1	Clarks Fork r. U.S.A.
134 C1	Clarkston U.S.A.
140 B1	Clarksville AR U.S.A.
140 C1	Clarksville TN U.S.A.
154 B1	Claro r. Brazil
143 C1	Claude U.S.A.
143 C1	Clayton U.S.A.
97 B3	Clear, Cape Ireland
126 C3	Cleare, Cape U.S.A.
137 E2	Clear Lake U.S.A.
135 B3	Clear Lake l. U.S.A.
128 C2	Clearwater Can.
129 C2	Clearwater r. Can.
141 D3	Clearwater U.S.A.
134 C1	Clearwater r. U.S.A.
143 D2	Cleburne U.S.A.
101 E1	Clenze Ger.
51 D2	Clermont Austr.
105 C2	Clermont-Ferrand France
100 C2	Clervaux Lux.
52 A2	Cleve Austr.
140 B2	Cleveland MS U.S.A.
138 C2	Cleveland OH U.S.A.
141 D1	Cleveland TN U.S.A.
134 D1	Cleveland, Mount U.S.A.
154 B3	Clevelândia Brazil
97 B2	Clew Bay Ireland
141 D3	Clewiston U.S.A.
97 A2	Clifden Ireland
53 D1	Clifton Austr.
142 B2	Clifton U.S.A.
142 B1	Clines Corners U.S.A.
128 B2	Clinton Can.
137 E2	Clinton IA U.S.A.
137 E3	Clinton MO U.S.A.
143 D1	Clinton OK U.S.A.
125 H8	Clipperton, Île terr. N. Pacific Ocean
96 A2	Clisham h. U.K.
98 B3	Clitheroe U.K.
97 B3	Clonakilty Ireland
51 D2	Cloncurry Austr.
97 C1	Clones Ireland
97 C2	Clonmel Ireland
101 D1	Cloppenburg Ger.
137 E1	Cloquet U.S.A.
136 B2	Cloud Peak U.S.A.
135 C3	Clovis CA U.S.A.
143 C2	Clovis NM U.S.A.
	Cluain Meala Ireland see Clonmel
129 D2	Cluff Lake Mine Can.
110 B1	Cluj-Napoca Romania
51 C2	Cluny Austr.
105 D2	Cluses France
54 A3	Clutha r. N.Z.
96 B3	Clyde r. U.K.
96 B3	Clyde, Firth of est. U.K.
96 B3	Clydebank U.K.
127 H2	Clyde River Can.
144 B3	Coalcomán Mex.
128 C3	Coaldale Can.
135 C3	Coaldale U.S.A.
128 B2	Coal River Can.
150 C3	Coari Brazil
150 C3	Coari r. Brazil
141 C2	Coastal Plain U.S.A.
128 B2	Coast Mountains Can.
134 B2	Coast Ranges mts U.S.A.
96 B3	Coatbridge U.K.
129 F1	Coats Island Can.
55 C2	Coats Land reg. Antarctica
145 C3	Coatzacoalcos Mex.
146 A3	Cobán Guat.
53 C2	Cobar Austr.
97 B3	Cobh Ireland
152 B2	Cobija Bol.
	Coblenz Ger. see Koblenz
130 C3	Cobourg Can.
50 C1	Cobourg Peninsula Austr.
53 C3	Cobram Austr.
101 E2	Coburg Ger.
152 B1	Cochabamba Bol.
100 C2	Cochem Ger.
	Cochin India see Kochi
128 C2	Cochrane Alta. Can.
130 B3	Cochrane Ont. Can.
153 A4	Cochrane Chile
52 A2	Cockaleechie Austr.
52 B2	Cockburn Austr.

69 E1 Daqing China
4 A3 Dara Senegal
80 B2 Dar'ā Syria
81 D3 Dārāb Iran
15 D1 Daraj Libya
81 D2 Dārān Iran
Đaravica Kosovo see Gjeravicë
75 C2 Darbhanga India
Dardo China see Kangding
49 D3 Dar es Salaam Tanz.
17 A3 Darfur reg. Sudan
74 B1 Dargai Pak.
54 B1 Dargaville N.Z.
63 C3 Dargo Austr.
58 D1 Darhan Mongolia
60 B2 Darién, Golfo del g. Col.
Darjeeling India see Darjiling
75 C2 Darjiling India
58 C2 Darlag China
53 C1 Darling r. Austr.
53 C1 Darling Downs hills Austr.
50 A3 Darling Range hills Austr.
98 C2 Darlington U.K.
53 C2 Darlington Point Austr.
103 D1 Darłowo Pol.
101 F3 Darmstadt Ger.
115 E1 Darnah Libya
52 B2 Darnick Austr.
55 H3 Darnley, Cape Antarctica
107 C1 Daroca Spain
99 D4 Dartford U.K.
99 A4 Dartmoor hills U.K.
131 D3 Dartmouth Can.
99 B4 Dartmouth U.K.
59 D3 Daru P.N.G.
59 C2 Daruba Indon.
50 C1 Darwin Austr.
79 C2 Darwin Falkland Is
65 A1 Dārzīn Iran
65 A1 Dashiqiao China
Dashkhovuz Turkm. see Daşoguz
74 A2 Dasht r. Pak.
76 B2 Daşoguz Turkm.
61 C1 Datadian Indon.
111 C3 Datça Turkey
66 D2 Date Japan
71 B3 Datian China
70 B1 Datong China
70 B1 Datu Piang Phil.
74 B1 Daud Khel Pak.
88 B2 Daugava r. Latvia
88 C2 Daugavpils Latvia
100 C2 Daun Ger.
129 D2 Dauphin Can.
129 F7 Dauphin Lake Can.
23 B3 Davangere India
64 B3 Davao Phil.
64 B3 Davao Gulf Phil.
137 E2 Davenport U.S.A.
99 C3 Daventry U.K.
123 C2 Daveyton S. Africa
46 B4 David Panama
29 D2 Davidson Can.
26 F3 Davidson Lake Can.
35 B3 Davis U.S.A.
31 D2 Davis Inlet (abandoned) Can.
55 I3 Davis Sea Antarctica
60 P3 Davis Strait str. Can./Greenland
05 D2 Davos Switz.
88 C3 Davyd-Haradok Belarus
78 A2 Dawmat al Jandal Saudi Arabia
79 C3 Dawqah Oman
78 B3 Dawqah Saudi Arabia
26 C2 Dawson r. Austr.
41 D2 Dawson U.S.A.
28 B2 Dawson Creek Can.
28 B2 Dawsons Landing Can.
68 C2 Dawu China
Dawukou China see Shizuishan
79 C2 Dawwah Oman
04 B3 Dax France
Daxian China see Dazhou
68 C2 Da Xueshan mts China
52 B3 Daylesford Austr.
Dayong China see Zhangjiajie
81 C2 Dayr az Zawr Syria
38 C3 Dayton U.S.A.
41 D3 Daytona Beach U.S.A.
71 B3 Dayu China

Da Yunhe canal China see Jinghang Yunhe
79 C2 Dayyer Iran
70 A2 Dazhou China
122 B3 De Aar S. Africa
141 D3 Deadman Bay U.S.A.
80 B2 Dead Sea salt l. Asia
99 D4 Deal U.K.
71 B3 De'an China
152 B3 Deán Funes Arg.
128 B2 Dease Lake Can.
126 E2 Dease Strait Can.
135 C3 Death Valley depr. U.S.A.
104 C2 Deauville France
61 C1 Debak Malaysia
111 B2 Debar Macedonia
103 E1 Dębica Pol.
103 E1 Dęblin Pol.
114 B3 Débo, Lac l. Mali
103 E2 Debrecen Hungary
117 B3 Debre Markos Eth.
117 B3 Debre Sīna Eth.
117 B3 Debre Tabor Eth.
117 B4 Debre Zeyit Eth.
140 C2 Decatur AL U.S.A.
138 B3 Decatur IL U.S.A.
73 B3 Deccan plat. India
53 D1 Deception Bay Austr.
71 A3 Dechang China
102 C1 Děčín Czech Rep.
137 E2 Decorah U.S.A.
154 C2 Dedo de Deus mt. Brazil
88 C2 Dedovichi Rus. Fed.
121 C2 Dedza Malawi
99 B3 Dee r. England/Wales U.K.
96 C2 Dee r. Scotland U.K.
130 C3 Deep River Can.
53 D1 Deepwater Austr.
131 E3 Deer Lake Can.
134 D1 Deer Lodge U.S.A.
138 C2 Defiance U.S.A.
68 C2 Dêgê China
117 C4 Degeh Bur Eth.
139 F1 Dégelis Can.
102 C2 Deggendorf Ger.
91 E2 Degtevo Rus. Fed.
81 C2 Dehlorān Iran
74 B1 Dehra Dun India
75 C2 Dehri India
69 C2 Dehui China
100 A2 Deinze Belgium
110 B1 Dej Romania
138 B2 De Kalb U.S.A.
116 B3 Dekemhare Eritrea
118 C3 Dekese Dem. Rep. Congo
118 B2 Dékoa C.A.R.
141 D3 De Land U.S.A.
135 C3 Delano U.S.A.
135 D3 Delano Peak U.S.A.
48 I3 Delap-Uliga-Djarrit Marshall Is
75 C2 Delārām Afgh.
123 C2 Delareyville S. Africa
129 D2 Delaronde Lake Can.
138 C2 Delaware U.S.A.
139 D3 Delaware r. U.S.A.
139 D3 Delaware state U.S.A.
139 D3 Delaware Bay U.S.A.
53 C3 Delegate Austr.
118 C2 Délembé C.A.R.
105 D2 Delémont Switz.
100 B1 Delft Neth.
100 C1 Delfzijl Neth.
121 D2 Delgado, Cabo c. Moz.
68 C1 Delgerhaan Mongolia
68 C2 Delhi China
74 B2 Delhi India
60 B2 Deli i. Indon.
128 B1 Déline Can.
Delingha China see Delhi
101 E2 Delitzsch Ger.
107 D2 Dellys Alg.
135 C4 Del Mar U.S.A.
101 D1 Delmenhorst Ger.
109 C1 Delnice Croatia
136 B3 Del Norte U.S.A.
83 L1 De-Longa, Ostrova is Rus. Fed.
De Long Islands is Rus. Fed. see De-Longa, Ostrova
De Long Strait sea chan. Rus. Fed. see Longa, Proliv
129 D3 Deloraine Can.
111 B3 Delphi tourist site Greece
141 D3 Delray Beach U.S.A.
143 C3 Del Rio U.S.A.

136 B3 Delta CO U.S.A.
135 D3 Delta UT U.S.A.
126 C2 Delta Junction U.S.A.
109 D3 Delvinë Albania
106 C1 Demanda, Sierra de la mts Spain
Demavend mt. Iran see Damāvand, Kūh-e
118 C3 Demba Dem. Rep. Congo
119 D1 Dembech'a Eth.
117 B4 Dembī Dolo Eth.
Demerara Guyana see Georgetown
91 C3 Demerdzhi mt. Ukr.
89 D2 Demidov Rus. Fed.
142 B2 Deming U.S.A.
111 C3 Demirci Turkey
111 C2 Demirköy Turkey
102 C1 Demmin Ger.
140 C2 Demopolis U.S.A.
61 C2 Dempo, Gunung vol. Indon.
89 D2 Demyansk Rus. Fed.
122 B3 De Naawte S. Africa
117 C3 Denakil reg. Africa
98 B3 Denbigh U.K.
100 B1 Den Burg Neth.
62 B2 Den Chai Thai.
60 B2 Dendang Indon.
100 B2 Dendermonde Belgium
100 B2 Dendre r. Belgium
Dengjiabu China see Yujiang
70 A1 Dengkou China
Dengxian China see Dengzhou
70 B2 Dengzhou China
Dengzhou China see Penglai
Den Haag Neth. see The Hague
100 B1 Denham Austr.
100 B1 Den Helder Neth.
107 D2 Dénia Spain
52 B3 Deniliquin Austr.
134 C2 Denio U.S.A.
137 D2 Denison IA U.S.A.
143 D2 Denison TX U.S.A.
111 C3 Denizli Turkey
53 D2 Denman Austr.
50 A3 Denmark Austr.
93 E4 Denmark country Europe
84 B2 Denmark Strait Greenland/Iceland
77 C3 Denov Uzbek.
61 C2 Denpasar Indon.
143 D2 Denton U.S.A.
50 A3 D'Entrecasteaux, Point Austr.
59 E3 D'Entrecasteaux Islands P.N.G.
141 D2 Dentsville U.S.A.
136 B3 Denver U.S.A.
75 C2 Deogarh Orissa India
74 B2 Deogarh Rajasthan India
75 C2 Deoghar India
138 B2 De Pere U.S.A.
83 K2 Deputatskiy Rus. Fed.
62 A1 Dêqên China
140 B2 De Queen U.S.A.
74 A2 Dera Bugti Pak.
74 B1 Dera Ghazi Khan Pak.
74 B1 Dera Ismail Khan Pak.
87 D4 Derbent Rus. Fed.
50 B1 Derby Austr.
99 C3 Derby U.K.
137 D3 Derby U.S.A.
99 D3 Dereham U.K.
97 B2 Derg, Lough l. Ireland
91 D1 Derhachi Ukr.
140 B2 De Ridder U.S.A.
91 D2 Derkul r. Rus. Fed./Ukr.
75 B1 Dêrub China
116 B3 Derudeb Sudan
122 B3 De Rust S. Africa
109 C2 Derventa Bos.-Herz.
98 C3 Derwent r. England U.K.
98 C3 Derwent r. England U.K.
98 B2 Derwent Water l. U.K.
77 C1 Derzhavinsk Kazakh.
Derzhavinskiy Kazakh. see Derzhavinsk
152 B1 Desaguadero r. Bol.
49 M5 Désappointement, Îles du is Fr. Polynesia
129 D2 Deschambault Lake Can.
134 B1 Deschutes r. U.S.A.
117 B3 Desē Eth.
153 B4 Deseado Arg.

153 B4 Deseado r. Arg.
142 A2 Desemboque Mex.
137 E2 Des Moines U.S.A.
137 E2 Des Moines r. U.S.A.
91 C1 Desna r. Rus. Fed./Ukr.
89 D3 Desnogorsk Rus. Fed.
101 F2 Dessau Ger.
Dessye Eth. see Desē
128 A1 Destruction Bay Can.
149 C4 Desventuradas, Islas is S. Pacific Ocean
128 C1 Detah Can.
120 B2 Dete Zimbabwe
101 D2 Detmold Ger.
138 C2 Detroit U.S.A.
137 D1 Detroit Lakes U.S.A.
Dett Zimbabwe see Dete
100 B2 Deurne Neth.
110 B1 Deva Romania
100 C1 Deventer Neth.
96 C2 Deveron r. U.K.
103 D2 Devět skal h. Czech Rep.
137 D1 Devil's Lake U.S.A.
128 A2 Devil's Paw mt. U.S.A.
99 C4 Devizes U.K.
74 B2 Devli India
110 C2 Devnya Bulg.
128 C2 Devon Can.
126 F1 Devon Island Can.
51 D4 Devonport Austr.
74 B2 Dewas India
137 F3 Dexter U.S.A.
70 A2 Deyang China
59 D3 Deyong, Tanjung pt Indon.
81 C2 Dezfūl Iran
70 B2 Dezhou China
79 C2 Dhahran Saudi Arabia
75 D2 Dhaka Bangl.
78 B3 Dhamār Yemen
75 C2 Dhamtari India
75 C2 Dhanbad India
74 B2 Dhandhuka India
75 C2 Dhankuta Nepal
74 B2 Dhar India
75 D2 Dharmanagar India
73 B3 Dharmapuri India
75 C2 Dharmjaygarh India
114 B3 Dhar Oualâta hills Maur.
114 B3 Dhar Tîchît hills Maur.
73 B3 Dharwad India
Dharwar India see Dharwad
74 B2 Dhasa India
75 C2 Dhaulagiri mt. Nepal
78 B3 Dhubāb Yemen
74 B2 Dhule India
Dhulia India see Dhule
117 C4 Dhuusa Marreeb Somalia
144 A1 Diablo, Picacho del mt. Mex.
142 B2 Diablo Plateau U.S.A.
121 C2 Diaca Moz.
51 C2 Diamantina watercourse Austr.
155 D1 Diamantina Brazil
151 E4 Diamantina, Chapada plat. Brazil
159 F6 Diamantina Deep sea feature Indian Ocean
151 D4 Diamantino Mato Grosso Brazil
154 B1 Diamantino Mato Grosso Brazil
71 B3 Dianbai China
151 E4 Dianópolis Brazil
114 B4 Dianra Côte d'Ivoire
114 C3 Diapaga Burkina
79 C2 Dibā al Ḩiṣn U.A.E.
79 C2 Dibab Oman
118 C3 Dibaya Dem. Rep. Congo
122 B2 Dibeng S. Africa
72 D2 Dibrugarh India
136 C1 Dickinson U.S.A.
140 C1 Dickson U.S.A.
Dicle r. Turkey see Tigris
105 D2 Die France
Diedenhofen France see Thionville
114 B3 Diéma Mali
101 D2 Diemel r. Ger.
62 B1 Điên Biên Phu Vietnam
62 B2 Điên Châu Vietnam
101 D1 Diepholz Ger.
104 C2 Dieppe France
100 B2 Diest Belgium
115 D3 Diffa Niger

E

El Asnam Alg. *see* Chlef
111 B3 Elassona Greece
    Elat Israel *see* Eilat
80 B2 Elazığ Turkey
108 B2 Elba, Isola d' *i.* Italy
150 B2 El Banco Col.
142 B2 El Barreal *l.* Mex.
109 D2 Elbasan Albania
150 C2 El Baúl Venez.
114 C1 El Bayadh Alg.
101 D1 Elbe *r.* Ger.
136 B3 Elbert, Mount U.S.A.
141 D2 Elberton U.S.A.
104 C2 Elbeuf France
80 B2 Elbistan Turkey
103 D1 Elbląg Pol.
153 A4 El Bolsón Arg.
87 D4 El'brus *mt.* Rus. Fed.
81 C2 Elburz Mountains *mts* Iran
150 C2 El Callao Venez.
143 D3 El Campo U.S.A.
135 C4 El Centro U.S.A.
152 B1 El Cerro Bol.
107 C2 Elche-Elx Spain
145 C3 El Chichónal *vol.* Mex.
107 C2 Elda Spain
119 D2 Eldama Ravine Kenya
137 E3 Eldon U.S.A.
154 B3 Eldorado Arg.
154 C2 Eldorado Brazil
144 B2 El Dorado Mex.
140 B2 El Dorado AR U.S.A.
137 D3 El Dorado KS U.S.A.
134 D1 Electric Peak U.S.A.
114 B2 El Eglab *plat.* Alg.
106 C2 El Ejido Spain
89 E2 Elektrostal' Rus. Fed.
150 B3 El Encanto Col.
146 C2 Eleuthera *i.* Bahamas
116 A3 El Fasher Sudan
144 B2 El Fuerte Mex.
117 A3 El Fula Sudan
116 A3 El Geneina Sudan
116 B3 El Geteina Sudan
96 C2 Elgin U.K.
138 B2 Elgin U.S.A.
83 K2 El'ginskiy Rus. Fed.
    El Gîza Egypt *see* Giza
115 C1 El Goléa Alg.
144 A1 El Golfo de Santa Clara Mex.
119 D2 Elgon, Mount Kenya/Uganda
108 A3 El Hadjar Alg.
114 A2 El Hammâmi *reg.* Maur.
114 B2 El Hank *esc.* Mali/Maur.
114 A2 El Hierro *i.* Islas Canarias
145 C2 El Higo Mex.
114 C2 El Homr Alg.
    Elichpur India *see* Achalpur
126 B2 Élisabethville
    Dem. Rep. Congo *see* Lubumbashi
    El Iskandarîya Egypt *see* Alexandria
87 D4 Elista Rus. Fed.
139 E2 Elizabeth U.S.A.
141 E1 Elizabeth City U.S.A.
138 B3 Elizabethtown U.S.A.
114 B1 El Jadida Morocco
103 E1 Ełk Pol.
108 A3 El Kala Alg.
88 C2 Elkas kalns *h.* Latvia
143 D1 Elk City U.S.A.
114 B1 El Kelaâ des Srarhna Morocco
128 C2 Elkford Can.
138 B2 Elkhart U.S.A.
    El Khartûm Sudan *see* Khartoum
110 C2 Elkhovo Bulg.
139 D3 Elkins U.S.A.
128 C3 Elko Can.
134 C2 Elko U.S.A.
129 C2 Elk Point U.S.A.
137 E1 Elk River U.S.A.
126 F1 Ellef Ringnes Island Can.
137 D1 Ellendale U.S.A.
134 B1 Ellensburg U.S.A.
54 B2 Ellesmere, Lake N.Z.
127 G1 Ellesmere Island Can.
98 B3 Ellesmere Port U.K.
126 F2 Ellice *r.* Can.
    Ellice Islands *country* S. Pacific Ocean *see* Tuvalu
123 C3 Elliotdale S. Africa
138 C1 Elliot Lake Can.

96 C2 Ellon U.K.
139 F2 Ellsworth U.S.A.
55 R2 Ellsworth Mountains Antarctica
111 C3 Elmalı Turkey
115 C1 El Meghaïer Alg.
139 D2 Elmira U.S.A.
107 C2 El Moral Spain
101 D1 Elmshorn Ger.
117 A3 El Muglad Sudan
150 B2 El Nevado, Cerro *mt.* Col.
64 A2 El Nido Phil.
116 B3 El Obeid Sudan
144 B2 El Oro Mex.
115 C1 El Oued Alg.
142 A2 Eloy U.S.A.
    El Paso U.S.A. *see* Derby
142 B2 El Paso U.S.A.
145 D3 El Pinalón, Cerro *mt.* Guat.
144 B1 El Porvenir Mex.
107 D1 El Prat de Llobregat Spain
146 B3 El Progreso Hond.
106 B2 El Puerto de Santa María Spain
    El Qâhira Egypt *see* Cairo
    El Quds Israel/West Bank *see* Jerusalem
143 D1 El Reno U.S.A.
128 A1 Elsa Can.
145 B2 El Salado Mex.
144 B2 El Salto Mex.
146 B3 El Salvador *country* Central America
152 B2 El Salvador Chile
145 B2 El Salvador Mex.
142 B3 El Sauz Mex.
144 A1 El Socorro Mex.
142 B3 El Sueco Mex.
    El Suweis Egypt *see* Suez
108 A3 El Tarf Alg.
106 B1 El Teleno *mt.* Spain
145 C2 El Temascal Mex.
150 C2 El Tigre Venez.
147 D4 El Tocuyo Venez.
    El Uqsur Egypt *see* Luxor
73 C3 Eluru India
88 C2 Elva Estonia
106 B2 Elvas Port.
93 F3 Elverum Norway
119 E2 El Wak Kenya
99 D3 Ely U.K.
137 E1 Ely MN U.S.A.
135 D3 Ely NV U.S.A.
81 D2 Emāmrūd Iran
93 G4 Emån *r.* Sweden
76 B2 Emba Kazakh.
123 C2 Embalenhle S. Africa
118 B2 Embondo Dem. Rep. Congo
154 C1 Emborcação, Represa de *resr* Brazil
119 D3 Embu Kenya
100 C1 Emden Ger.
51 D2 Emerald Austr.
129 E3 Emerson Can.
111 C3 Emet Turkey
123 D2 eMgwenya S. Africa
115 D3 Emi Koussi *mt.* Chad
145 C3 Emiliano Zapata Mex.
110 C2 Emine, Nos *pt* Bulg.
80 B2 Emirdağ Turkey
123 D2 eMjindini S. Africa
88 B2 Emmaste Estonia
100 B1 Emmeloord Neth.
100 C2 Emmelshausen Ger.
100 C1 Emmen Neth.
143 C3 Emory Peak U.S.A.
144 A2 Empalme Mex.
123 D2 Empangeni S. Africa
156 D2 Emperor Trough N. Pacific Ocean
108 B2 Empoli Italy
111 C3 Emponas Greece
137 D3 Emporia KS U.S.A.
139 D3 Emporia VA U.S.A.
    Empty Quarter *des.* Saudi Arabia *see* Rub' al Khālī
100 C1 Ems *r.* Ger.
100 C1 Emsdetten Ger.
88 C2 Emumägi *h.* Estonia
123 C2 eMzinoni S. Africa
59 D3 Enarotali Indon.
144 B2 Encarnación Mex.
152 C2 Encarnación Para.
155 D1 Encruzilhada Brazil
61 D2 Ende Indon.
55 G2 Enderby Land *reg.* Antarctica

126 B2 Endicott Mountains U.S.A.
50 A2 Eneabba Austr.
91 C2 Enerhodar Ukr.
111 C2 Enez Turkey
97 C2 Enfield Ireland
87 D3 Engel's Rus. Fed.
60 B2 Enggano *i.* Indon.
99 C3 England *admin. div.* U.K.
130 C3 Englehart Can.
141 D3 Englewood U.S.A.
130 A2 English *r.* Can.
    English Bazar India *see* Ingraj Bazar
95 C4 English Channel France/U.K.
143 D1 Enid U.S.A.
    Enkeldoorn Zimbabwe *see* Chivhu
100 B1 Enkhuizen Neth.
93 G4 Enköping Sweden
108 B3 Enna Italy
129 D1 Ennadai Lake Can.
116 A3 En Nahud Sudan
115 E3 Ennedi, Massif *mts* Chad
53 C1 Enngonia Austr.
97 B2 Ennis Ireland
143 D2 Ennis U.S.A.
97 C2 Enniscorthy Ireland
97 C1 Enniskerry Ireland
97 C1 Enniskillen U.K.
97 B2 Ennistymon Ireland
102 C2 Enns Austria
102 C2 Enns *r.* Austria
92 J3 Eno Fin.
92 H2 Enontekiö Fin.
53 C3 Ensay Austr.
100 C1 Enschede Neth.
144 A1 Ensenada Mex.
70 A2 Enshi China
119 D2 Entebbe Uganda
128 C1 Enterprise Can.
140 C2 Enterprise AL U.S.A.
134 C1 Enterprise OR U.S.A.
152 B2 Entre Ríos Bol.
106 B2 Entroncamento Port.
115 C4 Enugu Nigeria
150 B3 Envira Brazil
118 B2 Epéna Congo
105 C2 Épernay France
135 D3 Ephraim U.S.A.
134 C1 Ephrata U.S.A.
105 D2 Épinal France
99 D4 Epping U.K.
99 C4 Epsom U.K.
118 A2 Equatorial Guinea *country* Africa
104 B2 Équeurdreville-Hainneville France
101 D3 Erbach Ger.
101 F3 Erbendorf Ger.
100 C3 Erbeskopf *h.* Ger.
81 C2 Erciş Turkey
80 B2 Erciyes Dağı *mt.* Turkey
103 D2 Érd Hungary
    Erdaobaihe China *see* Baihe
65 B1 Erdao Jiang *r.* China
111 C2 Erdek Turkey
80 B2 Erdemli Turkey
154 B2 Eré, Campos *hills* Brazil
55 M2 Erebus, Mount *vol.* Antarctica
152 C2 Erechim Brazil
69 D1 Ereentsav Mongolia
80 B2 Ereğli Konya Turkey
80 B1 Ereğli Zonguldak Turkey
69 D2 Erenhot China
    Erevan Armenia *see* Yerevan
101 E2 Erfurt Ger.
80 B2 Ergani Turkey
114 B2 'Erg Chech *des.* Alg./Mali
111 C2 Ergene *r.* Turkey
78 A3 Erheib Sudan
96 B2 Ericht, Loch *l.* U.K.
138 C2 Erie U.S.A.
138 C2 Erie, Lake Can./U.S.A.
66 D2 Erimo-misaki *c.* Japan
116 B3 Eritrea *country* Africa
101 E3 Erlangen Ger.
50 C2 Erldunda Austr.
123 C2 Ermelo S. Africa
80 B2 Ermenek Turkey
111 B3 Ermoupoli Greece
73 B4 Ernakulam India
97 B1 Erne *r.* Ireland/U.K.
73 B3 Erode India
100 B2 Erp Neth.
114 B1 Er Rachidia Morocco
116 B3 Er Rahad Sudan

97 B1 Errigal *h.* Ireland
97 A1 Erris Head *hd* Ireland
48 H5 Erromango *i.* Vanuatu
109 D2 Ersekë Albania
62 B1 Ertan Reservoir China
89 F3 Ertil' Rus. Fed.
101 D2 Erwitte Ger.
101 F2 Erzgebirge *mts* Czech Rep./Ger.
80 B2 Erzincan Turkey
81 C2 Erzurum Turkey
66 D2 Esashi Japan
93 E4 Esbjerg Denmark
135 D3 Escalante U.S.A.
135 D3 Escalante Desert U.S.A.
144 B2 Escalón Mex.
138 B1 Escanaba U.S.A.
145 C3 Escárcega Mex.
107 C1 Escatrón Spain
100 A2 Escaut *r.* Belgium/France
101 E1 Eschede Ger.
100 B3 Esch-sur-Alzette Lux.
101 E2 Eschwege Ger.
100 C2 Eschweiler Ger.
135 C4 Escondido U.S.A.
144 B2 Escuinapa Mex.
111 C3 Eşen Turkey
81 D2 Eşfahān Iran
123 D2 Eshowe S. Africa
123 D2 Esikhawini S. Africa
98 B2 Esk *r.* U.K.
131 D2 Esker Can.
92 □C2 Eskifjörður Iceland
93 G4 Eskilstuna Sweden
    Eskimo Point Can. *see* Arviat
111 D3 Eskişehir Turkey
81 C2 Eslāmābād-e Gharb Iran
111 C3 Esler Dağı *mt.* Turkey
111 C3 Eşme Turkey
146 C2 Esmeralda Cuba
150 B2 Esmeraldas Ecuador
107 D2 Es Mercadal Spain
79 D2 Espakeh Iran
105 C3 Espalion France
130 B3 Espanola Can.
142 B1 Espanola U.S.A.
50 B3 Esperance Austr.
153 A5 Esperanza Arg.
144 B2 Esperanza Mex.
106 B2 Espichel, Cabo *c.* Port.
154 B3 Espigão, Serra do *mts* Brazil
143 C3 Espinazo Mex.
155 D1 Espinhaço, Serra do *mts* Brazil
151 E4 Espinosa Brazil
155 D1 Espírito Santo *state* Brazil
48 H5 Espíritu Santo *i.* Vanuatu
144 A2 Espíritu Santo, Isla *i.* Mex.
93 H3 Espoo Fin.
153 A4 Esquel Arg.
114 B1 Essaouira Morocco
114 A2 Es Semara Western Sahara
100 C2 Essen Ger.
150 D2 Essequibo *r.* Guyana
139 E2 Essex Junction U.S.A.
83 L3 Esso Rus. Fed.
106 B1 Estaca de Bares, Punta de *pt* Spain
153 B5 Estados, Isla de los *i.* Arg.
81 D3 Eştahbān Iran
151 F4 Estância Brazil
123 C3 Estcourt S. Africa
107 C1 Estella Spain
106 B2 Estepona Spain
106 C1 Esteras de Medinaceli Spain
129 D2 Esterhazy Can.
152 B2 Esteros Para.
136 C2 Estes Park U.S.A.
129 D3 Estevan Can.
129 E2 Estherville U.S.A.
129 D2 Eston Can.
88 C2 Estonia *country* Europe
    Estonskaya S.S.R. *see* Estonia Europe *see* Estonia
106 B1 Estrela, Serra da *mts* Port.
106 B2 Estremoz Port.
52 A1 Etadunna Austr.
104 C2 Étampes France
104 C2 Étaples France
75 B2 Etawah India
123 D2 Ethandakukhanya S. Africa
122 B2 E'Thembini S. Africa
117 B4 Ethiopia *country* Africa
    Etna, Monte *vol.* Italy *see* Etna, Mount

Florence

Ghisonaccia

105 D3	Ghisonaccia France	
74 B2	Ghotaru India	
74 A2	Ghotki Pak.	
75 C2	Ghugri r. India	
76 C3	Ghūriān Afgh.	
91 E3	Giaginskaya Rus. Fed.	
97 C1	Giant's Causeway lava field U.K.	
61 C2	Gianyar Indon.	
109 C3	Giarre Italy	
108 A1	Giaveno Italy	
122 A2	Gibeon Namibia	
106 B2	Gibraltar Gibraltar	
106 B2	Gibraltar, Strait of Morocco/Spain	
140 B2	Gibsland U.S.A.	
50 B2	Gibson Desert Austr.	
68 C1	Gichgeniyn Nuruu mts Mongolia	
117 B4	Gīdolē Eth.	
105 C2	Gien France	
101 D2	Gießen Ger.	
101 E1	Gifhorn Ger.	
128 C2	Gift Lake Can.	
67 C3	Gifu Japan	
96 B3	Gigha i. U.K.	
76 C2	G'ijduvon Uzbek.	
106 B1	Gijón-Xixón Spain	
142 A2	Gila r. U.S.A.	
142 A2	Gila Bend U.S.A.	
51 D1	Gilbert r. Austr.	
48 I4	Gilbert Islands is Kiribati	
156 D5	Gilbert Ridge Pacific Ocean	
151 E3	Gilbués Brazil	
134 D1	Gildford U.S.A.	
	Gilf Kebir Plateau plat. Egypt see Haḍabat al Jilf al Kabīr	
53 C2	Gilgandra Austr.	
74 B1	Gilgit Pak.	
74 B1	Gilgit r. Pak.	
53 C2	Gilgunnia Austr.	
129 E2	Gillam Can.	
136 B2	Gillette U.S.A.	
99 D4	Gillingham U.K.	
130 C2	Gilmour Island Can.	
135 B3	Gilroy U.S.A.	
129 E2	Gimli Can.	
64 B3	Gingoog Phil.	
117 C4	Gīnīr Eth.	
109 C2	Ginosa Italy	
109 C2	Gioia del Colle Italy	
53 C3	Gippsland reg. Austr.	
74 A2	Girdar Dhor r. Pak.	
79 D1	Gīrdī Iran	
80 B1	Giresun Turkey	
	Girgenti Italy see Agrigento	
53 C2	Girilambone Austr.	
	Giron Sweden see Kiruna	
107 D1	Girona Spain	
104 B2	Gironde est. France	
53 C2	Girral Austr.	
96 B3	Girvan U.K.	
54 C1	Gisborne N.Z.	
93 F4	Gislaved Sweden	
119 C3	Gitarama Rwanda	
119 C3	Gitega Burundi	
	Giuba r. Somalia see Jubba	
108 B2	Giulianova Italy	
110 C2	Giurgiu Romania	
110 C1	Giuvala, Pasul pass Romania	
105 C2	Givors France	
123 D1	Giyani S. Africa	
119 D2	Giyon Eth.	
116 B2	Giza Egypt	
103 E1	Giżycko Pol.	
109 D2	Gjakovë Kosovo	
109 D2	Gjeravicë mt. Kosovo	
110 D2	Gjilan Kosovo	
109 D2	Gjirokastër Albania	
126 F2	Gjoa Haven Can.	
93 F3	Gjøvik Norway	
131 E3	Glace Bay Can.	
134 B1	Glacier Peak vol. U.S.A.	
143 E2	Gladewater U.S.A.	
51 E2	Gladstone Qld Austr.	
52 A2	Gladstone S.A. Austr.	
92 □A2	Gláma mts Iceland	
109 C2	Glamoč Bos.-Herz.	
100 C3	Glan r. Ger.	
97 B2	Glanaruddery Mountains hills Ireland	
96 B3	Glasgow U.K.	
138 B3	Glasgow KY U.S.A.	
136 B1	Glasgow MT U.S.A.	
99 B4	Glastonbury U.K.	
101 F2	Glauchau Ger.	
86 E3	Glazov Rus. Fed.	
89 E3	Glazunovka Rus. Fed.	
123 D2	Glencoe S. Africa	
96 B2	Glen Coe val. U.K.	
142 A2	Glendale AZ U.S.A.	
138 B2	Glendale WI U.S.A.	
53 D2	Glen Davis Austr.	
51 D2	Glenden Austr.	
136 C1	Glendive U.S.A.	
52 B3	Glenelg r. Austr.	
96 B2	Glenfinnan U.K.	
53 D1	Glen Innes Austr.	
96 B2	Glen More val. U.K.	
53 C1	Glenmorgan Austr.	
126 C2	Glennallen U.S.A.	
134 C2	Glenns Ferry U.S.A.	
136 B2	Glenrock U.S.A.	
96 C2	Glenrothes U.K.	
139 E2	Glens Falls U.S.A.	
96 C2	Glen Shee val. U.K.	
97 B1	Glenties Ireland	
142 B2	Glenwood U.S.A.	
136 B3	Glenwood Springs U.S.A.	
101 E1	Glinde Ger.	
103 D1	Gliwice Pol.	
142 A2	Globe U.S.A.	
103 D1	Głogów Pol.	
92 F2	Glomfjord Norway	
93 F4	Glomma r. Norway	
53 D2	Gloucester Austr.	
99 B4	Gloucester U.K.	
131 E3	Glovertown Can.	
101 F1	Glöwen Ger.	
77 E1	Glubokoye Kazakh.	
101 D1	Glückstadt Ger.	
103 C2	Gmünd Austria	
102 C2	Gmunden Austria	
101 D1	Gnarrenburg Ger.	
103 D1	Gniezno Pol.	
	Gnjilane Kosovo see Gjilan	
75 D2	Goalpara India	
96 B3	Goat Fell h. U.K.	
117 C4	Goba Eth.	
122 A1	Gobabis Namibia	
153 A4	Gobernador Gregores Arg.	
152 C2	Gobernador Virasoro Arg.	
68 D2	Gobi des. China/Mongolia	
67 C4	Gobō Japan	
100 C2	Goch Ger.	
122 A1	Gochas Namibia	
74 B3	Godavari r. India	
73 C3	Godavari, Mouths of the India	
75 C2	Godda India	
117 C4	Godē Eth.	
130 B3	Goderich Can.	
	Godhavn Greenland see Qeqertarsuaq	
74 B2	Godhra India	
129 E2	Gods r. Can.	
129 E2	Gods Lake Can.	
	Godthåb Greenland see Nuuk	
	Godwin-Austen, Mount mt. China/Pakistan see K2	
	Goedgegun Swaziland see Nhlangano	
130 C3	Goéland, Lac au l. Can.	
131 D2	Goélands, Lac aux l. Can.	
100 A2	Goes Neth.	
138 B1	Gogebic Range hills U.S.A.	
88 C1	Gogland, Ostrov i. Rus. Fed.	
	Gogra r. India see Ghaghara	
119 C2	Gogrial Sudan	
154 C1	Goiandira Brazil	
154 B1	Goianésia Brazil	
154 C1	Goiânia Brazil	
154 B1	Goiás Brazil	
154 B1	Goiás state Brazil	
154 B2	Goiatuba Brazil	
154 B2	Goio-Erê Brazil	
111 C2	Gökçeada i. Turkey	
111 C3	Gökçedağ Turkey	
121 B2	Gokwe Zimbabwe	
93 E3	Gol Norway	
62 A1	Golaghat India	
111 C2	Gölcük Turkey	
103 E1	Gołdap Pol.	
101 F1	Goldberg Ger.	
	Gold Coast country Africa see Ghana	
53 D1	Gold Coast Austr.	
114 B4	Gold Coast coastal area Ghana	
128 C2	Golden Can.	
54 B2	Golden Bay N.Z.	
134 B1	Goldendale U.S.A.	
128 B3	Golden Hinde mt. Can.	
97 B2	Golden Vale lowland Ireland	
135 C3	Goldfield U.S.A.	
128 B3	Gold River Can.	
141 E1	Goldsboro U.S.A.	
103 C1	Goleniów Pol.	
135 C4	Goleta U.S.A.	
	Golfe du St-Laurent g. Can. see St Lawrence, Gulf of	
111 C3	Gölhisar Turkey	
	Gollel Swaziland see Lavumisa	
75 D1	Golmud China	
81 D2	Golpāyegān Iran	
96 C2	Golspie U.K.	
	Golyshi Rus. Fed. see Vetluzhskiy	
119 C3	Goma Dem. Rep. Congo	
75 C2	Gomati r. India	
115 D3	Gombe Nigeria	
115 D3	Gombi Nigeria	
	Gomel' Belarus see Homyel'	
144 B2	Gómez Palacio Mex.	
81 D2	Gomīshān Iran	
75 C1	Gomo China	
147 C3	Gonaïves Haiti	
147 C3	Gonâve, Île de la i. Haiti	
81 D2	Gonbad-e Kāvūs Iran	
74 B2	Gondal India	
	Gondar Eth. see Gonder	
116 B3	Gonder Eth.	
75 C2	Gondia India	
111 C2	Gönen Turkey	
62 A1	Gongga China	
70 A3	Gongga Shan mt. China	
68 C2	Gonghe China	
115 D4	Gongola r. Nigeria	
53 C2	Gongolgon Austr.	
	Gongtang China see Damxung	
123 C3	Gonubie S. Africa	
145 C2	Gonzáles Mex.	
143 D3	Gonzales U.S.A.	
122 A3	Good Hope, Cape of S. Africa	
134 D2	Gooding U.S.A.	
136 C3	Goodland U.S.A.	
53 C1	Goodooga Austr.	
98 C2	Goole U.K.	
53 C2	Goolgowi Austr.	
52 A3	Goolwa Austr.	
53 D1	Goondiwindi Austr.	
134 B2	Goose Lake U.S.A.	
102 B2	Göppingen Ger.	
75 C2	Gorakhpur India	
109 C2	Goražde Bos.-Herz.	
111 C3	Gördes Turkey	
89 D3	Gordeyevka Rus. Fed.	
136 C2	Gordon U.S.A.	
51 D4	Gordon, Lake Austr.	
115 D4	Goré Chad	
117 B4	Gore Eth.	
54 A3	Gore N.Z.	
97 C2	Gorey Ireland	
81 D2	Gorgān Iran	
81 C1	Gori Georgia	
100 B2	Gorinchem Neth.	
108 B1	Gorizia Italy	
	Gor'kiy Rus. Fed. see Nizhniy Novgorod	
89 D2	Gor'kovskoye Vodokhranilishche resr Rus. Fed.	
103 E2	Gorlice Pol.	
103 C1	Görlitz Ger.	
	Gorna Dzhumaya Bulg. see Blagoevgrad	
109 D2	Gornji Milanovac Serbia	
109 C2	Gornji Vakuf Bos.-Herz.	
77 E1	Gorno-Altaysk Rus. Fed.	
86 F2	Gornopravdinsk Rus. Fed.	
110 C2	Gornotrakiyska Nizina lowland Bulg.	
66 D1	Gornozavodsk Rus. Fed.	
77 E1	Gornyak Rus. Fed.	
59 D3	Goroka P.N.G.	
52 B3	Goroke Austr.	
114 B3	Gorom Gorom Burkina	
121 C2	Gorongosa Moz.	
61 D1	Gorontalo Indon.	
89 E3	Goshchenskoye Rus. Fed.	
97 B3	Gorumna Island Ireland	
91 D3	Goryachiy Klyuch Rus. Fed.	
103 C1	Gorzów Wielkopolski Pol.	
59 E3	Goschen Strait P.N.G.	
53 D2	Gosford Austr.	
74 A2	Goshanak Pak.	
66 D2	Goshogawara Japan	
101 E2	Goslar Ger.	
109 C2	Gospić Croatia	
99 C4	Gosport U.K.	
111 B2	Gostivar Macedonia	
	Göteborg Sweden see Gothenburg	
101 E2	Gotha Ger.	
93 F4	Gothenburg Sweden	
136 C2	Gothenburg U.S.A.	
93 G4	Gotland i. Sweden	
111 B2	Gotse Delchev Bulg.	
93 G4	Gotska Sandön i. Sweden	
67 B4	Gōtsu Japan	
101 D2	Göttingen Ger.	
128 B2	Gott Peak Can.	
	Gottwaldow Czech Rep. see Zlín	
	Gotval'd Ukr. see Zmiyiv	
100 B1	Gouda Neth.	
114 A3	Goudiri Senegal	
115 D3	Goudoumaria Niger	
158 E7	Gough Island S. Atlantic Ocean	
130 C3	Gouin, Réservoir resr Can.	
53 C2	Goulburn Austr.	
53 D2	Goulburn r. N.S.W. Austr.	
53 C3	Goulburn r. Vic. Austr.	
114 B3	Goundam Mali	
107 D2	Gouraya Alg.	
114 B3	Gourcy Burkina	
104 C3	Gourdon France	
115 D3	Gouré Niger	
122 B3	Gourits r. S. Africa	
114 B3	Gourma-Rharous Mali	
53 C3	Gourock Range mts Austr.	
155 D1	Governador Valadares Brazil	
141 E3	Governor's Harbour Bahamas	
68 C2	Govĭ Altayn Nuruu mts Mongolia	
75 C2	Govind Ballash Pant Sagar resr India	
99 A3	Gower pen. U.K.	
152 C2	Goya Arg.	
81 C1	Göyçay Azer.	
80 B1	Göynük Turkey	
115 E3	Goz-Beïda Chad	
75 C1	Gozha Co salt l. China	
122 B3	Graaff-Reinet S. Africa	
122 A3	Graafwater S. Africa	
101 E1	Grabow Ger.	
109 C2	Gračac Croatia	
87 E3	Grachevka Rus. Fed.	
109 C2	Gradačac Bos.-Herz.	
104 B3	Gradignan France	
101 F2	Gräfenhainichen Ger.	
53 D1	Grafton Austr.	
137 D1	Grafton U.S.A.	
143 D2	Graham U.S.A.	
142 B2	Graham, Mount U.S.A.	
	Graham Bell Island i. Rus. Fed. see Greem-Bell, Ostrov	
128 A2	Graham Island Can.	
55 A3	Graham Land pen. Antarctica	
123 C3	Grahamstown S. Africa	
97 C2	Graiguenamanagh Ireland	
151 E3	Grajaú Brazil	
103 E1	Grajewo Pol.	
111 B2	Grammos mt. Greece	
96 B2	Grampian Mountains U.K.	
146 B3	Granada Nic.	
106 C2	Granada Spain	
97 C2	Granard Ireland	
139 E1	Granby Can.	
114 A2	Gran Canaria i. Islas Canarias	
152 B3	Gran Chaco reg. Arg./Para.	
136 C3	Grand r. MO U.S.A.	
136 C1	Grand r. SD U.S.A.	
146 C2	Grand Bahama i. Bahamas	
131 E3	Grand Bank Can.	
158 D2	Grand Banks of Newfoundland N. Atlantic Ocean	
	Grand Canal canal China see Jinghang Yunhe	
114 A2	Grand Canary i. Islas Canarias see Gran Canaria	
142 A1	Grand Canyon U.S.A.	
142 A1	Grand Canyon gorge U.S.A.	
146 B3	Grand Cayman i. Cayman Is	
129 C2	Grand Centre Can.	
134 C1	Grand Coulee U.S.A.	
152 B1	Grande r. Bol.	
154 B2	Grande r. Brazil	
153 B5	Grande, Bahía b. Arg.	

55 D2	Grande, Ilha i. Brazil	
150 C2	Grande, Serra mt. Brazil	
128 C2	Grande Cache Can.	
	Grande Comore i. Comoros see Ngazidja	
28 C2	Grande Prairie Can.	
14 B1	Grand Erg Occidental des. Alg.	
115 C2	Grand Erg Oriental des. Alg.	
31 D3	Grande-Rivière i. Can.	
152 B3	Grandes, Salinas salt flat Arg.	
131 D3	Grand Falls N.B. Can.	
131 E3	Grand Falls-Windsor Nfld. and Lab. Can.	
128 C3	Grand Forks Can.	
37 D1	Grand Forks U.S.A.	
138 B2	Grand Haven U.S.A.	
28 C1	Grandin, Lac l. Can.	
37 D2	Grand Island U.S.A.	
40 B3	Grand Isle U.S.A.	
36 B3	Grand Junction U.S.A.	
14 B4	Grand-Lahou Côte d'Ivoire	
31 D3	Grand Lake N.B. Can.	
31 E3	Grand Lake Nfld. and Lab. Can.	
37 E1	Grand Marais U.S.A.	
30 C3	Grand-Mère Can.	
06 B2	Grândola Port.	
29 E2	Grand Rapids Can.	
38 B2	Grand Rapids MI U.S.A.	
137 E1	Grand Rapids MN U.S.A.	
36 A2	Grand Teton mt. U.S.A.	
47 C2	Grand Turk Turks and Caicos Is	
29 D2	Grandview Can.	
34 C1	Grandview U.S.A.	
34 C1	Grangeville U.S.A.	
28 B2	Granisle Can.	
37 D2	Granite Falls U.S.A.	
34 E1	Granite Peak MT U.S.A.	
34 C2	Granite Peak NV U.S.A.	
08 B3	Granitola, Capo c. Italy	
93 F4	Gränna Sweden	
101 F1	Gransee Ger.	
99 C3	Grantham U.K.	
96 C2	Grantown-on-Spey U.K.	
42 B1	Grants U.S.A.	
34 B2	Grants Pass U.S.A.	
04 B2	Granville France	
29 D2	Granville Lake Can.	
55 D1	Grão Mogol Brazil	
23 D1	Graskop S. Africa	
05 D3	Grasse France	
50 B3	Grass Patch Austr.	
07 D1	Graus Spain	
92 F2	Gravdal Norway	
04 B2	Grave, Pointe de pt France	
29 D3	Gravelbourg Can.	
30 C3	Gravenhurst Can.	
53 D1	Gravesend Austr.	
99 D4	Gravesend U.K.	
05 D2	Gray France	
41 D1	Gray U.S.A.	
38 C2	Grayling U.S.A.	
99 D4	Grays U.K.	
03 D2	Graz Austr.	
46 C1	Great Abaco i. Bahamas	
50 B3	Great Australian Bight g. Austr.	
46 C2	Great Bahama Bank Bahamas	
54 C1	Great Barrier Island N.Z.	
51 D1	Great Barrier Reef Austr.	
35 C3	Great Basin U.S.A.	
28 C1	Great Bear Lake Can.	
93 F4	Great Belt sea chan. Denmark	
37 D3	Great Bend U.S.A.	
95 C3	Great Britain i. Europe	
63 A2	Great Coco Island Cocos Is	
53 B3	Great Dividing Range mts Austr.	
	Great Eastern Erg des. Alg. see Grand Erg Oriental	
46 C2	Greater Antilles is Caribbean Sea	
	Greater Khingan Mountains mts China see Da Hinggan Ling	
58 A3	Greater Sunda Islands Indon.	
34 D1	Great Falls U.S.A.	
23 C3	Great Fish r. S. Africa	
23 C3	Great Fish Point S. Africa	
47 C3	Great Inagua i. Bahamas	
22 B3	Great Karoo plat. S. Africa	
23 C3	Great Kei r. S. Africa	

99 B3	Great Malvern U.K.	
122 A2	Great Namaqualand reg. Namibia	
73 D4	Great Nicobar i. India	
98 B3	Great Ormes Head hd U.K.	
99 D3	Great Ouse r. U.K.	
136 C2	Great Plains U.S.A.	
119 D3	Great Rift Valley Africa	
119 D3	Great Ruaha r. Tanz.	
134 D2	Great Salt Lake U.S.A.	
134 D2	Great Salt Lake Desert U.S.A.	
116 A2	Great Sand Sea des. Egypt/Libya	
50 B2	Great Sandy Desert Austr.	
128 C1	Great Slave Lake Can.	
141 D1	Great Smoky Mountains U.S.A.	
99 A4	Great Torrington U.K.	
50 B2	Great Victoria Desert Austr.	
70 B1	Great Wall tourist site China	
	Great Western Erg des. Alg. see Grand Erg Occidental	
99 D3	Great Yarmouth U.K.	
92 □A3	Grebenkovskiy Ukr. see Hrebinka	
106 B1	Gredos, Sierra de mts Spain	
111 B3	Greece country Europe	
136 C2	Greeley U.S.A.	
82 F1	Greem-Bell, Ostrov i. Rus. Fed.	
138 F1	Green r. Can.	
138 B3	Green r. KY U.S.A.	
136 B3	Green r. WY U.S.A.	
138 B2	Green Bay U.S.A.	
138 B1	Green Bay b. U.S.A.	
138 C3	Greenbrier r. U.S.A.	
138 B3	Greencastle U.S.A.	
141 D1	Greeneville U.S.A.	
139 E2	Greenfield U.S.A.	
129 D2	Green Lake Can.	
160 L2	Greenland terr. N. America	
	Greenland Basin Arctic Ocean	
160 R2	Greenland Sea sea Greenland/Svalbard	
96 B3	Greenock U.K.	
97 C1	Greenore Ireland	
135 D3	Green River UT U.S.A.	
136 B2	Green River WY U.S.A.	
138 B3	Greensburg IN U.S.A.	
139 D2	Greensburg PA U.S.A.	
141 E2	Green Swamp U.S.A.	
142 A2	Green Valley U.S.A.	
114 B4	Greenville Liberia	
140 C2	Greenville AL U.S.A.	
139 F1	Greenville ME U.S.A.	
140 B2	Greenville MS U.S.A.	
141 E1	Greenville NC U.S.A.	
141 D2	Greenville SC U.S.A.	
143 D2	Greenville TX U.S.A.	
53 D2	Greenwell Point Austr.	
131 D3	Greenwood Can.	
140 B2	Greenwood MS U.S.A.	
141 D2	Greenwood SC U.S.A.	
50 B2	Gregory, Lake imp. l. Austr.	
51 D1	Gregory Range hills Austr.	
102 C1	Greifswald Ger.	
101 F2	Greiz Ger.	
86 C2	Gremikha Rus. Fed.	
93 F4	Grenaa Denmark	
140 C2	Grenada U.S.A.	
147 D3	Grenada country West Indies	
104 C3	Grenade France	
93 F4	Grenen spit Denmark	
53 C2	Grenfell Austr.	
129 D2	Grenfell Can.	
105 D2	Grenoble France	
51 D1	Grenville, Cape Austr.	
134 B1	Gresham U.S.A.	
140 B3	Gretna LA U.S.A.	
140 C3	Gretna VA U.S.A.	
100 C1	Greven Ger.	
111 B2	Grevena Greece	
100 C2	Grevenbroich Ger.	
101 E1	Grevesmühlen Ger.	
136 B2	Greybull U.S.A.	
128 A1	Grey Hunter Peak Can.	
131 E2	Grey Islands Can.	
54 B2	Greymouth N.Z.	
52 B1	Grey Range hills Austr.	
97 C2	Greystones Ireland	
123 D2	Greytown S. Africa	
91 E1	Gribanovskiy Rus. Fed.	
118 B2	Gribingui r. C.A.R.	
101 D3	Grieskirchen Ger.	
141 D2	Griffin U.S.A.	
53 C2	Griffith Austr.	

88 C3	Grigiškės Lith.	
118 C2	Grimari C.A.R.	
101 F2	Grimma Ger.	
102 C1	Grimmen Ger.	
98 C3	Grimsby U.K.	
128 C2	Grimshaw Can.	
92 □B2	Grímsstaðir Iceland	
93 E4	Grimstad Norway	
92 □A3	Grindavík Iceland	
93 E4	Grindsted Denmark	
137 E2	Grinnell U.S.A.	
123 C3	Griqualand East reg. S. Africa	
122 B2	Griqualand West reg. S. Africa	
122 B2	Griquatown S. Africa	
127 G1	Grise Fiord Can.	
	Grishino Ukr. see Krasnoarmiys'k	
99 D4	Gris Nez, Cap c. France	
96 C1	Gritley U.K.	
123 C2	Groblersdal S. Africa	
122 B2	Groblershoop S. Africa	
51 C1	Groote Eylandt i. Austr.	
120 A2	Grootfontein Namibia	
122 A2	Groot Karas Berg plat. Namibia	
122 B3	Groot Swartberge mts S. Africa	
122 B2	Grootvloer salt pan S. Africa	
123 C3	Groot Winterberg mt. S. Africa	
101 D2	Großenlüder Ger.	
101 E2	Großer Beerberg h. Ger.	
102 C2	Großer Rachel mt. Ger.	
103 C2	Grosser Speikkogel mt. Austria	
108 B2	Grosseto Italy	
101 D3	Groß-Gerau Ger.	
102 C2	Großglockner mt. Austria	
100 C1	Groß-Hesepe Ger.	
101 E2	Großröhrsdorf Ger.	
122 A1	Gross Ums Namibia	
136 A2	Gros Ventre Range mts U.S.A.	
131 E2	Groswater Bay Can.	
130 B3	Groundhog r. Can.	
135 B3	Grover Beach U.S.A.	
140 B3	Groves U.S.A.	
139 E2	Groveton U.S.A.	
87 D4	Groznyy Rus. Fed.	
109 C1	Grubišno Polje Croatia	
	Grudovo Bulg. see Sredets	
103 D1	Grudziądz Pol.	
122 A2	Grünau Namibia	
92 □A3	Grundarfjörður Iceland	
	Gruzinskaya S.S.R. country Asia see Georgia	
89 E3	Gryazi Rus. Fed.	
89 F2	Gryazovets Rus. Fed.	
103 D1	Gryfice Pol.	
102 C1	Gryfino Pol.	
153 E5	Grytviken S. Georgia	
146 C2	Guacanayabo, Golfo de b. Cuba	
144 B2	Guadalajara Mex.	
106 C1	Guadalajara Spain	
106 C1	Guadalajara reg. Spain	
48 H4	Guadalcanal i. Solomon Is	
107 C1	Guadalope r. Spain	
106 B2	Guadalquivir r. Spain	
132 B4	Guadalupe i. Mex.	
106 B2	Guadalupe, Sierra de mts Spain	
142 C2	Guadalupe Peak U.S.A.	
144 B2	Guadalupe Victoria Mex.	
144 B2	Guadalupe y Calvo Mex.	
106 C1	Guadarrama, Sierra de mts Spain	
147 D3	Guadeloupe terr. West Indies	
147 D3	Guadeloupe Passage Caribbean Sea	
106 B2	Guadiana r. Port./Spain	
106 C2	Guadix Spain	
154 B2	Guaíra Brazil	
147 C3	Guajira, Península de la pen. Col.	
150 B3	Gualaceo Ecuador	

59 D2	Guam terr. N. Pacific Ocean	
144 B2	Guamúchil Mex.	
144 B2	Guanacevi Mex.	
151 E4	Guanambi Brazil	
150 C2	Guanare Venez.	
146 B2	Guane Cuba	
70 A2	Guang'an China	
71 B3	Guangchang China	
71 B3	Guangdong prov. China	
	Guanghua China see Laohekou	
71 A3	Guangxi Zhuangzu Zizhiqu aut. reg. China	
70 A2	Guangyuan China	
71 B3	Guangzhou China	
155 D1	Guanhães Brazil	
147 D4	Guanipa r. Venez.	
71 A3	Guanling China	
65 A1	Guanshui China	
	Guansuo China see Guanling	
147 C2	Guantánamo Cuba	
155 C2	Guapé Brazil	
150 C4	Guaporé r. Bol./Brazil	
155 D2	Guarapari Brazil	
154 B3	Guarapuava Brazil	
154 C3	Guaraqueçaba Brazil	
155 C2	Guaratinguetá Brazil	
154 C3	Guaratuba Brazil	
106 B1	Guarda Port.	
	Guardafui, Cape c. Somalia see Gwardafuy, Gees	
154 C1	Guarda Mor Brazil	
106 C1	Guardo Spain	
155 C2	Guarujá Brazil	
144 B2	Guasave Mex.	
146 A3	Guatemala country Central America	
	Guatemala Guat. see Guatemala City	
157 G5	Guatemala Basin N. Pacific Ocean	
146 A3	Guatemala City Guat.	
150 C2	Guaviare r. Col.	
155 C2	Guaxupé Brazil	
150 B3	Guayaquil Ecuador	
150 A3	Guayaquil, Golfo de g. Ecuador	
152 B1	Guayaramerín Bol.	
144 A2	Guaymas Mex.	
68 C?	Guazhou China	
117 B3	Guba Eth.	
86 E1	Guba Dolgaya Rus. Fed.	
108 B2	Gubbio Italy	
89 E3	Gubkin Rus. Fed.	
73 C3	Gudivada India	
105 D2	Guebwiller France	
	Guecho Spain see Algorta	
114 A2	Guelb er Rîchât h. Maur.	
115 C1	Guelma Alg.	
114 A2	Guelmine Morocco	
130 B3	Guelph Can.	
145 C2	Guémez Mex.	
104 B2	Guérande France	
115 E2	Guerende Libya	
104 C2	Guéret France	
95 C4	Guernsey i. Channel Is	
95 C4	Guernsey terr. Channel Is	
144 A2	Guerrero Negro Mex.	
131 D2	Guers, Lac l. Can.	
158 D4	Guiana Basin N. Atlantic Ocean	
150 C2	Guiana Highlands mts Guyana/Venez.	
	Guichi China see Chizhou	
114 B2	Guider Cameroon	
108 B2	Guidonia-Montecelio Italy	
71 A3	Guigang China	
100 A3	Guignicourt France	
123 D1	Guija Moz.	
99 C4	Guildford U.K.	
71 B3	Guilin China	
130 C2	Guillaume-Delisle, Lac l. Can.	
106 B1	Guimarães Port.	
114 A3	Guinea country Africa	
113 D5	Guinea, Gulf of Africa	
114 A3	Guinea-Bissau country Africa	
104 B2	Guingamp France	
104 B2	Guipavas France	
154 B1	Guiratinga Brazil	
150 C1	Güiria Venez.	
151 D2	Guisanbourg Fr. Guiana	
98 C2	Guisborough U.K.	
100 A3	Guise France	
64 B2	Guiuan Phil.	
71 A3	Guiyang China	

H

117 C4 **Hobyo** Somalia
Hô Chi Minh Vietnam *see*
Ho Chi Minh City
63 B2 **Ho Chi Minh City** Vietnam
114 B3 **Hōd** *reg.* Maur.
117 D3 **Hodda** *mt.* Somalia
78 B3 **Hodeidah** Yemen
103 E2 **Hódmezóvásárhely**
Hungary
Hoek van Holland Neth. *see*
Hook of Holland
65 B2 **Hoeyang** N. Korea
101 E2 **Hof** Ger.
101 E2 **Hofheim in Unterfranken**
Ger.
92 □B3 **Höfn** Iceland
92 □A2 **Höfn** Iceland
92 □B3 **Hofsjökull** Iceland
67 B4 **Höfu** Japan
93 G4 **Högsby** Sweden
93 E3 **Høgste Breakulen** *mt.*
Norway
101 D2 **Hohe Rhön** *mts* Ger.
100 C2 **Hohe Venn** *moorland*
Belgium
70 B1 **Hohhot** China
75 C1 **Hoh Xil Shan** *mts* China
63 B2 **Hôi An** Vietnam
119 D2 **Hoima** Uganda
75 D2 **Hojai** India
54 B2 **Hokitika** N.Z.
66 D2 **Hokkaidō** *i.* Japan
91 C2 **Hola Prystan'** Ukr.
128 B2 **Holberg** Can.
53 C3 **Holbrook** Austr.
142 A2 **Holbrook** U.S.A.
137 D2 **Holdrege** U.S.A.
146 C2 **Holguín** Cuba
92 □B3 **Hóll** Iceland
103 D2 **Hollabrunn** Austria
138 B2 **Holland** U.S.A.
Hollandia Indon. *see*
Jayapura
135 B3 **Hollister** U.S.A.
103 E2 **Hollóháza** Hungary
93 I3 **Hollola** Fin.
100 B1 **Hollum** Neth.
140 C2 **Holly Springs** U.S.A.
134 C4 **Hollywood** U.S.A.
141 D3 **Hollywood** U.S.A.
92 F2 **Holm** Norway
Holman Can. *see*
Ulukhaktok
92 H3 **Holmsund** Sweden
122 A2 **Holoog** Namibia
93 E4 **Holstebro** Denmark
141 D1 **Holston** *r.* U.S.A.
98 A3 **Holyhead** U.K.
98 C2 **Holy Island** *England* U.K.
98 A3 **Holy Island** *Wales* U.K.
136 C2 **Holyoke** U.S.A.
Holy See Europe *see*
Vatican City
101 D2 **Holzminden** Ger.
62 A1 **Homalin** Myanmar
101 D2 **Homberg (Efze)** Ger.
114 B3 **Hombori** Mali
100 C3 **Homburg** Ger.
127 H2 **Home Bay** Can.
140 B2 **Homer** U.S.A.
141 D3 **Homestead** U.S.A.
92 F3 **Hommelvik** Norway
141 D3 **Homosassa Springs** U.S.A.
80 B2 **Homs** Syria
89 D3 **Homyel'** Belarus
Honan prov. China *see*
Henan
122 A3 **Hondeklipbaai** S. Africa
145 D3 **Hondo** *r.* Belize/Mex.
142 B2 **Hondo** *NM* U.S.A.
143 D3 **Hondo** *TX* U.S.A.
146 B3 **Honduras** *country*
Central America
146 B3 **Honduras, Gulf of** Belize/
Hond.
93 F3 **Hønefoss** Norway
135 B3 **Honey Lake** U.S.A.
104 C2 **Honfleur** France
70 B3 **Honghu** China
71 A3 **Hongjiang** China
71 B3 **Hong Kong** China
71 B3 **Hong Kong** *aut. reg.* China
Hongqizhen China *see*
Wuzhishan
131 D3 **Honguedo, Détroit d'**
sea chan. Can.
65 B1 **Hongwŏn** N. Korea
70 B2 **Hongze Hu** *l.* China

48 H4 **Honiara** Solomon Is
99 B4 **Honiton** U.K.
66 D3 **Honjō** Japan
92 I1 **Honningsvåg** Norway
49 L1 **Honolulu** U.S.A.
67 B3 **Honshū** *i.* Japan
134 B1 **Hood, Mount** *vol.* U.S.A.
50 A3 **Hood Point** Austr.
134 B1 **Hood River** U.S.A.
100 B2 **Hoogerheide** Neth.
100 C1 **Hoogeveen** Neth.
100 C1 **Hoogezand-Sappemeer**
Neth.
100 C2 **Hoog-Keppel** Neth.
100 B2 **Hook of Holland** Neth.
128 A2 **Hoonah** U.S.A.
123 C2 **Hoopstad** S. Africa
100 B1 **Hoorn** Neth.
49 J5 **Hoorn, Îles de** *is*
Wallis and Futuna Is
128 B3 **Hope** Can.
140 B2 **Hope** U.S.A.
83 N2 **Hope, Point** U.S.A.
131 D2 **Hopedale** Can.
Hopei prov. China *see*
Hebei
145 D3 **Hopelchén** Mex.
131 D2 **Hope Mountains** Can.
Hopes Advance Bay Can.
see Aupaluk
52 B3 **Hopetoun** Austr.
122 B2 **Hopetown** S. Africa
139 D3 **Hopewell** U.S.A.
130 C2 **Hopewell Islands** Can.
50 B2 **Hopkins, Lake** *imp. l.*
Austr.
138 B3 **Hopkinsville** U.S.A.
134 B1 **Hoquiam** U.S.A.
81 C1 **Horasan** Turkey
93 F4 **Hörby** Sweden
89 D3 **Horki** Belarus
91 D2 **Horlivka** Ukr.
79 D2 **Hormak** Iran
79 C2 **Hormuz, Strait of** Iran/
Oman
103 D2 **Horn** Austria
92 □A2 **Horn** *c.* Iceland
153 B5 **Horn, Cape** Chile
139 D2 **Hornell** U.S.A.
130 B3 **Hornepayne** Can.
Hornos, Cabo de *c.* Chile
see Horn, Cape
53 D2 **Hornsby** Austr.
98 C3 **Hornsea** U.K.
90 B2 **Horodenka** Ukr.
91 C1 **Horodnya** Ukr.
90 B2 **Horodok** Ukr.
90 A2 **Horodok** Ukr.
90 A1 **Horokhiv** Ukr.
Horqin Youyi Qianqi China
see Ulanhot
131 E2 **Horse Islands** Can.
52 B3 **Horsham** Austr.
99 C4 **Horsham** U.K.
93 F4 **Horten** Norway
126 D2 **Horton** *r.* Can.
117 B4 **Hosa'ina** Eth.
74 A2 **Hoshab** Pak.
74 B1 **Hoshiarpur** India
142 B1 **Hosta Butte** *mt.* U.S.A.
75 C1 **Hotan** China
122 B2 **Hotazel** S. Africa
92 G3 **Hoting** Sweden
Hot Springs *AR* U.S.A. *see*
Truth or Consequences
140 B2 **Hot Springs** *AR* U.S.A.
Hot Springs *NM* U.S.A. *see*
Truth or Consequences
136 C2 **Hot Springs** *SD* U.S.A.
128 C1 **Hottah Lake** Can.
62 B1 **Houayxay** Laos
100 B2 **Houffalize** Belgium
138 B1 **Houghton** U.S.A.
139 F1 **Houlton** U.S.A.
140 B3 **Houma** U.S.A.
70 B2 **Houma** China
128 B2 **Houston** Can.
143 D3 **Houston** U.S.A.
50 A2 **Houtman Abrolhos** *is*
Austr.
122 B3 **Houwater** S. Africa
68 C1 **Hovd** Mongolia
99 C4 **Hove** U.K.
90 A2 **Hoverla, Hora** *mt.* Ukr.
68 C1 **Hövsgöl Nuur** *l.* Mongolia
116 A3 **Howar, Wadi** *watercourse*
Sudan
98 C3 **Howden** U.K.
53 C3 **Howe, Cape** Austr.
123 D2 **Howick** S. Africa

49 J3 **Howland Island** *terr.*
N. Pacific Ocean
53 C3 **Howlong** Austr.
Howrah India *see* Haora
140 B1 **Hoxie** U.S.A.
101 D2 **Höxter** Ger.
96 C1 **Hoy** *i.* U.K.
93 E3 **Høyanger** Norway
102 C1 **Hoyerswerda** Ger.
62 A2 **Hpapun** Myanmar
103 D1 **Hradec Králové** Czech Rep.
109 C2 **Hrasnica** Bos.-Herz.
92 □B2 **Hraun** Iceland
91 C1 **Hrebinka** Ukr.
88 B3 **Hrodna** Belarus
62 A1 **Hsi-hseng** Myanmar
71 C3 **Hsinchu** Taiwan
71 C3 **Hsinying** Taiwan
62 A1 **Hsipaw** Myanmar
70 A2 **Huachi** China
150 B4 **Huacho** Peru
70 B1 **Huade** China
65 B1 **Huadian** China
70 B2 **Huai'an** China
70 B2 **Huaibei** China
70 B2 **Huai He** *r.* China
71 A3 **Huaihua** China
70 B2 **Huainan** China
70 B2 **Huaiyang** China
145 C3 **Huajuápan de León** Mex.
59 C3 **Huaki** Indon.
71 C3 **Hualien** Taiwan
150 B3 **Huallaga** *r.* Peru
120 A2 **Huambo** Angola
150 B4 **Huancavelica** Peru
150 B4 **Huancayo** Peru
Huangcaoba China *see*
Xingyi
□B2 **Huangchuan** China
Huang Hai *sea*
N. Pacific Ocean *see*
Yellow Sea
Huang He *r.* China *see*
Yellow River
71 A4 **Huangliu** China
70 B2 **Huangshan** China
70 B2 **Huangshi** China
70 A2 **Huangtu Gaoyuan** *plat.*
China
71 C3 **Huangyan** China
70 A2 **Huangyuan** China
65 B1 **Huanren** China
150 B3 **Huánuco** Peru
152 B1 **Huanuni** Bol.
150 B4 **Huaral** Peru
150 B4 **Huaráz** Peru
150 B4 **Huarmey** Peru
152 A2 **Huasco** Chile
152 A2 **Huasco** *r.* Chile
144 B2 **Huatabampo** Mex.
145 C3 **Huatusco** Mex.
71 A3 **Huayuan** China
70 B2 **Hubei** *prov.* China
73 B3 **Hubli** India
100 C2 **Hückelhoven** Ger.
99 C3 **Hucknall** U.K.
98 C3 **Huddersfield** U.K.
93 G3 **Hudiksvall** Sweden
139 E2 **Hudson** *r.* U.S.A.
129 D2 **Hudson Bay** Can.
127 G3 **Hudson Bay** *sea* Can.
128 B2 **Hudson's Hope** Can.
127 H2 **Hudson Strait** Can.
63 B2 **Huế** Vietnam
146 A3 **Huehuetenango** Guat.
144 B2 **Huehueto, Cerro** *mt.* Mex.
145 C2 **Huejutla** Mex.
106 B2 **Huelva** Spain
107 C2 **Huércal-Overa** Spain
107 C1 **Huesca** Spain
106 C2 **Huéscar** Spain
51 D2 **Hughenden** Austr.
50 B3 **Hughes (abandoned)** Austr.
75 C2 **Hugli** *r. mouth* India
143 D2 **Hugo** U.S.A.
Huhehot China *see*
Hohhot
122 B2 **Huhudi** S. Africa
122 A2 **Huib-Hoch Plateau**
Namibia
71 B3 **Huichang** China
Huicheng China *see*
Huilai
65 B1 **Huich'ŏn** N. Korea
120 A2 **Huíla, Planalto da** Angola
71 B3 **Huilai** China
71 A3 **Huili** China
70 B2 **Huimin** China

69 E2 **Huinan** China
Huinan China *see* Nanhui
93 H3 **Huittinen** Fin.
145 C3 **Huixtla** Mex.
71 A3 **Huize** China
78 B2 **Hujr** Saudi Arabia
122 B1 **Hukuntsi** Botswana
78 B2 **Ḥulayfah** Saudi Arabia
66 B1 **Hulin** China
130 C3 **Hull** Can.
70 C1 **Huludao** China
69 D1 **Hulun Buir** China
69 D1 **Hulun Nur** *l.* China
91 D2 **Hulyaypole** Ukr.
69 E1 **Huma** China
150 C3 **Humaitá** Brazil
122 B3 **Humansdorp** S. Africa
98 C2 **Humber** *est.* U.K.
143 D3 **Humble** U.S.A.
129 D2 **Humboldt** Can.
135 C2 **Humboldt** *NV* U.S.A.
140 C1 **Humboldt** *TN* U.S.A.
135 C2 **Humboldt** *r.* U.S.A.
103 E2 **Humenné** Slovakia
53 C3 **Hume Reservoir** Austr.
142 A1 **Humphreys Peak** U.S.A.
115 D2 **Hūn** Libya
92 □A2 **Húnaflói** *b.* Iceland
71 B3 **Hunan** *prov.* China
65 C1 **Hunchun** China
110 B1 **Hunedoara** Romania
101 D2 **Hünfeld** Ger.
103 D2 **Hungary** *country* Europe
52 B1 **Hungerford** Austr.
65 B2 **Hŭngnam** N. Korea
65 A1 **Hun He** *r.* China
Hunjiang China *see*
Baishan
99 D3 **Hunstanton** U.K.
101 D1 **Hunte** *r.* Ger.
48 I6 **Hunter Island**
S. Pacific Ocean
51 D4 **Hunter Islands** Austr.
99 C3 **Huntingdon** U.K.
138 B2 **Huntington** *IN* U.S.A.
138 C3 **Huntington** *WV* U.S.A.
135 C4 **Huntington Beach**
U.S.A.
54 C1 **Huntly** N.Z.
96 C1 **Huntly** U.K.
130 C3 **Huntsville** Can.
140 C2 **Huntsville** *AL* U.S.A.
143 D2 **Huntsville** *TX* U.S.A.
Hunyani *r.* Moz./Zimbabwe
see Manyame
59 D3 **Huon Peninsula** P.N.G.
Huoxian China *see*
Huozhou
70 B2 **Huozhou** China
Hupeh *prov.* China *see*
Hubei
Hurghada Egypt *see*
Al Ghurdaqah
137 D2 **Huron** U.S.A.
138 C2 **Huron, Lake** Can./U.S.A.
135 D3 **Hurricane** U.S.A.
100 C3 **Hürth** Ger.
92 □B2 **Húsavík** Iceland
110 C1 **Huşi** Romania
126 B2 **Huslia** U.S.A.
78 B3 **Ḥusn Äl 'Abr** Yemen
102 B1 **Husum** Ger.
68 C1 **Hutag-Öndör** Mongolia
60 A1 **Hutanopan** Indon.
137 D3 **Hutchinson** U.S.A.
141 D3 **Hutchinson Island** U.S.A.
71 A3 **Hutuo He** *r.* China
100 B2 **Huy** Belgium
70 C2 **Huzhou** China
92 □C3 **Hvalnes** Iceland
92 □B3 **Hvannadalshnúkur** *vol.*
Iceland
109 C2 **Hvar** Croatia
109 C2 **Hvar** *i.* Croatia
91 C2 **Hvardiys'ke** Ukr.
120 B2 **Hwange** Zimbabwe
Hwang Ho *r.* China *see*
Yellow River
136 C2 **Hyannis** U.S.A.
68 C1 **Hyargas Nuur** *salt l.*
Mongolia
50 A3 **Hyden** Austr.
73 B3 **Hyderabad** India
74 A2 **Hyderabad** Pak.
Hydra *i.* Greece *see* Ydra
105 D3 **Hyères** France
105 D3 **Hyères, Îles d'** *is* France
65 B1 **Hyesan** N. Korea

28 B2 Hyland Post Can.
67 B3 Hyöno-sen mt. Japan
99 D4 Hythe U.K.
67 B4 Hyūga Japan
93 H3 Hyvinkää Fin.

I

14 B2 Iabès, Erg des. Alg.
50 C3 Iaco r. Brazil
10 C2 Ialomiţa r. Romania
10 C1 Ianca Romania
10 C1 Iaşi Romania
64 A2 Iba Phil.
15 C4 Ibadan Nigeria
50 B2 Ibagué Col.
54 B2 Ibaiti Brazil
50 B2 Ibarra Ecuador
78 B3 Ibb Yemen
00 C1 Ibbenbüren Ger.
15 C4 Ibi Nigeria
07 C2 Ibi Spain
55 C1 Ibiá Brazil
55 D1 Ibiaí Brazil
55 D1 Ibiraçu Brazil
07 D2 Ibiza Spain
07 D2 Ibiza i. Spain
51 E4 Ibotirama Brazil
79 C2 Ibrā' Oman
79 C2 Ibrī Oman
50 B4 Ica Peru
Icaria i. Greece see Ikaria
İçel Turkey see Mersin
92 □B2 Iceland country Europe
60 M4 Iceland Basin N. Atlantic Ocean
60 L3 Icelandic Plateau N. Atlantic Ocean
66 D3 Ichinoseki Japan
91 C1 Ichnya Ukr.
65 B2 Ich'ŏn N. Korea
51 F3 Icó Brazil
55 D2 Iconha Brazil
43 E2 Idabel U.S.A.
15 C4 Idah Nigeria
34 D2 Idaho state U.S.A.
34 D2 Idaho Falls U.S.A.
00 C3 Idar-Oberstein Ger.
68 C1 Ideriyn Gol r. Mongolia
16 B2 Idfū Egypt
Idi Amin Dada, Lake Dem. Rep. Congo/Uganda see Edward, Lake
18 B3 Idiofa Dem. Rep. Congo
80 B2 Idlib Syria
23 C3 Idutywa S. Africa
88 B2 Iecava Latvia
54 B2 Iepê Brazil
00 A2 Ieper Belgium
11 C3 Ierapetra Greece
19 D3 Ifakara Tanz.
15 C4 Ife Nigeria
14 C3 Ifôghas, Adrar des hills Mali
18 C3 Ifumo Dem. Rep. Congo
61 C1 Igan Malaysia
19 D2 Iganga Uganda
54 A2 Igarapava Brazil
82 G2 Igarka Rus. Fed.
74 B3 Igatpuri India
81 C2 Iğdır Turkey
08 A3 Iglesias Italy
27 G2 Igloolik Can.
Igluligaarjuk Can. see Chesterfield Inlet
30 A3 Ignace Can.
88 C2 Ignalina Lith.
10 C2 İğneada Turkey
10 C2 İğneada Burnu pt Turkey
15 C3 Igoumenitsa Greece
86 E3 Igra Rus. Fed.
86 F2 Igrim Rus. Fed.
54 B3 Iguaçu r. Brazil
54 B3 Iguaçu Falls Arg./Brazil
45 C3 Iguala Mex.
07 D1 Igualada Spain
54 C2 Iguape Brazil
54 B2 Iguatemi Brazil
54 B2 Iguatemi r. Brazil
51 F3 Iguatu Brazil
18 A3 Iguéla Gabon
54 B2 Iguidi, Erg des. Alg./Maur.
19 D3 Igunga Tanz.
21 □D2 Iharaña Madag.
21 □D3 Ihosy Madag.
92 I2 Iijoki r. Fin.

92 I3 Iisalmi Fin.
67 B4 Iizuka Japan
115 C4 Ijebu-Ode Nigeria
100 B1 IJmuiden Neth.
100 B1 IJssel r. Neth.
100 B1 IJsselmeer l. Neth.
152 C2 Ijuí Brazil
Ikaahuk Can. see Sachs Harbour
123 C2 Ikageleng S. Africa
123 C2 Ikageng S. Africa
111 C3 Ikaria i. Greece
118 C3 Ikela Dem. Rep. Congo
110 B2 Ikhtiman Bulg.
67 A4 Iki-shima i. Japan
118 A2 Ikom Nigeria
121 □D3 Ikongo Madag.
65 B2 Iksan S. Korea
119 D3 Ikungu Tanz.
114 C2 Ilaferh, Oued watercourse Alg.
64 B2 Ilagan Phil.
81 C2 Īlām Iran
75 C2 Ilam Nepal
103 D1 Iława Pol.
79 C2 Ilazārān, Kūh-e mt. Iran
129 D2 Île-à-la-Crosse Can.
129 D2 Île-à-la-Crosse, Lac l. Can.
118 C3 Ilebo Dem. Rep. Congo
119 D2 Ileret Kenya
99 D4 Ilford U.K.
99 A4 Ilfracombe U.K.
155 C2 Ilhabela Brazil
155 D2 Ilha Grande, Baía da b. Brazil
154 B2 Ilha Grande, Represa resr Brazil
154 B2 Ilha Solteíra, Represa resr Brazil
106 B1 Ílhavo Port.
151 F4 Ilhéus Brazil
Ili Kazakh. see Kapshagay
126 B3 Iliamna Lake U.S.A.
64 B3 Iligan Phil.
Iliysk Kazakh. see Kapshagay
98 C3 Ilkley U.K.
152 A3 Illapel Chile
90 C2 Illichivs'k Ukr.
138 A3 Illinois r. U.S.A.
138 B3 Illinois state U.S.A.
90 B2 Illintsi Ukr.
115 C2 Illizi Alg.
89 D2 Il'men', Ozero l. Rus. Fed.
101 E2 Ilmenau Ger.
150 B4 Ilo Peru
64 B2 Iloilo Phil.
92 J3 Ilomantsi Fin.
115 C4 Ilorin Nigeria
87 D4 Ilovlya Rus. Fed.
53 D1 Iluka Austr.
127 I2 Ilulissat Greenland
Iman Rus. Fed. see Dal'nerechensk
66 B1 Iman r. Rus. Fed.
67 A4 Imari Japan
93 I3 Imatra Fin.
imeni Petra Stuchki Latvia see Aizkraukle
117 C3 Īmī Eth.
65 B2 Imjin-gang r. N. Korea/S. Korea
141 D3 Immokalee U.S.A.
108 B2 Imola Italy
151 E3 Imperatriz Brazil
108 A2 Imperia Italy
136 C2 Imperial U.S.A.
118 B2 Impfondo Congo
72 D2 Imphal India
111 C2 İmroz Turkey
67 C3 Ina Japan
150 C4 Inambari r. Peru
128 C3 In Aménas Alg.
54 A3 Inangahua Junction N.Z.
59 C3 Inanwatan Indon.
92 I2 Inari Fin.
92 I2 Inarijärvi l. Fin.
67 D3 Inawashiro-ko l. Japan
80 B1 İnce Burun pt Turkey
65 B2 Inch'ŏn S. Korea
123 C3 Incomati r. Moz.
116 B3 Inda Silasē Eth.
144 B2 Indé Mex.
135 C3 Independence CA U.S.A.
137 E2 Independence IA U.S.A.
137 D3 Independence KS U.S.A.

137 E3 Independence MO U.S.A.
134 C2 Independence Mountains U.S.A.
76 B2 Inderbor Kazakh.
72 B2 India country Asia
139 D2 Indiana U.S.A.
138 B2 Indiana state U.S.A.
138 B3 Indianapolis U.S.A.
129 D2 Indian Head Can.
159 Indian Ocean
137 E2 Indianola IA U.S.A.
140 B2 Indianola MS U.S.A.
135 D3 Indian Peak U.S.A.
135 C3 Indian Springs U.S.A.
86 D2 Indiga Rus. Fed.
83 K2 Indigirka r. Rus. Fed.
109 D1 Inđija Serbia
135 C4 Indio U.S.A.
58 B3 Indonesia country Asia
74 B2 Indore India
60 B2 Indramayu, Tanjung pt Indon.
Indrapura, Gunung vol. Indon. see Kerinci, Gunung
75 C3 Indravati r. India
104 C2 Indre r. France
74 A2 Indus r. China/Pak.
74 A2 Indus, Mouths of the Pak.
159 E2 Indus Cone Indian Ocean
80 B1 İnebolu Turkey
111 C2 İnegöl Turkey
Infantes Spain see Villanueva de los Infantes
144 B3 Infiernillo, Presa resr Mex.
51 D1 Ingham Austr.
53 D1 Inglewood Austr.
102 C2 Ingolstadt Ger.
75 C2 Ingraj Bazar India
123 D2 Ingwavuma S. Africa
120 B2 Ingwe Zambia
123 D2 Inhaca Moz.
121 C3 Inhambane Moz.
121 C2 Inhaminga Moz.
151 D2 Inini Fr. Guiana
Inis Ireland see Ennis
97 A2 Inishbofin i. Ireland
97 B2 Inishmore i. Ireland
97 C1 Inishowen pen. Ireland
54 B2 Inland Kaikoura Range mts N.Z.
Inland Sea sea Japan see Seto-naikai
102 C2 Inn r. Europe
127 H1 Innaanganeq c. Greenland
52 B1 Innamincka Austr.
70 A1 Inner Mongolia aut. reg China
96 B2 Inner Sound sea chan. U.K.
51 D1 Innisfail Austr.
102 C2 Innsbruck Austria
97 C2 Inny r. Ireland
154 B1 Inocência Brazil
118 B3 Inongo Dem. Rep. Congo
111 D3 İnönü Turkey
103 D1 Inowrocław Pol.
114 C2 In Salah Alg.
62 A2 Insein Myanmar
110 C2 Însurăţei Romania
86 F2 Inta Rus. Fed.
105 D2 Interlaken Switz.
137 E1 International Falls U.S.A.
63 A2 Interview Island India
130 C2 Inukjuak Can.
126 D2 Inuvik Can.
96 B2 Inveraray U.K.
96 C2 Inverbervie U.K.
54 A3 Invercargill N.Z.
53 D1 Inverell Austr.
96 B2 Invergordon U.K.
96 B2 Invermere Can.
96 C2 Inverness Can.
96 C2 Inverness U.K.
96 C2 Inverurie U.K.
159 F4 Investigator Ridge Indian Ocean
52 A3 Investigator Strait Austr.
77 E1 Inya Rus. Fed.
121 C2 Inyanga Zimbabwe see Nyanga
119 D3 Inyonga Tanz.
87 D3 Inza Rus. Fed.
137 D3 Iola U.S.A.
96 A2 Iona i. U.K.
111 B3 Ionian Islands Greece
109 C3 Ionian Sea Greece/Italy

Ionioi Nisoi is Greece see Ionian Islands
111 C3 Ios i. Greece
137 E2 Iowa state U.S.A.
137 E2 Iowa City U.S.A.
154 C1 Ipameri Brazil
155 D1 Ipatinga Brazil
81 C1 Ipatovo Rus. Fed.
123 C2 Ipelegeng S. Africa
150 B2 Ipiales Col.
151 F4 Ipiaú Brazil
154 B3 Ipiranga Brazil
150 B3 Ipixuna Brazil
60 B1 Ipoh Malaysia
154 B1 Iporá Brazil
118 C2 Ippy C.A.R.
111 C2 Ipsala Turkey
53 D1 Ipswich Austr.
99 D3 Ipswich U.K.
127 H2 Iqaluit Can.
152 A2 Iquique Chile
150 B3 Iquitos Peru
Irakleio Greece see Iraklion
111 C3 Iraklion Greece
81 D2 Iran country Asia
61 C1 Iran, Pegunungan mts Indon.
79 D2 Īrānshahr Iran
144 B2 Irapuato Mex.
81 C2 Iraq country Asia
154 B3 Irati Brazil
88 B2 Irbe Strait Estonia/Latvia
80 B2 Irbid Jordan
86 F3 Irbit Rus. Fed.
151 E4 Irecê Brazil
97 C2 Ireland country Europe
118 C3 Irema Dem. Rep. Congo
Iri S. Korea see Iksan
Irian Jaya reg. Indon. see Papua
115 C3 Iriba Chad
114 B3 Irigui reg. Mali/Maur.
119 D3 Iringa Tanz.
151 D3 Iriri r. Brazil
Irish Free State country Europe see Ireland
95 B3 Irish Sea Ireland/U.K.
68 C1 Irkutsk Rus. Fed.
160 M4 Irminger Basin N. Atlantic Ocean
139 D2 Irondequoit U.S.A.
110 B2 Iron Gates gorge Romania/Serbia
52 A2 Iron Knob Austr.
138 B1 Iron Mountain U.S.A.
138 C1 Ironton U.S.A.
138 A1 Ironwood U.S.A.
130 B3 Iroquois Falls Can.
64 B2 Irosin Phil.
67 C1 Irō-zaki pt Japan
90 C1 Irpin' Ukr.
62 A2 Irrawaddy r. Myanmar
63 A2 Irrawaddy, Mouths of the Myanmar
77 D1 Irtysh r. Kazakh./Rus. Fed.
107 C1 Irun Spain
96 B3 Irvine U.K.
143 D2 Irving U.S.A.
64 B3 Isabela Phil.
146 B3 Isabela, Cordillera mts Nic.
92 □A2 Ísafjarðardjúp est. Iceland
92 □A2 Ísafjörður Iceland
67 B4 Isahaya Japan
102 C2 Isar r. Ger.
96 □ Isbister U.K.
108 B2 Ischia, Isola d' i. Italy
67 C4 Ise Japan
118 C2 Isengi Dem. Rep. Congo
105 D2 Isère r. France
100 C2 Iserlohn Ger.
101 D1 Isernhagen Ger.
67 C4 Ise-wan b. Japan
114 C4 Iseyin Nigeria
Isfahan Iran see Eşfahān
66 D3 Ishikari-wan b. Japan
82 F3 Ishim Rus. Fed.
67 D3 Ishinomaki Japan
67 D3 Ishioka Japan
67 B4 Ishizuchi-san mt. Japan
74 B1 Ishkoshim Tajik.
138 B1 Ishpeming U.S.A.
111 C2 Işıklar Dağı mts Turkey
111 C3 Işıklı Turkey
123 D2 Isipingo S. Africa
119 C2 Isiro Dem. Rep. Congo
80 B2 Iskenderun Turkey
82 G3 Iskitim Rus. Fed.
110 B2 Iskŭr r. Bulg.

70 A2 Jiangyou China
70 B3 Jianli China
70 B2 Jianqiao China
71 B3 Jianyang Fujian China
70 A2 Jianyang Sichuan China
69 E2 Jiaohe China
Jiaojiang China see Taizhou
70 C2 Jiaozhou China
70 B2 Jiaozuo China
Jiashan China see Mingguang
70 C2 Jiaxing China
68 C2 Jiayuguan China
Jiddah Saudi Arabia see Jeddah
92 G2 Jiehkkevárri mt. Norway
70 B2 Jiexiu China
70 A2 Jigzhi China
103 D2 Jihlava Czech Rep.
115 C1 Jijel Alg.
117 C4 Jijiga Eth.
117 C4 Jilib Somalia
65 B1 Jilin China
65 B1 Jilin prov. China
70 B1 Jilin Hada Ling mts China
117 B4 Jīma Eth.
110 B1 Jimbolia Romania
144 B2 Jiménez Chihuahua Mex.
143 C3 Jiménez Coahuila Mex.
70 B2 Jinan China
70 A2 Jinchang China
70 B2 Jincheng China
Jinchuan China see Jinchang
Jinchuan China
53 C3 Jindabyne Austr.
103 D2 Jindřichův Hradec Czech Rep.
Jin'e China see Longchang
70 A2 Jingbian China
71 B3 Jingdezhen China
62 B1 Jingdong China
53 C3 Jingellic Austr.
70 B2 Jinghang Yunhe canal China
62 B1 Jinghong China
70 B2 Jingmen China
70 A2 Jingning China
Jingsha China see Jingzhou
70 A2 Jingtai China
71 A3 Jingxi China
Jingxian China see Jingzhou
65 B1 Jingyu China
70 A2 Jingyuan China
70 B2 Jingzhou Hubei China
70 B2 Jingzhou Hubei China
71 A3 Jingzhou Hunan China
71 B3 Jinhua China
70 B2 Jining Nei Mongol China
70 B2 Jining Shandong China
119 D2 Jinja Uganda
117 B4 Jinka Eth.
146 B3 Jinotega Nic.
146 B3 Jinotepe Nic.
71 A3 Jinping China
Jinsha Jiang r. China see Yangtze
70 B3 Jinshi China
Jinshi China see Xinning
Jinxi China see Lianshan
70 B2 Jinzhong China
70 C1 Jinzhou China
75 C1 Jiparaná r. Brazil
75 D2 Jirang China
116 B2 Jirjā Egypt
79 C3 Jīroft Iran
79 C2 Jirwān Saudi Arabia
71 A3 Jishou China
110 B2 Jiu r. Romania
70 A2 Jiuding Shan mt. China
70 B3 Jiujiang China
Jiulian China see Mojiang
79 D2 Jiwani Pak.
66 B1 Jixi China
78 B3 Jīzān Saudi Arabia
77 C2 Jizzax Uzbek.
155 D1 Joaíma Brazil
João Belo Moz. see Xai-Xai
151 F3 João Pessoa Brazil
155 C1 João Pinheiro Brazil
74 B2 Jodhpur India
92 I3 Joensuu Fin.
67 C3 Jōetsu Japan
121 C3 Jofane Moz.
88 C2 Jõgeva Estonia
Jogjakarta Indon. see Yogyakarta
123 C2 Johannesburg S. Africa
134 C2 John Day U.S.A.
134 B1 John Day r. U.S.A.

128 C2 John D'Or Prairie Can.
141 E1 John H. Kerr Reservoir U.S.A.
96 C1 John o'Groats U.K.
141 D1 Johnson City U.S.A.
128 A1 Johnson's Crossing Can.
50 B3 Johnston, Lake imp. l. Austr.
49 J2 Johnston Atoll N. Pacific Ocean
96 B3 Johnstone U.K.
Johnstone Lake l. Can. see Old Wives Lake
88 B2 Johnstown U.S.A.
150 C3 Johor Bahru Malaysia
88 C2 Jõhvi Estonia
154 C3 Joinville Brazil
105 D2 Joinville France
154 B1 Joinville Island Antarctica
92 G2 Jokkmokk Sweden
92 □B2 Jökulsá á Fjöllum r. Iceland
138 B2 Joliet U.S.A.
130 C3 Joliette Can.
64 B3 Jolo Phil.
64 B3 Jolo i. Phil.
61 C2 Jombang Indon.
75 C2 Jomsom Nepal
88 B2 Jonava Lith.
140 B1 Jonesboro AR U.S.A.
140 B2 Jonesboro LA U.S.A.
139 F2 Jonesport U.S.A.
127 G1 Jones Sound sea chan. Can.
93 F4 Jönköping Sweden
131 C3 Jonquière Can.
145 C3 Jonuta Mex.
137 E3 Joplin U.S.A.
80 B2 Jordan country Asia
80 B2 Jordan r. Asia
134 E1 Jordan U.S.A.
155 D1 Jordânia Brazil
134 C2 Jordan Valley U.S.A.
72 D2 Jorhat India
101 D1 Jork Ger.
93 E4 Jørpeland Norway
115 C4 Jos Nigeria
119 C3 José Cardel Mex.
118 C3 Joseph, Lac l. Can.
50 B1 Joseph Bonaparte Gulf Austr.
115 C4 Jos Plateau Nigeria
93 E3 Jotunheimen mts Norway
122 B3 Joubertina S. Africa
123 C2 Jouberton S. Africa
104 C2 Joué-lès-Tours France
93 I3 Joutseno Fin.
134 B1 Juan de Fuca Strait Can./U.S.A.
Juanshui China see Tongcheng
145 B2 Juárez Mex.
144 A1 Juárez, Sierra de mts Mex.
151 E3 Juazeiro Brazil
151 F3 Juazeiro do Norte Brazil
117 B4 Juba Sudan
78 B2 Jubba r. Somalia
Jubbah Saudi Arabia
Jubbulpore India see Jabalpur
145 C3 Juchitán Mex.
155 E1 Jucuruçu Brazil
102 C2 Judenburg Austria
155 E1 Juerana Brazil
101 D2 Jühnde Ger.
146 B3 Juigalpa Nic.
150 D4 Juína Brazil
100 C1 Juist i. Ger.
155 D2 Juiz de Fora Brazil
136 C2 Julesburg U.S.A.
150 B4 Juliaca Peru
Julianatop mt. Indon. see Mandala, Puncak
151 D2 Juliana Top mt. Suriname
Jullundur India see Jalandhar
107 C2 Jumilla Spain
75 C2 Jumla Nepal
Jumna r. India see Yamuna
74 B2 Junagadh India
143 D2 Junction U.S.A.
137 D3 Junction City U.S.A.
154 C2 Jundiaí Brazil
128 A2 Juneau U.S.A.
53 C2 Junee Austr.
105 D2 Jungfrau mt. Switz.
139 D2 Juniata r. U.S.A.
153 B3 Junín Arg.
92 G3 Junsele Sweden
134 C2 Juntura U.S.A.
Junxi China see Datian

Junxian China see Danjiangkou
154 B2 Jupiá, Represa resr Brazil
141 D3 Jupiter U.S.A.
154 C2 Juquiá Brazil
117 A4 Jur r. Sudan
105 D2 Jura mts France/Switz.
96 B2 Jura i. U.K.
96 B3 Jura, Sound of sea chan. U.K.
88 B2 Jurbarkas Lith.
88 B2 Jūrmala Latvia
150 C3 Juruá r. Brazil
54 C1 Juruena r. Brazil
154 C2 Jurumirim, Represa de resr Brazil
151 D3 Juruti Brazil
154 B1 Jussara Brazil
54 B1 Jutaí r. Brazil
74 B2 Jüterbog Ger.
115 D1 Juti Brazil
145 D3 Jutiapa Guat.
93 E4 Jutland pen. Denmark
146 B2 Juventud, Isla de la i. Cuba
70 B2 Juxian China
81 D3 Jüyom Iran
122 B1 Jwaneng Botswana
93 E4 Jylland pen. Denmark see Jutland
93 I3 Jyväskylä Fin.

K

74 B1 K2 mt. China/Pakistan
Kaakhka Turkm. see Kaka
92 I2 Kaamanen Fin.
61 D2 Kabaena i. Indon.
119 C3 Kabalo Dem. Rep. Congo
119 C3 Kabambare Dem. Rep. Congo
119 C3 Kabare Dem. Rep. Congo
119 C3 Kabemba Dem. Rep. Congo
130 B3 Kabinakagami Lake Can.
118 C3 Kabinda Dem. Rep. Congo
118 B2 Kabo C.A.R.
120 B2 Kabompo Zambia
119 C3 Kabongo Dem. Rep. Congo
77 C3 Kābul Afgh.
64 B3 Kaburuang i. Indon.
121 B2 Kabwe Zambia
109 D2 Kaçanik Kosovo
74 A2 Kachchh, Gulf of India
74 B2 Kachchh, Rann of marsh India
83 I3 Kachug Rus. Fed.
81 C1 Kaçkar Dağı mt. Turkey
111 C2 Kadıköy Turkey
52 A2 Kadina Austr.
114 B3 Kadiolo Mali
Kadiyevka Ukr. see Stakhanov
73 B3 Kadmat atoll India
89 F2 Kadnikov Rus. Fed.
121 B2 Kadoma Zimbabwe
63 A2 Kadonkani Myanmar
117 A3 Kadugli Sudan
115 C3 Kaduna Nigeria
89 E2 Kaduy Rus. Fed.
86 E2 Kadzherom Rus. Fed.
114 A3 Kaédi Maur.
118 B1 Kaélé Cameroon
65 B2 Kaesŏng N. Korea
118 C3 Kafakumba Dem. Rep. Congo
114 A3 Kaffrine Senegal
80 B2 Kafr ash Shaykh Egypt
121 B2 Kafue Zambia
120 B2 Kafue r. Zambia
67 C3 Kaga Japan
118 B2 Kaga Bandoro C.A.R.
91 E2 Kagal'nitskaya Rus. Fed.
Kaganovichi Pervyye Ukr. see Polis'ke
60 A2 Kagologolo Indon.
67 B4 Kagoshima Japan
Kagul Moldova see Cahul
90 C2 Kaharlyk Ukr.
61 C2 Kahayan r. Indon.
118 B3 Kahemba Dem. Rep. Congo
101 E2 Kahla Ger.
79 C2 Kahnūj Iran
92 H2 Kahperusvaarat mts Fin.
80 B2 Kahramanmaraş Turkey
79 C2 Kahūrak Iran
59 C3 Kai, Kepulauan is Indon.

115 C4 Kaiama Nigeria
54 B2 Kaiapoi N.Z.
59 C3 Kai Besar i. Indon.
70 B2 Kaifeng China
Kaihua China see Wenshan
122 B2 Kaiingveld reg. S. Africa
59 C3 Kai Kecil i. Indon.
54 B2 Kaikoura N.Z.
114 A4 Kailahun Sierra Leone
Kailas Range mts China see Gangdisê Shan
71 A3 Kaili China
59 C3 Kaimana Indon.
54 C1 Kaimanawa Mountains N.Z.
72 C2 Kaimur Range hills India
88 B2 Käina Estonia
67 C4 Kainan Japan
115 C3 Kainji Reservoir Nigeria
54 B1 Kaipara Harbour N.Z.
74 B2 Kairana India
115 D1 Kairouan Tunisia
100 C3 Kaiserslautern Ger.
55 L2 Kaiser Wilhelm II Land reg. Antarctica
54 B1 Kaitaia N.Z.
54 C1 Kaitawa N.Z.
Kaitong China see Tongyu
59 C3 Kaiwatu Indon.
65 A1 Kaiyuan Liaoning China
71 A3 Kaiyuan Yunnan China
92 I3 Kajaani Fin.
51 D2 Kajabbi Austr.
53 C1 Kajarabie, Lake Austr.
76 B3 Kaka Turkm.
122 B2 Kakamas S. Africa
119 D2 Kakamega Kenya
114 A4 Kakata Liberia
91 C2 Kakhovka Ukr.
91 C2 Kakhovs'ke Vodoskhovyshche resr Ukr.
Kakhul Moldova see Cahul
73 C3 Kakinada India
128 C1 Kakisa Can.
67 B4 Kakogawa Japan
119 C3 Kakoswa Dem. Rep. Congo
126 C2 Kaktovik U.S.A.
Kalaallit Nunaat terr. N. America see Greenland
59 C3 Kalabahi Indon.
120 B2 Kalabo Zambia
91 E1 Kalach Rus. Fed.
119 D2 Kalacha Dida Kenya
87 D4 Kalach-na-Donu Rus. Fed.
62 A1 Kaladan r. India/Myanmar
120 B3 Kalahari Desert Africa
92 H3 Kalajoki Fin.
123 C1 Kalamare Botswana
111 B3 Kalamaria Greece
111 B3 Kalamata Greece
138 B2 Kalamazoo U.S.A.
111 B3 Kalampaka Greece
88 B2 Kalana Estonia
91 C2 Kalanchak Ukr.
115 E2 Kalanshiyū ar Ramlī al Kabīr, Sarīr des. Libya
61 D2 Kalao i. Indon.
61 D2 Kalaotoa i. Indon.
63 B2 Kalasin Thai.
79 C2 Kalāt Iran
74 A2 Kalat Pak.
50 A2 Kalbarri Austr.
111 C3 Kale Turkey
80 B1 Kalecik Turkey
118 C3 Kalema Dem. Rep. Congo
119 C3 Kalemie Dem. Rep. Congo
62 A1 Kalemyo Myanmar
86 C2 Kalevala Rus. Fed.
Kalgan China see Zhangjiakou
50 B3 Kalgoorlie Austr.
109 C2 Kali Croatia
110 C2 Kaliakra, Nos pt Bulg.
60 A2 Kaliet Indon.
119 C3 Kalima Dem. Rep. Congo
61 C2 Kalimantan reg. Indon.
Kalinin Rus. Fed. see Tver'
88 B3 Kaliningrad Rus. Fed.
91 D2 Kalininskaya Rus. Fed.
88 C3 Kalinkavichy Belarus
134 D1 Kalispell U.S.A.
103 D1 Kalisz Pol.
91 E2 Kalitva r. Rus. Fed.
92 H2 Kalix Sweden
92 H2 Kalixälven r. Sweden
111 C3 Kalkan Turkey
120 A3 Kalkfeld Namibia
100 C2 Kall Ger.
92 I3 Kallavesi l. Fin.

Kallsjön

92 F3 **Kallsjön** *l.* Sweden
93 G4 **Kalmar** Sweden
93 G4 **Kalmarsund** *sea chan.* Sweden
73 C4 **Kalmunai** Sri Lanka
119 C3 **Kalole** Dem. Rep. Congo
120 B2 **Kalomo** Zambia
128 B2 **Kalone Peak** Can.
74 B1 **Kalpa** India
73 B3 **Kalpeni** *atoll* India
75 B2 **Kalpi** India
126 B2 **Kaltag** U.S.A.
101 D1 **Kaltenkirchen** Ger.
118 B2 **Kaltungo** Nigeria
89 E3 **Kaluga** Rus. Fed.
93 F4 **Kalundborg** Denmark
74 B1 **Kalur Kot** Pak.
90 A2 **Kalush** Ukr.
74 B3 **Kalyan** India
89 E2 **Kalyazin** Rus. Fed.
111 C3 **Kalymnos** Greece
111 C3 **Kalymnos** *i.* Greece
119 C3 **Kama** Dem. Rep. Congo
62 A2 **Kama** Myanmar
86 E3 **Kama** *r.* Rus. Fed.
66 D3 **Kamaishi** Japan
80 B2 **Kaman** Turkey
120 A2 **Kamanjab** Namibia
78 B3 **Kamarān** *i.* Yemen
Kamaran Island *i.* Yemen *see* **Kamarān**
74 A2 **Kamarod** Pak.
50 B3 **Kambalda** Austr.
119 C4 **Kambove** Dem. Rep. Congo
160 C4 **Kamchatka Basin** Bering Sea
83 L3 **Kamchatka Peninsula** Rus. Fed.
110 C2 **Kamchiya** *r.* Bulg.
108 B2 **Kamenjak, Rt** *pt* Croatia
86 D2 **Kamenka** *Arkhangel'skaya Oblast'* Rus. Fed.
87 D3 **Kamenka** *Penzenskaya Oblast'* Rus. Fed.
66 C2 **Kamenka** *Primorskiy Kray* Rus. Fed.
91 D1 **Kamenka** *Voronezhskaya Oblast'* Rus. Fed.
Kamenka-Strumilovskaya Ukr. *see* **Kam"yanka-Buz'ka**
91 E3 **Kamennomostskiy** Rus. Fed.
91 E2 **Kamenolomni** Rus. Fed.
Kamenongue Angola *see* **Camanongue**
83 M2 **Kamenskoye** Rus. Fed.
Kamenskoye Ukr. *see* **Dniprodzerzhyns'k**
91 E2 **Kamensk-Shakhtinskiy** Rus. Fed.
86 F3 **Kamensk-Ural'skiy** Rus. Fed.
89 F2 **Kameshkovo** Rus. Fed.
72 C1 **Kamet** *mt.* China/India
75 B1 **Kamet** *mt.* China/India
122 A3 **Kamiesberge** *mts* S. Africa
122 A3 **Kamieskroon** S. Africa
129 D1 **Kamilukuak Lake** Can.
119 C3 **Kamina** Dem. Rep. Congo
129 E1 **Kaminak Lake** Can.
90 A1 **Kamin'-Kashyrs'kyy** Ukr.
119 C3 **Kamituga** Dem. Rep. Congo
128 B2 **Kamloops** Can.
54 B1 **Kamo** N.Z.
116 B3 **Kamob Sanha** Sudan
118 C3 **Kamonia** Dem. Rep. Congo
119 D2 **Kampala** Uganda
60 B1 **Kampar** *r.* Indon.
60 B1 **Kampar** Malaysia
100 B1 **Kampen** Neth.
119 C3 **Kampene** Dem. Rep. Congo
63 A2 **Kamphaeng Phet** Thai.
63 B2 **Kâmpóng Cham** Cambodia
63 B2 **Kâmpóng Chhnăng** Cambodia
Kâmpóng Saôm Cambodia *see* **Sihanoukville**
63 B2 **Kâmpóng Spœ** Cambodia
63 B2 **Kâmpôt** Cambodia
Kampuchea *country* Asia *see* **Cambodia**
129 D2 **Kamsack** Can.
86 E3 **Kamskoye Vodokhranilishche** *resr* Rus. Fed.
117 C4 **Kamsuuma** Somalia
90 B2 **Kam"yanets'-Podil's'kyy** Ukr.
90 A1 **Kam"yanka-Buz'ka** Ukr.
88 B3 **Kamyanyets** Belarus
91 D2 **Kamyshevatskaya** Rus. Fed.
87 D3 **Kamyshin** Rus. Fed.
135 D3 **Kanab** U.S.A.
118 C3 **Kananga** Dem. Rep. Congo
87 D3 **Kanash** Rus. Fed.
138 C3 **Kanawha** *r.* U.S.A.
67 C3 **Kanazawa** Japan
62 A1 **Kanbalu** Myanmar
63 A2 **Kanchanaburi** Thai.
73 B3 **Kanchipuram** India
77 C3 **Kandahār** Afgh.
86 C2 **Kandalaksha** Rus. Fed.
61 C2 **Kandangan** Indon.
74 A2 **Kandh Kot** Pak.
114 C3 **Kandi** Benin
74 A2 **Kandiaro** Pak.
74 B2 **Kandla** India
53 C2 **Kandos** Austr.
121 □D2 **Kandreho** Madag.
73 C4 **Kandy** Sri Lanka
76 B2 **Kandyagash** Kazakh.
127 H1 **Kane Bassin** *b.* Greenland
91 D2 **Kanevskaya** Rus. Fed.
122 B1 **Kang** Botswana
127 I2 **Kangaatsiaq** Greenland
114 B3 **Kangaba** Mali
80 B2 **Kangal** Turkey
79 C2 **Kangān** Iran
60 B1 **Kangar** Malaysia
52 A3 **Kangaroo Island** Austr.
93 H3 **Kangasala** Fin.
81 C2 **Kangāvar** Iran
75 C2 **Kangchenjunga** *mt.* India/Nepal
70 A2 **Kangding** China
65 B2 **Kangdong** N. Korea
61 C2 **Kangean, Kepulauan** *is* Indon.
119 D2 **Kangen** *r.* Sudan
127 J2 **Kangeq** *c.* Greenland
127 I2 **Kangerlussuaq** *inlet* Greenland
127 J2 **Kangerlussuaq** *inlet* Greenland
127 I2 **Kangersuatsiaq** Greenland
65 B1 **Kanggye** N. Korea
131 D2 **Kangiqsualujjuaq** Can.
127 H2 **Kangiqsujuaq** Can.
131 C1 **Kangirsuk** Can.
75 C2 **Kangmar** China
65 B2 **Kangnŭng** S. Korea
65 A1 **Kangping** China
72 D2 **Kangto** *mt.* China/India
62 A1 **Kani** Myanmar
118 C3 **Kaniama** Dem. Rep. Congo
61 C1 **Kanibongan** Malaysia
86 D2 **Kanin, Poluostrov** *pen.* Rus. Fed.
86 D2 **Kanin Nos** Rus. Fed.
86 D2 **Kanin Nos, Mys** *c.* Rus. Fed.
91 C2 **Kaniv** Ukr.
52 B3 **Kaniva** Austr.
93 H3 **Kankaanpää** Fin.
138 B2 **Kankakee** U.S.A.
114 B3 **Kankan** Guinea
75 C2 **Kanker** India
73 B3 **Kannur** India
115 C3 **Kano** Nigeria
122 B3 **Kanonpunt** *pt* S. Africa
67 B4 **Kanoya** Japan
75 C2 **Kanpur** India
136 C3 **Kansas** *r.* U.S.A.
137 D3 **Kansas** *state* U.S.A.
137 E3 **Kansas City** *KS* U.S.A.
137 E3 **Kansas City** *MO* U.S.A.
83 H3 **Kansk** Rus. Fed.
Kansu *prov.* China *see* **Gansu**
63 B2 **Kantaralak** Thai.
114 C3 **Kantchari** Burkina
91 D2 **Kantemirovka** Rus. Fed.
49 J4 **Kanton** *atoll* Kiribati
97 B2 **Kanturk** Ireland
123 D2 **Kanyamazane** S. Africa
123 C1 **Kanye** Botswana
71 C3 **Kaohsiung** Taiwan
120 A2 **Kaokoveld** *plat.* Namibia
114 A3 **Kaolack** Senegal
120 B2 **Kaoma** Zambia
118 C3 **Kapanga** Dem. Rep. Congo
88 C3 **Kapatkyevichy** Belarus
100 B2 **Kapellen** Belgium
121 B2 **Kapiri Mposhi** Zambia
127 I2 **Kapisillit** Greenland
130 B2 **Kapiskau** *r.* Can.
61 C1 **Kapit** Malaysia
63 A3 **Kapoe** Thai.
117 B4 **Kapoeta** Sudan
103 D2 **Kaposvár** Hungary
102 B1 **Kappeln** Ger.
65 B1 **Kapsan** N. Korea
77 D2 **Kapshagay** Kazakh.
77 D2 **Kapshagay, Vodokhranilishche** *resr* Kazakh.
Kapsukas Lith. *see* **Marijampolė**
61 B2 **Kapuas** *r.* Indon.
52 A2 **Kapunda** Austr.
130 B3 **Kapuskasing** Can.
53 D2 **Kaputar** *mt.* Austr.
103 D2 **Kapuvár** Hungary
88 C3 **Kapyl'** Belarus
77 D3 **Kaqung** China
114 C4 **Kara** Togo
111 C3 **Kara Ada** *i.* Turkey
77 D2 **Kara-Balta** Kyrg.
76 C1 **Karabalyk** Kazakh.
81 D1 **Karabaur, Uval** *hills* Kazakh./Uzbek.
Kara-Bogaz-Gol Turkm. *see* **Garabogazköl**
80 B1 **Karabük** Turkey
76 C2 **Karabutak** Kazakh.
111 C2 **Karacabey** Turkey
111 C2 **Karacaköy** Turkey
81 C1 **Karachayevsk** Rus. Fed.
89 D3 **Karachev** Rus. Fed.
74 A2 **Karachi** Pak.
77 D2 **Karagandy** Kazakh.
77 D2 **Karagayly** Kazakh.
83 L3 **Karaginskiy Zaliv** *b.* Rus. Fed.
81 D2 **Karaj** Iran
Kara-Kala Turkm. *see* **Magtymguly**
64 B3 **Karakelong** *i.* Indon.
Karaklis Armenia *see* **Vanadzor**
77 D2 **Kara-Köl** Kyrg.
77 D2 **Karakol** Kyrg.
74 B1 **Karakoram Range** *mts* Asia
117 B3 **Kara K'orē** Eth.
Karakum, Peski *des.* Kazakh. *see* **Karakum Desert**
76 B2 **Karakum Desert** *des.* Kazakh.
76 C3 **Karakum Desert** Turkm.
Karakumy, Peski *des.* Turkm. *see* **Karakum Desert**
80 B2 **Karaman** Turkey
77 E2 **Karamay** China
54 B2 **Karamea** N.Z.
54 B2 **Karamea Bight** *b.* N.Z.
80 B2 **Karapınar** Turkey
122 A2 **Karasburg** Namibia
86 F1 **Kara Sea** Rus. Fed.
92 I2 **Karasjok** Norway
Kara Strait *str.* Rus. Fed. *see* **Karskiye Vorota, Proliv**
111 D2 **Karasu** Turkey
Karasubazar Ukr. *see* **Bilohirs'k**
77 D1 **Karasuk** Rus. Fed.
77 D2 **Karatau** Kazakh.
77 C2 **Karatau, Khrebet** *mts* Kazakh.
86 F2 **Karatayka** Rus. Fed.
67 A4 **Karatsu** Japan
111 B3 **Karavas** Greece
60 B2 **Karawang** Indon.
81 C2 **Karbalā'** Iraq
81 D2 **Karbūsh, Kūh-e** *mt.* Iran
103 E2 **Karcag** Hungary
Kardeljevo Croatia *see* **Ploče**
111 B3 **Karditsa** Greece
88 B2 **Kärdla** Estonia
122 B3 **Kareeberge** *mts* S. Africa
75 B2 **Kareli** India
88 C3 **Karelichy** Belarus
92 H2 **Karesuando** Sweden
Karghalik China *see* **Yecheng**
74 B1 **Kargil** India
Kargilik China *see* **Yecheng**
86 C2 **Kargopol'** Rus. Fed.
118 B3 **Kari** Nigeria
121 B2 **Kariba** Zimbabwe
121 B2 **Kariba, Lake** *resr* Zambia/Zimbabwe
60 B2 **Karimata, Pulau-pulau** *is* Indon.
60 B2 **Karimata, Selat** *str.* Indon.
73 B3 **Karimnagar** India
61 C2 **Karimunjawa, Pulau-pulau** *is* Indon.
91 C2 **Karkinits'ka Zatoka** *g.* U
91 D2 **Karlivka** Ukr.
Karl-Marx-Stadt Ger. *see* **Chemnitz**
109 C1 **Karlovac** Croatia
102 C1 **Karlovy Vary** Czech Rep.
Karlsburg Romania *see* **Alba Iulia**
93 F4 **Karlshamn** Sweden
93 F4 **Karlskoga** Sweden
93 G4 **Karlskrona** Sweden
102 B2 **Karlsruhe** Ger.
93 F4 **Karlstad** Sweden
101 D3 **Karlstadt** Ger.
89 D3 **Karma** Belarus
93 E4 **Karmøy** *i.* Norway
75 D2 **Karnafuli Reservoir** Ban
74 B2 **Karnal** India
110 C2 **Karnobat** Bulg.
74 A2 **Karodi** Pak.
121 B2 **Karoi** Zimbabwe
121 C1 **Karonga** Malawi
116 B3 **Karora** Eritrea
111 C3 **Karpathos** Greece
111 C3 **Karpathos** *i.* Greece
111 B3 **Karpenisi** Greece
Karpilovka Belarus *see* **Aktsyabrski**
86 D2 **Karpogory** Rus. Fed.
50 A2 **Karratha** Austr.
81 C1 **Kars** Turkey
88 C2 **Kārsava** Latvia
Karshi Uzbek. *see* **Qarsh**
111 C3 **Karşıyaka** Turkey
86 E2 **Karskiye Vorota, Proliv** *s* Rus. Fed.
Karskoye More *sea* Rus. Fed. *see* **Kara Sea**
101 E1 **Karstädt** Ger.
111 C2 **Kartal** Turkey
87 F3 **Kartaly** Rus. Fed.
81 C2 **Kārūn, Rūd-e** *r.* Iran
73 B3 **Karwar** India
83 I3 **Karymskoye** Rus. Fed.
111 B3 **Karystos** Greece
111 C3 **Kaş** Turkey
130 B2 **Kasabonika Lake** Can.
118 C3 **Kasaï, Plateau du** Dem. Rep. Congo
118 C4 **Kasaji** Dem. Rep. Congo
121 C2 **Kasama** Zambia
120 B2 **Kasane** Botswana
118 B3 **Kasangulu** Dem. Rep. Congo
73 B3 **Kasaragod** India
129 D1 **Kasba Lake** Can.
120 B2 **Kasempa** Zambia
119 C4 **Kasenga** Dem. Rep. Con
119 C3 **Kasese** Dem. Rep. Congo
119 D2 **Kasese** Uganda
Kasevo Rus. Fed. *see* **Neftekamsk**
81 D2 **Kāshān** Iran
Kashgar China *see* **Kashi**
77 D3 **Kashi** China
67 D3 **Kashima-nada** *b.* Japan
89 E2 **Kashin** Rus. Fed.
89 E3 **Kashira** Rus. Fed.
89 E3 **Kashirskoye** Rus. Fed.
67 C3 **Kashiwazaki** Japan
76 B3 **Kāshmar** Iran
Kashmir *terr.* Asia *see* **Jammu and Kashmir**
74 A2 **Kashmore** Pak.
119 C3 **Kashyukulu** Dem. Rep. Congo
89 F3 **Kasimov** Rus. Fed.
138 B3 **Kaskaskia** *r.* U.S.A.
93 H3 **Kaskinen** Fin.
119 C3 **Kasongo** Dem. Rep. Con
118 B3 **Kasongo-Lunda** Dem. Rep. Congo
111 C3 **Kasos** *i.* Greece
Kaspiyskiy Rus. Fed. *see* **Lagan'**
116 B3 **Kassala** Sudan
101 D2 **Kassel** Ger.
115 C1 **Kasserine** Tunisia
80 B1 **Kastamonu** Turkey
Kastellorizon *i.* Greece *s* **Megisti**
111 B2 **Kastoria** Greece
89 D3 **Kastsyukovichy** Belarus
119 D3 **Kasulu** Tanz.
121 C2 **Kasungu** Malawi
139 F1 **Katahdin, Mount** U.S.A.
118 C3 **Katako-Kombe** Dem. Rep. Congo

49 D2	Katakwi Uganda
50 A3	Katanning Austr.
53 A3	Katchall i. India
41 B2	Katerini Greece
49 D3	Katesh Tanz.
28 A2	Kate's Needle mt. Can./U.S.A.
21 C2	Katete Zambia
52 A1	Katha Myanmar
50 C1	Katherine Austr.
50 C1	Katherine r. Austr.
52 B2	Kathiawar pen. India
75 C2	Kathmandu Nepal
22 B2	Kathu S. Africa
74 B1	Kathua India
48 B3	Kati Mali
75 C2	Katihar India
54 C1	Katikati N.Z.
23 C3	Katikati S. Africa
20 B2	Katima Mulilo Namibia
44 B4	Katiola Côte d'Ivoire
23 C2	Katlehong S. Africa
	Katmandu Nepal see Kathmandu
41 B3	Kato Achaïa Greece
19 C3	Katompi Dem. Rep. Congo
53 D2	Katoomba Austr.
09 C2	Katowice Pol.
80 B3	Kātrīnā, Jabal mt. Egypt
93 G4	Katrineholm Sweden
45 C3	Katsina Nigeria
45 C3	Katsina-Ala Nigeria
67 D3	Katsuura Japan
77 C3	Kattaqo'rg'on Uzbek.
93 F4	Kattegat str. Denmark/Sweden
00 B1	Katwijk aan Zee Neth.
01 D3	Katzenbuckel h. Ger.
49 L1	Kaua'i i. U.S.A.
93 H3	Kauhajoki Fin.
88 B3	Kaunas Lith.
15 C3	Kaura-Namoda Nigeria
	Kaushany Moldova see Căuşeni
92 H2	Kautokeino Norway
09 D2	Kavadarci Macedonia
09 C2	Kavajë Albania
09 C2	Kavala Greece
66 C2	Kavalerovo Rus. Fed.
73 C3	Kavali India
53 C2	Kavaratti atoll India
10 C2	Kavarna Bulg.
59 C3	Kavieng P.N.G.
77 C3	Kavīr, Dasht-e des. Iran
67 C3	Kawagoe Japan
54 B1	Kawakawa N.Z.
21 B1	Kawambwa Zambia
66 C2	Kawanishi Japan
30 C3	Kawartha Lakes Can.
67 C3	Kawasaki Japan
54 C1	Kawerau N.Z.
63 A2	Kawkareik Myanmar
62 A1	Kawlin Myanmar
63 A2	Kawmapyin Myanmar
46 B2	Kawm Umbū Egypt
63 A2	Kawthaung Myanmar
	Kaxgar China see Kashi
77 D3	Kaxgar He r. China
48 B3	Kaya Burkina
81 C3	Kayacı Dağı h. Turkey
21 C1	Kayambi Zambia
61 C1	Kayan r. Indon.
36 B2	Kaycee U.S.A.
	Kaydanovo Belarus see Dzyarzhynsk
42 A1	Kayenta U.S.A.
14 A3	Kayes Mali
77 D2	Kaynar Kazakh.
81 C1	Kayseri Turkey
34 D2	Kaysville U.S.A.
60 B2	Kayuagung Indon.
	Kazakhskaya S.S.R. country Asia see Kazakhstan
76 B2	Kazakhskiy Zaliv b. Kazakh.
76 C2	Kazakhstan country Asia
	Kazakhstan Kazakh. see Aksay
87 D4	Kazan' Rus. Fed.
	Kazandzhik Turkm. see Bereket
10 C2	Kazanlŭk Bulg.
	Kazan-rettō is Japan see Volcano Islands
76 A2	Kazbek mt. Georgia/Rus. Fed.
81 D3	Kāzerūn Iran
23 E2	Kazincbarcika Hungary
18 C3	Kazumba Dem. Rep. Congo
66 D2	Kazuno Japan
86 F2	Kazymskiy Mys Rus. Fed.
111 B3	Kea i. Greece
97 C1	Keady U.K.
137 D2	Kearney U.S.A.
142 A2	Kearny U.S.A.
115 C1	Kebili Tunisia
116 A3	Kebkabiya Sudan
92 G2	Kebnekaise mt. Sweden
117 C4	K'ebrī Dehar Eth.
60 B2	Kebumen Indon.
128 B2	Kechika r. Can.
111 D3	Keçiborlu Turkey
103 D2	Kecskemét Hungary
88 B2	Kėdainiai Lith.
114 A3	Kédougou Senegal
103 D1	Kędzierzyn-Koźle Pol.
128 B1	Keele r. Can.
128 A1	Keele Peak Can.
	Keelung Taiwan see Chilung
139 E2	Keene U.S.A.
122 A2	Keetmanshoop Namibia
129 E3	Keewatin Can.
	Kefallonia i. Greece see Cephalonia
59 C3	Kefamenanu Indon.
92 □A3	Keflavík Iceland
77 D2	Kegen Kazakh.
128 C2	Keg River Can.
88 C2	Kehra Estonia
62 A1	Kehsi Mansam Myanmar
98 C3	Keighley U.K.
88 B2	Keila Estonia
116 S.	Keiskammahoek S. Africa
92 I3	Keitele l. Fin.
52 B3	Keith Austr.
96 C2	Keith U.K.
128 B1	Keith Arm b. Can.
134 B2	Keizer U.S.A.
103 E2	Kékes mt. Hungary
117 C4	K'elafo Eth.
	Kelang Malaysia see Klang
92 J2	Keles-Uayv, Gora h. Rus. Fed.
102 C2	Kelheim Ger.
76 C3	Kelif Uzboýy marsh Turkm.
80 B1	Kelkit r. Turkey
128 B1	Keller Lake Can.
134 C1	Kellogg U.S.A.
92 I2	Kelloselkä Fin.
97 C2	Kells Ireland
88 B2	Kelmė Lith.
115 D4	Kélo Chad
128 C3	Kelowna Can.
96 C3	Kelso U.K.
134 B1	Kelso U.S.A.
60 B1	Keluang Malaysia
129 D2	Kelvington Can.
86 C2	Kem' Rus. Fed.
	Ke Macina Mali see Macina
128 B2	Kemano (abandoned) Can.
118 C2	Kembé C.A.R.
111 C3	Kemer Turkey
82 G3	Kemerovo Rus. Fed.
92 H2	Kemi Fin.
92 I2	Kemijärvi Fin.
92 I2	Kemijärvi l. Fin.
92 I2	Kemijoki r. Fin.
136 A2	Kemmerer U.S.A.
92 I3	Kempele Fin.
55 G2	Kemp Land reg. Antarctica
55 A2	Kemp Peninsula Antarctica
53 D2	Kempsey Austr.
130 C3	Kempt, Lac l. Can.
102 C2	Kempten (Allgäu) Ger.
123 C2	Kempton Park S. Africa
61 C2	Kemujan i. Indon.
126 B2	Kenai U.S.A.
129 D2	Kenaston Can.
98 B2	Kendal U.K.
141 D3	Kendall U.S.A.
61 C2	Kendari Indon.
61 B2	Kendawangan Indon.
115 D3	Kendégué Chad
114 A4	Kenema Sierra Leone
118 B3	Kenge Dem. Rep. Congo
62 A1	Kengtung Myanmar
122 B2	Kenhardt S. Africa
114 B1	Kenitra Morocco
97 B3	Kenmare Ireland
136 C1	Kenmare U.S.A.
97 A3	Kenmare River inlet Ireland
100 C3	Kenn Ger.
143 C2	Kenna U.S.A.
139 F2	Kennebec r. U.S.A.
	Kennedy, Cape c. U.S.A. see Canaveral, Cape
140 B3	Kenner U.S.A.
99 C4	Kennet r. U.K.
137 E3	Kennett U.S.A.
134 C1	Kennewick U.S.A.
130 A3	Kenora Can.
138 B2	Kenosha U.S.A.
142 C2	Kent U.S.A.
77 C2	Kentau Kazakh.
138 B3	Kentucky r. U.S.A.
138 C3	Kentucky state U.S.A.
138 B3	Kentucky Lake U.S.A.
140 B2	Kentwood U.S.A.
119 D2	Kenya country Africa
119 D3	Kenya, Mount mt. Kenya
60 B1	Kenyir, Tasik resr Malaysia
137 C2	Keokuk U.S.A.
75 C2	Keonjhar India
111 C3	Kepsut Turkey
52 B3	Kerang Austr.
91 D2	Kerch Ukr.
59 D3	Kerema P.N.G.
128 C3	Keremeos Can.
116 B3	Keren Eritrea
81 C2	Kerend Iran
159 E7	Kerguelen, Îles is Indian Ocean
159 E7	Kerguelen Plateau Indian Ocean
119 D3	Kericho Kenya
54 B1	Kerikeri N.Z.
60 B2	Kerinci, Gunung vol. Indon.
	Kerintji vol. Indon. see Kerinci, Gunung
100 C2	Kerkrade Neth.
111 A3	Kerkyra Greece
	Kerkyra i. Greece see Corfu
116 B3	Kerma Sudan
49 J7	Kermadec Islands S. Pacific Ocean
79 C1	Kermān Iran
81 C2	Kermānshāh Iran
	Kermine Uzbek. see Navoiy
143 C2	Kermit U.S.A.
135 C3	Kern r. U.S.A.
114 B4	Kérouané Guinea
100 C2	Kerpen Ger.
129 D2	Kerrobert Can.
143 D2	Kerrville U.S.A.
97 B2	Kerry Head hd Ireland
	Keryneia Cyprus see Kyrenia
130 B2	Kesagami Lake Can.
111 C2	Keşan Turkey
66 D3	Kesennuma Japan
74 B2	Keshod India
100 C2	Kessel Neth.
98 B2	Keswick U.K.
103 D2	Keszthely Hungary
82 G3	Ket' r. Rus. Fed.
60 C2	Ketapang Indon.
128 A2	Ketchikan U.S.A.
134 D2	Ketchum U.S.A.
114 B4	Kete Krachi Ghana
118 B2	Kétté Cameroon
99 C3	Kettering U.K.
138 C3	Kettering U.S.A.
134 C1	Kettle River Range mts U.S.A.
93 H3	Keuruu Fin.
100 C2	Kevelaer Ger.
138 B2	Kewanee U.S.A.
138 B1	Keweenaw Bay U.S.A.
138 B1	Keweenaw Peninsula U.S.A.
141 D3	Key Largo U.S.A.
99 B4	Keynsham U.K.
141 D3	Keyser U.S.A.
141 D4	Key West U.S.A.
123 C2	Kgotsong S. Africa
69 F1	Khabarovsk Rus. Fed.
91 D3	Khadyzhensk Rus. Fed.
75 D2	Khagrachari Bangl.
74 A2	Khairpur Pak.
122 B1	Khakhea Botswana
76 B3	Khalīlābād Iran
86 F2	Khal'mer"yu Rus. Fed.
68 C1	Khamar-Daban, Khrebet mts Rus. Fed.
74 B2	Khambhat India
74 B3	Khambhat, Gulf of India
74 B3	Khamgaon India
79 C2	Khamīr Iran
78 B3	Khamir Yemen
78 B3	Khamis Mushayṭ Saudi Arabia
77 C3	Khānābād Afgh.
74 B2	Khandwa India
83 K2	Khandyga Rus. Fed.
74 B1	Khanewal Pak.
	Khan Hung Vietnam see Soc Trăng
83 J3	Khani Rus. Fed.
66 B2	Khanka, Lake China/Rus. Fed.
115 C2	Khannfoussa h. Alg.
74 B2	Khanpur Pak.
77 D2	Khantau Kazakh.
83 H2	Khantayskoye, Ozero l. Rus. Fed.
86 F2	Khanty-Mansiysk Rus. Fed.
63 A3	Khao Chum Thong Thai.
63 A2	Khao Laem, Ang Kep Nam Thai.
74 B1	Khaplu Pak.
87 D4	Kharabali Rus. Fed.
75 C2	Kharagpur India
79 C2	Khārān r. Iran
	Kharga Oasis oasis Egypt see Wāḥāt al Khārijah
74 B2	Khargon India
91 D2	Kharkiv Ukr.
	Khar'kov Ukr. see Kharkiv
111 C2	Kharmanli Bulg.
89 F2	Kharovsk Rus. Fed.
116 B3	Khartoum Sudan
87 D4	Khasavyurt Rus. Fed.
79 D2	Khāsh Iran
86 F2	Khashgort Rus. Fed.
78 A3	Khashm el Girba Sudan
78 A3	Khashm el Girba Dam Sudan
81 C1	Khashuri Georgia
75 D2	Khasi Hills India
111 C2	Khaskovo Bulg.
83 H2	Khatanga Rus. Fed.
123 C3	Khayamnandi S. Africa
78 A2	Khaybar Saudi Arabia
122 A3	Khayelitsha S. Africa
107 D2	Khemis Miliana Alg.
63 B2	Khemmarat Thai.
115 C1	Khenchela Alg.
81 D3	Kherämeh Iran
91 C2	Kherson Ukr.
83 H2	Kheta r. Rus. Fed.
69 D1	Khilok Rus. Fed.
89 E2	Khimki Rus. Fed.
74 A2	Khipro Pak.
89 E3	Khlevnoye Rus. Fed.
63 B2	Khlung Thai.
90 B2	Khmel'nyts'kyy Ukr.
	Khmer Republic country Asia see Cambodia
	Khodzheyli Uzbek. see Xo'jayli
89 E3	Khokhol'skiy Rus. Fed.
74 B3	Khokhropar Pak.
74 A1	Kholm Afgh.
89 D2	Kholm Rus. Fed.
89 D2	Kholm-Zhirkovskiy Rus. Fed.
122 A1	Khomas Highland hills Namibia
89 E3	Khomutovo Rus. Fed.
79 C2	Khonj Iran
63 B2	Khon Kaen Thai.
62 A1	Khonsa India
83 K2	Khonuu Rus. Fed.
86 E2	Khoreyver Rus. Fed.
69 D1	Khorinsk Rus. Fed.
120 A3	Khorixas Namibia
66 B2	Khorol Rus. Fed.
91 C2	Khorol Ukr.
81 C2	Khorramābād Iran
81 C2	Khorramshahr Iran
77 D3	Khorugh Tajik.
77 C3	Khorugh Tajik.
	Khotan China see Hotan
90 B2	Khotyn Ukr.
114 B1	Khouribga Morocco
88 C3	Khoyniki Belarus
62 A1	Khreum Myanmar
81 B1	Khromtau Kazakh.
75 C2	Khulna Bangl.
	Khūnīnshahr Iran see Khorramshahr
79 B2	Khurays Saudi Arabia
74 B1	Khushab Pak.
90 A2	Khust Ukr.
123 C2	Khutsong S. Africa
74 A2	Khuzdar Pak.
81 D2	Khvānsär Iran
81 D3	Khvormūj Iran
81 C2	Khvoy Iran

Khvoynaya

89 D2 **Khvoynaya** Rus. Fed.
77 D3 **Khyber Pass** Afgh./Pak.
53 D2 **Kiama** Austr.
64 B3 **Kiamba** Phil.
119 C3 **Kiambi** Dem. Rep. Congo
Kiangsi *prov.* China *see* **Jiangxi**
Kiangsu *prov.* China *see* **Jiangsu**
119 D3 **Kibaha** Tanz.
119 D3 **Kibaya** Tanz.
119 D3 **Kibiti** Tanz.
119 C3 **Kibombo** Dem. Rep. Congo
119 D3 **Kibondo** Tanz.
119 D2 **Kibre Mengist** Eth.
119 D3 **Kibungo** Rwanda
111 B2 **Kičevo** Macedonia
114 C3 **Kidal** Mali
99 B3 **Kidderminster** U.K.
114 A3 **Kidira** Senegal
74 B1 **Kidmang** India
54 C1 **Kidnappers, Cape** N.Z.
102 C1 **Kiel** Ger.
103 E1 **Kielce** Pol.
98 B2 **Kielder Water** *resr* U.K.
119 C4 **Kienge** Dem. Rep. Congo
90 C1 **Kiev** Ukr.
114 A3 **Kiffa** Maur.
119 D3 **Kigali** Rwanda
119 C3 **Kigoma** Tanz.
88 B2 **Kihnu** *i.* Estonia
92 I2 **Kiiminki** Fin.
67 B4 **Kii-suidō** *sea chan.* Japan
109 D1 **Kikinda** Serbia
119 C3 **Kikondja** Dem. Rep. Congo
59 D3 **Kikori** P.N.G.
59 D3 **Kikori** *r.* P.N.G.
118 B3 **Kikwit** Dem. Rep. Congo
65 B1 **Kilchu** N. Korea
97 C2 **Kilcock** Ireland
97 C2 **Kildare** Ireland
118 B3 **Kilembe** Dem. Rep. Congo
143 E2 **Kilgore** U.S.A.
119 D3 **Kilifi** Kenya
119 D3 **Kilimanjaro** *vol.* Tanz.
119 D3 **Kilindoni** Tanz.
80 B2 **Kilis** Turkey
90 B2 **Kiliya** Ukr.
97 B2 **Kilkee** Ireland
97 D1 **Kilkeel** U.K.
97 C2 **Kilkenny** Ireland
111 B2 **Kilkis** Greece
97 B1 **Killala** Ireland
97 B1 **Killala Bay** Ireland
97 B2 **Killaloe** Ireland
128 C2 **Killam** Can.
97 B2 **Killarney** Ireland
143 D2 **Killeen** U.S.A.
96 B2 **Killin** U.K.
131 D1 **Killiniq** Can.
97 B2 **Killorglin** Ireland
97 B1 **Killybegs** Ireland
96 B3 **Kilmarnock** U.K.
53 B3 **Kilmore** Austr.
119 D3 **Kilosa** Tanz.
97 B2 **Kilrush** Ireland
119 C3 **Kilwa** Dem. Rep. Congo
119 D3 **Kilwa Masoko** Tanz.
119 D3 **Kimambi** Tanz.
52 A2 **Kimba** Austr.
136 C2 **Kimball** U.S.A.
59 E3 **Kimbe** P.N.G.
128 C3 **Kimberley** Can.
122 B2 **Kimberley** S. Africa
50 B1 **Kimberley Plateau** Austr.
65 B1 **Kimch'aek** N. Korea
65 B2 **Kimch'ŏn** S. Korea
65 B2 **Kimhae** S. Korea
127 H2 **Kimmirut** Can.
89 E3 **Kimovsk** Rus. Fed.
118 C3 **Kimpanga** Dem. Rep. Congo
118 B3 **Kimpese** Dem. Rep. Congo
89 E2 **Kimry** Rus. Fed.
61 C1 **Kinabalu, Gunung** *mt.* Malaysia
128 C2 **Kinbasket Lake** Can.
96 C1 **Kinbrace** U.K.
130 B3 **Kincardine** Can.
62 A1 **Kinchang** Myanmar
119 C3 **Kinda** Dem. Rep. Congo
98 C3 **Kinder Scout** *h.* U.K.
129 D2 **Kindersley** Can.
114 A3 **Kindia** Guinea
119 C3 **Kindu** Dem. Rep. Congo
89 F2 **Kineshma** Rus. Fed.
118 B3 **Kingandu** Dem. Rep. Congo
51 E2 **Kingaroy** Austr.
135 B3 **King City** U.S.A.
130 C2 **King George Islands** Can.
88 C2 **Kingisepp** Rus. Fed.
51 D3 **King Island** Austr.
Kingisseppa Estonia *see* **Kuressaare**
50 B1 **King Leopold Ranges** *hills* Austr.
142 A1 **Kingman** U.S.A.
135 B3 **Kings** *r.* U.S.A.
52 A3 **Kingscote** Austr.
97 C2 **Kingscourt** Ireland
99 D3 **King's Lynn** U.K.
50 B1 **King Sound** *b.* Austr.
134 D2 **Kings Peak** U.S.A.
141 D1 **Kingsport** U.S.A.
51 D4 **Kingston** Austr.
130 C3 **Kingston** Can.
146 C3 **Kingston** Jamaica
139 E2 **Kingston** U.S.A.
52 A3 **Kingston South East** Austr.
98 C3 **Kingston upon Hull** U.K.
147 D3 **Kingstown** St Vincent
143 D3 **Kingsville** U.S.A.
99 B4 **Kingswood** U.K.
96 B2 **Kingussie** U.K.
126 F2 **King William Island** Can.
123 C3 **King William's Town** S. Africa
67 D3 **Kinka-san** *i.* Japan
96 B2 **Kinlochleven** U.K.
93 F4 **Kinna** Sweden
97 B3 **Kinsale** Ireland
118 B3 **Kinshasa** Dem. Rep. Congo
141 E1 **Kinston** U.S.A.
88 B2 **Kintai** Lith.
114 B4 **Kintampo** Ghana
96 C2 **Kintore** U.K.
96 B3 **Kintyre** *pen.* U.K.
62 A1 **Kin-U** Myanmar
119 D3 **Kiomboi** Tanz.
130 C3 **Kipawa, Lac** *l.* Can.
119 D3 **Kipembawe** Tanz.
119 D3 **Kipengere Range** *mts* Tanz.
129 D2 **Kipling** Can.
Kipling Station Can. *see* **Kipling**
119 C4 **Kipushi** Dem. Rep. Congo
119 C4 **Kipushia** Dem. Rep. Congo
101 D2 **Kirchhain** Ger.
101 D3 **Kirchheim-Bolanden** Ger.
83 I3 **Kirenga** *r.* Rus. Fed.
83 I3 **Kirensk** Rus. Fed.
89 E3 **Kireyevsk** Rus. Fed.
Kirghizia *country* Asia *see* **Kyrgyzstan**
77 D2 **Kirghiz Range** *mts* Kazakh./Kyrg.
Kirgizskaya S.S.R. *country* Asia *see* **Kyrgyzstan**
49 J4 **Kiribati** *country* Pacific Ocean
80 B2 **Kırıkkale** Turkey
89 E2 **Kirillov** Rus. Fed.
Kirin China *see* **Jilin**
Kirin *prov.* China *see* **Jilin**
Kirinyaga *mt.* Kenya *see* **Kenya, Mount**
89 D2 **Kirishi** Rus. Fed.
48 L3 **Kiritimati** *atoll* Kiribati
111 C3 **Kırkağaç** Turkey
98 B3 **Kirkby** U.K.
98 B2 **Kirkby Stephen** U.K.
96 C2 **Kirkcaldy** U.K.
96 B3 **Kirkcudbright** U.K.
92 J2 **Kirkenes** Norway
88 B1 **Kirkkonummi** Fin.
130 B3 **Kirkland Lake** Can.
111 C2 **Kırklareli** Turkey
137 E2 **Kirksville** U.S.A.
81 C2 **Kirkūk** Iraq
96 C1 **Kirkwall** U.K.
Kīrov Kazakh. *see* **Balpyk Bi**
89 D3 **Kirov** *Kaluzhskaya Oblast'* Rus. Fed.
86 D3 **Kirov** *Kirovskaya Oblast'* Rus. Fed.
Kirovabad Azer. *see* **Gäncä**
Kirovakan Armenia *see* **Vanadzor**
Kirovo Ukr. *see* **Kirovohrad**
86 E3 **Kirovo-Chepetsk** Rus. Fed.
Kirovo-Chepetskiy Rus. Fed. *see* **Kirovo-Chepetsk**
91 C2 **Kirovohrad** Ukr.
86 C2 **Kirovsk** Rus. Fed.
91 D2 **Kirovs'ke** Ukr.
Kirovskiy Kazakh. *see* **Balpyk Bi**
66 B1 **Kirovskiy** Rus. Fed.
96 C2 **Kirriemuir** U.K.
86 E3 **Kirs** Rus. Fed.
87 D3 **Kirsanov** Rus. Fed.
80 B2 **Kırşehir** Turkey
74 A2 **Kirthar Range** *mts* Pak.
92 H2 **Kiruna** Sweden
67 C3 **Kiryū** Japan
89 E2 **Kirzhach** Rus. Fed.
119 D3 **Kisaki** Tanz.
119 C2 **Kisangani** Dem. Rep. Congo
118 B3 **Kisantu** Dem. Rep. Congo
60 A1 **Kisaran** Indon.
82 G3 **Kiselevsk** Rus. Fed.
75 C2 **Kishanganj** India
115 C4 **Kishi** Nigeria
Kishinev Moldova *see* **Chişinău**
67 C4 **Kishiwada** Japan
77 D1 **Kishkenekol'** Kazakh.
75 D2 **Kishoreganj** Bangl.
74 B1 **Kishtwar** India
119 D3 **Kisii** Kenya
103 D2 **Kiskunfélegyháza** Hungary
103 D2 **Kiskunhalas** Hungary
87 D4 **Kislovodsk** Rus. Fed.
117 C5 **Kismaayo** Somalia
Kismayu Somalia *see* **Kismaayo**
119 C3 **Kisoro** Uganda
111 B3 **Kissamos** Greece
114 A4 **Kissidougou** Guinea
141 D3 **Kissimmee** U.S.A.
141 D3 **Kissimmee, Lake** U.S.A.
129 D2 **Kississing Lake** Can.
Kistna *r.* India *see* **Krishna**
119 D3 **Kisumu** Kenya
103 E2 **Kisvárda** Hungary
Kisykkamys Kazakh. *see* **Zhanakala**
114 B3 **Kita** Mali
67 D3 **Kitaibaraki** Japan
66 D3 **Kitakami** Japan
66 D3 **Kitakami-gawa** *r.* Japan
67 B4 **Kita-Kyūshū** Japan
119 D2 **Kitale** Kenya
66 D2 **Kitami** Japan
130 B3 **Kitchener** Can.
93 J3 **Kitee** Fin.
119 D2 **Kitgum** Uganda
128 B2 **Kitimat** Can.
118 B3 **Kitona** Dem. Rep. Congo
92 H2 **Kittilä** Fin.
141 E1 **Kitty Hawk** U.S.A.
119 D3 **Kitunda** Tanz.
128 B2 **Kitwanga** Can.
121 B2 **Kitwe** Zambia
102 C2 **Kitzbühel** Austria
101 E3 **Kitzingen** Ger.
59 D3 **Kiunga** P.N.G.
92 I3 **Kiuruvesi** Fin.
92 I2 **Kivalo** *ridge* Fin.
90 B1 **Kivertsi** Ukr.
88 C2 **Kiviõli** Estonia
91 D2 **Kivsharivka** Ukr.
119 C3 **Kivu, Lake** Dem. Rep. Congo/Rwanda
111 C2 **Kıyıköy** Turkey
86 E3 **Kizel** Rus. Fed.
111 C3 **Kızılca Dağ** *mt.* Turkey
80 B1 **Kızılırmak** *r.* Turkey
87 D4 **Kizlyar** Rus. Fed.
Kizyl-Arbat Turkm. *see* **Serdar**
92 I1 **Kjøllefjord** Norway
92 G2 **Kjøpsvik** Norway
102 C1 **Kladno** Czech Rep.
102 C2 **Klagenfurt** Austria
88 B2 **Klaipėda** Lith.
94 B1 **Klaksvík** Faroe Is
134 B2 **Klamath** *r.* U.S.A.
134 B2 **Klamath Falls** U.S.A.
134 B2 **Klamath Mountains** U.S.A.
60 B1 **Klang** Malaysia
102 C2 **Klatovy** Czech Rep.
122 A3 **Klawer** S. Africa
128 A2 **Klawock** U.S.A.
128 B2 **Kleena Kleene** Can.
122 B2 **Kleinbegin** S. Africa
122 A2 **Klein Karas** Namibia
122 A2 **Kleinsee** S. Africa
123 C2 **Klerksdorp** S. Africa
90 B1 **Klesiv** Ukr.
89 D3 **Kletnya** Rus. Fed.
100 C2 **Kleve** Ger.
88 C3 **Klichaw** Belarus
89 D3 **Klimavichy** Belarus
89 D3 **Klimovo** Rus. Fed.
89 E2 **Klimovsk** Rus. Fed.
89 E2 **Klin** Rus. Fed.
101 F2 **Klingenthal** Ger.
101 F2 **Klínovec** *mt.* Czech Rep.
93 G4 **Klintehamn** Sweden
89 D3 **Klintsy** Rus. Fed.
109 C2 **Ključ** Bos.-Herz.
103 D1 **Kłodzko** Pol.
100 C1 **Kloosterhaar** Neth.
103 D2 **Klosterneuburg** Austria
101 E1 **Klötze (Altmark)** Ger.
128 A1 **Kluane Lake** Can.
Kluang Malaysia *see* **Keluang**
103 D1 **Kluczbork** Pol.
Klukhori Rus. Fed. *see* **Karachayevsk**
128 A2 **Klukwan** U.S.A.
89 F2 **Klyaz'ma** *r.* Rus. Fed.
88 C3 **Klyetsk** Belarus
83 L3 **Klyuchi** Rus. Fed.
98 C2 **Knaresborough** U.K.
93 F3 **Knästen** *h.* Sweden
129 E2 **Knee Lake** Can.
101 E1 **Knesebeck** Ger.
101 E3 **Knetzgau** Ger.
109 C2 **Knin** Croatia
103 C2 **Knittelfeld** Austria
109 D2 **Knjaževac** Serbia
Knob Lake Can. *see* **Schefferville**
97 B3 **Knockboy** *h.* Ireland
100 A2 **Knokke-Heist** Belgium
141 D1 **Knoxville** U.S.A.
127 H1 **Knud Rasmussen Land** *r* Greenland
122 B3 **Knysna** S. Africa
60 B2 **Koba** Indon.
76 B1 **Kobda** Kazakh.
67 C4 **Kōbe** Japan
København Denmark *see* **Copenhagen**
100 C2 **Koblenz** Ger.
59 C3 **Kobroör** *i.* Indon.
88 B3 **Kobryn** Belarus
Kocaeli Turkey *see* **İzmit**
111 B2 **Kočani** Macedonia
111 C2 **Kocasu** *r.* Turkey
109 B1 **Kočevje** Slovenia
75 C2 **Koch Bihar** India
89 F3 **Kochetovka** Rus. Fed.
73 B4 **Kochi** India
67 B4 **Kōchi** Japan
87 D4 **Kochubey** Rus. Fed.
75 C2 **Kodarma** India
126 B3 **Kodiak** U.S.A.
126 B3 **Kodiak Island** U.S.A.
123 C1 **Kodibeleng** Botswana
117 B4 **Kodok** Sudan
90 B2 **Kodyma** Ukr.
111 C2 **Kodzhaele** *mt.* Bulg./Gree
122 A2 **Koës** Namibia
122 C2 **Koffiefontein** S. Africa
114 B4 **Koforidua** Ghana
67 C3 **Kōfu** Japan
131 D2 **Kogaluk** *r.* Can.
114 B3 **Kogoni** Mali
74 B1 **Kohat** Pak.
72 D2 **Kohima** India
88 C2 **Kohtla-Järve** Estonia
128 A1 **Koidern** Can.
Kokand Uzbek. *see* **Qo'qo**
88 B2 **Kökar** Fin.
Kokchetav Kazakh. *see* **Kokshetau**
122 A2 **Kokerboom** Namibia
88 C3 **Kokhanava** Belarus
89 F2 **Kokhma** Rus. Fed.
92 H3 **Kokkola** Fin.
88 C2 **Koknese** Latvia
138 B2 **Kokomo** U.S.A.
122 B1 **Kokong** Botswana
123 C2 **Kokosi** S. Africa
77 E2 **Kokpekty** Kazakh.
77 C1 **Kokshetau** Kazakh.
131 D2 **Koksoak** *r.* Can.
123 C3 **Kokstad** S. Africa
Koktokay China *see* **Fuyu**
61 D2 **Kolaka** Indon.
86 C2 **Kola Peninsula** Rus. Fed.
92 H2 **Kolari** Fin.
Kolarovgrad Bulg. *see* **Shumen**
114 A3 **Kolda** Senegal
93 E4 **Kolding** Denmark

49 C2	Kole Dem. Rep. Congo
07 D2	Koléa Alg.
86 D2	Kolguyev, Ostrov i. Rus. Fed.
73 B3	Kolhapur India
68 B2	Kolkasrags pt Latvia
75 C2	Kolkata India
73 B4	Kollam India
00 C1	Kollum Neth.
	Köln Ger. see Cologne
03 D1	Koło Pol.
03 D1	Kołobrzeg Pol.
14 B3	Kolokani Mali
89 E2	Kolomna Rus. Fed.
90 B2	Kolomyya Ukr.
14 B3	Kolondiéba Mali
61 D2	Kolonedale Indon.
22 B2	Kolonkwaneng Botswana
82 G3	Kolpashevo Rus. Fed.
89 E3	Kolpny Rus. Fed.
	Kol'skiy Poluostrov pen. Rus. Fed. see Kola Peninsula
78 B3	Koluli Eritrea
92 F3	Kolvereid Norway
19 C4	Kolwezi Dem. Rep. Congo
83 L2	Kolyma r. Rus. Fed.
	Kolyma Lowland lowland Rus. Fed. see Kolymskaya Nizmennost'
	Kolyma Range mts Rus. Fed. see Kolymskaya Nagor'ye
83 L2	Kolymskaya Nizmennost' lowland Rus. Fed.
83 M2	Kolymskoye Nagor'ye mts Rus. Fed.
22 A2	Komaggas S. Africa
67 C3	Komaki Japan
83 M3	Komandorskiye Ostrova is Rus. Fed.
03 D2	Komárno Slovakia
23 D2	Komati r. S. Africa/ Swaziland
23 D2	Komatipoort S. Africa
67 C3	Komatsu Japan
20 A2	Kombat Namibia
19 C3	Kombe Dem. Rep. Congo
	Komintern Ukr. see Marhanets'
90 C2	Kominternivs'ke Ukr.
09 C2	Komiža Croatia
03 D2	Komló Hungary
	Kommunarsk Ukr. see Alchevs'k
18 B3	Komono Congo
11 C7	Komotini Greece
	Kompong Som Cambodia see Sihanoukville
22 B3	Komsberg mts S. Africa
83 H1	Komsomolets, Ostrov i. Rus. Fed.
89 F2	Komsomol'sk Rus. Fed.
91 C2	Komsomol's'k Ukr.
	Komsomol'skiy Rus. Fed. see Yugorsk
83 M2	Komsomol'skiy Chukotskiy Avtonomnyy Okrug Rus. Fed.
87 D4	Komsomol'skiy Respublika Kalmykiya-Khalm'g-Tangch Rus. Fed.
83 K3	Komsomol'sk-na-Amure Rus. Fed.
89 E2	Konakovo Rus. Fed.
75 D3	Kondagaon India
86 F2	Kondinskoye Rus. Fed. Kondinskoye Rus. Fed. see Oktyabr'skoye
19 D3	Kondoa Tanz.
86 C2	Kondopoga Rus. Fed.
89 E3	Kondrovo Rus. Fed.
27 J2	Kong Christian IX Land reg. Greenland
27 K2	Kong Christian X Land reg. Greenland
27 J2	Kong Frederik VI Kyst coastal area Greenland
65 B2	Kongju S. Korea
19 C3	Kongolo Dem. Rep. Congo
93 E4	Kongsberg Norway
93 F3	Kongsvinger Norway
77 D3	Kongur Shan mt. China
00 C2	Königswinter Ger.
03 D1	Konin Pol.
09 C2	Konjic Bos.-Herz.
22 A2	Konkiep watercourse Namibia
86 D2	Konosha Rus. Fed.
91 C1	Konotop Ukr.
103 E1	Końskie Pol.
	Konstantinograd Ukr. see Krasnohrad
102 B2	Konstanz Ger.
115 C3	Kontagora Nigeria
63 B2	Kon Tum Vietnam
63 B2	Kon Tum, Cao Nguyên Vietnam
80 B2	Konya Turkey
77 D2	Konyrat Kazakh.
100 C3	Konz Ger.
86 E3	Konzhakovskiy Kamen', Gora mt. Rus. Fed.
134 C1	Kooskia U.S.A.
128 C3	Kootenay Lake Can.
53 D2	Kootingal Austr.
122 B3	Kootjieskolk S. Africa
92 □B2	Kópasker Iceland
108 B1	Koper Slovenia
93 G4	Köping Sweden
123 C1	Kopong Botswana
93 G4	Kopparberg Sweden
109 C1	Koprivnica Croatia
89 F3	Korablino Rus. Fed.
73 C3	Koraput India
75 C2	Korba India
101 D2	Korbach Ger.
109 C2	Korçë Albania
109 C2	Korčula Croatia
109 C2	Korčula i. Croatia
70 C2	Korea Bay g. China/ N. Korea
65 B1	Korea, North country Asia
65 B2	Korea, South country Asia
65 B3	Korea Strait Japan/S. Korea
89 D3	Korenevo Rus. Fed.
91 D2	Korenovsk Rus. Fed.
	Korenovskaya Rus. Fed. see Korenovsk
90 B1	Korets' Ukr.
111 C2	Körfez Turkey
114 B4	Korhogo Côte d'Ivoire
	Korinthos Greece see Corinth
103 D2	Kőris-hegy h. Hungary
109 D2	Koritnik mt. Albania/Kosovo
	Koritsa Albania see Korçë
67 D3	Kōriyama Japan
87 F3	Korkino Rus. Fed.
111 D3	Korkuteli Turkey
	Korla China
103 D2	Körmend Hungary
49 I5	Koro i. Fiji
114 B3	Koro Mali
131 D2	Koroc r. Can.
91 D1	Korocha Rus. Fed.
119 D3	Korogwe Tanz.
59 C2	Koror Palau
103 E2	Körös r. Hungary
90 B1	Korosten' Ukr.
90 B1	Korostyshiv Ukr.
115 D3	Koro Toro Chad
93 H3	Korpo Fin.
91 D2	Korsakov Rus. Fed.
91 C2	Korsun'-Shevchenkivs'kyy Ukr.
103 E1	Korsze Pol.
116 B3	Korti Sudan
100 A2	Kortrijk Belgium
83 L3	Koryakskaya, Sopka vol. Rus. Fed.
83 M2	Koryakskoye Nagor'ye mts Rus. Fed.
86 D2	Koryazhma Rus. Fed.
65 B2	Koryŏng S. Korea
91 C1	Koryukivka Ukr.
111 C3	Kos Greece
111 C3	Kos i. Greece
91 D2	Kosa Biryuchyy Ostriv i. Ukr.
65 B2	Kosan N. Korea
103 D1	Kościan Pol.
	Kosciusko, Mount mt. Austr. see Kosciuszko, Mount
53 C3	Kosciuszko, Mount Austr.
77 E2	Kosh-Agach Rus. Fed.
67 A4	Koshikijima-rettō is Japan
103 E2	Košice Slovakia
92 H2	Koskullskulle Sweden
65 B2	Kosŏng N. Korea
109 D2	Kosovo country Europe
109 D2	Kosovska Mitrovica Kosovo see Mitrovicë
48 H3	Kosrae atoll Micronesia
114 B4	Kossou, Lac de l. Côte d'Ivoire
76 C1	Kostanay Kazakh.
110 B2	Kostenets Bulg.
123 C2	Koster S. Africa
116 B3	Kosti Sudan
92 J3	Kostomuksha Rus. Fed.
90 B1	Kostopil' Ukr.
89 F2	Kostroma Rus. Fed.
89 F2	Kostroma r. Rus. Fed.
102 C1	Kostrzyn Pol.
91 D2	Kostyantynivka Ukr.
103 D1	Koszalin Pol.
103 D2	Kőszeg Hungary
74 B2	Kota India
60 B2	Kotaagung Indon.
61 C2	Kotabaru Indon.
61 C1	Kota Belud Malaysia
60 B1	Kota Bharu Malaysia
60 B2	Kotabumi Indon.
61 C1	Kota Kinabalu Malaysia
75 C3	Kotaparh India
61 C1	Kota Samarahan Malaysia
86 D3	Kotel'nich Rus. Fed.
87 D4	Kotel'nikovo Rus. Fed.
83 K1	Kotel'nyy, Ostrov i. Rus. Fed.
91 C1	Kotel'va Ukr.
101 E2	Köthen (Anhalt) Ger.
119 D2	Kotido Uganda
93 I3	Kotka Fin.
86 D2	Kotlas Rus. Fed.
126 B2	Kotlik U.S.A.
109 C2	Kotor Varoš Bos.-Herz.
87 D3	Kotovo Rus. Fed.
91 E1	Kotovsk Rus. Fed.
90 B2	Kotovs'k Ukr.
73 C3	Kottagudem India
118 C2	Kotto r. C.A.R.
83 H2	Kotuy r. Rus. Fed.
126 B2	Kotzebue U.S.A.
126 B2	Kotzebue Sound sea chan. U.S.A.
114 A3	Koubia Guinea
100 A2	Koudekerke Neth.
114 B3	Koudougou Burkina
122 B3	Kougaberge mts S. Africa
118 B3	Koulamoutou Gabon
114 B3	Koulikoro Mali
118 B2	Koum Cameroon
118 B2	Koumra Chad
114 A3	Koundâra Guinea
	Kounradskiy Kazakh. see Konyrat
131 D2	Kourou Fr. Guiana
114 B3	Kouroussa Guinea
115 D3	Kousséri Cameroon
114 B3	Koutiala Mali
93 I3	Kouvola Fin.
109 D1	Kovačica Serbia
92 J2	Kovdor Rus. Fed.
90 A1	Kovel' Ukr.
	Kovno Lith. see Kaunas
89 F2	Kovrov Rus. Fed.
51 D1	Kowanyama Austr.
54 B2	Kowhitirangi N.Z.
111 C3	Köyceğiz Turkey
86 D2	Koyda Rus. Fed.
126 B2	Koyukuk r. U.S.A.
111 B2	Kozani Greece
90 C1	Kozelets' Ukr.
89 E3	Kozel'sk Rus. Fed.
73 B3	Kozhikode India
90 B2	Kozyatyn Ukr.
114 C4	Kpalimé Togo
63 A2	Kra, Isthmus of Myanmar/ Thai.
63 A3	Krabi Thai.
63 A2	Kra Buri Thai.
63 B2	Krâchéh Cambodia
93 E4	Kragerø Norway
100 B1	Kraggenburg Neth.
109 D2	Kragujevac Serbia
60 B2	Krakatau i. Indon.
103 D1	Kraków Pol.
109 D2	Kraljevo Serbia
91 D2	Kramators'k Ukr.
93 G3	Kramfors Sweden
111 B3	Kranidi Greece
102 C2	Kranj Slovenia
123 D2	Krasino Tanz.
86 E1	Krasino Rus. Fed.
88 C3	Kráslava Latvia
103 F2	Kraslice Czech Rep.
89 D3	Krasnapollye Belarus
89 D3	Krasnaya Gora Rus. Fed.
89 F2	Krasnaya Gorbatka Rus. Fed.
	Krasnoarmeysk Kazakh. see Taiynsha
87 D3	Krasnoarmeysk Rus. Fed.
	Krasnoarmeyskaya Rus. Fed. see Poltavskaya
91 D2	Krasnoarmiys'k Ukr.
86 D2	Krasnoborsk Rus. Fed.
91 D2	Krasnodar Rus. Fed.
91 D2	Krasnodon Ukr.
88 C2	Krasnogorodskoye Rus. Fed.
91 D1	Krasnogvardeyskoye Rus. Fed.
91 D2	Krasnohrad Ukr.
91 C2	Krasnohvardiys'ke Ukr.
86 E3	Krasnokamsk Rus. Fed.
89 D2	Krasnomayskiy Rus. Fed.
91 C2	Krasnoperekops'k Ukr.
87 D3	Krasnoslobodsk Rus. Fed.
86 F3	Krasnotur'insk Rus. Fed.
86 E3	Krasnoufimsk Rus. Fed.
86 E2	Krasnovishersk Rus. Fed.
	Krasnovodsk Turkm. see Türkmenbaşy
83 H3	Krasnoyarsk Rus. Fed.
89 E3	Krasnoye Rus. Fed.
83 M2	Krasnoye, Ozero l. Rus. Fed.
89 F2	Krasnoye-na-Volge Rus. Fed.
103 E1	Krasnystaw Pol.
89 D3	Krasnyye Baki Rus. Fed.
	Krasnyy Kamyshanik Rus. Fed. see Komsomol'skiy
89 E2	Krasnyy Kholm Rus. Fed.
91 D2	Krasnyy Luch Ukr.
91 E2	Krasnyy Sulin Rus. Fed.
90 B2	Krasyliv Ukr.
	Kraulshavn Greenland see Nuussuaq
100 C2	Krefeld Ger.
91 C2	Kremenchuk Ukr.
91 C2	Kremenchuts'ke Vodoskhovyshche resr Ukr.
90 B1	Kremenets' Ukr.
103 D2	Křemešník h. Czech Rep.
	Kremges Ukr. see Svitlovods'k
91 D2	Kreminna Ukr.
136 D2	Kremmling U.S.A.
103 D2	Krems an der Donau Austria
89 D2	Kresttsy Rus. Fed.
88 B2	Kretinga Lith.
100 C2	Kreuzau Ger.
101 C2	Kreuztal Ger.
118 A2	Kribi Cameroon
123 C2	Kriel S. Africa
111 B3	Krikellos Greece
66 D1	Kril'on, Mys c. Rus. Fed.
111 D3	Krios, Akrotirio pt Greece
73 C3	Krishna r. India
73 C3	Krishna, Mouths of the India
75 C2	Krishnanagar India
93 E4	Kristiansand Norway
93 F4	Kristianstad Sweden
92 E3	Kristiansund Norway
93 F4	Kristinehamn Sweden
	Kristinopol' Ukr. see Chervonohrad
111 C3	Kriti i. Greece see Crete
110 B2	Kritiko Pelagos sea Greece
	Kriva Palanka Macedonia
	Krivoy Rog Ukr. see Kryvyy Rih
109 C1	Križevci Croatia
108 B1	Krk i. Croatia
92 F3	Krokom Sweden
91 C1	Krolevets' Ukr.
89 E3	Kromy Rus. Fed.
101 E2	Kronach Ger.
63 B2	Krŏng Kaôh Kŏng Cambodia
127 J2	Kronprins Frederik Bjerge nunataks Greenland
123 C2	Kroonstad S. Africa
91 E2	Kropotkin Rus. Fed.
103 E2	Krosno Pol.
103 D1	Krotoszyn Pol.
60 B2	Krui Indon.
122 B3	Kruisfontein S. Africa
109 C2	Krujë Albania
111 C2	Krumovgrad Bulg.
	Krung Thep Thai. see Bangkok
88 C3	Krupki Belarus
109 C2	Kruševac Serbia
101 E2	Krušné hory mts Czech Rep.
128 A2	Kruzof Island U.S.A.
88 C3	Krychaw Belarus
91 D2	Krylovskaya Rus. Fed.
91 D3	Krymsk Rus. Fed.

Krymskaya Rus. Fed. see
Krymsk
Kryms'kyy Pivostriv pen.
Ukr. see Crimea
Krystynopol Ukr. see
Chervonohrad
91 C2 Kryvyy Rih Ukr.
90 B2 Kryzhopil' Ukr.
114 B2 Ksabi Alg.
107 D2 Ksar el Boukhari Alg.
114 B1 Ksar el Kebir Morocco
Ksar-es-Souk Morocco see
Er Rachidia
89 E3 Kshenskiy Rus. Fed.
78 B2 Kū', Jabal al h. Saudi Arabia
61 C1 Kuala Belait Brunei
Kuala Dungun Malaysia see
Dungun
60 B1 Kuala Kangsar Malaysia
60 B1 Kuala Kerai Malaysia
60 B1 Kuala Lipis Malaysia
60 B1 Kuala Lumpur Malaysia
61 C2 Kualapembuang Indon.
60 B1 Kuala Terengganu Malaysia
60 B2 Kualatungal Indon.
61 C1 Kuamut Malaysia
65 A1 Kuandian China
60 B1 Kuantan Malaysia
91 D2 Kuban' r. Rus. Fed.
89 E2 Kubenskoye, Ozero l.
Rus. Fed.
110 C2 Kubrat Bulg.
60 B2 Kubu Indon.
61 C1 Kubuang Indon.
60 C1 Kuching Malaysia
109 C2 Kuçovë Albania
61 C1 Kudat Malaysia
61 C2 Kudus Indon.
102 C2 Kufstein Austria
127 G2 Kugaaruk Can.
91 D2 Kugey Rus. Fed.
126 E2 Kugluktuk Can.
126 D2 Kugmallit Bay Can.
92 I3 Kuhmo Fin.
79 C2 Kührān, Küh-e mt. Iran
122 A1 Kuis Namibia
Kuitin China see Kuytun
120 A2 Kuito Angola
92 I2 Kuivaniemi Fin.
65 B2 Kujang N. Korea
66 D2 Kuji Japan
67 B4 Kujū-san vol. Japan
109 D2 Kukës Albania
76 B3 Kükürtli Turkm.
111 C3 Kula Turkey
75 D2 Kula Kangri mt. Bhutan/
China
76 B2 Kulandy Kazakh.
88 B2 Kuldiga Latvia
122 B1 Kule Botswana
101 E2 Kulmbach Ger.
77 C3 Kūlob Tajik.
76 B2 Kul'sary Kazakh.
111 D3 Kulübe Tepe mt. Turkey
77 D1 Kulunda Rus. Fed.
77 D1 Kulundinskoye, Ozero salt l.
Rus. Fed.
127 J2 Kulusuk Greenland
67 C3 Kumagaya Japan
67 B4 Kumamoto Japan
67 C4 Kumano Japan
110 B2 Kumanovo Macedonia
114 B4 Kumasi Ghana
118 A2 Kum-Dag Turkm. see
Gumdag
78 B2 Kumdah Saudi Arabia
87 E3 Kumertau Rus. Fed.
65 B2 Kumi S. Korea
119 D2 Kumi Uganda
111 C3 Kumkale Turkey
93 G4 Kumla Sweden
115 D3 Kumo Nigeria
62 A1 Kumon Range mts
Myanmar
62 B2 Kumphawapi Thai.
Kumul China see Hami
66 D2 Kunashir, Ostrov i. Rus. Fed.
120 A2 Kunene r. Angola/Namibia
77 D2 Kungei Alatau mts Kazakh./
Kyrg.
93 F4 Kungsbacka Sweden
118 B2 Kungu Dem. Rep. Congo
86 E3 Kungur Rus. Fed.
62 A1 Kunhing Myanmar
62 A1 Kunlong Myanmar
75 B1 Kunlun Shan mts China

71 A3 Kunming China
65 B2 Kunsan S. Korea
50 B1 Kununurra Austr.
101 D3 Künzelsau Ger.
92 I3 Kuopio Fin.
109 C1 Kupa r. Croatia/Slovenia
59 C3 Kupang Indon.
88 B2 Kupiškis Lith.
111 C2 Küplü Turkey
128 A2 Kupreanof Island U.S.A.
91 D2 Kup"yans'k Ukr.
77 E2 Kuqa China
81 C2 Kür r. Azer.
67 B4 Kurashiki Japan
75 C2 Kurasia India
67 B3 Kurayoshi Japan
89 E3 Kurchatov Rus. Fed.
111 C2 Kürdzhali Bulg.
67 B4 Kure Japan
57 T7 Kure Atoll U.S.A.
88 B2 Kuressaare Estonia
87 F3 Kurgan Rus. Fed.
Kuria Muria Islands is Oman
see Ḩalāniyāt, Juzur al
93 H3 Kurikka Fin.
156 C2 Kuril Basin Sea of Okhotsk
69 F1 Kuril Islands is Rus. Fed.
69 F1 Kuril'sk Rus. Fed.
Kuril'skiye Ostrova is
Rus. Fed. see Kuril Islands
156 C3 Kuril Trench
N. Pacific Ocean
89 E3 Kurkino Rus. Fed.
Kurmashkino Kazakh. see
Kurshim
117 B3 Kurmuk Sudan
73 B3 Kurnool India
67 D3 Kuroiso Japan
53 D2 Kurri Kurri Austr.
88 B2 Kuršėnai Lith.
77 E2 Kurshim Kazakh.
78 B2 Kursh, Jabal mt.
Saudi Arabia
89 E3 Kursk Rus. Fed.
109 D2 Kuršumlija Serbia
122 B2 Kuruman S. Africa
122 B2 Kuruman watercourse
S. Africa
67 B4 Kurume Japan
83 I3 Kurumkan Rus. Fed.
73 C4 Kurunegala Sri Lanka
81 D1 Kuryk Kazakh.
111 C3 Kuşadası Turkey
111 C3 Kuşadası Körfezi b. Turkey
111 C2 Kuş Gölü l. Turkey
91 D2 Kushchevskaya Rus. Fed.
66 D2 Kushiro Japan
Kushka Turkm. see
Serhetabat
75 C2 Kushtia Bangl.
126 B2 Kuskokwim r. U.S.A.
126 B2 Kuskokwim Mountains
U.S.A.
76 C1 Kusmuryn Kazakh.
66 D2 Kussharo-ko l. Japan
Kustanay Kazakh. see
Kostanay
63 B2 Kut, Ko i. Thai.
111 C3 Kütahya Turkey
81 C1 K'ut'aisi Georgia
Kutaraja Indon. see
Banda Aceh
109 C1 Kutjevo Croatia
103 D1 Kutno Pol.
118 B3 Kutu Dem. Rep. Congo
116 A3 Kutum Sudan
126 E2 Kuujjua r. Can.
131 D2 Kuujjuaq Can.
130 C2 Kuujjuarapik Can.
92 I2 Kuusamo Fin.
120 A2 Kuvango Angola
89 D2 Kuvshinovo Rus. Fed.
78 B2 Kuwait country Asia
78 B2 Kuwait Kuwait
82 G3 Kuybyshev Rus. Fed.
Kuybyshev Rus. Fed. see
Samara
91 D2 Kuybysheve Ukr.
Kuybyshevka-Vostochnaya
Rus. Fed. see Belogorsk
87 D3 Kuybyshevskoye
Vodokhranilishche resr
Rus. Fed.
77 E2 Kuytun China
111 C3 Kuyucak Turkey
87 D3 Kuznetsk Rus. Fed.

90 B1 Kuznetsovs'k Ukr.
86 C2 Kuzomen' Rus. Fed.
92 H1 Kvalsund Norway
123 D2 KwaMashu S. Africa
61 D1 Kwandang Indon.
Kwangchow China see
Guangzhou
65 B2 Kwangju S. Korea
Kwangtung prov. China see
Guangdong
65 B1 Kwanmo-bong mt. N. Korea
123 C3 Kwanobuhle S. Africa
123 C3 KwaNojoli S. Africa
122 B3 KwaNonzame S. Africa
115 C3 Kwatarkwashi Nigeria
123 C3 Kwatinidubu S. Africa
123 C2 KwaZamokuhle S. Africa
122 B3 KwaZamukucinga S. Africa
123 C2 KwaZanele S. Africa
123 D2 KwaZulu-Natal prov.
S. Africa
Kweichow prov. China see
Guizhou
Kweiyang China see
Guiyang
121 B2 Kwekwe Zimbabwe
118 B3 Kwenge r. Dem. Rep. Congo
123 C3 Kwezi-Naledi S. Africa
103 D1 Kwidzyn Pol.
59 D3 Kwikila P.N.G.
118 B3 Kwilu r. Angola/
Dem. Rep. Congo
59 C3 Kwoka mt. Indon.
118 B2 Kyabé Chad
53 C3 Kyabram Austr.
62 A2 Kyaikto Myanmar
63 A2 Kya-in Seikkyi Myanmar
68 D1 Kyakhta Rus. Fed.
52 A2 Kyancutta Austr.
62 A1 Kyaukpadaung Myanmar
62 A2 Kyaukpyu Myanmar
88 B3 Kybartai Lith.
62 A2 Kyebogyi Myanmar
54 B3 Kyeburn N.Z.
62 A2 Kyeintali Myanmar
74 B1 Kyelang India
Kyiv Ukr. see Kiev
90 C1 Kyivs'ke Vodoskhovyshche
resr Ukr.
Kyklades is Greece see
Cyclades
129 D2 Kyle Can.
96 B2 Kyle of Lochalsh U.K.
100 C3 Kyll r. Ger.
111 B3 Kyllini mt. Greece
111 B3 Kymi Greece
52 B3 Kyneton Austr.
119 D2 Kyoga, Lake Uganda
53 D1 Kyogle Austr.
65 B2 Kyŏnggi-man b. S. Korea
65 B2 Kyŏngju S. Korea
67 C4 Kyōto Japan
111 B3 Kyparissia Greece
111 B3 Kyparissiakos Kolpos b.
Greece
111 B3 Kyra Panagia i. Greece
80 B2 Kyrenia Cyprus
77 D2 Kyrgyzstan country Asia
101 F1 Kyritz Ger.
93 H3 Kyrönjoki r. Fin.
86 E2 Kyrta Rus. Fed.
86 D2 Kyssa Rus. Fed.
83 J2 Kytalyktakh Rus. Fed.
111 B3 Kythira i. Greece
111 B3 Kythnos i. Greece
128 B2 Kyuquot Can.
67 B4 Kyūshū i. Japan
110 B2 Kyustendil Bulg.
62 A2 Kywebwe Myanmar
92 H3 Kyyjärvi Fin.
68 C1 Kyzyl Rus. Fed.
76 C2 Kyzyl Kum desert Kazakh./
Uzbek.
77 C2 Kyzylorda Kazakh.
Kyzyl-Orda Kazakh. see
Kyzylorda
Kyzyltu Kazakh. see
Kishkenekol'

L

145 C3 La Angostura, Presa de resr
Mex.
117 C4 Laascaanood Somalia
117 C3 Laasgoray Somalia
150 C1 La Asunción Venez.

114 A2 Laâyoune Western Sahara
87 D4 Laba r. Rus. Fed.
144 B2 La Babia Mex.
61 D2 Labala Indon.
152 B2 La Banda Arg.
61 C1 Labang Malaysia
104 B3 La Baule-Escoublac France
102 C1 Labe r. Czech Rep.
114 A3 Labé Guinea
128 A1 Laberge, Lake Can.
128 C2 La Biche, Lac l. Can.
108 B1 Labin Croatia
87 D4 Labinsk Rus. Fed.
64 B2 Labo Phil.
104 B3 Labouheyre France
153 B3 Laboulaye Arg.
131 D2 Labrador reg. Can.
131 D2 Labrador City Can.
127 I2 Labrador Sea Can./
Greenland
150 C3 Lábrea Brazil
61 C1 Labuan Malaysia
61 C2 Labuhanbajo Indon.
60 B1 Labuhanbilik Indon.
59 C3 Labuna Indon.
63 A2 Labutta Myanmar
86 F2 Labytnangi Rus. Fed.
109 C2 Laç Albania
107 D2 La Cabaneta Spain
La Calle Alg. see El Kala
105 C2 La Capelle France
73 B3 Laccadive Islands India
129 E2 Lac du Bonnet Can.
146 B3 La Ceiba Hond.
52 A3 Lacepede Bay Austr.
134 B1 Lacey U.S.A.
105 D2 La Chaux-de-Fonds Switz.
53 B2 Lachlan r. Austr.
146 C4 La Chorrera Panama
139 E1 Lachute Can.
105 D3 La Ciotat France
128 C2 Lac La Biche Can.
Lac la Martre Can. see
Whatì
139 E1 Lac-Mégantic Can.
128 C2 Lacombe Can.
146 B4 La Concepción Panama
145 C3 La Concordia Mex.
108 A3 Laconi Italy
139 E2 Laconia U.S.A.
128 C2 La Crete Can.
138 A2 La Crosse U.S.A.
144 B2 La Cruz Mex.
144 B2 La Cuesta Mex.
155 D1 Ladainha Brazil
74 B1 Ladakh Range mts
India/Pak.
122 B3 Ladismith S. Africa
79 D2 Lādīz Iran
89 D1 Ladoga, Lake Rus. Fed.
Ladozhskoye Ozero l.
Rus. Fed. see Ladoga, Lake
141 D2 Ladson U.S.A.
123 C3 Lady Grey S. Africa
123 B3 Ladysmith S. Africa
123 C2 Ladysmith S. Africa
59 D3 Lae P.N.G.
143 C3 La Encantada, Sierra mts
Mex.
152 B2 La Esmeralda Bol.
93 F4 Læsø i. Denmark
Lafayette Alg. see Bougaa
141 C2 La Fayette U.S.A.
138 B2 Lafayette IN U.S.A.
140 B2 Lafayette LA U.S.A.
115 C4 Lafia Nigeria
104 B2 La Flèche France
141 D1 La Follette U.S.A.
130 C2 Laforge Can.
79 C2 Läft Iran
108 A3 La Galite i. Tunisia
89 D1 Lagan' Rus. Fed.
151 F4 Lagarto Brazil
118 B2 Lagdo, Lac de l. Cameroon
115 C1 Laghouat Alg.
155 D1 Lagoa Santa Brazil
114 A2 La Gomera i. Islas Canarias
114 C4 Lagos Nigeria
106 B2 Lagos Port.
134 C1 La Grande U.S.A.
130 C2 La Grande 3, Réservoir resr
Can.
130 C2 La Grande 4, Réservoir resr
Can.
50 B1 La Grange Austr.
141 C2 La Grange U.S.A.
150 C2 La Gran Sabana plat. Venez.
152 D2 Laguna Brazil

Leiden

100 B1   Leiden Neth.
100 A2   Leie *r.* Belgium
52 A2   Leigh Creek Austr.
97 C2   Leighlinbridge Ireland
99 C4   Leighton Buzzard U.K.
93 E3   Leikanger Norway
101 D1   Leine *r.* Ger.
50 B2   Leinster Austr.
97 C2   Leinster *reg.* Ireland
97 C2   Leinster, Mount *h.* Ireland
101 F2   Leipzig Ger.
92 F2   Leiranger Norway
106 B2   Leiria Port.
93 E4   Leirvik Norway
97 C2   Leixlip Ireland
71 B3   Leiyang China
71 B3   Leizhou China
71 A3   Leizhou Bandao *pen.* China
118 B3   Lékana Congo
122 A2   Lekkersing S. Africa
140 B2   Leland U.S.A.
      Leli China *see* Tianlin
100 B1   Lelystad Neth.
153 B5   Le Maire, Estrecho de
      sea chan. Arg.
      Léman, Lac *l.* France/Switz.
      see Geneva, Lake
104 C2   Le Mans France
137 D2   Le Mars U.S.A.
102 B2   Lemberg *mt.* Ger.
154 C2   Leme Brazil
      Lemesos Cyprus *see*
      Limassol
101 D1   Lemförde Ger.
127 H2   Lemieux Islands Can.
136 C1   Lemmon U.S.A.
135 C3   Lemoore U.S.A.
131 D2   Le Moyne, Lac *l.* Can.
62 A1   Lemro *r.* Myanmar
109 C2   Le Murge *hills* Italy
83 J2   Lena *r.* Rus. Fed.
100 C1   Lengerich Ger.
70 A2   Lenglong Ling *mts* China
71 B3   Lengshuijiang China
71 B3   Lengshuitan China
      Leninabad Tajik. *see*
      Khŭjand
      Leninakan Armenia *see*
      Gyumri
91 D2   Lenine Ukr.
      Leningrad Rus. Fed. *see*
      St Petersburg
91 D2   Leningradskaya Rus. Fed.
      Leninobod Tajik. *see*
      Khŭjand
77 D3   Lenin Peak Kyrg./Tajik.
      Leninsk Kazakh. *see*
      Baykonyr
89 E3   Leninskiy Rus. Fed.
53 D1   Lennox Head Austr.
141 D1   Lenoir U.S.A.
100 A2   Lens Belgium
105 C1   Lens France
83 J2   Lensk Rus. Fed.
103 D2   Lenti Hungary
109 C3   Lentini Italy
114 B3   Léo Burkina
103 D2   Leoben Austria
      Leodhais, Eilean *i.* U.K. *see*
      Lewis, Isle of
99 B3   Leominster U.K.
144 B2   León Mex.
146 B3   León Nic.
106 B1   León Spain
122 A1   Leonardville Namibia
108 B3   Leonforte Italy
53 C3   Leongatha Austr.
50 B2   Leonora Austr.
      Léopold II, Lac *l.*
      Dem. Rep. Congo *see*
      Mai-Ndombe, Lac
155 D2   Leopoldina Brazil
      Léopoldville Dem. Rep.
      Congo *see* Kinshasa
90 B2   Leova Moldova
      Leovo Moldova *see* Leova
123 C1   Lephalale S. Africa
123 C1   Lephepe Botswana
123 C3   Lephoi S. Africa
71 B3   Leping China
77 D2   Lepsi Kazakh.
105 C2   Le Puy-en-Velay France
123 C1   Lerala Botswana
123 C2   Leratswana S. Africa
150 B3   Lérida Col.
      Lérida Spain *see* Lleida
106 C1   Lerma Spain
111 C3   Leros *i.* Greece

130 C2   Le Roy, Lac *l.* Can.
93 F4   Lerum Sweden
96 □   Lerwick U.K.
111 C3   Lesbos *i.* Greece
147 C3   Les Cayes Haiti
104 C3   Les Escaldes Andorra
139 F1   Les Escoumins Can.
107 D1   Le Seu d'Urgell Spain
70 A3   Leshan China
104 B2   Les Herbiers France
86 D2   Leshukonskoye Rus. Fed.
      Leskhimstroy Ukr. *see*
      Syeverodonets'k
109 D2   Leskovac Serbia
104 B2   Lesneven France
      Lesnoy Rus. Fed. *see* Umba
89 E2   Lesnoye Rus. Fed.
83 H3   Lesosibirsk Rus. Fed.
123 C2   Lesotho *country* Africa
66 B1   Lesozavodsk Rus. Fed.
104 B2   Les Sables-d'Olonne
      France
147 D3   Lesser Antilles *is*
      Caribbean Sea
81 C1   Lesser Caucasus *mts* Asia
      Lesser Khingan Mountains
      mts China *see*
      Xiao Hinggan Ling
128 C2   Lesser Slave Lake Can.
58 B3   Lesser Sunda Islands
      Indon.
105 C3   Les Vans France
      Lesvos *i.* Greece *see* Lesbos
103 D1   Leszno Pol.
123 D1   Letaba S. Africa
99 C3   Letchworth Garden City
      U.K.
128 C3   Lethbridge Can.
150 D2   Lethem Guyana
59 C3   Leti, Kepulauan *is* Indon.
150 C3   Leticia Col.
120 B3   Letlhakane Botswana
123 C1   Letlhakeng Botswana
104 C1   Le Touquet-Paris-Plage
      France
99 D4   Le Tréport France
123 D1   Letsitele S. Africa
63 A2   Letsok-aw Kyun *i.* Myanmar
123 C2   Letsopa S. Africa
97 C1   Letterkenny Ireland
105 C3   Leucate, Étang de *l.* France
      Leukas Greece *see* Lefkada
60 A1   Leuser, Gunung *mt.* Indon.
100 B3   Leuven Belgium
92 F3   Levanger Norway
143 C2   Levelland U.S.A.
50 B1   Lévêque, Cape Austr.
96 A2   Leverburgh U.K.
100 C2   Leverkusen Ger.
103 D2   Levice Slovakia
54 C2   Levin N.Z.
131 C3   Lévis Can.
139 E2   Levittown U.S.A.
89 E3   Lev Tolstoy Rus. Fed.
99 D4   Lewes U.K.
96 A1   Lewis, Isle of *i.* U.K.
139 D2   Lewisburg PA U.S.A.
140 C1   Lewisburg TN U.S.A.
138 C3   Lewisburg WV U.S.A.
134 D1   Lewis Range *mts* U.S.A.
134 C1   Lewiston ID U.S.A.
139 E2   Lewiston ME U.S.A.
134 E1   Lewistown U.S.A.
138 C3   Lexington KY U.S.A.
136 D2   Lexington NE U.S.A.
143 D1   Lexington OK U.S.A.
139 D3   Lexington VA U.S.A.
64 B2   Leyte *i.* Phil.
109 C2   Lezhë Albania
89 F2   Lezhnevo Rus. Fed.
89 E3   L'gov Rus. Fed.
75 C2   Lhagoi Kangri *mt.* China
75 D1   Lharigarbo China
75 D2   Lhasa China
75 C2   Lhazê China
60 A1   Lhokseumawe Indon.
62 A1   Lhünzê China
111 B3   Liakoura *mt.* Greece
67 B3   Liancourt Rocks *is*
      N. Pacific Ocean
      Liangzhou China *see*
      Wuwei
70 B2   Liangzi Hu *l.* China
      Lianhe China *see*
      Qianjiang
71 B3   Lianhua China
71 B3   Lianjiang China
      Lianran China *see* Anning

70 C1   Lianshan China
      Liantang China *see*
      Nanchang
      Lianxian China *see*
      Lianzhou
70 B2   Lianyungang China
71 B3   Lianzhou China
      Lianzhou China *see* Hepu
70 B2   Liaocheng China
70 C1   Liaodong Bandao *pen.*
      China
70 C1   Liaodong Wan *b.* China
65 A1   Liao He *r.* China
70 C1   Liaoning *prov.* China
70 C1   Liaoyang China
65 B1   Liaoyuan China
128 B1   Liard *r.* Can.
134 C1   Libby U.S.A.
118 B2   Libenge Dem. Rep. Congo
136 C3   Liberal U.S.A.
103 D1   Liberec Czech Rep.
114 B4   Liberia *country* Africa
146 B3   Liberia Costa Rica
150 C2   Libertad Venez.
137 E3   Liberty U.S.A.
100 B3   Libin Belgium
64 B2   Libmanan Phil.
71 A3   Libo China
123 C3   Libode S. Africa
119 E2   Liboi Kenya
104 B3   Libourne France
100 B3   Libramont Belgium
142 A3   Libres, Sierra *mts* Mex.
118 A2   Libreville Gabon
115 D2   Libya *country* Africa
115 E2   Libyan Desert Egypt/Libya
116 A1   Libyan Plateau Egypt/Libya
108 B3   Licata Italy
      Licheng China *see* Lipu
99 C3   Lichfield U.K.
121 C2   Lichinga Moz.
101 E2   Lichte Ger.
123 C2   Lichtenburg S. Africa
101 E2   Lichtenfels Ger.
88 C3   Lida Belarus
93 F4   Lidköping Sweden
50 C2   Liebig, Mount Austr.
105 D2   Liechtenstein *country* Europe
100 B2   Liège Belgium
92 J3   Lieksa Fin.
119 C2   Lienart Dem. Rep. Congo
63 B2   Liên Nghia Vietnam
102 C2   Lienz Austria
88 B2   Liepāja Latvia
100 B2   Lier Belgium
102 C2   Liezen Austria
97 C2   Liffey *r.* Ireland
97 C1   Lifford Ireland
53 C1   Lightning Ridge Austr.
121 C2   Ligonha *r.* Moz.
105 D3   Ligurian Sea France/Italy
119 C4   Likasi Dem. Rep. Congo
118 C2   Likati Dem. Rep. Congo
128 B2   Likely Can.
      Likhachevo Ukr. *see*
      Pervomays'kyy
      Likhachyovo Ukr. *see*
      Pervomays'kyy
89 E2   Likhoslavl' Rus. Fed.
60 B1   Liku Indon.
105 D3   L'Île-Rousse France
71 B3   Liling China
93 F4   Lilla Edet Sweden
100 B3   Lille Belgium
105 C1   Lille France
      Lille Bælt *sea chan.*
      Denmark *see* Little Belt
93 F3   Lillehammer Norway
93 F4   Lillestrøm Norway
128 B2   Lillooet Can.
121 C2   Lilongwe Malawi
64 B3   Liloy Phil.
150 B4   Lima Peru
138 C2   Lima U.S.A.
155 D2   Lima Duarte Brazil
79 C2   Limah Oman
80 B2   Limassol Cyprus
97 C1   Limavady U.K.
153 B3   Limay *r.* Arg.
101 F2   Limbach-Oberfrohna Ger.
88 B2   Limbaži Latvia
118 A2   Limbe Cameroon
101 D2   Limburg an der Lahn Ger.
122 B2   Lime Acres S. Africa
154 C2   Limeira Brazil
97 B2   Limerick Ireland
93 E4   Limfjorden *sea chan.*
      Denmark

92 I3   Liminka Fin.
111 C3   Limnos *i.* Greece
104 C2   Limoges France
136 C3   Limon U.S.A.
104 C2   Limousin, Plateaux du
      France
104 C3   Limoux France
123 C1   Limpopo *prov.* S. Africa
121 C3   Limpopo *r.* S. Africa/
      Zimbabwe
64 A2   Linapacan *i.* Phil.
153 A3   Linares Chile
145 C2   Linares Mex.
106 C2   Linares Spain
108 A3   Linas, Monte *mt.* Italy
62 B1   Lincang China
      Linchuan China *see* Fuzhou
98 C3   Lincoln U.K.
138 B2   Lincoln IL U.S.A.
139 F1   Lincoln ME U.S.A.
137 D2   Lincoln NE U.S.A.
134 B2   Lincoln City U.S.A.
151 D2   Linden Guyana
140 C1   Linden U.S.A.
119 C2   Lindi *r.* Dem. Rep. Congo
119 D3   Lindi Tanz.
      Lindisfarne *i.* U.K. *see*
      Holy Island
111 C3   Lindos Greece
130 C3   Lindsay Can.
48 K3   Line Islands Kiribati
70 B2   Linfen China
64 B2   Lingayen Phil.
70 B2   Lingbao China
      Lingcheng China *see*
      Lingshan
      Lingcheng China *see*
      Lingshui
123 C3   Lingelethu S. Africa
123 C3   Lingelihle S. Africa
100 C1   Lingen (Ems) Ger.
60 B1   Lingga *i.* Indon.
60 B2   Lingga, Kepulauan *is* Indon.
71 A3   Lingshan China
71 A4   Lingshui China
114 A3   Linguère Senegal
155 D1   Linhares Brazil
70 A1   Linhe China
      Linjiang China *see*
      Shanghang
71 B3   Linjiang China
93 G4   Linköping Sweden
66 B1   Linkou China
71 B3   Linli China
96 B2   Linnhe, Loch *inlet* U.K.
70 B2   Linqing China
154 C2   Lins Brazil
136 C1   Linton U.S.A.
69 D2   Linxi China
70 A2   Linxia China
70 B2   Linyi Shandong China
70 B2   Linyi Shandong China
70 B2   Linying China
102 C2   Linz Austria
105 C3   Lion, Golfe du *g.* France
      Lions, Gulf of *g.* France *see*
      Lion, Golfe du
109 B3   Lipari Italy
108 B3   Lipari, Isole *is* Italy
89 E3   Lipetsk Rus. Fed.
110 B1   Lipova Romania
101 D2   Lippstadt Ger.
53 C3   Liptrap, Cape Austr.
71 B3   Lipu China
119 D2   Lira Uganda
108 B2   Liri *r.* Italy
76 C1   Lisakovsk Kazakh.
118 C2   Lisala Dem. Rep. Congo
      Lisboa Port. *see* Lisbon
106 B2   Lisbon Port.
97 C1   Lisburn U.K.
97 B2   Liscannor Bay Ireland
97 B2   Lisdoonvarna Ireland
      Lishi China *see* Dingnan
71 B3   Lishui China
104 C2   Lisieux France
99 A4   Liskeard U.K.
89 E3   Liski Rus. Fed.
53 D1   Lismore Austr.
97 C1   Lismore Ireland
97 C1   Lisnaskea U.K.
97 B1   Listowel Ireland
71 A3   Litang Guangxi China
68 C2   Litang Sichuan China
138 B3   Litchfield IL U.S.A.
137 E1   Litchfield MN U.S.A.
53 D2   Lithgow Austr.
111 B3   Lithino, Akrotirio *pt* Greece

88 B2 Lithuania *country* Europe
111 B2 Litochoro Greece
102 C1 Litoměřice Czech Rep.
Litovskaya S.S.R. *country* Europe *see* Lithuania
146 C2 Little Abaco *i.* Bahamas
73 D3 Little Andaman *i.* India
147 E3 Little Bahama Bank *sea feature* Bahamas
93 E4 Little Belt *sea chan.* Denmark
146 B3 Little Cayman *i.* Cayman Is
142 A1 Little Colorado *r.* U.S.A.
138 C1 Little Current Can.
137 E1 Little Falls U.S.A.
143 C2 Littlefield U.S.A.
99 C4 Littlehampton U.K.
122 A2 Little Karas Berg *plat.* Namibia
122 B3 Little Karoo *plat.* S. Africa
96 A2 Little Minch *sea chan.* U.K.
136 C1 Little Missouri *r.* U.S.A.
73 D4 Little Nicobar *i.* India
140 B2 Little Rock U.S.A.
139 E2 Littleton U.S.A.
121 C2 Litunde Moz.
90 B2 Lityn Ukr.
Liuchow China *see* Liuzhou
70 B2 Liujiachang China
Liupanshui China *see* Lupanshui
121 C2 Liupo Moz.
71 A3 Liuzhou China
111 B3 Livadeia Greece
88 C2 Līvāni Latvia
141 D2 Live Oak U.S.A.
50 B1 Liveringa Austr.
142 C2 Livermore, Mount U.S.A.
53 D2 Liverpool Austr.
131 D3 Liverpool Can.
98 B3 Liverpool U.K.
127 G2 Liverpool, Cape Can.
134 C1 Liverpool Range *mts* Austr.
136 C3 Livingston *MT* U.S.A.
143 C1 Livingston *TX* U.S.A.
143 D2 Livingston, Lake U.S.A.
120 B2 Livingstone Zambia
55 A3 Livingston Island Antarctica
109 C2 Livno Bos.-Herz.
89 E3 Livny Rus. Fed.
138 C2 Livonia U.S.A.
108 B2 Livorno Italy
119 D3 Liwale Tanz.
99 A5 Lizard Point U.K
108 B1 Ljubljana Slovenia
93 G3 Ljungan *r.* Sweden
93 F4 Ljungby Sweden
93 G3 Ljusdal Sweden
93 G3 Ljusnan *r.* Sweden
99 B4 Llandeilo U.K.
99 B3 Llandovery U.K.
99 B3 Llandrindod Wells U.K.
98 B3 Llandudno U.K.
99 A4 Llanelli U.K.
106 C1 Llanes Spain
98 A3 Llangefni U.K.
99 B3 Llangollen U.K.
99 B3 Llangurig U.K.
143 C2 Llano Estacado *plain* U.S.A.
150 C2 Llanos *plain* Col./Venez.
107 D1 Lleida Spain
99 A3 Lleyn Peninsula U.K.
107 C2 Llíria Spain
106 C1 Llodio Spain
128 B2 Lloyd George, Mount Can.
129 D2 Lloyd Lake Can.
129 C2 Lloydminster Can.
152 B2 Llullaillaco, Volcán *vol.* Chile
154 B2 Loanda Brazil
123 C2 Lobatse Botswana
103 D1 Łobez Pol.
120 A2 Lobito Angola
101 F1 Loburg Ger.
96 B2 Lochaber *reg.* U.K.
96 B2 Lochaline U.K.
Loch Baghasdail U.K. *see* Lochboisdale
96 A2 Lochboisdale U.K.
104 C2 Loches France
96 B2 Lochgilphead U.K.
96 B1 Lochinver U.K.
96 A2 Lochmaddy U.K.
96 C2 Lochnagar *mt.* U.K.
Loch nam Madadh U.K. *see* Lochmaddy

96 B3 Lochranza U.K.
52 A2 Lock Austr.
96 C3 Lockerbie U.K.
53 C3 Lockhart Austr.
143 D3 Lockhart U.S.A.
51 D1 Lockhart River Austr.
139 D2 Lock Haven U.S.A.
139 D2 Lockport U.S.A.
63 B2 Lôc Ninh Vietnam
105 C3 Lodève France
115 D2 Lodeynoye Pole Rus. Fed.
74 B2 Lodhran Pak.
108 A1 Lodi Italy
135 B3 Lodi U.S.A.
92 F2 Løding Norway
92 G2 Lødingen Norway
118 C3 Lodja Dem. Rep. Congo
119 D2 Lodwar Kenya
103 D1 Łódź Pol.
62 B2 Loei Thai.
122 A3 Loeriesfontein S. Africa
92 F2 Lofoten *is* Norway
134 D2 Logan U.S.A.
128 A1 Logan, Mount Can.
138 B2 Logansport U.S.A.
108 B1 Logatec Slovenia
118 B2 Logone *r.* Africa
106 C1 Logroño Spain
93 H3 Lohja Fin.
101 D1 Löhne Ger.
101 D1 Lohne (Oldenburg) Ger.
62 A2 Loikaw Myanmar
93 H3 Loimaa Fin.
104 B2 Loire *r.* France
150 B3 Loja Ecuador
106 C2 Loja Spain
92 I2 Lokan tekojärvi *resr* Fin.
100 B2 Lokeren Belgium
122 B1 Lokgwabe Botswana
91 C1 Lokhvytsya Ukr.
119 D2 Lokichar Kenya
119 D2 Lokichokio Kenya
93 E4 Løkken Denmark
89 D2 Loknya Rus. Fed.
115 C4 Lokoja Nigeria
89 D3 Lokot' Rus. Fed.
88 C2 Loksa Estonia
127 H2 Loks Land *i.* Can.
114 B4 Lola Guinea
93 F5 Lolland *i.* Denmark
119 D3 Lollondo Tanz.
118 C2 Lolo Dem. Rep. Congo
122 B2 Lolwane S. Africa
110 B2 Lom Bulg.
93 E3 Lom Norway
119 C2 Lomami *r.* Dem. Rep. Congo
153 C3 Lomas de Zamora Arg.
50 B1 Lombardina Austr.
61 C2 Lombok *i.* Indon.
61 C2 Lombok, Selat *sea chan.* Indon.
114 C4 Lomé Togo
118 C3 Lomela *r.* Dem. Rep. Congo
100 B2 Lommel Belgium
96 B2 Lomond, Loch *l.* U.K.
88 C2 Lomonosov Rus. Fed.
160 A1 Lomonosov Ridge Arctic Ocean
61 C2 Lompobattang, Gunung *mt.* Indon.
135 B4 Lompoc U.S.A.
63 B2 Lom Sak Thai.
103 E1 Łomża Pol.
130 B3 London Can.
99 C4 London U.K.
138 C3 London U.S.A.
97 C1 Londonderry U.K.
50 B1 Londonderry, Cape Austr.
154 C2 Londrina Brazil
135 C3 Lone Pine U.S.A.
83 M2 Longa, Proliv *sea chan.* Rus. Fed.
61 C1 Long Akah Malaysia
141 E2 Long Bay U.S.A.
135 C4 Long Beach U.S.A.
71 A3 Longchang China
99 C3 Long Eaton U.K.
97 C2 Longford Ireland
96 C1 Longhope U.K.
119 D3 Longido Tanz.
61 C2 Longiram Indon.
147 C2 Long Island Bahamas
130 C2 Long Island Can.
59 D3 Long Island P.N.G.
139 E2 Long Island U.S.A.
130 B3 Longlac Can.
130 B3 Long Lake Can.

71 A3 Longli China
71 A3 Longming China
136 B2 Longmont U.S.A.
70 A2 Longnan China
Longping China *see* Luodian
138 C2 Long Point Can.
71 B3 Longquan China
131 E3 Long Range Mountains Can.
51 D2 Longreach Austr.
Longshan China *see* Longli
99 D3 Long Stratton U.K.
98 B2 Longtown U.K.
105 D2 Longuyon France
143 D2 Longview *TX* U.S.A.
134 B1 Longview *WA* U.S.A.
61 C1 Longwai Indon.
70 A2 Longxi China
Longxian China *see* Wengyuan
71 B3 Longxi Shan *mt.* China
63 B2 Long Xuyên Vietnam
71 B3 Longyan China
82 C1 Longyearbyen Svalbard
108 B1 Lonigo Italy
100 C1 Löningen Ger.
105 D2 Lons-le-Saunier France
141 E2 Lookout, Cape U.S.A.
97 B2 Loop Head *hd* Ireland
Lopasnya Rus. Fed. *see* Chekhov
63 B2 Lop Buri Thai.
64 B2 Lopez Phil.
118 A3 Lopez, Cap *c.* Gabon
68 C2 Lop Nur *salt flat* China
118 B2 Lopori *r.* Dem. Rep. Congo
92 H1 Lopphavet *b.* Norway
74 A2 Lora, Hāmūn-i- *dry lake* Afgh./Pak.
106 B2 Lora del Río Spain
138 C2 Lorain U.S.A.
74 A1 Loralai Pak.
107 C2 Lorca Spain
51 E3 Lord Howe Island Austr.
142 B2 Lordsburg U.S.A.
155 C2 Lorena Brazil
59 D3 Lorengau P.N.G.
59 D3 Lorentz *r.* Indon.
152 B1 Loreto Bol.
144 A2 Loreto Mex.
104 B2 Lorient France
96 B2 Lorn, Firth of *est.* U.K.
52 B3 Lorne Austr.
105 D2 Lorraine *reg.* France
142 B1 Los Alamos U.S.A.
143 D3 Los Aldamas Mex.
153 A3 Los Ángeles Chile
135 C4 Los Angeles U.S.A.
135 B3 Los Banos U.S.A.
152 B2 Los Blancos Arg.
89 F3 Losevo Rus. Fed.
108 B2 Lošinj *i.* Croatia
144 B2 Los Mochis Mex.
118 B2 Losombo Dem. Rep. Congo
106 B2 Los Pedroches *plat.* Spain
147 D3 Los Roques, Islas *is* Venez.
96 C2 Lossiemouth U.K.
150 C2 Los Teques Venez.
59 E3 Losuia P.N.G.
152 A3 Los Vilos Chile
104 C3 Lot *r.* France
96 C1 Loth U.K.
134 D1 Lothair U.S.A.
Lothringen *reg.* France *see* Lorraine
119 D2 Lotikipi Plain Kenya/Sudan
118 C3 Loto Dem. Rep. Congo
89 E2 Lotoshino Rus. Fed.
62 B1 Louangnamtha Laos
62 B2 Louangphabang Laos
118 B3 Loubomo Congo
104 B2 Loudéac France
71 B3 Loudi China
118 B3 Loudima Congo
114 A3 Louga Senegal
99 C3 Loughborough U.K.
97 B2 Loughrea Ireland
105 D2 Louhans France
97 B2 Louisburgh Ireland
51 E1 Louisiade Archipelago *is* P.N.G.
140 B2 Louisiana *state* U.S.A.
123 C1 Louis Trichardt S. Africa
138 B3 Louisville *KY* U.S.A.

140 C2 Louisville *MS* U.S.A.
86 C2 Loukhi Rus. Fed.
118 B3 Loukoléla Congo
106 B2 Loulé Port.
118 A2 Loum Cameroon
130 C2 Loups Marins, Lacs des *lakes* Can.
104 B3 Lourdes France
151 D2 Lourenço Brazil
Lourenço Marques Moz. *see* Maputo
106 B1 Lousã Port.
53 C2 Louth Austr.
98 C3 Louth U.K.
Louvain Belgium *see* Leuven
122 A1 Louwater-Suid Namibia
89 D2 Lovat' *r.* Rus. Fed.
110 B2 Lovech Bulg.
136 B2 Loveland U.S.A.
136 B2 Lovell U.S.A.
135 C2 Lovelock U.S.A.
88 C1 Loviisa Fin.
143 C2 Lovington U.S.A.
86 C2 Lovozero Rus. Fed.
119 C3 Lowa Dem. Rep. Congo
139 E2 Lowell U.S.A.
119 D2 Lowelli Sudan
128 C3 Lower Arrow Lake Can.
Lower California *pen.* Mex. *see* Baja California
54 B2 Lower Hutt N.Z.
97 C1 Lower Lough Erne *l.* U.K.
128 B2 Lower Post Can.
137 E1 Lower Red Lake U.S.A.
Lower Tunguska *r.* Rus. Fed. *see* Nizhnyaya Tunguska
99 D3 Lowestoft U.K.
103 D1 Łowicz Pol.
139 D2 Lowville U.S.A.
52 B2 Loxton Austr.
Loyang China *see* Luoyang
48 H6 Loyauté, Îles New Caledonia
89 D3 Loyew Belarus
92 F2 Løypskardtinden *mt.* Norway
109 C2 Loznica Serbia
91 D2 Lozova Ukr.
120 B2 Luacano Angola
70 B2 Lu'an China
120 A1 Luanda Angola
63 B3 Luang, Thale *lag.* Thai.
121 C2 Luangwa *r.* Zambia
121 B2 Luanshya Zambia
Luao Angola *see* Luau
106 B1 Luarca Spain
120 B2 Luau Angola
103 E1 Lubaczów Pol.
120 A2 Lubango Angola
119 C3 Lubao Dem. Rep. Congo
103 E1 Lubartów Pol.
101 D1 Lübbecke Ger.
102 C1 Lübben Ger.
143 C2 Lubbock U.S.A.
101 E1 Lübeck Ger.
69 E2 Lubei China
76 B1 Lubenka Kazakh.
119 C3 Lubero Dem. Rep. Congo
103 D1 Lubin Pol.
103 E1 Lublin Pol.
91 C1 Lubny Ukr.
61 C1 Lubok Antu Malaysia
101 E1 Lübtheen Ger.
101 E1 Lübz Ger.
119 C3 Lubudi Dem. Rep. Congo
60 B2 Lubuklinggau Indon.
119 C4 Lubumbashi Dem. Rep. Congo
120 B2 Lubungu Zambia
119 C3 Lubutu Dem. Rep. Congo
120 A1 Lucala Angola
120 B1 Lucapa Angola
97 C2 Lucan Ireland
108 B2 Lucca Italy
96 B3 Luce Bay U.K.
154 B2 Lucélia Brazil
64 B2 Lucena Phil.
106 C2 Lucena Spain
103 D2 Lučenec Slovakia
109 C2 Lucera Italy
105 D2 Lucerne Switz.
66 B1 Luchegorsk Rus. Fed.
101 E1 Lüchow Ger.
120 A2 Lucira Angola
Łuck Ukr. *see* Luts'k

08 A1 **Maggiore, Lake** l. Italy
16 B2 **Maghāghah** Egypt
97 C1 **Magherafelt** U.K.
87 E3 **Magnitogorsk** Rus. Fed.
40 B2 **Magnolia** U.S.A.
21 C2 **Magoé** Moz.
30 C3 **Magog** Can.
31 D2 **Magpie, Lac** l. Can.
44 A3 **Magta' Lahjar** Maur.
81 D2 **Magtymguly** Turkm.
19 D3 **Magu** Tanz.
51 E3 **Maguarinho, Cabo** c. Brazil
23 D2 **Magude** Moz.
90 B2 **Măgura, Dealul** h. Moldova
62 A1 **Magwe** Myanmar
81 C2 **Mahābād** Iran
21 A1 **Mahajan** India
21 □D2 **Mahajanga** Madag.
21 A1 **Mahakam** r. Indon.
23 C1 **Mahalapye** Botswana
21 □D2 **Mahalevona** Madag.
75 C2 **Mahanadi** r. India
21 □D2 **Mahanoro** Madag.
74 B3 **Maharashtra** state India
63 B2 **Maha Sarakham** Thai.
21 □D2 **Mahavavy** r. Madag.
68 B3 **Mahbubnagar** India
78 B2 **Mahd adh Dhahab** Saudi Arabia
07 D2 **Mahdia** Alg.
50 D2 **Mahdia** Guyana
13 I6 **Mahé** i. Seychelles
75 C3 **Mahendragiri** mt. India
19 D3 **Mahenge** Tanz.
54 B3 **Maheno** N.Z.
74 B2 **Mahesana** India
74 B1 **Mahi** r. India
54 C1 **Mahia Peninsula** N.Z.
89 D3 **Mahilyow** Belarus
07 D2 **Mahón** Spain
14 B3 **Mahou** Mali
Mahsana India see **Mahesana**
74 B2 **Mahuva** India
11 C? **Mahya Daği** mt. Turkey
06 B1 **Maia** Port.
Maiaia Moz. see **Nacala**
47 C3 **Maicao** Col.
74 A1 **Maīdān Shahr** Afgh.
29 D2 **Maidstone** Can.
99 D4 **Maidstone** U.K.
15 D3 **Maiduguri** Nigeria
75 D2 **Maijdi** Bangl.
75 C2 **Mailani** India
01 D2 **Mainz** Ger.
18 B3 **Mai-Ndombe, Lac** l. Dem. Rep. Congo
01 E3 **Main-Donau-Kanal** canal Ger.
39 F1 **Maine** state U.S.A.
31 D3 **Maine, Gulf of** Can./U.S.A.
62 A1 **Maingkwan** Myanmar
96 C1 **Mainland** i. Scotland U.K.
96 □ **Mainland** i. Scotland U.K.
21 □D2 **Maintirano** Madag.
01 D2 **Mainz** Ger.
50 C1 **Maiquetía** Venez.
20 B3 **Maitengwe** Botswana
53 D2 **Maitland** N.S.W. Austr.
52 A2 **Maitland** S.A. Austr.
46 B3 **Maíz, Islas del** is Nic.
67 C3 **Maizuru** Japan
09 C2 **Maja Jezercë** mt. Albania
61 C2 **Majene** Indon.
19 D2 **Majī** Eth.
07 D2 **Majorca** i. Spain
Majunga Madag. see **Mahajanga**
23 C2 **Majwemasweu** S. Africa
18 B3 **Makabana** Congo
61 C2 **Makale** Indon.
18 C3 **Makamba** Burundi
77 E2 **Makanshy** Kazakh.
18 B3 **Makanza** Dem. Rep. Congo
90 B1 **Makariv** Ukr.
69 F1 **Makarov** Rus. Fed.
60 B1 **Makarov Basin** Arctic Ocean
09 C2 **Makarska** Croatia
61 C2 **Makassar** Indon.
61 C2 **Makassar, Selat** Indon.
61 C2 **Makat** Kazakh.
19 D3 **Makatapora** Tanz.
Makatini Flats lowland S. Africa
14 A4 **Makeni** Sierra Leone
20 B3 **Makgadikgadi** depr. Botswana

87 D4 **Makhachkala** Rus. Fed.
76 B2 **Makhambet** Kazakh.
119 D3 **Makindu** Kenya
77 D1 **Makinsk** Kazakh.
91 D2 **Makiyivka** Ukr.
Makkah Saudi Arabia see **Mecca**
131 E2 **Makkovik** Can.
103 E2 **Makó** Hungary
118 B2 **Makokou** Gabon
119 D3 **Makongolosi** Tanz.
122 B2 **Makopong** Botswana
119 C2 **Makoro** Dem. Rep. Congo
118 B3 **Makoua** Congo
111 B3 **Makrakomi** Greece
79 D2 **Makran** reg. Iran/Pak.
74 A2 **Makran Coast Range** mts Pak.
89 E2 **Maksatikha** Rus. Fed.
81 C2 **Mākū** Iran
62 A1 **Makum** India
67 B4 **Makurazaki** Japan
115 C4 **Makurdi** Nigeria
92 G2 **Malá** Sweden
146 B4 **Mala, Punta** pt Panama
73 B3 **Malabar Coast** India
118 A2 **Malabo** Equat. Guinea
155 D1 **Malacacheta** Brazil
Malacca Malaysia see **Melaka**
60 A1 **Malacca, Strait of** Indon./ Malaysia
134 D2 **Malad City** U.S.A.
88 C3 **Maladzyechna** Belarus
106 C2 **Málaga** Spain
Malagasy Republic country Africa see **Madagascar**
121 □D3 **Malaimbandy** Madag.
97 B1 **Málainn Mhóir** Ireland
48 H4 **Malaita** i. Solomon Is
117 B4 **Malakal** Sudan
48 H5 **Malakula** i. Vanuatu
61 D2 **Malamala** Indon.
61 C2 **Malang** Indon.
Malange Angola see **Malanje**
120 A1 **Malanje** Angola
93 G4 **Mälaren** l. Sweden
153 B3 **Malargüe** Arg.
130 C3 **Malartic** Can.
88 B3 **Malaryta** Belarus
80 B2 **Malatya** Turkey
121 C2 **Malawi** country Africa
Malawi, Lake l. Africa see **Nyasa, Lake**
110 A3 **Malaya Vishera** Rus. Fed.
64 B3 **Malaybalay** Phil.
81 C2 **Maläyer** Iran
60 B1 **Malaysia** country Asia
81 C2 **Malazgirt** Turkey
103 D1 **Malbork** Pol.
101 F1 **Malchin** Ger.
100 A2 **Maldegem** Belgium
48 L4 **Malden Island** Kiribati
56 C5 **Maldives** country Indian Ocean
99 D4 **Maldon** U.K.
56 I9 **Male** Maldives
111 B3 **Maleas, Akrotirio** pt Greece
103 D2 **Malé Karpaty** hills Slovakia
119 C3 **Malela** Dem. Rep. Congo
116 A3 **Malha** Sudan
134 C2 **Malheur Lake** U.S.A.
114 B3 **Mali** country Africa
114 A3 **Mali** Guinea
59 C3 **Maliana** East Timor
58 C3 **Malili** Indon.
97 C1 **Malin** Ireland
119 E3 **Malindi** Kenya
97 C1 **Malin Head** hd Ireland
Malin More Ireland see **Málainn Mhór**
111 C2 **Malkara** Turkey
88 C3 **Mal'kavichy** Belarus
110 C2 **Malko Tŭrnovo** Bulg.
53 C3 **Mallacoota** Austr.
53 C3 **Mallacoota Inlet** b. Austr.
96 B2 **Mallaig** U.K.
116 B2 **Mallawi** Egypt
129 E1 **Malley Lake** Can.
Mallorca i. Spain see **Majorca**
97 B2 **Mallow** Ireland
92 F3 **Malm** Norway
92 H2 **Malmberget** Sweden
100 C2 **Malmédy** Belgium
122 A3 **Malmesbury** S. Africa
93 F4 **Malmö** Sweden

71 A3 **Malong** China
118 C4 **Malonga** Dem. Rep. Congo
86 C2 **Maloshuyka** Rus. Fed.
93 E3 **Måløy** Norway
89 E2 **Maloyaroslavets** Rus. Fed.
89 E2 **Maloye Borisovo** Rus. Fed.
86 D2 **Malozemel'skaya Tundra** lowland Rus. Fed.
125 J9 **Malpelo, Isla de** i. N. Pacific Ocean
84 F5 **Malta** country Europe
88 C2 **Malta** Latvia
134 E1 **Malta** U.S.A.
122 A1 **Maltahöhe** Namibia
98 C2 **Malton** U.K.
Maluku is Indon. see **Moluccas**
93 F3 **Malung** Sweden
123 C2 **Maluti Mountains** Lesotho
73 B3 **Malvan** India
140 B2 **Malvern** U.S.A.
117 B4 **Malwal** Sudan
90 B1 **Malyn** Ukr.
83 L2 **Malyy Anyuy** r. Rus. Fed.
Malyy Kavkaz mts Asia see **Lesser Caucasus**
83 K2 **Malyy Lyakhovskiy, Ostrov** i. Rus. Fed.
151 F3 **Mamafubedu** S. Africa
151 F3 **Mamanguape** Brazil
64 B3 **Mambajao** Phil.
119 C2 **Mambasa** Dem. Rep. Congo
118 B2 **Mambéré** r. C.A.R.
64 B2 **Mamburao** Phil.
123 C2 **Mamelodi** S. Africa
118 A2 **Mamfe** Cameroon
135 C3 **Mammoth Lakes** U.S.A.
88 A3 **Mamonovo** Rus. Fed.
150 C4 **Mamoré** r. Bol./Brazil
114 A3 **Mamou** Guinea
114 B4 **Mampong** Ghana
61 C2 **Mamuju** Indon.
114 B4 **Man** Côte d'Ivoire
150 C3 **Manacapuru** Brazil
107 D2 **Manacor** Spain
59 C2 **Manado** Indon.
146 B3 **Managua** Nic.
121 □D3 **Manakara** Madag.
78 B3 **Manākhah** Yemen
79 C2 **Manama** Bahrain
59 D3 **Manam Island** P.N.G.
121 □D3 **Mananara** r. Madag.
121 □D2 **Mananara Avaratra** Madag.
121 □D3 **Mananjary** Madag.
114 A3 **Manantali, Lac de** l. Mali
54 A3 **Manapouri, Lake** l. N.Z.
77 E2 **Manas Hu** l. China
75 C2 **Manaslu** mt. Nepal
Manastir Macedonia see **Bitola**
59 C3 **Manatuto** East Timor
62 A2 **Man-aung Kyun** Myanmar
150 C3 **Manaus** Brazil
80 B2 **Manavgat** Turkey
116 A3 **Manawashei** Sudan
98 B3 **Manchester** U.K.
139 E2 **Manchester** CT U.S.A.
139 E2 **Manchester** NH U.S.A.
140 C1 **Manchester** TN U.S.A.
81 D3 **Mand, Rūd-e** r. Iran
117 A4 **Manda, Jebel** mt. Sudan
121 □D3 **Mandabe** Madag.
93 E4 **Mandal** Norway
59 D3 **Mandala, Puncak** mt. Indon.
62 A1 **Mandalay** Myanmar
68 D1 **Mandalgovĭ** Mongolia
136 C1 **Mandan** U.S.A.
118 B1 **Mandara Mountains** Cameroon/Nigeria
108 A3 **Mandas** Italy
119 E2 **Mandera** Kenya
100 C2 **Manderscheid** Ger.
74 B1 **Mandi** India
114 B3 **Mandiana** Guinea
Mandidzudze Zimbabwe see **Chimanimani**
75 C2 **Mandla** India
121 □D2 **Mandritsara** Madag.
74 B2 **Mandsaur** India
50 A3 **Mandurah** Austr.
73 B3 **Mandya** India
108 B1 **Manerbio** Italy
90 B1 **Manevychi** Ukr.
109 C2 **Manfredonia** Italy
109 C2 **Manfredonia, Golfo di** g. Italy
114 B3 **Manga** Burkina

118 B3 **Mangai** Dem. Rep. Congo
49 L6 **Mangaia** i. Cook Is
54 C1 **Mangakino** N.Z.
110 C2 **Mangalia** Romania
73 B3 **Mangalore** India
123 C2 **Manganui** S. Africa
60 B2 **Manggar** Indon.
Mangghyshlaq Kazakh. see **Mangistau**
76 B2 **Mangistau** Kazakh.
61 C1 **Mangkalihat, Tanjung** pt Indon.
68 C2 **Mangnai** China
121 C2 **Mangochi** Malawi
121 □D3 **Mangoky** r. Madag.
59 C3 **Mangole** i. Indon.
54 B1 **Mangonui** N.Z.
Mangshi China see **Luxi**
106 B1 **Mangualde** Port.
154 B3 **Mangueirinha** Brazil
69 E1 **Mangui** China
Mangyshlak Kazakh. see **Mangistau**
137 D3 **Manhattan** U.S.A.
121 C3 **Manhica** Moz.
155 D2 **Manhuaçu** Brazil
121 □D2 **Mania** r. Madag.
108 B1 **Maniago** Italy
121 C2 **Maniamba** Moz.
150 C3 **Manicoré** Brazil
131 D3 **Manicouagan** r. Can.
131 D2 **Manicouagan, Petit Lac** l. Can.
131 D2 **Manicouagan, Réservoir** resr Can.
79 B2 **Manīfah** Saudi Arabia
49 K5 **Manihiki** atoll Cook Is
64 B2 **Manila** Phil.
53 D2 **Manilla** Austr.
Manipur India see **Imphal**
111 C3 **Manisa** Turkey
107 C2 **Manises** Spain
98 A2 **Man, Isle of** i. Irish Sea
138 B2 **Manistee** U.S.A.
138 B1 **Manistique** U.S.A.
129 E2 **Manitoba** prov. Can.
129 E2 **Manitoba, Lake** Can.
138 B1 **Manitou Islands** U.S.A.
130 B3 **Manitoulin Island** Can.
136 C3 **Manitou Springs** U.S.A.
130 B3 **Manitouwadge** Can.
138 B2 **Manitowoc** U.S.A.
130 C3 **Maniwaki** Can.
150 B2 **Manizales** Col.
121 □D3 **Manja** Madag.
121 C3 **Manjacaze** Moz.
137 E2 **Mankato** U.S.A.
114 B4 **Mankono** Côte d'Ivoire
129 D3 **Mankota** Can.
73 C4 **Mankulam** Sri Lanka
74 B2 **Manmad** India
52 A2 **Mannahill** Austr.
73 B4 **Mannar** Sri Lanka
73 B4 **Mannar, Gulf of** India/ Sri Lanka
101 D3 **Mannheim** Ger.
128 C2 **Manning** Can.
52 A2 **Mannum** Austr.
129 C2 **Mannville** Can.
59 C3 **Manokwari** Indon.
119 C3 **Manono** Dem. Rep. Congo
63 A2 **Manoron** Myanmar
105 D3 **Manosque** France
131 C2 **Manouane, Lac** l. Can.
65 B1 **Manp'o** N. Korea
107 D1 **Manresa** Spain
121 B2 **Mansa** Zambia
127 G2 **Mansel Island** Can.
92 I2 **Mansel'kya** ridge Fin./ Rus. Fed.
53 C3 **Mansfield** Austr.
98 C3 **Mansfield** U.K.
140 B1 **Mansfield** AR U.S.A.
140 B2 **Mansfield** LA U.S.A.
138 C2 **Mansfield** OH U.S.A.
139 D2 **Mansfield** PA U.S.A.
150 A3 **Manta** Ecuador
64 A3 **Mantalingajan, Mount** Phil.
155 D1 **Mantena** Brazil
141 E1 **Manteo** U.S.A.
104 C2 **Mantes-la-Jolie** France
Mantiqueira, Serra da mts Brazil
Mantova Italy see **Mantua**
88 C1 **Mäntsälä** Fin.
108 B1 **Mantua** Italy
86 D3 **Manturovo** Rus. Fed.
151 D3 **Manuelzinho** Brazil

61 D2	Manui i. Indon.
54 B1	Manukau N.Z.
59 D3	Manus Island P.N.G.
140 B2	Many U.S.A.
121 C2	Manyame r. Moz./ Zimbabwe
119 D3	Manyara, Lake salt l. Tanz.
	Manyas Gölü l. Turkey see Kuş Gölü
87 D4	Manych-Gudilo, Ozero l. Rus. Fed.
142 B1	Many Farms U.S.A.
119 D3	Manyoni Tanz.
106 C2	Manzanares Spain
146 C2	Manzanillo Cuba
144 B3	Manzanillo Mex.
119 C3	Manzanza Dem. Rep. Congo
69 D1	Manzhouli China
123 D2	Manzini Swaziland
115 D3	Mao Chad
	Maó Spain see Mahón
59 D3	Maoke, Pegunungan mts Indon.
123 C2	Maokeng S. Africa
65 A1	Maokui Shan mt. China
70 A2	Maomao Shan mt. China
71 B3	Maoming China
121 C3	Mapai Moz.
75 C1	Mapam Yumco l. China
61 D2	Mapane Indon.
145 C3	Mapastepec Mex.
144 B2	Mapimí Mex.
121 C3	Mapinhane Moz.
129 D3	Maple Creek Can.
156 D4	Mapmakers Seamounts N. Pacific Ocean
59 D3	Maprik P.N.G.
123 D1	Mapulanguene Moz.
121 C3	Maputo Moz.
123 D2	Maputo r. Moz./S. Africa
123 C2	Maputsoe Lesotho
114 A2	Maqteïr reg. Maur.
75 C2	Maquan He r. China
120 A1	Maquela do Zombo Angola
153 B4	Maquinchao Arg.
137 E2	Maquoketa U.S.A.
123 C1	Mara S. Africa
150 C3	Maraã Brazil
151 E3	Marabá Brazil
151 D2	Maracá, Ilha de i. Brazil
150 B1	Maracaibo Venez.
	Maracaibo, Lago de inlet Venez. see Maracaibo, Lake
150 B2	Maracaibo, Lake inlet Venez.
154 A2	Maracaju Brazil
154 A2	Maracaju, Serra de hills Brazil
150 C1	Maracay Venez.
115 C3	Marādah Libya
115 C3	Maradi Niger
81 C2	Marāgheh Iran
150 C2	Marahuaca, Cerro mt. Venez.
151 E3	Marajó, Baía de est. Brazil
151 D3	Marajó, Ilha de i. Brazil
79 C2	Marākī Iran
119 D2	Maralal Kenya
	Maralbashi China see Bachu
50 C3	Maralinga Austr.
	Maralwexi China see Bachu
142 A2	Marana U.S.A.
81 C2	Marand Iran
	Marandellas Zimbabwe see Marondera
150 B3	Marañón r. Peru
	Maraş Turkey see Kahramanmaraş
110 C1	Mărăşeşti Romania
130 B3	Marathon Can.
141 D4	Marathon U.S.A.
106 C2	Marbella Spain
50 A2	Marble Bar Austr.
123 C1	Marble Hall S. Africa
123 D3	Marburg S. Africa
101 D2	Marburg an der Lahn Ger.
103 D2	Marcali Hungary
137 E3	Marceline U.S.A.
99 D3	March U.K.
100 B2	Marche-en-Famenne Belgium
106 B2	Marchena Spain
152 B3	Mar Chiquita, Laguna l. Arg.
150 B4	Marcona Peru
139 E2	Marcy, Mount U.S.A.
74 B1	Mardan Pak.
153 C3	Mar del Plata Arg.
81 C2	Mardin Turkey
96 B2	Maree, Loch l. U.K.
51 D1	Mareeba Austr.
108 B3	Marettimo, Isola i. Italy
89 D2	Marevo Rus. Fed.
142 C2	Marfa U.S.A.
	Margao India see Madgaon
50 A3	Margaret River Austr.
150 C1	Margarita, Isla de i. Venez.
123 D3	Margate S. Africa
99 D4	Margate U.K.
62 A1	Margherita India
	Margherita, Lake l. Eth. see Lake Abaya
119 C2	Margherita Peak Dem. Rep. Congo/Uganda
76 C3	Märgow, Dasht-e des. Afgh.
91 C2	Marhanets' Ukr.
62 A1	Mari Myanmar
152 B1	María Elena Chile
156 C5	Mariana Trench N. Pacific Ocean
	Mariánica, Cordillera mts Spain see Morena, Sierra
140 B2	Marianna AR U.S.A.
141 C2	Marianna FL U.S.A.
102 C2	Mariánské Lázně Czech Rep.
144 B2	Marías, Islas is Mex.
78 B3	Ma'rib Yemen
109 C1	Maribor Slovenia
	Maricourt Can. see Kangiqsujuaq
119 C2	Maridi Sudan
117 A4	Maridi watercourse Sudan
55 P2	Marie Byrd Land reg. Antarctica
147 D3	Marie-Galante i. Guadeloupe
93 G3	Mariehamn Fin.
122 A1	Mariental Namibia
93 F4	Mariestad Sweden
141 D2	Marietta GA U.S.A.
138 C3	Marietta OH U.S.A.
105 D3	Marignane France
83 K3	Marii, Mys pt Rus. Fed.
88 B3	Marijampolė Lith.
154 C2	Marília Brazil
106 B1	Marín Spain
135 B3	Marina U.S.A.
109 C3	Marina di Gioiosa Ionica Italy
88 C3	Mar"ina Horka Belarus
138 B1	Marinette U.S.A.
154 B2	Maringá Brazil
106 B2	Marinha Grande Port.
138 B2	Marion IN U.S.A.
138 C2	Marion OH U.S.A.
141 E2	Marion SC U.S.A.
138 C3	Marion VA U.S.A.
141 D2	Marion, Lake U.S.A.
52 A3	Marion Bay Austr.
152 B2	Mariscal José Félix Estigarribia Para.
105 D3	Maritime Alps mts France/Italy
110 C2	Maritsa r. Bulg.
91 D2	Mariupol' Ukr.
117 C4	Marka Somalia
68 C3	Markam China
123 C1	Marken S. Africa
100 B1	Markermeer l. Neth.
98 C3	Market Rasen U.K.
98 C3	Market Weighton U.K.
83 I2	Markha r. Rus. Fed.
139 D2	Markham Can.
91 D2	Markivka Ukr.
118 B2	Markounda C.A.R.
140 B2	Marksville U.S.A.
101 D3	Marktheidenfeld Ger.
101 F2	Marktredwitz Ger.
100 C2	Marl Ger.
51 C2	Marla Austr.
105 C2	Marle France
143 D2	Marlin U.S.A.
53 C3	Marlo Austr.
104 C3	Marmande France
111 C2	Marmara, Sea of g. Turkey
	Marmara Denizi g. Turkey see Marmara, Sea of
111 C3	Marmaris Turkey
105 C2	Marne r. France
105 C2	Marne-la-Vallée France
118 B2	Maro Chad
121 □D2	Maroantsetra Madag.
54 B1	Marokopa N.Z.
101 E2	Maroldsweisach Ger.
121 □D2	Maromokotro mt. Madag.
121 C2	Marondera Zimbabwe
151 D2	Maroni r. Fr. Guiana
51 E2	Maroochydore Austr.
61 C2	Maros Indon.
49 M6	Marotiri is Fr. Polynesia
118 B1	Maroua Cameroon
121 □D2	Marovoay Madag.
49 N4	Marquesas Islands Fr. Polynesia
141 D4	Marquesas Keys is U.S.A.
155 D2	Marquês de Valença Brazil
138 B1	Marquette U.S.A.
116 A3	Marra, Jebel mt. Sudan
116 A3	Marra, Jebel Sudan
123 D2	Marracuene Moz.
114 B1	Marrakech Morocco
	Marrakesh Morocco see Marrakech
52 A1	Marree Austr.
86 F2	Marresale Rus. Fed.
121 C2	Marromeu Moz.
121 C2	Marrupa Moz.
116 B2	Marsá al 'Alam Egypt
115 D1	Marsa al Burayqah Libya
119 D2	Marsabit Kenya
78 A2	Marsa Delwein Sudan
108 B3	Marsala Italy
116 A1	Marsá Maţrūḩ Egypt
101 D2	Marsberg Ger.
108 B2	Marsciano Italy
53 C2	Marsden Austr.
100 B1	Marsdiep sea chan. Neth.
105 D3	Marseille France
	Marseilles France see Marseille
140 B1	Marshall AR U.S.A.
137 D2	Marshall MN U.S.A.
137 E3	Marshall MO U.S.A.
143 E2	Marshall TX U.S.A.
48 H2	Marshall Islands country N. Pacific Ocean
137 E2	Marshalltown U.S.A.
138 A2	Marshfield U.S.A.
146 C2	Marsh Harbour Bahamas
140 B3	Marsh Island U.S.A.
93 G4	Märsta Sweden
	Martaban, Gulf of g. Myanmar see Mottama, Gulf of
61 C2	Martapura Indon.
60 B2	Martapura Indon.
139 E2	Martha's Vineyard i. U.S.A.
105 D2	Martigny Switz.
103 D2	Martin Slovakia
136 C2	Martin U.S.A.
145 C2	Martínez Mex.
141 D2	Martinez U.S.A.
155 C1	Martinho Campos Brazil
147 D3	Martinique terr. West Indies
147 D3	Martinique Passage Dominica/Martinique
139 D3	Martinsburg U.S.A.
138 D3	Martinsville U.S.A.
76 B1	Martok Kazakh.
54 C2	Marton N.Z.
107 D1	Martorell Spain
106 C2	Martos Spain
81 D3	Marv Dasht Iran
105 C3	Marvejols France
76 C3	Mary Turkm.
51 E2	Maryborough Austr.
122 B2	Marydale S. Africa
139 D3	Maryland state U.S.A.
98 B2	Maryport U.K.
137 D3	Marysville MO U.S.A.
137 E2	Marysville U.S.A.
141 D1	Maryville TN U.S.A.
119 D3	Masai Steppe plain Tanz.
119 D3	Masaka Uganda
61 C2	Masalembu Besar i. Indon.
61 D2	Masamba Indon.
65 B2	Masan S. Korea
119 D4	Masasi Tanz.
64 B2	Masbate Phil.
64 B2	Masbate i. Phil.
107 D2	Mascara Alg.
155 E1	Mascote Brazil
123 C2	Maseru Lesotho
	Mashaba Zimbabwe see Mashava
121 C3	Mashava Zimbabwe
76 B3	Mashhad Iran
123 D2	Mashishing S. Africa
74 A2	Mashkel, Hamun-i- salt flat Pak.
123 C3	Masibambane S. Africa
79 C3	Masīlah, Wādī al watercourse Yemen
123 C2	Masilo S. Africa
118 B3	Masi-Manimba Dem. Rep. Congo
119 D2	Masindi Uganda
122 B3	Masinyusane S. Africa
79 C3	Masira, Gulf of b. Oman see Maşīrah, Khalīj
81 C2	Maşīrah, Khalīj
97 B2	Maşīrah, Khalīj b. Oman
	Masjed-e Soleymān Iran
121 □E2	Mask, Lough l. Ireland
137 E2	Masoala, Tanjona c. Madag.
	Mason City U.S.A.
	Masqaţ Oman see Muscat
108 B2	Massa Italy
139 E2	Massachusetts state U.S.A.
139 E2	Massachusetts Bay U.S.A.
115 D3	Massaguet Chad
115 D3	Massakory Chad
121 C3	Massangena Moz.
120 A1	Massango Angola
116 B3	Massawa Eritrea
139 E2	Massena U.S.A.
115 D3	Massenya Chad
128 A2	Masset Can.
105 C2	Massif Central mts France
138 C2	Massillon U.S.A.
121 C3	Massinga Moz.
123 D1	Massingir Moz.
78 A2	Mastābah Saudi Arabia
54 C2	Masterton N.Z.
74 B1	Mastuj Pak.
74 A2	Mastung Pak.
78 A2	Mastūrah Saudi Arabia
88 B3	Masty Belarus
67 B4	Masuda Japan
	Masuku Gabon see Franceville
121 C3	Masvingo Zimbabwe
119 D3	Maswa Tanz.
69 D1	Matad Mongolia
118 B3	Matadi Dem. Rep. Congo
146 B3	Matagalpa Nic.
130 C3	Matagami Can.
130 C3	Matagami, Lac l. Can.
143 D3	Matagorda Island U.S.A.
143 D3	Matagorda Peninsula U.S.A.
120 A2	Matala Angola
114 A3	Matam Senegal
54 C1	Matamata N.Z.
144 B2	Matamoros Chihuahua Mex.
145 C2	Matamoros Tamaulipas Mex.
119 D3	Matandu r. Tanz.
131 D3	Matane Can.
146 B2	Matanzas Cuba
	Matapan, Cape c. Greece see Tainaro, Akra
73 C4	Matara Sri Lanka
61 C2	Mataram Indon.
50 C1	Mataranka Austr.
107 D1	Mataró Spain
123 C3	Matatiele S. Africa
54 A3	Mataura N.Z.
54 A3	Mataura r. N.Z.
54 C1	Matawai N.Z.
152 B1	Mategua Bol.
145 B2	Matehuala Mex.
109 C2	Matera Italy
108 A3	Mateur Tunisia
129 E2	Matheson Island Can.
143 D3	Mathis U.S.A.
74 B2	Mathura India
64 B3	Mati Phil.
145 C3	Matías Romero Mex.
98 C3	Matlock U.K.
150 D4	Mato Grosso Brazil
154 B1	Mato Grosso state Brazil
154 B1	Mato Grosso do Sul state Brazil
123 D2	Matola Moz.
106 B1	Matosinhos Port.
	Matou China see Pingguo
79 C2	Maţraḩ Oman
67 B3	Matsue Japan
66 D2	Matsumae Japan
67 C3	Matsumoto Japan
67 C4	Matsusaka Japan
71 C3	Matsu Tao i. Taiwan
67 B4	Matsuyama Japan
130 B2	Mattagami r. Can.
130 C3	Mattawa Can.
105 D2	Matterhorn mt. Italy/Switz.
134 C2	Matterhorn mt. U.S.A.
141 D1	Matthews U.S.A.
79 C2	Maţţi, Sabkhat salt pan Saudi Arabia

38	B3	Mattoon U.S.A.
50	C2	Maturín Venez.
91	D2	Matveyev Kurgan Rus. Fed.
23	C2	Matwabeng S. Africa
05	C1	Maubeuge France
04	C3	Maubourguet France
55	E3	Maud Seamount sea feature S. Atlantic Ocean
49	L1	Maui i. U.S.A.
41	D2	Mauldin U.S.A.
38	C2	Maumee r. U.S.A.
61	D2	Maumere Indon.
20	B2	Maun Botswana
75	C2	Maunath Bhanjan India
23	C1	Maunatlala Botswana
62	A1	Maungdaw Myanmar
75	B2	Mau Ranipur India
50	C2	Maurice, Lake imp. l. Austr.
14	A3	Mauritania country Africa
13	I8	Mauritius country Indian Ocean
20	B2	Mavinga Angola
23	C3	Mavuya S. Africa
78	B2	Māwān, Khashm mt. Saudi Arabia
18	B3	Mawanga Dem. Rep. Congo
71	B3	Mawei China
62	A1	Mawkmai Myanmar
62	A1	Mawlaik Myanmar
63	A2	Mawlamyaing Myanmar
78	B2	Mawqaq Saudi Arabia
55	L3	Mawson Peninsula Antarctica
78	B2	Mawza Yemen
08	A3	Maxia, Punta mt. Italy
121	C3	Maxixe Moz.
54	B1	Maxwell N.Z.
83	J2	Maya r. Rus. Fed.
47	C2	Mayaguana i. Bahamas
147	D3	Mayagüez Puerto Rico
81	D2	Mayāmey Iran
96	B3	Maybole U.K.
116	B3	Maych'ew Eth.
117	C3	Maydh Somalia
?00	C1	Mayen Ger.
04	B2	Mayenne France
08	B2	Mayenne r. France
28	C2	Mayerthorpe Can.
38	B3	Mayfield U.S.A.
91	E3	Maykop Rus. Fed.
76	C3	Maymanah Afgh.
42	A1	Mayo Can.
18	B3	Mayo r. Can.
		Mayoko Congo
		Mayo Landing Can. see Mayo
64	B2	Mayon vol. Phil.
121	D2	Mayotte terr. Africa
83	J3	Mayskiy Rus. Fed.
138	C3	Maysville U.S.A.
118	B3	Mayumba Gabon
137	D1	Mayville U.S.A.
120	D1	Mazabuka Zambia
		Mazagan Morocco see El Jadida
151	D3	Mazagão Brazil
104	C3	Mazamet France
74	B1	Mazar China
108	B3	Mazara del Vallo Italy
77	C3	Mazār-e Sharīf Afgh.
107	C2	Mazarrón Spain
107	C2	Mazarrón, Golfo de b. Spain
144	B2	Mazatlán Mex.
146	A3	Mazatenango Guat.
144	B2	Mazatlán Mex.
88	B2	Mažeikiai Lith.
88	B2	Mazsalaca Latvia
103	E1	Mazowiecka, Nizina lowland Pol.
121	B3	Mazunga Zimbabwe
103	E1	Mazurskie, Pojezierze reg. Pol.
88	C3	Mazyr Belarus
123	D2	Mbabane Swaziland
118	B2	Mbaïki C.A.R.
118	B2	Mbakaou, Lac de l. Cameroon
121	C1	Mbala Zambia
121	B3	Mbalabala Zimbabwe
119	D2	Mbale Uganda
118	B2	Mbalmayo Cameroon
118	B3	Mbandaka Dem. Rep. Congo
118	B2	Mbandjok Cameroon
120	A1	M'banza Congo Angola
118	B3	Mbanza-Ngungu Dem. Rep. Congo
119	D3	Mbarara Uganda
118	B2	Mbé Cameroon
119	D3	Mbeya Tanz.
119	D4	Mbinga Tanz.
119	D3	Mbizi Mountains Tanz.
118	B2	Mboki C.A.R.
118	B2	Mbomo Congo
118	B2	Mbouda Cameroon
114	A3	Mbour Senegal
114	A3	Mbout Maur.
118	C3	Mbuji-Mayi Dem. Rep. Congo
119	D3	Mbuyuni Tanz.
139	F1	McAdam Can.
143	D2	McAlester U.S.A.
143	D3	McAllen U.S.A.
128	B2	McBride Can.
134	C2	McCall U.S.A.
143	C2	McCamey U.S.A.
126	F2	McClintock Channel Can.
126	E2	McClure Strait Can.
140	B2	McComb U.S.A.
136	C2	McConaughy, Lake U.S.A.
136	C2	McCook U.S.A.
134	C2	McDermitt U.S.A.
159	E7	McDonald Islands Indian Ocean
134	D1	McDonald Peak U.S.A.
135	D3	McGill U.S.A.
126	B2	McGrath U.S.A.
134	D1	McGuire, Mount U.S.A.
121	C2	Mchinji Malawi
140	C1	McKenzie r. Can.
51	D2	McKinlay Austr.
134	B2	McKinleyville U.S.A.
128	C2	McLennan Can.
128	B2	McLeod Lake Can.
134	B1	McMinnville OR U.S.A.
140	C1	McMinnville TN U.S.A.
137	D3	McPherson U.S.A.
128	C1	McTavish Arm b. Can.
123	C3	Mdantsane S. Africa
107	E2	M'Doukal Alg.
135	D3	Mead, Lake resr U.S.A.
136	C3	Meade U.S.A.
129	D2	Meadow Lake Can.
138	C2	Meadville U.S.A.
66	D2	Meaken-dake vol. Japan
106	B1	Mealhada Port.
131	E2	Mealy Mountains Can.
128	C2	Meander River Can.
78	A2	Mecca Saudi Arabia
139	D3	Mechanicsville U.S.A.
100	B2	Mechelen Belgium
100	B2	Mechelen Neth.
100	C2	Mechernich Ger.
100	C2	Meckenheim Ger.
101	E1	Mecklenburgische Seenplatte reg. Ger.
106	B1	Meda Port.
60	A1	Medan Indon.
153	B4	Medanosa, Punta pt Arg.
73	C4	Medawachchiya Sri Lanka
107	D2	Médéa Alg.
150	B2	Medellín Col.
115	D1	Medenine Tunisia
134	B2	Medford U.S.A.
110	C1	Medgidia Romania
110	B1	Mediaş Romania
136	B2	Medicine Bow Mountains U.S.A.
136	B2	Medicine Bow Peak U.S.A.
129	C2	Medicine Hat Can.
137	D3	Medicine Lodge U.S.A.
155	D1	Medina Brazil
78	A2	Medina Saudi Arabia
106	C1	Medinaceli Spain
106	C1	Medina del Campo Spain
106	B1	Medina de Rioseco Spain
75	C2	Medinipur India
84	E5	Mediterranean Sea
129	C2	Medley Can.
87	E3	Mednogorsk Rus. Fed.
62	A1	Médog China
88	B2	Medvégalio kalnas h. Lith.
83	L2	Medvezh'i, Ostrova is Rus. Fed.
86	C2	Medvezh'yegorsk Rus. Fed.
50	A2	Meekatharra Austr.
136	B2	Meeker U.S.A.
74	B2	Meerut India
100	A2	Meetkerke Belgium
119	D2	Méga Eth.
60	D2	Mega i. Indon.
119	D2	Mega Escarpment Eth./Kenya
111	B3	Megalopoli Greece
75	C2	Meghalaya state India
75	C2	Meghasani mt. India
111	C3	Megisti i. Greece
92	I1	Mehamn Norway
50	A2	Meharry, Mount Austr.
137	E3	Mehlville U.S.A.
79	C2	Mehrān watercourse Iran
74	B1	Mehtar Lām Afgh.
119	D3	Meia Meia Tanz.
154	C1	Meia Ponte r. Brazil
118	B2	Meiganga Cameroon
65	B1	Meihekou China
		Meijiang China see Ningdu
62	A1	Meiktila Myanmar
101	E2	Meiningen Ger.
102	C1	Meißen Ger.
		Meixian China see Meizhou
71	B3	Meizhou China
152	B2	Mejicana mt. Arg.
152	A2	Mejillones Chile
118	B2	Mékambo Gabon
116	B3	Mek'elē Eth.
114	C2	Mekerrhane, Sebkha salt pan Alg.
114	B1	Meknès Morocco
63	B2	Mekong r. Asia
63	B3	Mekong, Mouths of the Vietnam
60	B1	Melaka Malaysia
156	D6	Melanesia is Pacific Ocean
156	D5	Melanesian Basin Pacific Ocean
53	B3	Melbourne Austr.
141	D3	Melbourne U.S.A.
108	A2	Mele, Capo c. Italy
		Melekess Rus. Fed. see Dimitrovgrad
89	F2	Melenki Rus. Fed.
131	C2	Mélèzes, Rivière aux r. Can.
115	D3	Melfi Chad
109	C2	Melfi Italy
129	D2	Melfort Can.
92	F3	Melhus Norway
106	B1	Melide Spain
114	B1	Melilla N. Africa
129	D3	Melita Can.
91	D2	Melitopol' Ukr.
119	D2	Melka Guba Eth.
101	D1	Melle Ger.
93	F4	Mellerud Sweden
101	E2	Mellrichstadt Ger.
101	D1	Mellum i. Ger.
123	D2	Melmoth S. Africa
152	C3	Melo Uru.
115	C1	Melrhir, Chott salt l. Alg.
96	C3	Melrose U.K.
		Melsetter Zimbabwe see Chimanimani
52	B3	Melton Austr.
99	C3	Melton Mowbray U.K.
105	C2	Melun France
129	D2	Melville Can.
51	D1	Melville, Cape Austr.
131	E2	Melville, Lake Can.
50	C1	Melville Island Austr.
126	E1	Melville Island Can.
127	G2	Melville Peninsula Can.
61	C2	Memboro Indon.
102	C2	Memmingen Ger.
60	B1	Mempawah Indon.
80	B3	Memphis tourist site Egypt
140	B1	Memphis TN U.S.A.
143	C2	Memphis TX U.S.A.
91	C1	Mena Ukr.
140	B2	Mena U.S.A.
121	□D3	Menabe mts Madag.
114	C3	Ménaka Mali
		Mènam Khong r. Laos/Thai. see Mekong
105	C3	Méndez Mex.
117	B4	Mendi Eth.
59	D3	Mendi P.N.G.
99	B4	Mendip Hills U.K.
138	B2	Mendota U.S.A.
153	B3	Mendoza Arg.
111	C3	Menemen Turkey
100	A2	Menen Belgium
70	B2	Mengcheng China
60	B2	Menggala Indon.
71	A3	Mengzi China
131	D2	Menihek Can.
52	B2	Menindee Austr.
52	B2	Menindee Lake Austr.
52	A3	Meningie Austr.
104	C2	Mennecy France
138	B1	Menominee U.S.A.
120	A2	Menongue Angola
		Menorca i. Spain see Minorca
61	C1	Mensalong Indon.
60	A2	Mentawai, Kepulauan is Indon.
60	B2	Mentok Indon.
61	C1	Menyapa, Gunung mt. Indon.
108	A3	Menzel Bourguiba Tunisia
50	B2	Menzies Austr.
144	B2	Meoqui Mex.
100	C1	Meppel Neth.
100	C1	Meppen Ger.
121	C3	Mepuze Moz.
123	C2	Meqheleng S. Africa
138	B2	Mequon U.S.A.
108	B1	Merano Italy
59	D3	Merauke Indon.
52	B2	Merbein Austr.
		Merca Somalia see Marka
135	B3	Merced U.S.A.
152	C2	Mercedes Arg.
143	D3	Mercedes U.S.A.
127	H2	Mercy, Cape Can.
143	C1	Meredith, Lake U.S.A.
91	D2	Merefa Ukr.
116	A3	Merga Oasis Sudan
63	A2	Mergui Archipelago is Myanmar
110	C2	Meriç r. Greece/Turkey
111	C2	Meriç Turkey
145	D2	Mérida Mex.
106	B2	Mérida Spain
150	B2	Mérida Venez.
147	C4	Mérida, Cordillera de mts Venez.
134	C2	Meridian ID U.S.A.
140	C2	Meridian MS U.S.A.
143	D2	Meridian TX U.S.A.
104	B3	Mérignac France
93	H3	Merikarvia Fin.
53	C3	Merimbula Austr.
88	B3	Merkinė Lith.
116	B3	Merowe Sudan
50	A3	Merredin Austr.
96	B3	Merrick h. U.K.
138	B1	Merrill U.S.A.
138	B2	Merrillville U.S.A.
136	C2	Merriman U.S.A.
128	B2	Merritt Can.
53	C2	Merrygoen Austr.
116	C3	Mersa Fatma Eritrea
100	C3	Mersch Lux.
101	E2	Merseburg (Saale) Ger.
98	B2	Mersey r. U.K.
80	B2	Mersin Turkey
60	B1	Mersing Malaysia
99	D4	Mers-les-Bains France
74	B2	Merta India
99	B4	Merthyr Tydfil U.K.
119	D2	Merti Plateau Kenya
106	B2	Mértola Port.
76	B2	Mertvyy Kultuk, Sor dry lake Kazakh.
119	D2	Meru vol. Tanz.
122	B3	Merweville S. Africa
80	B1	Merzifon Turkey
100	C3	Merzig Ger.
142	A2	Mesa AZ U.S.A.
142	C2	Mesa NM U.S.A.
137	E1	Mesabi Range hills U.S.A.
109	C2	Mesagne Italy
142	B2	Mescalero U.S.A.
142	C2	Mescalero Ridge U.S.A.
101	D2	Meschede Ger.
89	E3	Meshchovsk Rus. Fed.
		Meshed Iran see Mashhad
91	E2	Meshkovskaya Rus. Fed.
142	B2	Mesilla U.S.A.
111	B2	Mesimeri Greece
111	B3	Mesolongi Greece
115	C1	Messaad Alg.
121	D2	Messalo r. Moz.
109	C3	Messina Italy
109	C3	Messina, Strait of str. Italy
		Messina, Stretta di str. Italy see Messina, Strait of
111	B3	Messini Greece
111	B3	Messiniakos Kolpos g. Greece
111	B2	Mesta r. Bulg.
		Mesta r. Greece see Nestos
150	C2	Meta r. Col./Venez.
130	C3	Métabetchouan Can.
127	H2	Meta Incognita Peninsula Can.

Moero, Lake *l.* Dem. Rep. Congo/Zambia *see* Mweru, Lake
00 C2 Moers Ger.
96 C3 Moffat U.K.
17 C4 Mogadishu Somalia
Mogador Morocco *see* Essaouira
06 B1 Mogadouro, Serra de *mts* Port.
23 C1 Mogalakwena *r.* S. Africa
62 A1 Mogaung Myanmar
Mogilev Belarus *see* Mahilyow
54 C2 Mogi-Mirim Brazil
83 I3 Mogocha Rus. Fed.
23 C1 Mogoditshane Botswana
62 A1 Mogok Myanmar
42 A2 Mogollon Plateau U.S.A.
03 D2 Mohács Hungary
23 C3 Mohale's Hoek Lesotho
07 C1 Mohammadia Alg.
42 A2 Mohave Mountains U.S.A.
39 E2 Mohawk *r.* U.S.A.
19 D3 Mohnyin Myanmar
19 D3 Mohoro Tanz.
92 F2 Mohyliv-Podil's'kyy Ukr.
23 C1 Moijabana Botswana
10 C1 Moineşti Romania
Mointy Kazakh. *see* Moyynty
75 C2 Mo i Rana Norway
92 F2 Mo i Rana Norway
88 C2 Mõisaküla Estonia
04 C3 Moissac France
35 C3 Mojave U.S.A.
35 C3 Mojave *r.* U.S.A.
35 C3 Mojave Desert U.S.A.
62 B1 Mojiang China
55 C2 Moji das Cruzes Brazil
54 C2 Moji-Guaçu *r.* Brazil
09 C2 Mojkovac Montenegro
54 B1 Mokau N.Z.
23 C2 Mokhotlong Lesotho
83 J2 Mokhsogollokh Rus. Fed.
18 B1 Mokolo Cameroon
23 C1 Mokopane S. Africa
65 B3 Mokp'o S. Korea
09 C2 Mola di Bari Italy
45 C2 Molango Mex.
Moldavia *country* Europe *see* Moldova
Moldavskaya S.S.R. *country* Europe *see* Moldova
93 E3 Molde Norway
90 B2 Moldova *country* Europe
10 B2 Moldova Nouǎ Romania
10 B1 Moldoveanu, Vârful *mt.* Romania
10 B1 Moldovei, Podişul *plat.* Romania
90 B2 Moldovei Centrale, Podişul *plat.* Moldova
23 C1 Molepolole Botswana
88 C2 Molėtai Lith.
09 C2 Molfetta Italy
Molière Alg. *see* Bordj Bounaama
07 C1 Molina de Aragón Spain
07 C2 Molina de Segura Spain
19 D3 Moliro Dem. Rep. Congo
50 B4 Mollendo Peru
93 F4 Mölnlycke Sweden
91 D2 Molochna *r.* Ukr.
89 E2 Molokovo Rus. Fed.
53 C2 Molong Austr.
22 B2 Molopo *watercourse* Botswana/S. Africa
Molotov Rus. Fed. *see* Perm'
Molotovsk Rus. Fed. *see* Severodvinsk
Molotovsk Rus. Fed. *see* Nolinsk
18 B2 Moloundou Cameroon
59 C3 Moluccas *is* Indon.
Molucca Sea *sea* Indon. *see* Laut Maluku
52 B2 Momba Austr.
19 D3 Mombasa Kenya
54 B1 Mombuca, Serra da *hills* Brazil
11 C2 Momchilgrad Bulg.
93 F4 Møn *i.* Denmark
05 D3 Monaco *country* Europe
96 B2 Monadhliath Mountains U.K.
97 C1 Monaghan Ireland
43 C2 Monahans U.S.A.
47 D3 Mona Passage Dom. Rep./Puerto Rico
20 A1 Mona Quimbundo Angola

Monastir Macedonia *see* Bitola
89 D3 Monastyrshchina Rus. Fed.
90 B2 Monastyryshche Ukr.
118 B2 Monatélé Cameroon
66 D2 Monbetsu Japan
108 A1 Moncalieri Italy
107 C1 Moncayo *mt.* Spain
86 C2 Monchegorsk Rus. Fed.
100 C2 Mönchengladbach Ger.
144 B2 Monclova Mex.
131 D3 Moncton Can.
106 B1 Mondego *r.* Port.
118 C2 Mondjamboli Dem. Rep. Congo
123 D2 Mondlo S. Africa
108 A2 Mondovì Italy
111 B3 Monemvasia Greece
66 D1 Moneron, Ostrov *i.* Rus. Fed.
139 D1 Monet Can.
137 E3 Monett U.S.A.
108 B1 Monfalcone Italy
106 B1 Monforte de Lemos Spain
119 D2 Mongbwalu Dem. Rep. Congo
62 B1 Mông Cai Vietnam
62 A1 Mong Hang Myanmar
Monghyr India *see* Munger
75 C2 Mongla Bangl.
62 A1 Mong Lin Myanmar
62 A1 Mong Nawng Myanmar
115 D3 Mongo Chad
68 C2 Mongolia *country* Asia
74 B1 Mongora Pak.
62 A1 Mong Pawk Myanmar
62 A1 Mong Ping Myanmar
120 B2 Mongu Zambia
135 C3 Monitor Range *mts* U.S.A.
103 E1 Mońki Pol.
53 D1 Monmouth U.K.
Mono *r.* Benin/Togo
135 C3 Mono Lake U.S.A.
109 C2 Monopoli Italy
107 C1 Monreal del Campo Spain
140 B2 Monroe LA U.S.A.
140 C2 Monroe MI U.S.A.
Monroe WI U.S.A.
140 C2 Monroeville U.S.A.
114 A4 Monrovia Liberia
100 A2 Mons Belgium
155 E1 Monsarás, Ponta de *pt* Brazil
100 C2 Montabaur Ger.
122 B3 Montagu S. Africa
109 C3 Montallo *mt.* Italy
110 B2 Montana Bulg.
134 E1 Montana *state* U.S.A.
105 C2 Montargis France
104 C3 Montauban France
139 E2 Montauk Point U.S.A.
123 C2 Mont-aux-Sources *mt.* Lesotho
105 C2 Montbard France
105 D2 Montbéliard France
105 C2 Montbrison France
100 B3 Montcornet France
104 B3 Mont-de-Marsan France
104 C2 Montdidier France
151 D3 Monte Alegre Brazil
154 C1 Monte Alegre de Minas Brazil
139 E1 Montebello Can.
105 D3 Montecarlo Arg.
Monte-Carlo Monaco
154 C1 Monte Carmelo Brazil
152 C3 Monte Caseros Arg.
123 C1 Monte Cristo S. Africa
108 B2 Montecristo, Isola di *i.* Italy
146 C3 Montego Bay Jamaica
105 C3 Montélimar France
109 C2 Montella Italy
145 C2 Montemorelos Mex.
104 B2 Montendre France
109 C2 Montenegro *country* Europe
121 C2 Montepuez Moz.
108 B2 Montepulciano Italy
135 B3 Monterey U.S.A.
135 B3 Monterey Bay U.S.A.
150 B2 Montería Col.
152 B1 Montero Bol.
145 B2 Monterrey Mex.
109 C2 Montesano sulla Marcellana Italy
109 C2 Monte Sant'Angelo Italy
151 F4 Monte Santo Brazil
108 A2 Monte Santu, Capo di *c.* Italy

155 D1 Montes Claros Brazil
153 C3 Montevideo Uru.
137 D2 Montevideo U.S.A.
136 B3 Monte Vista U.S.A.
140 C2 Montgomery U.S.A.
100 B3 Monthermé France
105 D2 Monthey Switz.
140 B2 Monticello AR U.S.A.
141 D2 Monticello FL U.S.A.
135 E3 Monticello UT U.S.A.
104 C2 Montignac France
100 B2 Montignies-le-Tilleul Belgium
105 D2 Montigny-le-Roi France
106 B2 Montijo Port.
106 B2 Montijo Spain
106 C2 Montilla Spain
154 B1 Montividiu Brazil
131 D3 Mont-Joli Can.
130 C3 Mont-Laurier Can.
104 C2 Montluçon France
131 C3 Montmagny Can.
104 C2 Montmorillon France
51 E2 Monto Austr.
134 D2 Montpelier ID U.S.A.
139 E2 Montpelier VT U.S.A.
105 C3 Montpellier France
130 C3 Montréal Can.
129 D2 Montreal Lake Can.
129 D2 Montreal Lake *l.* Can.
99 D4 Montreuil France
105 D2 Montreux Switz.
96 C2 Montrose U.K.
136 B3 Montrose U.S.A.
147 D3 Montserrat *terr.* West Indies
62 A1 Monywa Myanmar
108 A1 Monza Italy
107 D1 Monzón Spain
123 C1 Mookane Botswana
52 A1 Moolawatana Austr.
52 B1 Moomba Austr.
53 D1 Moonie Austr.
53 C1 Moonie *r.* Austr.
52 A2 Moonta Austr.
50 A3 Moora Austr.
50 A2 Moore, Lake *imp. l.* Austr.
137 D1 Moorhead U.S.A.
122 A3 Moorreesburg S. Africa
130 B2 Moose *r.* Can.
130 B2 Moose Factory Can.
139 F1 Moosehead Lake U.S.A.
129 D2 Moose Jaw Can.
137 E1 Moose Lake U.S.A.
130 B2 Moosonee Can.
52 B2 Mootwingee Austr.
123 C1 Mopane S. Africa
114 B3 Mopti Mali
150 B4 Moquegua Peru
103 D2 Mór Hungary
118 B1 Mora Cameroon
93 F3 Mora Sweden
137 E1 Mora U.S.A.
74 B2 Moradabad India
121 □D2 Morafenobe Madag.
121 □D2 Moramanga Madag.
136 A2 Moran U.S.A.
51 D2 Moranbah Austr.
103 D2 Morava *r.* Europe
96 B2 Moray Firth *b.* U.K.
100 C3 Morbach Ger.
74 B2 Morbi India
93 G4 Mörbylånga Sweden
104 B3 Morcenx France
69 E1 Mordaga China
129 E3 Morden Can.
89 F3 Mordovo Rus. Fed.
98 B2 Morecambe U.K.
98 B2 Morecambe Bay U.K.
53 C1 Moree Austr.
59 D3 Morehead P.N.G.
138 C3 Morehead U.S.A.
141 E2 Morehead City U.S.A.
145 B3 Morelia Mex.
107 C1 Morella Spain
74 B2 Morena India
106 B2 Morena, Sierra *mts* Spain
110 C2 Moreni Romania
142 A3 Moreno Mex.
128 A2 Moresby, Mount Can.
128 A2 Moresby Island Can.
53 D1 Moreton Island Austr.
52 A2 Morgan Austr.
140 B3 Morgan City U.S.A.
141 D1 Morganton U.S.A.
139 D3 Morgantown U.S.A.
105 D2 Morges Switz.

76 C3 Morghāb Afgh.
77 C3 Morghāb *r.* Afgh.
68 C2 Mori China
66 D2 Mori Japan
53 C1 Moriarty's Range *hills* Austr.
128 B2 Morice Lake Can.
66 D3 Morioka Japan
53 D2 Morisset Austr.
104 B2 Morlaix France
98 C3 Morley U.K.
157 G9 Mornington Abyssal Plain S. Atlantic Ocean
51 C1 Mornington Island Austr.
59 D3 Morobe P.N.G.
114 B1 Morocco *country* Africa
119 D3 Morogoro Tanz.
64 B3 Moro Gulf Phil.
122 B2 Morokweng S. Africa
121 □D3 Morombe Madag.
68 C1 Mörön Mongolia
121 □D3 Morondava Madag.
106 B2 Morón de la Frontera Spain
121 D2 Moroni Comoros
59 C2 Morotai *i.* Indon.
119 D2 Moroto Uganda
98 C2 Morpeth U.K.
140 B1 Morrilton U.S.A.
154 C1 Morrinhos Brazil
54 C1 Morrinsville N.Z.
129 E3 Morris Can.
137 D1 Morris U.S.A.
141 D1 Morristown U.S.A.
154 C2 Morro Agudo Brazil
155 E1 Morro d'Anta Brazil
55 K3 Morse, Cape Antarctica
87 D3 Morshanka Rus. Fed.
Morshansk Rus. Fed. *see* Morshanka
151 C3 Mortes, Rio das *r.* Brazil
52 B3 Mortlake Austr.
48 G3 Mortlock Islands Micronesia
138 B2 Morton U.S.A.
53 C2 Morundah Austr.
53 D3 Moruya Austr.
96 D2 Morvern *reg.* U.K.
Morvi India *see* Morbi
53 C3 Morwell Austr.
101 D3 Mosbach Ger.
89 E2 Moscow Rus. Fed.
134 C1 Moscow U.S.A.
100 C2 Mosel *r.* Ger.
122 B2 Moselebe *watercourse* Botswana
105 D2 Moselle *r.* France
134 C1 Moses Lake U.S.A.
92 □A3 Mosfellsbær Iceland
54 B3 Mosgiel N.Z.
88 C2 Moshchnyy, Ostrov *i.* Rus. Fed.
89 D2 Moshenskoye Rus. Fed.
119 D3 Moshi Tanz.
92 F2 Mosjøen Norway
Moskva Rus. Fed. *see* Moscow
89 E2 Moskva *r.* Rus. Fed.
103 D2 Mosonmagyaróvár Hungary
146 B4 Mosquitos, Golfo de los *b.* Panama
93 F4 Moss Norway
Mossâmedes Angola *see* Namibe
122 B3 Mossel Bay S. Africa
122 B3 Mossel Bay *b.* S. Africa
118 B3 Mossendjo Congo
52 B2 Mossgiel Austr.
51 D1 Mossman Austr.
151 F3 Mossoró Brazil
53 D2 Moss Vale Austr.
102 C1 Most Czech Rep.
114 C1 Mostaganem Alg.
109 C2 Mostar Bos.-Herz.
152 C3 Mostardas Brazil
106 C1 Móstoles Spain
81 C2 Mosul Iraq
145 D3 Motagua *r.* Guat.
93 G4 Motala Sweden
123 C2 Motetema S. Africa
96 C3 Motherwell U.K.
75 C2 Motihari India
107 C2 Motilla del Palancar Spain
122 B1 Motokwe Botswana
106 C2 Motril Spain
110 B2 Motru Romania
136 C1 Mott U.S.A.
63 A2 Mottama Myanmar
63 A2 Mottama, Gulf of Myanmar

Motueka

54	B2	Motueka N.Z.
145	D2	Motul Mex.
49	L5	Motu One atoll Fr. Polynesia
114	A3	Moudjéria Maur.
111	C3	Moudros Greece
118	B3	Mouila Gabon
52	B3	Moulamein Austr.
105	C2	Moulins France
141	D2	Moultrie U.S.A.
141	E2	Moultrie, Lake U.S.A.
138	B3	Mound City U.S.A.
115	D4	Moundou Chad
138	C3	Moundsville U.S.A.
137	E3	Mountain Grove U.S.A.
140	B1	Mountain Home AR U.S.A.
134	C2	Mountain Home ID U.S.A.
141	D1	Mount Airy U.S.A.
123	C3	Mount Ayliff S. Africa
52	A3	Mount Barker Austr.
53	C3	Mount Beauty Austr.
97	B2	Mountbellew Ireland
121	C2	Mount Darwin Zimbabwe
139	F2	Mount Desert Island U.S.A.
123	C3	Mount Fletcher S. Africa
123	C3	Mount Frere S. Africa
52	B3	Mount Gambier Austr.
59	D3	Mount Hagen P.N.G.
53	C2	Mount Hope Austr.
51	C2	Mount Isa Austr.
52	A3	Mount Lofty Range mts Austr.
50	A2	Mount Magnet Austr.
52	B2	Mount Manara Austr.
54	C1	Mount Maunganui N.Z.
97	C2	Mountmellick Ireland
111	B2	Mount Olympus mt. Greece
137	E2	Mount Pleasant IA U.S.A.
138	C2	Mount Pleasant MI U.S.A.
141	E2	Mount Pleasant SC U.S.A.
143	E2	Mount Pleasant TX U.S.A.
135	D3	Mount Pleasant UT U.S.A.
99	A4	Mount's Bay U.K.
134	B2	Mount Shasta U.S.A.
54	B2	Mount Somers N.Z.
138	B3	Mount Vernon IL U.S.A.
138	C2	Mount Vernon OH U.S.A.
134	B1	Mount Vernon WA U.S.A.
51	D2	Moura Austr.
106	B2	Moura Port.
115	E3	Mourdi, Dépression du depr. Chad
97	C1	Mourne Mountains hills U.K.
100	A2	Mouscron Belgium
115	D3	Moussoro Chad
61	D1	Moutong Indon.
115	C2	Mouydir, Monts du plat. Alg.
100	B3	Mouzon France
97	B1	Moy r. Ireland
117	B4	Moyale Eth.
		Moyen Congo country Africa see Congo
123	C3	Moyeni Lesotho
76	B2	Mo'ynoq Uzbek.
119	D2	Moyo Uganda
77	D2	Moyynkum Kazakh.
77	D2	Moyynty Kazakh.
121	C3	Mozambique country Africa
113	G8	Mozambique Channel Africa
81	C1	Mozdok Rus. Fed.
89	E2	Mozhaysk Rus. Fed.
119	D3	Mpanda Tanz.
121	C2	Mpika Zambia
118	B2	Mpoko r. C.A.R.
121	C1	Mporokoso Zambia
123	C2	Mpumalanga prov. S. Africa
119	D3	Mpwapwa Tanz.
62	A1	Mrauk-U Myanmar
115	D1	M'Saken Tunisia
119	D3	Msambweni Kenya
119	D3	Msata Tanz.
88	C2	Mshinskaya Rus. Fed.
115	C1	M'Sila Alg.
89	D2	Msta r. Rus. Fed.
89	D2	Mstinskiy Most Rus. Fed.
89	D3	Mstsislaw Belarus
123	C3	Mthatha S. Africa
		Mtoko Zimbabwe see Mutoko
89	E3	Mtsensk Rus. Fed.
123	D2	Mtubatuba S. Africa
119	E4	Mtwara Tanz.
151	E3	Muana Brazil
118	B3	Muanda Dem. Rep. Congo
62	B1	Muang Hiam Laos
62	B2	Muang Hinboun Laos
63	B2	Muang Không Laos
63	B2	Muang Khôngxédôn Laos
62	B1	Muang Ngoy Laos
62	B2	Muang Pakbeng Laos
63	B2	Muang Phalan Laos
62	B1	Muang Sing Laos
62	B2	Muang Vangviang Laos
60	B1	Muar Malaysia
60	B2	Muarabungo Indon.
60	B2	Muaradua Indon.
61	C2	Muaralaung Indon.
60	A2	Muarasiberut Indon.
60	B2	Muaratembesi Indon.
61	C2	Muarateweh Indon.
		Muara Tuang Malaysia see Kota Samarahan
119	D2	Mubende Uganda
115	D3	Mubi Nigeria
120	B2	Muconda Angola
120	A2	Mucope Angola
121	C2	Mucubela Moz.
155	E1	Mucuri Brazil
155	E1	Mucuri r. Brazil
66	A2	Mudanjiang China
66	A1	Mudan Jiang r. China
111	C2	Mudanya Turkey
136	B2	Muddy Gap U.S.A.
101	E1	Müden (Örtze) Ger.
53	C2	Mudgee Austr.
63	A2	Mudon Myanmar
80	B1	Mudurnu Turkey
121	C2	Mueda Moz.
121	B2	Mufulira Zambia
120	B2	Mufumbwe Zambia
111	C3	Muğla Turkey
116	B2	Muhammad Qol Sudan
		Muhammarah Iran see Khorramshahr
101	F2	Mühlberg Ger.
101	E2	Mühlhausen (Thüringen) Ger.
88	B2	Muhu i. Estonia
96	B3	Muirkirk U.K.
121	C2	Muite Moz.
65	B2	Muju S. Korea
		Mukačevo Ukr. see Mukacheve
90	A2	Mukacheve Ukr.
61	C1	Mukah Malaysia
79	B3	Mukalla Yemen
63	B2	Mukdahan Thai.
		Mukden China see Shenyang
		Mukhtuya Rus. Fed. see Lensk
50	A3	Mukinbudin Austr.
60	B2	Mukomuko Indon.
121	C2	Mulanje, Mount Malawi
101	F2	Mulde r. Ger.
119	D3	Muleba Tanz.
144	A2	Mulegé Mex.
143	C2	Muleshoe U.S.A.
106	C2	Mulhacén mt. Spain
100	C2	Mülheim an der Ruhr Ger.
105	D2	Mulhouse France
62	B1	Muli China
66	B2	Muling China
66	B1	Muling He r. China
96	B2	Mull i. U.K.
53	C2	Mullaley Austr.
136	C2	Mullen U.S.A.
61	C1	Muller, Pegunungan mts Indon.
50	A2	Mullewa Austr.
97	C2	Mullingar Ireland
96	B3	Mull of Galloway c. U.K.
96	B3	Mull of Kintyre hd U.K.
96	A3	Mull of Oa hd U.K.
53	D1	Mullumbimby Austr.
120	B2	Mulobezi Zambia
74	B1	Multan Pak.
86	F2	Mulym'ya Rus. Fed.
74	B3	Mumbai India
120	B2	Mumbeji Zambia
120	B2	Mumbwa Zambia
61	D2	Muna i. Indon.
145	D2	Muna Mex.
101	E2	Münchberg Ger.
		München Ger. see Munich
		München-Gladbach Ger. see Mönchengladbach
138	B2	Muncie U.S.A.
50	B3	Mundrabilla Austr.
138	B3	Munfordville U.S.A.
119	C2	Mungbere Dem. Rep. Congo
75	C2	Munger India
52	A1	Mungeranie Austr.
53	C1	Mungindi Austr.
102	C2	Munich Ger.
155	D2	Muniz Freire Brazil
101	E1	Münster Ger.
100	C2	Münster Ger.
97	B2	Munster reg. Ireland
100	C2	Münsterland reg. Ger.
62	B1	Mường Nhe Vietnam
92	H2	Muonio Fin.
92	H2	Muonioälven r. Fin./Sweden
		Muqdisho Somalia see Mogadishu
155	D2	Muqui Brazil
103	D2	Mur r. Austria
67	C3	Murakami Japan
119	C3	Muramvya Burundi
119	D3	Murang'a Kenya
86	D3	Murashi Rus. Fed.
81	B2	Murat r. Turkey
111	C2	Muratlı Turkey
67	D3	Murayama Japan
50	A2	Murchison watercourse Austr.
107	C2	Murcia Spain
107	C2	Murcia aut. comm. Spain
136	C2	Murdo U.S.A.
131	D3	Murdochville Can.
111	C2	Mürefte Turkey
110	B1	Mureşul r. Romania
104	C3	Muret France
140	C1	Murfreesboro U.S.A.
77	D3	Murghob Tajik.
155	D2	Muriaé Brazil
120	B1	Muriege Angola
101	F1	Müritz l. Ger.
92	J2	Murmansk Rus. Fed.
86	C2	Murmanskiy Bereg coastal area Rus. Fed.
87	D3	Murom Rus. Fed.
66	D2	Muroran Japan
106	B1	Muros Spain
67	B4	Muroto Japan
67	B4	Muroto-zaki pt Japan
141	D1	Murphy U.S.A.
53	C1	Murra Murra Austr.
52	A3	Murray r. Austr.
128	B2	Murray r. Can.
138	B3	Murray U.S.A.
59	D3	Murray, Lake P.N.G.
141	D2	Murray, Lake U.S.A.
52	A3	Murray Bridge Austr.
122	B3	Murraysburg S. Africa
52	B3	Murrayville Austr.
52	B2	Murrumbidgee r. Austr.
53	C2	Murrumburrah Austr.
121	C2	Murrupula Moz.
53	D2	Murrurundi Austr.
109	C1	Murska Sobota Slovenia
54	C1	Murupara N.Z.
49	N6	Mururoa atoll Fr. Polynesia
75	C2	Murwara India
53	D1	Murwillumbah Austr.
74	A2	Murzechirla Turkm.
115	D2	Murzūq Libya
115	D2	Murzūq, Idhān des. Libya
103	D2	Mürzzuschlag Austria
81	C2	Muş Turkey
110	B2	Musala mt. Bulg.
65	B1	Musan N. Korea
78	B3	Musaymir Yemen
79	C2	Muscat Oman
		Muscat and Oman country Asia see Oman
137	E2	Muscatine U.S.A.
50	C2	Musgrave Ranges mts Austr.
118	B3	Mushie Dem. Rep. Congo
60	B2	Musi r. Indon.
123	D1	Musina S. Africa
138	B2	Muskegon U.S.A.
138	B2	Muskegon r. U.S.A.
138	C3	Muskingum r. U.S.A.
143	D1	Muskogee U.S.A.
139	D1	Muskoka, Lake Can.
128	B2	Muskwa r. Can.
74	A1	Muslimbagh Pak.
116	B3	Musmar Sudan
119	D3	Musoma Tanz.
59	D3	Mussau Island P.N.G.
96	C3	Musselburgh U.K.
117	C4	Mustahīl Eth.
88	B2	Mustjala Estonia
53	D2	Muswellbrook Austr.
116	A2	Mūţ Egypt
121	C2	Mutare Zimbabwe
121	C2	Mutoko Zimbabwe
66	D2	Mutsu Japan
66	D2	Mutsu-wan b. Japan
121	C2	Mutuali Moz.
155	D1	Mutum Brazil
92	I2	Muurola Fin.
70	A2	Mu Us Shamo des. China
120	A1	Muxaluando Angola
86	C2	Muyezerskiy Rus. Fed.
119	D3	Muyinga Burundi
74	B1	Muzaffargarh Pak.
75	C2	Muzaffarpur India
123	D1	Muzamane Moz.
155	C2	Muzambinho Brazil
144	B2	Múzquiz Mex.
75	C1	Muz Tag mt. China
117	A4	Mvolo Sudan
119	C3	Mwanza Dem. Rep. Congo
119	D3	Mwanza Tanz.
118	C3	Mweka Dem. Rep. Congo
121	B2	Mwenda Zambia
118	C3	Mwene-Ditu Dem. Rep. Congo
121	C3	Mwenezi Zimbabwe
121	C3	Mwenezi r. Zimbabwe
119	C3	Mweru, Lake Dem. Rep. Congo/Zambia
121	B1	Mweru Wantipa, Lake Zambia
118	C3	Mwimba Dem. Rep. Congo
120	B2	Mwinilunga Zambia
88	C3	Myadzyel Belarus
62	A2	Myanaung Myanmar
62	A1	Myanmar country Asia
		Myanmar country Asia see Myanmar
63	A2	Myaungmya Myanmar
63	A2	Myeik Myanmar
		Myeik Kyunzu is Myanma see Mergui Archipelago
62	A1	Myingyan Myanmar
62	A1	Myitkyina Myanmar
90	A2	Mykolayiv L'vivs'ka Oblas Ukr.
91	C2	Mykolayiv Mykolayivs'ka Oblast' Ukr.
111	C3	Mykonos Greece
111	C3	Mykonos i. Greece
86	E2	Myla Rus. Fed.
75	D2	Mymensingh Bangl.
65	B1	Myŏnggan N. Korea
88	C2	Myory Belarus
92	□B3	Mýrdalsjökull Iceland
92	G2	Myre Norway
91	C2	Myrhorod Ukr.
111	C3	Myrina Greece
90	C2	Myronivka Ukr.
141	E2	Myrtle Beach U.S.A.
53	C3	Myrtleford Austr.
134	B2	Myrtle Point U.S.A.
89	E2	Myshkin Rus. Fed.
		Myshkino Rus. Fed. see Myshkin
103	C1	Myślibórz Pol.
73	B3	Mysore India
83	N2	Mys Shmidta Rus. Fed.
63	B2	My Tho Vietnam
111	C3	Mytilini Greece
89	E3	Mytishchi Rus. Fed.
123	C3	Mzamomhle S. Africa
121	C2	Mzimba Malawi
121	C2	Mzuzu Malawi

N

101	F3	Naab r. Ger.
100	B1	Naarden Neth.
97	C2	Naas Ireland
122	A2	Nababeep S. Africa
87	E3	Naberezhnyye Chelny Rus. Fed.
59	D3	Nabire Indon.
80	B2	Nāblus West Bank
123	C1	Naboomspruit S. Africa
121	D2	Nacala Moz.
119	D4	Nachingwea Tanz.
103	D1	Náchod Czech Rep.
73	D3	Nachuge India
143	E2	Nacogdoches U.S.A.
144	B1	Nacozari de García Mex.
		Nada China see Danzhou
74	B2	Nadiad India
90	A2	Nadvirna Ukr.
86	C2	Nadvoitsy Rus. Fed.
86	G2	Nadym Rus. Fed.
93	F4	Næstved Denmark
111	B3	Nafpaktos Greece
111	B3	Nafplio Greece

45 D1 **Nafūsah, Jabal** hills Libya
78 B2 **Nafy** Saudi Arabia
54 B2 **Naga** Phil.
30 B2 **Nagagami** r. Can.
57 C3 **Nagano** Japan
57 C3 **Nagaoka** Japan
75 D2 **Nagaon** India
74 B1 **Nagar** India
74 B2 **Nagar Parkar** Pak.
57 A4 **Nagasaki** Japan
57 B4 **Nagato** Japan
74 B2 **Nagercoil** India
73 B4 **Nagercoil** India
74 A2 **Nagha Kalat** Pak.
74 B2 **Nagina** India
67 C3 **Nagoya** Japan
75 B2 **Nagpur** India
75 D1 **Nagqu** China
41 E1 **Nags Head** U.S.A.
82 E1 **Nagsk** Rus. Fed.
03 D2 **Nagyatád** Hungary
03 D2 **Nagykanizsa** Hungary
28 B1 **Nahanni Butte** Can.
78 B3 **Nahāvand** Iran
01 E1 **Nahrendorf** Ger.
53 A4 **Nahuel Huapí, Lago** l. Arg.
41 D2 **Nahunta** U.S.A.
31 D2 **Nain** Can.
31 D2 **Nā'īn** Iran
21 C2 **Naiopué** Moz.
96 C2 **Nairn** U.K.
49 D3 **Nairobi** Kenya
     Naissus Serbia see **Niš**
19 D3 **Naivasha** Kenya
81 D2 **Najafābād** Iran
78 B2 **Najd** reg. Saudi Arabia
06 C1 **Nájera** Spain
65 C1 **Najin** N. Korea
78 B3 **Najrān** Saudi Arabia
19 D2 **Nakasongola** Uganda
67 C3 **Nakatsugawa** Japan
78 A3 **Nakfa** Eritrea
66 B2 **Nakhodka** Rus. Fed.
63 B2 **Nakhon Nayok** Thai.
63 B2 **Nakhon Pathom** Thai.
62 B2 **Nakhon Phanom** Thai.
63 B2 **Nakhon Ratchasima** Thai.
63 B2 **Nakhon Sawan** Thai.
63 A3 **Nakhon Si Thammarat** Thai.
     Nakhrachi Rus. Fed. see **Kondinskoye**
30 B2 **Nakina** Can.
26 B3 **Naknek** U.S.A.
21 C1 **Nakonde** Zambia
93 F5 **Nakskov** Denmark
19 D3 **Nakuru** Kenya
28 C2 **Nakusp** Can.
75 D2 **Nalbari** India
87 D4 **Nal'chik** Rus. Fed.
15 D1 **Nālūt** Libya
23 D2 **Namaacha** Moz.
23 C2 **Namahadi** S. Africa
81 D2 **Namak, Daryācheh-ye** imp. l. Iran
76 B3 **Namak, Kavīr-e** salt flat Iran
79 C1 **Namakzar-e Shadad** salt flat Iran
77 D2 **Namangan** Uzbek.
19 D3 **Namanyere** Tanz.
22 A2 **Namaqualand** reg. S. Africa
51 E2 **Nambour** Austr.
53 D2 **Nambucca Heads** Austr.
63 B3 **Năm Căn** Vietnam
75 D1 **Nam Co** salt l. China
62 B1 **Nam Đinh** Vietnam
23 C2 **Namialo** Moz.
20 A3 **Namib Desert** Namibia
20 A2 **Namibe** Angola
20 A3 **Namibia** country Africa
72 D2 **Namjagbarwa Feng** mt. China
59 C3 **Namlea** Indon.
62 B2 **Nam Ngum Reservoir** Laos
53 C2 **Namoi** r. Austr.
34 C2 **Nampa** U.S.A.
14 B3 **Nampala** Mali
65 B2 **Namp'o** N. Korea
21 C2 **Nampula** Moz.
72 D2 **Namrup** India
93 E3 **Namsang** Myanmar
92 F3 **Namsos** Norway
92 F3 **Namsskogan** Norway
63 A2 **Nam Tok** Thai.
83 J2 **Namtsy** Rus. Fed.
62 A1 **Namtu** Myanmar
21 C2 **Namuno** Moz.
00 B2 **Namur** Belgium

120 B2 **Namwala** Zambia
65 B2 **Namwŏn** S. Korea
62 A1 **Namya Ra** Myanmar
62 B2 **Nan** Thai.
128 B3 **Nanaimo** Can.
71 B3 **Nan'an** China
122 A1 **Nananib Plateau** Namibia
     Nan'ao China see **Dayu**
67 C3 **Nanao** Japan
71 B3 **Nanchang** Jiangxi China
71 B3 **Nanchang** Jiangxi China
71 B3 **Nancheng** China
70 A2 **Nanchong** China
63 A3 **Nancowry** i. India
105 D2 **Nancy** France
75 C1 **Nanda Devi** mt. India
71 A3 **Nandan** China
74 B3 **Nanded** India
     Nander India see **Nanded**
53 D2 **Nandewar Range** mts Austr.
74 B2 **Nandurbar** India
73 B3 **Nandyal** India
71 B3 **Nanfeng** China
62 A1 **Nang** China
118 B2 **Nanga Eboko** Cameroon
61 C2 **Nangahpinoh** Indon.
77 D3 **Nanga Parbat** mt. Pak.
61 C2 **Nangatayap** Indon.
63 A2 **Nangin** Myanmar
65 B1 **Nangnim-sanmaek** mts N. Korea
70 B2 **Nangong** China
119 D3 **Nangulangwa** Tanz.
70 C2 **Nanhui** China
70 B2 **Nanjing** China
     Nanking China see **Nanjing**
67 B4 **Nankoku** Japan
120 A2 **Nankova** Angola
70 B2 **Nanle** China
71 B3 **Nan Ling** mts China
71 A3 **Nanning** China
127 I2 **Nanortalik** Greenland
71 A3 **Nanpan Jiang** r. China
75 C2 **Nanpara** India
71 B3 **Nanping** China
     Nanpu China see **Pucheng**
     Nansei-shotō is Japan see **Ryukyu Islands**
160 I1 **Nansen Basin** Arctic Ocean
126 F1 **Nansen Sound** sea chan. Can.
104 B2 **Nantes** France
70 C2 **Nantong** China
139 E2 **Nantucket** U.S.A.
139 F2 **Nantucket Island** U.S.A.
99 B3 **Nantwich** U.K.
49 I4 **Nanumea** atoll Tuvalu
155 D1 **Nanuque** Brazil
64 B3 **Nanusa, Kepulauan** is Indon.
71 B3 **Nanxiong** China
70 B2 **Nanyang** China
119 D3 **Nanyuki** Kenya
70 B2 **Nanzhang** China
     Nanzhao China see **Zhao'an**
107 D2 **Nao, Cabo de la** c. Spain
131 C2 **Naocacane, Lac** l. Can.
88 C3 **Naozhou Dao** i. China
135 B3 **Napa** U.S.A.
126 E2 **Napaktulik Lake** Can.
139 D2 **Napanee** Can.
127 I2 **Napasoq** Greenland
137 F2 **Naperville** U.S.A.
54 C1 **Napier** N.Z.
108 B2 **Naples** Italy
141 D3 **Naples** U.S.A.
150 B3 **Napo** r. Ecuador/Peru
     Napoli Italy see **Naples**
     Napug China see **Gê'gyai**
114 B3 **Nara** Mali
88 C3 **Narach** Belarus
52 A3 **Naracoorte** Austr.
53 C2 **Naradhan** Austr.
145 C2 **Naranjos** Mex.
63 B3 **Narathiwat** Thai.
104 C3 **Narbonne** France
105 C3 **Narcondam Island** India
127 H1 **Nares Strait** Can./Greenland
103 E1 **Narew** r. Pol.
122 A1 **Narib** Namibia
87 D4 **Narimanov** Rus. Fed.
67 D3 **Narita** Japan
74 B2 **Narmada** r. India
74 B2 **Narnaul** India
108 B2 **Narni** Italy

86 F2 **Narodnaya, Gora** mt. Rus. Fed.
90 B1 **Narodychi** Ukr.
89 E2 **Naro-Fominsk** Rus. Fed.
53 D3 **Narooma** Austr.
88 C3 **Narowlya** Belarus
93 H3 **Närpes** Fin.
53 C2 **Narrabri** Austr.
53 C2 **Narrandera** Austr.
53 C2 **Narromine** Austr.
67 B4 **Naruto** Japan
88 C2 **Narva** Estonia
88 C2 **Narva Bay** Estonia/Rus. Fed.
92 G2 **Narvik** Norway
88 C2 **Narvskoye Vodokhranilishche** resr Estonia/Rus. Fed.
86 E2 **Nar'yan-Mar** Rus. Fed.
77 D2 **Naryn** Kyrg.
74 B2 **Nashik** India
139 E2 **Nashua** U.S.A.
140 C1 **Nashville** U.S.A.
137 E1 **Nashwauk** U.S.A.
109 C1 **Našice** Croatia
117 B4 **Nasir** Sudan
     Nasirabad Bangl. see **Mymensingh**
119 C4 **Nasondoye** Dem. Rep. Congo
76 B3 **Naşrābād** Iran
128 B2 **Nass** r. Can.
146 C2 **Nassau** Bahamas
116 B2 **Nasser, Lake** resr Egypt
93 F4 **Nässjö** Sweden
130 C2 **Nastapoca** r. Can.
130 C2 **Nastapoka Islands** Can.
89 D2 **Nasva** Rus. Fed.
120 B3 **Nata** Botswana
151 F3 **Natal** Brazil
60 A1 **Natal** Indon.
     Natal prov. S. Africa see **KwaZulu-Natal**
159 D6 **Natal Basin** Indian Ocean
143 D3 **Natalia** U.S.A.
131 D2 **Natashquan** Can.
131 D2 **Natashquan** r. Can.
140 B2 **Natchez** U.S.A.
140 B2 **Natchitoches** U.S.A.
53 C3 **Nathalia** Austr.
107 D1 **Nati, Punta** pt Spain
114 C3 **Natitingou** Benin
151 F4 **Natividade** Brazil
67 D3 **Natori** Japan
119 D3 **Natron, Lake** salt l. Tanz.
60 B1 **Natuna, Kepulauan** is Indon.
60 B1 **Natuna Besar** i. Indon.
122 A1 **Nauchas** Namibia
101 F1 **Nauen** Ger.
64 B2 **Naujan** Phil.
88 B2 **Naujoji Akmenė** Lith.
74 A2 **Naukot** Pak.
101 E2 **Naumburg (Saale)** Ger.
48 H34 **Nauru** country S. Pacific Ocean
150 B3 **Nauta** Peru
145 C2 **Nautla** Mex.
88 C3 **Navahrudak** Belarus
106 B2 **Navalmoral de la Mata** Spain
106 B2 **Navalvillar de Pela** Spain
97 C2 **Navan** Ireland
     Navangar India see **Jamnagar**
88 C2 **Navapolatsk** Belarus
83 M2 **Navarin, Mys** c. Rus. Fed.
153 B5 **Navarino, Isla** i. Chile
107 C1 **Navarra** aut. comm. Spain
     Navarre aut. comm. Spain see **Navarra**
96 B1 **Naver** r. U.K.
74 B2 **Navi Mumbai** India
89 D3 **Navlya** Rus. Fed.
110 C2 **Năvodari** Romania
77 C2 **Navoiy** Uzbek.
144 B2 **Navojoa** Mex.
144 B2 **Navolato** Mex.
74 A2 **Nawabshah** Pak.
75 C2 **Nawada** India
62 A1 **Nawnghkio** Myanmar
62 A1 **Nawngleng** Myanmar
81 C2 **Naxçıvan** Azer.
111 C3 **Naxos** Greece
111 C3 **Naxos** i. Greece
144 B2 **Nayar** Mex.
66 D2 **Nayoro** Japan
62 A2 **Nay Pyi Taw** Myanmar
     Nazareth Israel see **Nazerat**

144 B2 **Nazas** Mex.
144 B2 **Nazas** r. Mex.
150 B4 **Nazca** Peru
157 H7 **Nazca Ridge** S. Pacific Ocean
80 B2 **Nazerat** Israel
111 C3 **Nazilli** Turkey
117 B4 **Nazrēt** Eth.
79 C2 **Nazwá** Oman
121 B1 **Nchelenge** Zambia
122 B1 **Ncojane** Botswana
120 A1 **N'dalatando** Angola
118 C2 **Ndélé** C.A.R.
118 B3 **Ndendé** Gabon
115 D3 **Ndjamena** Chad
118 A3 **Ndogo, Lagune** lag. Gabon
121 B2 **Ndola** Zambia
97 C1 **Neagh, Lough** l. U.K.
50 C2 **Neale, Lake** imp. l. Austr.
111 B3 **Neapoli** Greece
111 B2 **Nea Roda** Greece
99 B4 **Neath** U.K.
119 D2 **Nebbi** Uganda
53 C1 **Nebine Creek** r. Austr.
150 C2 **Neblina, Pico da** mt. Brazil
136 C2 **Nebo, Mount** U.S.A.
137 D2 **Nebolchi** Rus. Fed.
136 C2 **Nebraska** state U.S.A.
137 D2 **Nebraska City** U.S.A.
108 B3 **Nebrodi, Monti** mts Italy
143 E3 **Neches** r. U.S.A.
156 E4 **Necker Island** U.S.A.
153 C3 **Necochea** Arg.
143 E3 **Nederland** U.S.A.
100 B2 **Neder Rijn** r. Neth.
130 C2 **Nedlouc, Lac** l. Can.
139 E2 **Needham** U.S.A.
135 D4 **Needles** U.S.A.
74 B2 **Neemuch** India
129 E2 **Neepawa** Can.
87 E3 **Neftekamsk** Rus. Fed.
82 F2 **Nefteyugansk** Rus. Fed.
108 A3 **Néfza** Tunisia
120 A1 **Negage** Angola
117 B4 **Negēlē** Eth.
109 D2 **Negotin** Serbia
111 B2 **Negotino** Macedonia
150 A3 **Negra, Punta** pt Peru
155 D1 **Negra, Serra** mts Brazil
63 A2 **Negrais, Cape** Myanmar
153 B4 **Negro** r. Arg.
150 D3 **Negro** r. S. America
152 C3 **Negro** r. Uru.
106 B2 **Negro, Cabo** c. Morocco
64 B3 **Negros** i. Phil.
79 D1 **Nehbandān** Iran
69 E1 **Nehe** China
70 A3 **Neijiang** China
129 D2 **Neilburg** Can.
150 B2 **Neiva** Col.
129 E2 **Nejanilini Lake** Can.
     Nejd reg. Saudi Arabia see **Najd**
117 B4 **Nek'emtē** Eth.
89 F2 **Nekrasovskoye** Rus. Fed.
89 D2 **Nelidovo** Rus. Fed.
73 B3 **Nellore** India
128 C3 **Nelson** Can.
129 E2 **Nelson** r. Can.
54 B2 **Nelson** N.Z.
52 B3 **Nelson, Cape** Austr.
53 D2 **Nelson Bay** Austr.
129 E2 **Nelson House** Can.
134 E1 **Nelson Reservoir** U.S.A.
123 D2 **Nelspruit** S. Africa
114 B3 **Néma** Maur.
88 B2 **Neman** Rus. Fed.
105 C2 **Nemours** France
66 D2 **Nemuro** Japan
90 B2 **Nemyriv** Ukr.
97 B2 **Nenagh** Ireland
99 D3 **Nene** r. U.K.
69 E1 **Nenjiang** China
137 E3 **Neosho** U.S.A.
     Nepal country Asia
75 C2 **Nepalganj** Nepal
75 C2 **Nepean** Can.
139 D1 **Nephi** U.S.A.
97 B1 **Nephin** h. Ireland
97 B1 **Nephin Beg Range** hills Ireland
131 C2 **Nepisiguit** r. Can.
119 C2 **Nepoko** r. Dem. Rep. Congo
139 E2 **Neptune City** U.S.A.
108 B2 **Nera** r. Italy
104 C2 **Nérac** France
53 D1 **Nerang** Austr.
69 D1 **Nerchinsk** Rus. Fed.

3 D2	Nondweni S. Africa
	Nonghui China *see*
	Guang'an
2 B2	Nong Khai Thai.
5 D2	Nongstoin India
2 A2	Nonning Austr.
4 B2	Nonoava Mex.
5 B2	Nonsan S. Korea
3 B2	Nonthaburi Thai.
2 B3	Nonzwakazi S. Africa
0 B1	Noordwijk-Binnen Neth.
7 C3	Norak Tajik.
2 C1	Nordaustlandet *i.* Svalbard
8 C2	Nordegg Can.
0 C1	Norden Ger.
3 H1	Nordenshel'da, Arkhipelag *is* Rus. Fed.
	Nordenskjold Archipelago *is* Rus. Fed. *see*
	Nordenshel'da, Arkhipelag
0 C1	Norderney Ger.
0 C1	Norderney *i.* Ger.
1 E1	Norderstedt Ger.
3 E3	Nordfjordeid Norway
	Nordfriesische Inseln *is* Ger. *see* North Frisian Islands
1 E2	Nordhausen Ger.
1 D1	Nordholz Ger.
0 C1	Nordhorn Ger.
	Nordkapp *c.* Norway *see* North Cape
2 F3	Nordli Norway
2 C2	Nördlingen Ger.
2 C2	Nordmaling Sweden
4 B1	Norðoyar *is* Faroe Is
7 C2	Nore *r.* Ireland
8 B3	Noreikiškės Lith.
7 D2	Norfolk NE U.S.A.
9 D3	Norfolk VA U.S.A.
8 H6	Norfolk Island *terr.* S. Pacific Ocean
3 E3	Norheimsund Norway
2 G2	Noril'sk Rus. Fed.
5 C2	Norkyung China
3 D1	Norman U.S.A.
	Normandes, Îles *is* English Chan. *see* Channel Islands
0 D2	Normandia Brazil
	Normandie *reg.* France *see* Normandy
4 B2	Normandy *reg.* France
1 D1	Normanton Austr.
8 B1	Norman Wells Can.
3 G4	Norrköping Sweden
3 G4	Norrtälje Sweden
0 B3	Norseman Austr.
2 G3	Norsjö Sweden
5 M2	North, Cape Antarctica
8 C2	Northallerton U.K.
0 A3	Northam Austr.
0 A2	Northampton U.K.
9 C3	Northampton U.K.
3 D3	North Andaman *i.* India
9 F4	North Australian Basin Indian Ocean
9 D2	North Battleford Can.
0 C3	North Bay Can.
0 C2	North Belcher Islands Can.
6 C2	North Berwick U.K.
	North Borneo *state* Malaysia *see* Sabah
2 I1	North Cape *c.* Norway
4 B1	North Cape N.Z.
0 A2	North Caribou Lake Can.
1 E1	North Carolina *state* U.S.A.
0 B3	North Channel *lake channel* Can.
6 A3	North Channel U.K.
1 E2	North Charleston U.S.A.
8 B3	North Cowichan Can.
6 C1	North Dakota *state* U.S.A.
9 C4	North Downs *hills* U.K.
7 E3	Northeast Pacific Basin N. Pacific Ocean
1 E3	Northeast Providence Channel Bahamas
1 D2	Northeim Ger.
2 A2	Northern Cape *prov.* S. Africa
1 D2	Northern Donets *r.* Rus. Fed./Ukr.
	Northern Dvina *r.* Rus. Fed. *see* Severnaya Dvina
9 E2	Northern Indian Lake Can.
7 C1	Northern Ireland *prov.* U.K.

59 D1	Northern Mariana Islands *terr.* N. Pacific Ocean
	Northern Rhodesia *country* Africa *see* Zambia
50 C1	Northern Territory *admin. div.* Austr.
	Northern Transvaal *prov.* S. Africa *see* Limpopo
96 C2	North Esk *r.* U.K.
137 E2	Northfield U.S.A.
99 D4	North Foreland *c.* U.K.
102 B1	North Frisian Islands *is* Ger.
160 P2	North Geomagnetic Pole (2009)
54 B1	North Island N.Z.
129 E2	North Knife Lake Can.
65 B1	North Korea *country* Asia
72 D2	North Lakhimpur India
	North Land *is* Rus. Fed. *see* Severnaya Zemlya
160 R1	North Magnetic Pole (2009)
128 B1	North Nahanni *r.* Can.
136 C2	North Platte U.S.A.
136 C2	North Platte *r.* U.S.A.
96 C1	North Ronaldsay *i.* U.K.
129 D2	North Saskatchewan *r.* Can.
94 D2	North Sea Europe
63 A2	North Sentinel Island India
130 A2	North Spirit Lake Can.
53 D1	North Stradbroke Island Austr.
131 D3	North Sydney Can.
54 B1	North Taranaki Bight *b.* N.Z.
130 C2	North Twin Island Can.
98 B2	North Tyne *r.* U.K.
96 A2	North Uist *i.* U.K.
131 D3	Northumberland Strait Can.
99 D3	North Walsham U.K.
123 C2	North West *prov.* S. Africa
158 D1	Northwest Atlantic Mid-Ocean Channel *sea chan.* N. Atlantic Ocean
50 A2	North West Cape Austr.
156 D3	Northwest Pacific Basin N. Pacific Ocean
141 E3	Northwest Providence Channel Bahamas
131 E2	North West River Can.
128 B1	Northwest Territories *admin. div.* Can.
98 C2	North York Moors *moorland* U.K.
138 C3	Norton U.S.A.
121 C2	Norton Zimbabwe
	Norton de Matos Angola *see* Balombo
126 B2	Norton Sound *sea chan.* U.S.A.
55 D2	Norvegia, Cape Antarctica
138 C2	Norwalk U.S.A.
93 F3	Norway *country* Europe
129 E2	Norway House Can.
160 L3	Norwegian Basin N. Atlantic Ocean
92 E2	Norwegian Sea N. Atlantic Ocean
99 D3	Norwich U.K.
139 E2	Norwich CT U.S.A.
139 D2	Norwich NY U.S.A.
66 D2	Noshiro Japan
91 C1	Nosivka Ukr.
122 B2	Nosop *watercourse* Africa
86 E2	Nosovaya Rus. Fed.
79 C2	Noşratābād Iran
103 D1	Noteć *r.* Pol.
93 E4	Notodden Norway
67 C3	Noto-hantō *pen.* Japan
131 D3	Notre-Dame, Monts *mts* Can.
131 E3	Notre Dame Bay Can.
130 C2	Nottaway *r.* Can.
99 C3	Nottingham U.K.
114 A2	Nouâdhibou Maur.
114 A3	Nouakchott Maur.
114 A3	Nouâmghâr Maur.
63 B2	Nouei Vietnam
48 H6	Nouméa New Caledonia
114 B3	Nouna Burkina
122 B3	Noupoort S. Africa
	Nouveau-Comptoir Can. *see* Wemindji
	Nouvelle Anvers Dem. Rep. Congo *see* Makanza

	Nouvelles Hébrides *country* S. Pacific Ocean *see* Vanuatu
	Nova Chaves Angola *see* Muconda
154 B2	Nova Esperança Brazil
	Nova Freixa Moz. *see* Cuamba
155 D2	Nova Friburgo Brazil
109 C1	Nova Gradiška Croatia
154 C2	Nova Granada Brazil
155 D2	Nova Iguaçu Brazil
91 C2	Nova Kakhovka Ukr.
155 D1	Nova Lima Brazil
	Nova Lisboa Angola *see* Huambo
154 B2	Nova Londrina Brazil
91 C2	Nova Odesa Ukr.
154 C1	Nova Ponte Brazil
108 A1	Novara Italy
151 E3	Nova Remanso Brazil
131 D3	Nova Scotia *prov.* Can.
155 D1	Nova Venécia Brazil
	Novaya Sibir', Ostrov *i.* Rus. Fed.
86 E1	Novaya Zemlya *is* Rus. Fed.
107 C2	Novelda Spain
103 D2	Nové Zámky Slovakia
	Novgorod Rus. Fed. *see* Velikiy Novgorod
91 C1	Novhorod-Sivers'kyy Ukr.
110 B2	Novi Iskŭr Bulg.
66 D1	Novikovo Rus. Fed.
108 A2	Novi Ligure Italy
109 D2	Novi Pazar Serbia
109 C1	Novi Sad Serbia
	Novoalekseyevka Kazakh. *see* Kobda
87 D3	Novoanninskiy Rus. Fed.
150 C3	Novo Aripuanã Brazil
91 D2	Novoazovs'k Ukr.
91 E2	Novocherkassk Rus. Fed.
89 D2	Novodugino Rus. Fed.
86 D2	Novodvinsk Rus. Fed.
	Novoekonomicheskoye Ukr. *see* Dymytrov
152 C2	Novo Hamburgo Brazil
154 C2	Novo Horizonte Brazil
90 B1	Novohrad-Volyns'kyy Ukr.
	Novokazalinsk Kazakh. *see* Ayteke Bi
91 E1	Novokhopersk Rus. Fed.
68 B1	Novokuznetsk Rus. Fed.
109 C1	Novo Mesto Slovenia
91 D3	Novomikhaylovskiy Rus. Fed.
89 E3	Novomoskovsk Rus. Fed.
91 D2	Novomoskovs'k Ukr.
91 C2	Novomyrhorod Ukr.
	Novonikolayevsk Rus. Fed. *see* Novosibirsk
91 C2	Novooleksiyivka Ukr.
150 C2	Novo Paraíso Brazil
91 E2	Novopokrovskaya Rus. Fed.
91 D2	Novopskov Ukr.
	Novo Redondo Angola *see* Sumbe
91 D3	Novorossiysk Rus. Fed.
88 C2	Novorzhev Rus. Fed.
91 D2	Novosergiyevka Rus. Fed.
91 D2	Novoshakhtinsk Rus. Fed.
82 G3	Novosibirsk Rus. Fed.
	Novosibirskiye Ostrova *is* Rus. Fed. *see* New Siberia Islands
89 E3	Novosil' Rus. Fed.
89 D2	Novosokol'niki Rus. Fed.
91 C2	Novotroyits'ke Ukr.
91 D2	Novovolyns'k Ukr.
89 E3	Novovoronezh Rus. Fed.
	Novovoronezhskiy Rus. Fed. *see* Novovoronezh
89 D3	Novozybkov Rus. Fed.
103 D2	Nový Jičín Czech Rep.
86 E2	Novyy Bor Rus. Fed.
91 C2	Novyy Buh Ukr.
	Novyy Donbass Ukr. *see* Dymytrov
	Novyye Petushki Rus. Fed. *see* Petushki
	Novyy Margelan Uzbek. *see* Farg'ona
89 E2	Novyy Nekouz Rus. Fed.
91 D1	Novyy Oskol Rus. Fed.
86 G2	Novyy Port Rus. Fed.
86 G2	Novyy Urengoy Rus. Fed.
69 E1	Novyy Urgal Rus. Fed.

	Novyy Uzen' Kazakh. *see* Zhanaozen
103 D1	Nowogard Pol.
	Nowoaradomsk Pol. *see* Radomsko
53 D2	Nowra Austr.
81 D2	Nowshahr Iran
74 B1	Nowshera Pak.
103 E2	Nowy Sącz Pol.
103 E2	Nowy Targ Pol.
82 G2	Noyabr'sk Rus. Fed.
105 C2	Noyon France
68 C2	Noyon Mongolia
121 C2	Nsanje Malawi
121 B2	Nsombo Zambia
118 B3	Ntandembele Dem. Rep. Congo
111 B3	Ntha S. Africa
111 B3	Ntoro, Kavo *pt* Greece
118 A2	Ntoum Gabon
119 D3	Ntungamo Uganda
	Nuanetsi *r.* Zimbabwe *see* Mwenezi
79 C2	Nu'aym *reg.* Oman
116 B2	Nubian Desert Sudan
150 B4	Nudo Coropuna *mt.* Peru
143 D3	Nueces *r.* U.S.A.
129 E1	Nueltin Lake Can.
150 B2	Nueva Loja Ecuador
153 A4	Nueva Lubecka Arg.
145 B2	Nueva Rosita Mex.
144 B1	Nuevo Casas Grandes Mex.
144 B2	Nuevo Ideal Mex.
145 C2	Nuevo Laredo Mex.
117 C4	Nugaal *watercourse* Somalia
105 C2	Nuits-St-Georges France
	Nu Jiang *r.* China/Myanmar *see* Salween
	Nukha Azer. *see* Şäki
49 J6	Nuku'alofa Tonga
49 M4	Nuku Hiva *i.* Fr. Polynesia
48 G4	Nukumanu Islands P.N.G.
76 B2	Nukus Uzbek.
50 B2	Nullagine Austr.
50 B3	Nullarbor Austr.
115 D4	Numan Nigeria
67 C3	Numazu Japan
51 C1	Numbulwar Austr.
93 E3	Numedal *val.* Norway
59 C3	Numfoor *i.* Indon.
53 C3	Numurkah Austr.
	Nunap Isua *c.* Greenland *see* Farewell, Cape
	Nunavut *admin. div.* Can.
99 C3	Nuneaton U.K.
126 A3	Nunivak Island U.S.A.
106 B1	Nuñomoral Spain
108 A2	Nuoro Italy
78 B2	Nuqrah Saudi Arabia
77 C1	Nura *r.* Kazakh.
101 E3	Nuremberg Ger.
74 B1	Nürestān Afgh.
52 A2	Nuriootpa Austr.
92 I3	Nurmes Fin.
	Nürnberg Ger. *see* Nuremberg
53 C2	Nurri, Mount *h.* Austr.
62 A1	Nu Shan *mts* China
74 A2	Nushki Pak.
127 I2	Nuuk Greenland
127 I2	Nuussuaq Greenland
127 I2	Nuussuaq *pen.* Greenland
80 B3	Nuwaybi' al Muzayyinah Egypt
122 A3	Nuwerus S. Africa
122 B3	Nuweveldberge *mts* S. Africa
86 F2	Nyagan' Rus. Fed.
75 D1	Nyainqêntanglha Feng *mt.* China
75 D2	Nyainqêntanglha Shan *mts* China
	Nyakh Rus. Fed. *see* Nyagan'
117 A3	Nyala Sudan
119 D4	Nyamtumbo Tanz.
	Nyande Zimbabwe *see* Masvingo
118 B3	Nyanga Congo
121 C2	Nyanga *r.* Angola
121 C2	Nyanga Zimbabwe
121 C1	Nyasa, Lake Africa
	Nyasaland *country* Africa *see* Malawi
88 C3	Nyasvizh Belarus

62	A2	Nyaunglebin Myanmar
93	F4	Nyborg Denmark
92	I1	Nyborg Norway
93	G4	Nybro Sweden
		Nyenchen Tanglha Range mts China see Nyainqêntanglha Shan
119	D3	Nyeri Kenya
68	C3	Nyingchi China
103	E2	Nyíregyháza Hungary
93	F5	Nykøbing Denmark
93	G4	Nyköping Sweden
53	C2	Nymagee Austr.
93	G4	Nynäshamn Sweden
53	C2	Nyngan Austr.
88	B3	Nyoman r. Belarus/Lith.
105	D3	Nyons France
86	E2	Nyrob Rus. Fed.
103	D1	Nysa Pol.
134	C2	Nyssa U.S.A.
119	C3	Nyunzu Dem. Rep. Congo
83	I2	Nyurba Rus. Fed.
91	C2	Nyzhni Sirohozy Ukr.
91	C2	Nyzhn'ohirs'kyy Ukr.
118	B3	Nzambi Congo
119	D3	Nzega Tanz.
114	B4	Nzérékoré Guinea
120	A1	N'zeto Angola

O

136	C2	Oahe, Lake U.S.A.
49	L1	O'ahu i. U.S.A.
52	B2	Oakbank Austr.
140	B2	Oakdale U.S.A.
53	D1	Oakey Austr.
138	B3	Oak Grove U.S.A.
99	C3	Oakham U.K.
134	B1	Oak Harbor U.S.A.
138	C3	Oak Hill U.S.A.
135	B3	Oakland CA U.S.A.
139	D3	Oakland MD U.S.A.
138	B2	Oak Lawn U.S.A.
136	C3	Oakley U.S.A.
50	B2	Oakover r. Austr.
134	B2	Oakridge U.S.A.
141	D1	Oak Ridge U.S.A.
54	B3	Oamaru N.Z.
64	B2	Oas Phil.
145	C3	Oaxaca Mex.
86	F2	Ob' r. Rus. Fed.
		Ob, Gulf of sea chan. Rus. Fed. see Obskaya Guba
88	C2	Obal' Belarus
118	B2	Obala Cameroon
96	B2	Oban U.K.
106	B1	O Barco Spain
		Obbia Somalia see Hobyo
136	C3	Oberlin U.S.A.
53	C3	Oberon Austr.
101	F3	Oberviechtach Ger.
59	C3	Obi i. Indon.
151	D3	Óbidos Brazil
66	D2	Obihiro Japan
69	E1	Obluch'ye Rus. Fed.
89	E2	Obninsk Rus. Fed.
119	C2	Obo C.A.R.
117	C3	Obock Djibouti
103	D1	Oborniki Pol.
118	B3	Obouya Congo
89	E3	Oboyan' Rus. Fed.
86	D2	Obozerskiy Rus. Fed.
144	B2	Obregón, Presa resr Mex.
109	D2	Obrenovac Serbia
134	B2	O'Brien U.S.A.
87	E3	Obshchiy Syrt hills Kazakh./Rus. Fed.
86	G2	Obskaya Guba sea chan. Rus. Fed.
114	B4	Obuasi Ghana
90	C1	Obukhiv Ukr.
86	D2	Ob''yachevo Rus. Fed.
141	D3	Ocala U.S.A.
144	B2	Ocampo Mex.
106	C2	Ocaña Spain
150	B2	Occidental, Cordillera mts Col.
150	B4	Occidental, Cordillera mts Peru
139	D3	Ocean City MD U.S.A.
139	E3	Ocean City NJ U.S.A.
128	B2	Ocean Falls Can.
135	C4	Oceanside U.S.A.
91	C2	Ochakiv Ukr.
86	E3	Ocher Rus. Fed.
101	E3	Ochsenfurt Ger.
110	B1	Ocna Mureş Romania
90	B2	Ocniţa Moldova
141	D2	Oconee r. U.S.A.
145	C3	Ocosingo Mex.
141	E1	Ocracoke Island U.S.A.
		October Revolution Island i. Rus. Fed. see Oktyabr'skoy Revolyutsii, Ostrov
59	C3	Ocussi enclave East Timor
116	B2	Oda, Jebel mt. Sudan
66	D2	Ōdate Japan
67	C3	Odawara Japan
93	E3	Odda Norway
106	B2	Odemira Port.
111	C3	Ödemiş Turkey
93	F4	Odense Denmark
101	D3	Odenwald reg. Ger.
102	C1	Oder r. Ger./Pol.
102	C1	Oderbucht b. Ger.
		Odesa Ukr. see Odessa
90	C2	Odessa Ukr.
143	C2	Odessa U.S.A.
114	B4	Odienné Côte d'Ivoire
89	E3	Odoyev Rus. Fed.
103	D1	Odra r. Ger./Pol.
151	E3	Oeiras Brazil
101	D2	Oelde Ger.
136	C2	Oelrichs U.S.A.
101	F2	Oelsnitz Ger.
100	B1	Oenkerk Neth.
137	E3	O'Fallon U.S.A.
109	C2	Ofanto r. Italy
101	D2	Offenbach am Main Ger.
102	B2	Offenburg Ger.
66	C3	Oga Japan
117	C4	Ogaden reg. Eth.
66	C3	Oga-hantō pen. Japan
67	C3	Ōgaki Japan
136	C2	Ogallala U.S.A.
		Ogasawara-shotō is Japan see Bonin Islands
115	C4	Ogbomosho Nigeria
134	D2	Ogden U.S.A.
139	D2	Ogdensburg U.S.A.
126	C2	Ogilvie r. Can.
126	C2	Ogilvie Mountains Can.
141	D2	Oglethorpe, Mount U.S.A.
130	B2	Ogoki r. Can.
130	B2	Ogoki Reservoir Can.
88	B2	Ogre Latvia
109	C1	Ogulin Croatia
81	D2	Ogurjaly Adasy i. Turkm.
115	C2	Ohanet Alg.
138	B3	Ohio r. U.S.A.
138	C2	Ohio state U.S.A.
101	E2	Ohrdruf Ger.
101	F2	Ohře r. Czech Rep.
111	B2	Ohrid Macedonia
109	D2	Ohrid, Lake l. Albania/Macedonia
151	D2	Oiapoque Brazil
139	D2	Oil City U.S.A.
100	A3	Oise r. France
67	B4	Ōita Japan
144	B2	Ojinaga Mex.
152	B2	Ojos del Salado, Nevado mt. Arg./Chile
89	F2	Oka r. Rus. Fed.
120	A3	Okahandja Namibia
120	A3	Okakarara Namibia
128	C3	Okanagan Falls Can.
128	C3	Okanagan Lake Can.
134	C1	Okanogan U.S.A.
134	C1	Okanogan r. U.S.A.
74	B1	Okara Pak.
120	B2	Okavango r. Africa
120	B2	Okavango Delta swamp Botswana
67	C3	Okaya Japan
67	B4	Okayama Japan
67	C4	Okazaki Japan
141	D3	Okeechobee U.S.A.
141	D3	Okeechobee, Lake U.S.A.
141	D2	Okefenokee Swamp U.S.A.
99	A4	Okehampton U.K.
115	C4	Okene Nigeria
101	E1	Oker r. Ger.
74	A2	Okha India
83	K3	Okha Rus. Fed.
75	C2	Okhaldhunga Nepal
83	K3	Okhotka r. Rus. Fed.
83	K3	Okhotsk Rus. Fed.
83	K3	Okhotsk, Sea of Japan/Rus. Fed.
		Okhotskoye More sea Japan/Rus. Fed. see Okhotsk, Sea of
91	C1	Okhtyrka Ukr.
69	E3	Okinawa i. Japan
67	B3	Oki-shotō is Japan
143	D1	Oklahoma state U.S.A.
143	D1	Oklahoma City U.S.A.
143	D1	Okmulgee U.S.A.
		Oknitsa Moldova see Ocniţa
78	A2	Oko, Wadi watercourse Sudan
118	B3	Okondja Gabon
128	C2	Okotoks Can.
89	D3	Okovskiy Les for. Rus. Fed.
118	B3	Okoyo Congo
92	H1	Øksfjord Norway
62	A2	Oktwin Myanmar
		Oktyabr' Kazakh. see Kandyagash
		Oktyabr'sk Kazakh. see Kandyagash
86	D2	Oktyabr'skiy Rus. Fed.
83	L3	Oktyabr'skiy Rus. Fed.
87	E3	Oktyabr'skiy Rus. Fed.
86	F2	Oktyabr'skoye Rus. Fed.
83	H1	Oktyabr'skoy Revolyutsii, Ostrov i. Rus. Fed.
89	D2	Okulovka Rus. Fed.
66	C2	Okushiri-tō i. Japan
92	□B2	Ólafsfjörður Iceland
92	□A3	Ólafsvík Iceland
88	B2	Olaine Latvia
93	G4	Öland i. Sweden
52	B2	Olary Austr.
136	B3	Olathe CO U.S.A.
137	E3	Olathe KS U.S.A.
153	B3	Olavarría Arg.
103	D1	Oława Pol.
108	A2	Olbia Italy
126	C2	Old Crow Can.
101	D1	Oldenburg Ger.
102	C1	Oldenburg in Holstein Ger.
100	C1	Oldenzaal Neth.
98	B3	Oldham U.K.
97	B3	Old Head of Kinsale hd Ireland
96	C2	Oldmeldrum U.K.
128	C2	Olds Can.
129	D2	Old Wives Lake Can.
139	D2	Olean U.S.A.
103	E1	Olecko Pol.
83	J2	Olekminsk Rus. Fed.
91	C2	Oleksandrivka Ukr.
		Oleksandrivs'k Ukr. see Zaporizhzhya
91	C2	Oleksandriya Ukr.
86	C2	Olenegorsk Rus. Fed.
83	I2	Olenek Rus. Fed.
83	I2	Olenek r. Rus. Fed.
89	D2	Olenino Rus. Fed.
		Olenivs'ki Kar"yery Ukr. see Dokuchayevs'k
		Olenya Rus. Fed. see Olenegorsk
		Oleshky Ukr. see Tsyurupyns'k
103	D1	Olesno Pol.
90	B1	Olevs'k Ukr.
106	B2	Olhão Port.
123	D1	Olifants r. Moz./S. Africa
122	A2	Olifants watercourse Namibia
123	D1	Olifants S. Africa
122	A3	Olifants r. S. Africa
122	B2	Olifantshoek S. Africa
154	C2	Olímpia Brazil
151	F3	Olinda Brazil
123	C1	Oliphants Drift Botswana
107	C2	Oliva Spain
155	D2	Oliveira Brazil
		Olivença Moz. see Lupilichi
106	B2	Olivenza Spain
140	B2	Olla U.S.A.
152	B2	Ollagüe Chile
77	C2	Olmaliq Uzbek.
106	C1	Olmedo Spain
105	D3	Olmeto France
150	B3	Olmos Peru
138	B3	Olney U.S.A.
103	D2	Olomouc Czech Rep.
86	C2	Olonets Rus. Fed.
64	B2	Olongapo Phil.
104	B3	Oloron-Ste-Marie France
107	D1	Olot Spain
69	D1	Olovyannaya Rus. Fed.
83	L2	Oloy r. Rus. Fed.
100	C2	Olpe Ger.
103	E1	Olsztyn Pol.
110	C2	Olteniţa Romania
143	C2	Olton U.S.A.
81	C1	Oltu Turkey
		Ol'viopol' Ukr. see Pervomays'k
111	B3	Olympia tourist site Greece
134	B1	Olympia U.S.A.
		Olympus, Mount mt. Greece see Mount Olympus
134	B1	Olympus, Mount U.S.A.
83	M2	Olyutorskiy Rus. Fed.
83	M3	Olyutorskiy, Mys c. Rus. Fed.
66	D2	Ōma Japan
97	C1	Omagh U.K.
137	D2	Omaha U.S.A.
79	C2	Oman country Asia
79	C2	Oman, Gulf of Asia
54	A2	Omarama N.Z.
120	A3	Omaruru Namibia
120	B2	Omatako watercourse Namibia
122	B2	Omaweneno Botswana
116	B3	Omdurman Sudan
53	C3	Omeo Austr.
145	C3	Ometepec Mex.
78	A3	Om Hajêr Eritrea
81	C2	Omīdiyeh Iran
128	B2	Omineca Mountains Can.
120	A3	Omitara Namibia
67	C3	Ōmiya Japan
100	C1	Ommen Neth.
83	L2	Omolon r. Rus. Fed.
100	B3	Omont France
82	F3	Omsk Rus. Fed.
83	L2	Omsukchan Rus. Fed.
110	C1	Omu, Vârful mt. Romania
67	A4	Ōmura Japan
138	A2	Onalaska U.S.A.
139	D3	Onancock U.S.A.
138	C1	Onaping Lake Can.
131	C3	Onatchiway, Lac l. Can.
63	A2	Onbingwin Myanmar
120	A2	Oncócua Angola
122	B3	Onderstedorings S. Africa
120	A2	Ondjiva Angola
69	D1	Öndörhaan Mongolia
122	B1	One Botswana
86	C2	Onega Rus. Fed.
86	C2	Onega r. Rus. Fed.
86	C2	Onega, Lake Rus. Fed.
139	D2	Oneida Lake U.S.A.
137	D2	O'Neill U.S.A.
139	D2	Oneonta U.S.A.
110	C1	Oneşti Romania
		Onezhskoye Ozero l. Rus. Fed. see Onega, Lake
122	B2	Ongers watercourse S. Africa
65	B2	Ongjin N. Korea
73	C3	Ongole India
121	□D3	Onilahy r. Madag.
115	C4	Onitsha Nigeria
120	A3	Onjati Mountain Namibia
67	C3	Ōno Japan
156	D6	Onotoa atoll Kiribati
122	A2	Onseepkans S. Africa
50	A2	Onslow Austr.
141	E2	Onslow Bay U.S.A.
130	A2	Ontario prov. Can.
134	C2	Ontario U.S.A.
139	D2	Ontario, Lake Can./U.S.A.
107	C2	Ontinyent Spain
51	C2	Oodnadatta Austr.
		Oostende Belgium see Ostend
100	B2	Oosterhout Neth.
100	B2	Oosterschelde est. Neth.
100	B1	Oost-Vlieland Neth.
128	B2	Ootsa Lake Can.
128	B2	Ootsa Lake l. Can.
118	C3	Opala Dem. Rep. Congo
130	C2	Opataca, Lac l. Can.
103	D2	Opava Czech Rep.
141	C2	Opelika U.S.A.
140	B2	Opelousas U.S.A.
117	A4	Opienge Dem. Rep. Congo
130	C2	Opinaca, Réservoir resr Can.
131	D2	Opiscotéo, Lac l. Can.
88	C2	Opochka Rus. Fed.
144	A2	Opodepe Mex.
103	D1	Opole Pol.
106	B1	Oporto Port.
54	C1	Opotiki N.Z.
93	E3	Oppdal Norway
134	C1	Opportunity U.S.A.
54	B1	Opunake N.Z.
120	A2	Opuwo Namibia
110	B1	Oradea Romania
		Orahovac Kosovo see Rahovec

4 B1	Oran Alg.	
2 B2	Orán Arg.	
5 B1	Ŏrang N. Korea	
3 C2	Orange Austr.	
5 C3	Orange France	
2 A2	Orange r. Namibia/S. Africa	
3 E2	Orange U.S.A.	
1 D2	Orangeburg U.S.A.	
	Orange Free State prov.	
	S. Africa see Free State	
1 D7	Orange Park U.S.A.	
8 C2	Orangeville Can.	
5 D3	Orange Walk Belize	
1 F1	Oranienburg Ger.	
2 A2	Oranjemund Namibia	
7 C3	Oranjestad Aruba	
3 B3	Orapa Botswana	
0 B1	Orăştie Romania	
	Oraşul Stalin Romania see	
	Braşov	
8 B2	Orbetello Italy	
3 C3	Orbost Austr.	
1 D3	Orchid Island U.S.A.	
1 B3	Orchomenos Greece	
0 B1	Ord, Mount h. Austr.	
6 B1	Ordes Spain	
0 B2	Ordos China	
0 B1	Ordu Turkey	
	Ordzhonikidze Rus. Fed. see	
	Vladikavkaz	
1 C2	Ordzhonikidze Ukr.	
3 G4	Örebro Sweden	
4 B2	Oregon state U.S.A.	
4 B1	Oregon City U.S.A.	
1 E1	Oregon Inlet U.S.A.	
7 C3	Orekhovo-Zuyevo Rus. Fed.	
9 E3	Orel Rus. Fed.	
3 K3	Orel', Ozero l. Rus. Fed.	
5 D2	Orem U.S.A.	
1 C3	Ören Turkey	
7 E3	Orenburg Rus. Fed.	
4 A3	Orepuki N.Z.	
1 C2	Orestiada Greece	
3 F4	Öresund str. Denmark/Sweden	
	Oretana, Cordillera mts	
	Spain see Toledo, Montes de	
9 D3	Orford Ness hd U.K.	
9 F2	Orgtrud Rus. Fed.	
4 A1	Orgūn Afgh.	
1 C3	Orhaneli Turkey	
1 C2	Orhangazi Turkey	
8 D1	Urhon Gol r. Mongolia	
0 B1	Oriental, Cordillera mts Bol.	
0 B2	Oriental, Cordillera mts Col.	
0 B4	Oriental, Cordillera mts Peru	
7 C2	Orihuela Spain	
1 D2	Orikhiv Ukr.	
1 C3	Orillia Can.	
3 I3	Orimattila Fin.	
2 C2	Orinoco r. Col./Venez.	
0 C2	Orinoco Delta Venez.	
5 C2	Orissa state India	
8 B2	Orissaare Estonia	
8 A3	Oristano Italy	
3 I3	Orivesi l. Fin.	
1 D3	Oriximiná Brazil	
5 C3	Orizaba Mex.	
5 C3	Orizaba, Pico de vol. Mex.	
4 C1	Orizona Brazil	
2 E3	Orkanger Norway	
3 F4	Örkelljunga Sweden	
3 E3	Orkla r. Norway	
6 C1	Orkney Islands U.K.	
4 C2	Orlândia Brazil	
1 D3	Orlando U.S.A.	
2 C2	Orléans France	
9 F2	Orleans U.S.A.	
9 E1	Orléans, Île d' i. Can.	
	Orléansville Alg. see Chlef	
4 A2	Ormara Pak.	
4 C2	Ormoc Phil.	
1 D3	Ormond Beach U.S.A.	
8 B3	Ormskirk U.K.	
4 B2	Orne r. France	
2 F2	Ørnes Norway	
2 G3	Örnsköldsvik Sweden	
4 B3	Orodara Burkina	
4 C1	Orofino U.S.A.	
9 F2	Orono U.S.A.	
	Oroqen Zizhiqi China see Alihe	
4 B3	Oroquieta Phil.	
8 A2	Orosei Italy	

108 A2	Orosei, Golfo di b. Italy	
103 E2	Orosháza Hungary	
135 B3	Oroville U.S.A.	
52 A2	Orroroo Austr.	
93 F3	Orsa Sweden	
89 D3	Orsha Belarus	
87 E3	Orsk Rus. Fed.	
110 B2	Orşova Romania	
93 E3	Ørsta Norway	
106 B1	Ortegal, Cabo c. Spain	
104 B3	Orthez France	
106 B1	Ortigueira Spain	
108 B1	Ortles mt. Italy	
108 B2	Ortona Italy	
137 D1	Ortonville U.S.A.	
83 J2	Orulgan, Khrebet mts Rus. Fed.	
	Orūmīyeh Iran see Urmia	
	Orūmīyeh, Daryācheh-ye salt l. Iran see Urmia, Lake	
152 B2	Oruro Bol.	
104 B2	Orvault France	
108 B2	Orvieto Italy	
93 F3	Os Norway	
146 B4	Osa, Península de pen. Costa Rica	
137 E3	Osage r. U.S.A.	
137 D3	Osage City U.S.A.	
67 C4	Ōsaka Japan	
77 D1	Osakarovka Kazakh.	
101 E1	Oschersleben (Bode) Ger.	
98 B2	Oschiri Italy	
138 C2	Oscoda U.S.A.	
89 E3	Osetr r. Rus. Fed.	
139 D1	Osgoode Can.	
77 D2	Osh Kyrg.	
120 A2	Oshakati Namibia	
130 C3	Oshawa Can.	
120 A2	Oshikango Namibia	
66 C2	Ō-shima i. Japan	
67 C4	Ō-shima i. Japan	
138 B2	Oshkosh U.S.A.	
81 C2	Oshnovīyeh Iran	
115 C4	Oshogbo Nigeria	
118 B3	Oshwe Dem. Rep. Congo	
109 C1	Osijek Croatia	
128 B2	Osilinka r. Can.	
108 B2	Osimo Italy	
	Osipenko Ukr. see Berdyans'k	
123 D2	Osizweni S. Africa	
137 F2	Oskaloosa U.S.A.	
93 G4	Oskarshamn Sweden	
89 E3	Oskol r. Rus. Fed.	
93 F4	Oslo Norway	
93 F4	Oslofjorden sea chan. Norway	
80 B1	Osmancık Turkey	
111 C2	Osmaneli Turkey	
80 B2	Osmaniye Turkey	
88 C2	Os'mino Rus. Fed.	
101 D1	Osnabrück Ger.	
153 A4	Osorno Chile	
106 C1	Osorno Spain	
128 C3	Osoyoos Can.	
100 B2	Oss Neth.	
51 D4	Ossa, Mount Austr.	
83 L3	Ossora Rus. Fed.	
89 D2	Ostashkov Rus. Fed.	
101 D1	Oste r. Ger.	
100 A2	Ostend Belgium	
101 E1	Osterburg (Altmark) Ger.	
93 F3	Österdalälven r. Sweden	
101 D1	Osterholz-Scharmbeck Ger.	
101 E2	Osterode am Harz Ger.	
92 F3	Östersund Sweden	
	Ostfriesische Inseln is Ger. see East Frisian Islands	
100 C1	Ostfriesland reg. Ger.	
93 G3	Östhammar Sweden	
103 D2	Ostrava Czech Rep.	
103 D1	Ostróda Pol.	
89 E3	Ostrogozhsk Rus. Fed.	
103 E1	Ostrołęka Pol.	
101 F2	Ostrov Czech Rep.	
88 C2	Ostrov Rus. Fed.	
	Ostrovets Pol. see Ostrowiec Świętokrzyski	
89 F2	Ostrovskoye Rus. Fed.	
103 E1	Ostrowiec Świętokrzyski Pol.	
103 E1	Ostrów Mazowiecka Pol.	
	Ostrowo Pol. see Ostrów Wielkopolski	
103 D1	Ostrów Wielkopolski Pol.	
109 C2	Ostuni Italy	
110 B2	Osŭm r. Bulg.	

67 B4	Ōsumi-kaikyō sea chan. Japan
67 B4	Ōsumi-shotō is Japan
106 B2	Osuna Spain
139 D2	Oswego U.S.A.
99 B3	Oswestry U.K.
67 C3	Ōta Japan
54 B3	Otago Peninsula N.Z.
54 C2	Otaki N.Z.
77 D2	Otar Kazakh.
66 D2	Otaru Japan
120 A2	Otavi Namibia
67 D3	Ōtawara Japan
92 G2	Oteren Norway
134 C1	Othello U.S.A.
120 A3	Otjiwarongo Namibia
109 C2	Otočac Croatia
	Otog Qi China see Ulan
117 B3	Otoro, Jebel mt. Sudan
	Otpor Rus. Fed. see Zabaykal'sk
93 E4	Otra r. Norway
109 C2	Otranto, Strait of Albania/Italy
67 C3	Ōtsu Japan
93 E3	Otta Norway
130 C3	Ottawa Can.
130 C3	Ottawa r. Can.
138 B2	Ottawa U.S.A.
137 D3	Ottawa KS U.S.A.
130 B2	Ottawa Islands Can.
98 B2	Otterburn U.K.
130 B2	Otter Rapids Can.
100 B2	Ottignies Belgium
137 E2	Ottumwa U.S.A.
150 B3	Otuzco Peru
52 B3	Otway, Cape Austr.
140 B2	Ouachita r. U.S.A.
140 B2	Ouachita, Lake U.S.A.
140 B2	Ouachita Mountains U.S.A.
118 C2	Ouadda C.A.R.
115 D3	Ouaddaï reg. Chad
114 B3	Ouagadougou Burkina
114 B3	Ouahigouya Burkina
114 B3	Oualâta Maur.
118 C2	Ouanda-Djallé C.A.R.
114 B2	Ouarâne reg. Maur.
115 C1	Ouargla Alg.
114 B1	Ouarzazate Morocco
100 A2	Oudenaarde Belgium
122 B3	Oudtshoorn S. Africa
107 C2	Oued Tlélat Alg.
114 B1	Oued Zem Morocco
104 A2	Ouessant, Île d' i. France
118 B2	Ouesso Congo
97 B2	Oughterard Ireland
118 B2	Ouham r. C.A.R./Chad
114 B1	Oujda Morocco
92 H3	Oulainen Fin.
107 D2	Ouled Farès Alg.
92 I2	Oulu Fin.
92 I3	Oulujärvi l. Fin.
108 A1	Oulx Italy
115 E3	Oum-Chalouba Chad
115 D3	Oum-Hadjer Chad
115 E3	Ounianga Kébir Chad
100 B2	Oupeye Belgium
100 C2	Our r. Ger./Lux.
106 B1	Ourense Spain
154 C2	Ourinhos Brazil
155 D2	Ouro Preto Brazil
100 B2	Ourthe r. Belgium
98 C3	Ouse r. U.K.
	Outaouais, Rivière des r. Can. see Ottawa
131 D3	Outardes, Rivière aux r. Can.
131 D2	Outardes Quatre, Réservoir resr Can.
96 A2	Outer Hebrides is U.K.
	Outer Mongolia country Asia see Mongolia
120 A3	Outjo Namibia
129 D2	Outlook Can.
92 I3	Outokumpu Fin.
52 B3	Ouyen Austr.
107 F2	Ovace, Punta d' mt. France
152 A3	Ovalle Chile
106 B1	Ovar Port.
92 H2	Överkalix Sweden
136 B3	Overland Park U.S.A.
92 H2	Övertorneå Sweden
106 B1	Oviedo Spain
141 D3	Oviedo U.S.A.
88 B2	Ovišrags hd Latvia
93 E3	Øvre Årdal Norway

93 F3	Øvre Rendal Norway
90 B1	Ovruch Ukr.
118 B3	Owando Congo
67 C4	Owase Japan
143 D1	Owasso U.S.A.
137 E2	Owatonna U.S.A.
139 D2	Owego U.S.A.
138 B3	Owensboro U.S.A.
135 C3	Owens Lake U.S.A.
130 B3	Owen Sound Can.
51 D1	Owen Stanley Range mts P.N.G.
115 C4	Owerri Nigeria
115 C4	Owo Nigeria
138 C2	Owosso U.S.A.
134 C2	Owyhee U.S.A.
134 C2	Owyhee r. U.S.A.
129 D3	Oxbow Can.
54 B2	Oxford N.Z.
99 C4	Oxford U.K.
140 C2	Oxford U.S.A.
129 E2	Oxford Lake Can.
145 D2	Oxkutzcab Mex.
52 B2	Oxley Austr.
97 B1	Ox Mountains hills Ireland
135 C4	Oxnard U.S.A.
67 C3	Oyama Japan
118 B2	Oyem Gabon
129 C2	Oyen Can.
105 D2	Oyonnax France
77 C2	Oyoqquduq Uzbek.
64 B3	Ozamiz Phil.
140 C2	Ozark AL U.S.A.
137 E3	Ozark MO U.S.A.
137 E3	Ozark Plateau U.S.A.
137 E3	Ozarks, Lake of the U.S.A.
83 L3	Ozernovskiy Rus. Fed.
88 B3	Ozersk Rus. Fed.
89 E3	Ozery Rus. Fed.
87 D3	Ozinki Rus. Fed.

P

127 I2	Paamiut Greenland
122 A3	Paarl S. Africa
103 D1	Pabianice Pol.
75 C2	Pabna Bangl.
88 C3	Pabradė Lith.
74 A2	Pab Range mts Pak.
150 B3	Pacasmayo Peru
142 B2	Pacheco Mex.
109 C3	Pachino Italy
145 C2	Pachuca Mex.
135 B3	Pacifica U.S.A.
157 E9	Pacific-Antarctic Ridge S. Pacific Ocean
156	Pacific Ocean
61 C2	Pacitan Indon.
52 B2	Packsaddle Austr.
103 D1	Paczków Pol.
60 B2	Padang Indon.
60 B1	Padang Endau Malaysia
60 B2	Padangpanjang Indon.
60 A1	Padangsidimpuan Indon.
101 D2	Paderborn Ger.
	Padova Italy see Padua
143 D3	Padre Island U.S.A.
99 A4	Padstow U.K.
52 B3	Padthaway Austr.
108 B1	Padua Italy
138 B3	Paducah U.S.A.
143 C2	Paducah U.S.A.
65 B1	Paegam N. Korea
	Paektu-san mt. China/N. Korea see Baitou Shan
65 A2	Paengnyŏng-do i. S. Korea
54 C1	Paeroa N.Z.
	Pafos Cyprus see Paphos
109 C2	Pag Croatia
109 C2	Pag i. Croatia
64 B3	Pagadian Phil.
60 B2	Pagai Selatan i. Indon.
60 B2	Pagai Utara i. Indon.
59 D1	Pagan i. N. Mariana Is
61 C2	Pagatan Indon.
142 A1	Page U.S.A.
88 B2	Pagėgiai Lith.
153 E5	Paget, Mount S. Georgia
136 B3	Pagosa Springs U.S.A.
88 C2	Paide Estonia
99 B4	Paignton U.K.
93 I3	Päijänne l. Fin.
75 C2	Paikū Co l. China
60 B2	Painan Indon.
138 C2	Painesville U.S.A.

Painted Desert

142 A1 **Painted Desert** U.S.A.
Paint Hills Can. *see* **Wemindji**
96 B3 **Paisley** U.K.
92 H2 **Pajala** Sweden
150 A3 **Paján** Ecuador
150 C2 **Pakaraima Mountains** *mts* S. America
150 D2 **Pakaraima Mountains** S. America
74 A2 **Pakistan** *country* Asia
62 A1 **Pakokku** Myanmar
88 B2 **Pakruojis** Lith.
103 D2 **Paks** Hungary
130 A2 **Pakwash Lake** Can.
62 B2 **Pakxan** Laos
63 B2 **Pakxé** Laos
115 D4 **Pala** Chad
60 B2 **Palabuhanratu, Teluk** *b.* Indon.
111 C3 **Palaikastro** Greece
111 B3 **Palaiochora** Greece
73 B3 **Palakkad** India
122 B1 **Palamakoloi** Botswana
107 D1 **Palamós** Spain
83 L3 **Palana** Rus. Fed.
64 B2 **Palanan** Phil.
61 C2 **Palangkaraya** Indon.
74 B2 **Palanpur** India
123 C1 **Palapye** Botswana
83 L2 **Palatka** Rus. Fed.
141 D3 **Palatka** U.S.A.
59 C2 **Palau** *country* N. Pacific Ocean
63 A2 **Palaw** Myanmar
64 A3 **Palawan** *i.* Phil.
64 A3 **Palawan Passage** *str.* Phil.
88 B2 **Paldiski** Estonia
89 F2 **Palekh** Rus. Fed.
60 B2 **Palembang** Indon.
106 C1 **Palencia** Spain
145 C3 **Palenque** Mex.
108 B3 **Palermo** Italy
143 D2 **Palestine** U.S.A.
62 A1 **Paletwa** Myanmar
Palghat India *see* **Palakkad**
74 B2 **Pali** India
48 G4 **Palikir** Micronesia
109 C2 **Palinuro, Capo** *c.* Italy
111 B3 **Paliouri, Akrotirio** *pt* Greece
100 B3 **Paliseul** Belgium
92 I3 **Paljakka** *h.* Fin.
88 C2 **Palkino** Rus. Fed.
73 B4 **Palk Strait** India/Sri Lanka
Palla Bianca *mt.* Austria/Italy *see* **Weißkugel**
54 C2 **Palliser, Cape** N.Z.
157 F7 **Palliser, Îles** *is* Fr. Polynesia
106 B2 **Palma del Río** Spain
107 D2 **Palma de Mallorca** Spain
154 B3 **Palmas** Brazil
151 E4 **Palmas** Brazil
154 B3 **Palmas, Campos de** *hills* Brazil
114 B4 **Palmas, Cape** Liberia
141 D3 **Palm Bay** U.S.A.
135 C4 **Palmdale** U.S.A.
154 C3 **Palmeira** Brazil
151 E3 **Palmeirais** Brazil
126 C2 **Palmer** U.S.A.
55 A2 **Palmer Land** *reg.* Antarctica
49 K5 **Palmerston** *atoll* Cook Is
54 C2 **Palmerston North** N.Z.
109 C3 **Palmi** Italy
145 C2 **Palmillas** Mex.
150 B2 **Palmira** Col.
154 B2 **Palmital** Brazil
135 C4 **Palm Springs** U.S.A.
Palmyra Syria *see* **Tadmur**
49 K3 **Palmyra Atoll** N. Pacific Ocean
135 B3 **Palo Alto** U.S.A.
117 B3 **Paloich** Sudan
145 C3 **Palomares** Mex.
61 D2 **Palopo** Indon.
107 C2 **Palos, Cabo de** *c.* Spain
92 I3 **Paltamo** Fin.
61 C2 **Palu** Indon.
83 M2 **Palyavaam** *r.* Rus. Fed.
150 B3 **Pamar** Col.
121 C3 **Pambarra** Moz.
104 C3 **Pamiers** France
77 D3 **Pamir** *mts* Asia
141 E1 **Pamlico Sound** *sea chan.* U.S.A.
152 B1 **Pampa Grande** Bol.
153 B3 **Pampas** *reg.* Arg.
150 B2 **Pamplona** Col.
107 C1 **Pamplona** Spain
111 D2 **Pamukova** Turkey
60 B2 **Panaitan** *i.* Indon.
73 B3 **Panaji** India
146 B4 **Panama** *country* Central America
Panamá Panama *see* **Panama City**
146 C4 **Panamá, Canal de** *canal* Panama
Panamá, Golfo de *g.* Panama *see* **Panama, Gulf of**
146 C4 **Panama, Gulf of** *g.* Panama
Panama Canal *canal* Panama *see* **Panamá, Canal de**
146 C4 **Panama City** Panama
140 C2 **Panama City** U.S.A.
135 C3 **Panamint Range** *mts* U.S.A.
60 B1 **Panarik** Indon.
64 B2 **Panay** *i.* Phil.
109 D2 **Pančevo** Serbia
64 B2 **Pandan** Phil.
64 B2 **Pandan** Phil.
75 C2 **Pandaria** India
73 B3 **Pandharpur** India
88 B2 **Panevėžys** Lith.
Panfilov Kazakh. *see* **Zharkent**
61 C2 **Pangkalanbuun** Indon.
60 A1 **Pangkalansusu** Indon.
60 B2 **Pangkalpinang** Indon.
61 D2 **Pangkalsiang, Tanjung** *pt* Indon.
127 H2 **Pangnirtung** Can.
86 G2 **Pangody** Rus. Fed.
89 F3 **Panino** Rus. Fed.
74 B2 **Panipat** India
74 A2 **Panjgur** Pak.
Panjim India *see* **Panaji**
118 A2 **Pankshin** Nigeria
65 C1 **Pan Ling** *mts* China
75 C2 **Panna** India
50 A2 **Pannawonica** Austr.
154 B2 **Panorama** Brazil
65 B1 **Panshi** China
152 C1 **Pantanal** *reg.* Brazil
145 C2 **Pánuco** Mex.
145 C2 **Pánuco** *r.* Mex.
71 A3 **Panxian** China
62 B1 **Panzhihua** China
109 C3 **Paola** Italy
118 B2 **Paoua** C.A.R.
63 B2 **Paôy Pêt** Cambodia
103 D2 **Pápa** Hungary
54 B1 **Papakura** N.Z.
145 C2 **Papantla** Mex.
96 □ **Papa Stour** *i.* U.K.
54 B1 **Papatoetoe** N.Z.
49 M5 **Papeete** Fr. Polynesia
100 C1 **Papenburg** Ger.
80 B2 **Paphos** Cyprus
137 D2 **Papillion** U.S.A.
59 D3 **Papua** *reg.* Indon.
59 D3 **Papua, Gulf of** P.N.G.
59 D3 **Papua New Guinea** *country* Oceania
89 F3 **Para** *r.* Rus. Fed.
50 A2 **Paraburdoo** Austr.
154 C1 **Paracatu** Brazil
155 C1 **Paracatu** *r.* Brazil
52 A2 **Parachilna** Austr.
109 D2 **Paraćin** Serbia
155 D1 **Pará de Minas** Brazil
151 D2 **Paradise** Guyana
135 B3 **Paradise** U.S.A.
140 B1 **Paragould** U.S.A.
151 D3 **Paraguai** *r.* Brazil
147 D3 **Paraguaná, Península de** *pen.* Venez.
152 C2 **Paraguay** *r.* Arg./Para.
152 C2 **Paraguay** *country* S. America
155 D2 **Paraíba do Sul** *r.* Brazil
154 B1 **Paraíso** Brazil
145 C3 **Paraíso** Mex.
114 C4 **Parakou** Benin
52 A2 **Parakylia** Austr.
151 D2 **Paramaribo** Suriname
83 L3 **Paramushir, Ostrov** *i.* Rus. Fed.
152 B3 **Paraná** Arg.
154 B2 **Paraná** *state* Brazil
154 A3 **Paraná** *r.* S. America
154 C1 **Paraná, Serra do** *hills* Brazil
154 C3 **Paranaguá** Brazil
154 B1 **Paranaíba** Brazil
154 B2 **Paranaíba** *r.* Brazil
154 B2 **Paranapanema** *r.* Brazil
154 B2 **Paranapiacaba, Serra** *mts* Brazil
154 B2 **Paranavaí** Brazil
90 A2 **Parângul Mare, Vârful** *mt.* Romania
54 B2 **Paraparaumu** N.Z.
155 D2 **Parati** Brazil
52 A2 **Paratoo** Austr.
151 D3 **Parauaquara, Serra** *h.* Brazil
154 B1 **Paraúna** Brazil
105 C2 **Paray-le-Monial** France
74 B2 **Parbati** *r.* India
74 B3 **Parbhani** India
101 E1 **Parchim** Ger.
103 E1 **Parczew** Pol.
155 E1 **Pardo** *r. Bahia* Brazil
154 B2 **Pardo** *r. Mato Grosso do Sul* Brazil
154 C2 **Pardo** *r. São Paulo* Brazil
103 D1 **Pardubice** Czech Rep.
152 C1 **Parecis, Serra dos** *hills* Brazil
130 C3 **Parent** Can.
130 C3 **Parent, Lac** *l.* Can.
61 C2 **Parepare** Indon.
89 D2 **Parfino** Rus. Fed.
111 B3 **Parga** Greece
109 C3 **Parghelia** Italy
147 D3 **Paria, Gulf of** Trin. and Tob./Venez.
150 C2 **Parima, Serra** *mts* Brazil
151 D3 **Parintins** Brazil
104 C2 **Paris** France
140 C1 **Paris** *TN* U.S.A.
143 D2 **Paris** *TX* U.S.A.
93 H3 **Parkano** Fin.
142 A2 **Parker** U.S.A.
138 C3 **Parkersburg** U.S.A.
53 C2 **Parkes** Austr.
138 A1 **Park Falls** U.S.A.
134 B1 **Parkland** U.S.A.
137 D1 **Park Rapids** U.S.A.
106 C1 **Parla** Spain
108 B2 **Parma** Italy
134 C2 **Parma** U.S.A.
151 E3 **Parnaíba** Brazil
151 E3 **Parnaíba** *r.* Brazil
Parnassus, Mount *mt.* Greece *see* **Parnassos**
54 B2 **Parnassus** N.Z.
111 B3 **Parnonas** *mts* Greece
88 B2 **Pärnu** Estonia
65 B2 **P'aro-ho** *l.* S. Korea
52 B2 **Paroo** *watercourse* Austr.
Paropamisus *mts* Afgh. *see* **Safīd Kūh**
111 C3 **Paros** *i.* Greece
135 D3 **Parowan** U.S.A.
153 A3 **Parral** Chile
53 D2 **Parramatta** Austr.
144 B2 **Parras** Mex.
126 D2 **Parry, Cape** Can.
126 E1 **Parry Islands** Can.
130 B3 **Parry Sound** Can.
137 D3 **Parsons** U.S.A.
108 B3 **Partanna** Italy
101 D2 **Partenstein** Ger.
104 B2 **Parthenay** France
108 B3 **Partinico** Italy
66 B2 **Partizansk** Rus. Fed.
97 B2 **Partry Mountains** *hills* Ireland
151 D3 **Paru** *r.* Brazil
131 E3 **Pasadena** Can.
135 C4 **Pasadena** *CA* U.S.A.
143 D3 **Pasadena** *TX* U.S.A.
62 A2 **Pasawng** Myanmar
140 C2 **Pascagoula** U.S.A.
110 C1 **Pașcani** Romania
134 C1 **Pasco** U.S.A.
155 E1 **Pascoal, Monte** *h.* Brazil
Pascua, Isla de *i.* S. Pacific Ocean *see* **Easter Island**
Pas de Calais *str.* France/U.K. *see* **Dover, Strait of**
102 C1 **Pasewalk** Ger.
129 D2 **Pasfield Lake** Can.
89 D1 **Pasha** Rus. Fed.
64 B2 **Pasig** Phil.
60 B1 **Pasir Putih** Malaysia
103 D1 **Pasłęk** Pol.
74 A2 **Pasni** Pak.
153 A4 **Paso Río Mayo** Arg.
135 B3 **Paso Robles** U.S.A.
97 B3 **Passage West** Ireland
155 D2 **Passa Tempo** Brazil
102 C2 **Passau** Ger.
152 C2 **Passo Fundo** Brazil
155 C2 **Passos** Brazil
88 C2 **Pastavy** Belarus
150 B3 **Pastaza** *r.* Peru
150 B2 **Pasto** Col.
74 B1 **Pasu** Pak.
61 C2 **Pasuruan** Indon.
88 B2 **Pasvalys** Lith.
103 D2 **Pásztó** Hungary
153 A5 **Patagonia** *reg.* Arg.
75 C2 **Patan** Nepal
54 B1 **Patea** N.Z.
139 E2 **Paterson** U.S.A.
74 B1 **Pathankot** India
Pathein Myanmar *see* **Bassein**
136 B2 **Pathfinder Reservoir** U.S
61 C2 **Pati** Indon.
74 B1 **Patiala** India
62 A1 **Patkai Bum** *mts* India/Myanmar
111 C3 **Patmos** *i.* Greece
75 C2 **Patna** India
81 C2 **Patnos** Turkey
154 B3 **Pato Branco** Brazil
152 C3 **Patos, Lagoa dos** *l.* Brazi
155 C1 **Patos de Minas** Brazil
152 B3 **Patquía** Arg.
Patra Greece *see* **Patras**
111 B3 **Patras** Greece
75 C2 **Patratu** India
154 C1 **Patrocínio** Brazil
63 B3 **Pattani** Thai.
63 B2 **Pattaya** Thai.
128 B2 **Pattullo, Mount** Can.
129 D2 **Patuanak** Can.
146 B3 **Patuca** *r.* Hond.
144 B3 **Pátzcuaro** Mex.
104 B3 **Pau** France
104 B3 **Pau, Gave de** *r.* France
104 B2 **Pauillac** France
150 C3 **Pauini** Brazil
62 A1 **Pauk** Myanmar
126 D2 **Paulatuk** Can.
Paulis Dem. Rep. Congo *see* **Isiro**
151 E3 **Paulistana** Brazil
151 F3 **Paulo Afonso** Brazil
123 D2 **Paulpietersburg** S. Africa
143 D2 **Pauls Valley** U.S.A.
62 A2 **Paungde** Myanmar
155 D1 **Pavão** Brazil
108 A1 **Pavia** Italy
88 B2 **Pāvilosta** Latvia
110 C2 **Pavlikeni** Bulg.
77 D1 **Pavlodar** Kazakh.
91 D2 **Pavlohrad** Ukr.
91 E1 **Pavlovsk** Rus. Fed.
91 D2 **Pavlovskaya** Rus. Fed.
139 E2 **Pawtucket** U.S.A.
111 B3 **Paxoi** *i.* Greece
60 B2 **Payakumbuh** Indon.
134 C2 **Payette** U.S.A.
134 C2 **Payette** *r.* U.S.A.
86 F2 **Pay-Khoy, Khrebet** *hills* Rus. Fed.
Payne Can. *see* **Kangirsu**
130 C2 **Payne, Lac** *l.* Can.
152 C3 **Paysandú** Uru.
81 C1 **Pazar** Turkey
110 B2 **Pazardzhik** Bulg.
111 C3 **Pazarköy** Turkey
108 B1 **Pazin** Croatia
63 A2 **Pe** Myanmar
128 C2 **Peace** *r.* Can.
128 C2 **Peace River** Can.
53 C2 **Peak Hill** *N.S.W.* Austr.
50 A2 **Peak Hill** *W.A.* Austr.
135 E3 **Peale, Mount** U.S.A.
140 C2 **Pearl** *r.* U.S.A.
71 B3 **Pearl River** *r.* China
143 D3 **Pearsall** U.S.A.
126 F1 **Peary Channel** Can.
121 C2 **Pebane** Moz.
Peć Kosovo *see* **Pejë**
155 D1 **Peçanha** Brazil
154 C3 **Peças, Ilha das** *i.* Brazil
92 J2 **Pechenga** Rus. Fed.
86 E2 **Pechora** Rus. Fed.
86 E2 **Pechora** *r.* Rus. Fed.
Pechora Sea *sea* Rus. Fed *see* **Pechorskoye More**
86 E2 **Pechorskoye More** *sea* Rus. Fed.
88 C2 **Pechory** Rus. Fed.
142 B1 **Pecos** *NM* U.S.A.
143 C2 **Pecos** *TX* U.S.A.

43 C3 Pecos r. U.S.A.
03 D2 Pécs Hungary
42 B3 Pedernales Mex.
55 D1 Pedra Azul Brazil
54 C2 Pedregulho Brazil
51 E3 Pedreiras Brazil
73 C4 Pedro, Point Sri Lanka
51 E3 Pedro Afonso Brazil
44 B1 Pedro de Valdivia Chile
54 B1 Pedro Gomes Brazil
52 C2 Pedro Juan Caballero Para.
06 B1 Pedroso Port.
96 C3 Peebles U.K.
41 E2 Pee Dee r. U.S.A.
26 D2 Peel r. Can.
98 A2 Peel Isle of Man
28 C2 Peerless Lake Can.
54 B2 Pegasus Bay N.Z.
01 E3 Pegnitz Ger.
62 A2 Pegu Myanmar
62 A2 Pegu Yoma mts Myanmar
53 B3 Pehuajó Arg.
01 E1 Peine Ger.
88 C2 Peipus, Lake Estonia/Rus. Fed.
Peiraias Greece see Piraeus
54 B2 Peixe r. Brazil
55 C2 Peixoto, Represa resr Brazil
51 D4 Peixoto de Azevedo Brazil
09 D2 Pejë Kosovo
53 C2 Peka Lesotho
60 B2 Pekalongan Indon.
60 B1 Pekan Malaysia
60 B1 Pekanbaru Indon.
Peking China see Beijing
30 B3 Pelee Island Can.
61 D2 Peleng i. Indon.
03 D2 Pelhřimov Czech Rep.
92 I2 Pelkosenniemi Fin.
22 A2 Pella S. Africa
37 E2 Pella U.S.A.
59 D3 Pelleluhu Islands P.N.G.
92 H2 Pello Fin.
28 N1 Pelly r. Can.
Pelly Bay Can. see Kugaaruk
28 A1 Pelly Mountains Can.
52 C2 Pelotas Brazil
52 C2 Pelotas, Rio das r. Brazil
39 F1 Pemadumcook Lake U.S.A.
60 B1 Pemangkat Indon.
60 A1 Pematangsiantar Indon.
21 D2 Pemba Moz.
20 B2 Pemba Zambia
49 D3 Pemba Island Tanz.
32 B2 Pemberton Can.
37 D1 Pembina r. Can.
37 D1 Pembina r. Can./U.S.A.
30 C3 Pembroke Can.
99 A4 Pembroke U.K.
41 D3 Pembroke Pines U.S.A.
06 C1 Peñalara mt. Spain
54 B2 Penápolis Brazil
06 B1 Peñaranda de Bracamonte Spain
07 C1 Peñarroya mt. Spain
06 B2 Peñarroya-Pueblonuevo Spain
06 B2 Peñas, Cabo de c. Spain
53 A4 Penas, Golfo de g. Chile
11 C2 Pendik Turkey
41 C1 Pendleton U.S.A.
28 B2 Pendleton Bay Can.
34 C1 Pend Oreille Lake U.S.A.
Penfro U.K. see Pembroke
74 B3 Penganga r. India
18 C3 Penge Dem. Rep. Congo
23 D1 Penge S. Africa
70 C2 Penglai China
71 A3 Pengshui China
06 B2 Peniche Port.
06 C1 Penicuik U.K.
60 B1 Peninsular Malaysia pen. Malaysia
08 B2 Penne Italy
52 A3 Penneshaw Austr.
98 B2 Pennines hills U.K.
39 D2 Pennsylvania state U.S.A.
27 H2 Penny Icecap Can.
89 D2 Peno Rus. Fed.
39 F2 Penobscot r. U.S.A.
52 A3 Penola Austr.
50 C3 Penong Austr.
57 E6 Penrhyn Basin S. Pacific Ocean
53 D2 Penrith Austr.
98 B2 Penrith U.K.
40 C2 Pensacola U.S.A.

55 B1 Pensacola Mountains Antarctica
61 C1 Pensiangan Malaysia
128 C3 Penticton Can.
96 C1 Pentland Firth sea chan. U.K.
99 B3 Penygadair h. U.K.
87 D3 Penza Rus. Fed.
99 A4 Penzance U.K.
83 L2 Penzhinskaya Guba b. Rus. Fed.
142 A2 Peoria AZ U.S.A.
138 B2 Peoria IL U.S.A.
107 C1 Perales del Alfambra Spain
111 B3 Perama Greece
131 D3 Percé Can.
50 B2 Percival Lakes imp. l. Austr.
51 E2 Percy Isles Austr.
107 D1 Perdido, Monte mt. Spain
86 F2 Perdizes Brazil
86 F2 Peregrebnoye Rus. Fed.
150 B2 Pereira Col.
154 B2 Pereira Barreto Brazil
145 D2 Pereira de Eça Angola see Ondjiva
90 A2 Peremyshlyany Ukr.
89 E2 Pereslavl'-Zalesskiy Rus. Fed.
91 C1 Pereyaslav-Khmel'nyts'kyy Ukr.
153 B3 Pergamino Arg.
92 H3 Perhonjoki r. Fin.
131 C2 Péribonka, Lac l. Can.
152 B2 Perico Arg.
144 B2 Pericos Mex.
104 C2 Périgueux France
150 B2 Perijá, Sierra de mts Venez.
111 B3 Peristeri Greece
153 A4 Perito Moreno Arg.
101 C1 Perleberg Ger.
86 E3 Perm' Rus. Fed.
109 D2 Përmet Albania
Pernambuco Brazil see Recife
52 A2 Pornatty Lagoon imp. l. Austr.
110 B2 Pernik Bulg.
Pernov Estonia see Pärnu
105 C2 Péronne France
105 C3 Perote Mex.
145 C3 Perpignan France
99 A4 Perranporth U.K.
Perrégaux Alg. see Mohammadia
141 D2 Perry FL U.S.A.
141 D2 Perry GA U.S.A.
137 E2 Perry IA U.S.A.
143 D1 Perry OK U.S.A.
138 C2 Perrysburg U.S.A.
143 C1 Perryton U.S.A.
137 F3 Perryville U.S.A.
Pershotravnevoye Ukr. see Pershotravens'k
99 B3 Pershore U.K.
91 D2 Pershotravens'k Ukr.
Persia country Asia see Iran
Persian Gulf g. Asia see The Gulf
50 A3 Perth Austr.
96 C2 Perth U.K.
159 F5 Perth Basin Indian Ocean
86 C2 Pertominsk Rus. Fed.
105 D3 Pertuis France
108 A2 Pertusato, Capo c. France
150 B3 Peru country S. America
138 B2 Peru U.S.A.
157 H6 Peru Basin S. Pacific Ocean
157 H7 Peru-Chile Trench S. Pacific Ocean
108 B2 Perugia Italy
154 B2 Peruíbe Brazil
100 A2 Péruwelz Belgium
90 C2 Pervomays'k Ukr.
91 C2 Pervomays'ke Ukr.
Pervomayskiy Rus. Fed. see Novodvinsk
89 F3 Pervomayskiy Rus. Fed.
91 D2 Pervomays'kyy Ukr.
108 B2 Pesaro Italy
108 B2 Pescara Italy
108 B2 Pescara r. Italy
74 B1 Peshawar Pak.
109 D2 Peshkopi Albania
108 C1 Pesnica Slovenia
104 B3 Pessac France
89 E2 Pestovo Rus. Fed.
140 C2 Petal U.S.A.
100 B3 Pétange Lux.

147 D3 Petare Venez.
144 B3 Petatlán Mex.
121 C2 Petauke Zambia
130 C3 Petawawa Can.
138 B2 Petenwell Lake U.S.A.
52 A2 Peterborough Austr.
130 C3 Peterborough Can.
99 C3 Peterborough U.K.
96 D2 Peterhead U.K.
55 R3 Peter I Island Antarctica
129 E1 Peter Lake Can.
50 B2 Petermann Ranges mts Austr.
129 D2 Peter Pond Lake Can.
128 A2 Petersburg AK U.S.A.
139 D3 Petersburg VA U.S.A.
101 D1 Petershagen Ger.
Peter the Great Bay b. Rus. Fed. see
131 E2 Petit Mécatina r. Can.
145 D2 Peto Mex.
138 C1 Petoskey U.S.A.
80 B2 Petra tourist site Jordan
66 B2 Petra Velikogo, Zaliv b. Rus. Fed.
111 B2 Petrich Bulg.
Petroaleksandrovsk Uzbek. see To'rtko'l
88 C2 Petrodvorets Rus. Fed.
Petrokov Pol. see Piotrków Trybunalski
151 E3 Petrolina Brazil
77 C1 Petropavlovsk Kazakh.
83 L3 Petropavlovsk-Kamchatskiy Rus. Fed.
110 B1 Petroşani Romania
Petrovskoye Rus. Fed. see Svetlograd
89 F3 Petrovskoye Rus. Fed.
89 E2 Petrovskoye Rus. Fed.
69 D1 Petrovsk-Zabaykal'skiy Rus. Fed.
86 C2 Petrozavodsk Rus. Fed.
123 C2 Petrusburg S. Africa
123 C2 Petrus Steyn S. Africa
122 B3 Petrusville S. Africa
Petsamo Rus. Fed. see Pechenga
87 F3 Petukhovo Rus. Fed.
89 E2 Petushki Rus. Fed.
60 A1 Peureula Indon.
83 M2 Pevek Rus. Fed.
102 B2 Pforzheim Ger.
102 C2 Pfunds Austria
101 D3 Pfungstadt Ger.
123 C1 Phagameng S. Africa
123 C2 Phahameng S. Africa
123 D1 Phalaborwa S. Africa
74 B2 Phalodi India
63 A3 Phangnga Thai.
62 B1 Phăng Xi Păng mt. Vietnam
63 B2 Phan Rang-Thap Cham Vietnam
63 B2 Phan Thiết Vietnam
63 B3 Phatthalung Thai.
62 A2 Phayao Thai.
129 D2 Phelps Lake Can.
141 C2 Phenix City U.S.A.
63 A2 Phet Buri Thai.
63 B2 Phetchabun Thai.
63 B2 Phichit Thai.
139 D3 Philadelphia U.S.A.
136 C2 Philip U.S.A.
Philip Atoll atoll Micronesia see Sorol
Philippeville Alg. see Skikda
100 B2 Philippeville Belgium
51 C2 Philippi, Lake imp. l. Austr.
156 C2 Philippine Basin N. Pacific Ocean
64 B2 Philippines country Asia
64 B2 Philippine Sea N. Pacific Ocean
126 C2 Philip Smith Mountains U.S.A.
122 B3 Philipstown S. Africa
53 C3 Phillip Island Austr.
137 D3 Phillipsburg U.S.A.
63 B2 Phimun Mangsahan Thai.
123 C2 Phiritona S. Africa
63 B2 Phitsanulok Thai.
63 B2 Phnom Penh Cambodia
Phnum Pénh Cambodia see Phnom Penh

142 A2 Phoenix U.S.A.
49 J4 Phoenix Islands Kiribati
63 B2 Phon Thai.
62 B2 Phong Nha Vietnam
62 B1 Phôngsali Laos
62 B1 Phong Thô Vietnam
62 B2 Phônsavan Laos
62 B2 Phrae Thai.
Phu Cuong Vietnam see Thu Dâu Môt
120 B3 Phuduhudu Botswana
63 A3 Phuket Thai.
63 B2 Phumĭ Kâmpóng Trâbêk Cambodia
62 A2 Phumĭphon, Khuan Thai.
63 B2 Phumĭ Sâmraông Cambodia
63 B2 Phu Quôc, Đao i. Vietnam
123 C2 Phuthaditjhaba S. Africa
Phu Vinh Vietnam see Tra Vinh
62 A2 Phyu Myanmar
108 A1 Piacenza Italy
108 B2 Pianosa, Isola i. Italy
110 C1 Piatra Neamţ Romania
151 E3 Piauí r. Brazil
108 B1 Piave r. Italy
117 B4 Pibor r. Sudan
117 B4 Pibor Post Sudan
Picardie reg. France see Picardy
104 C2 Picardy reg. France
140 C2 Picayune U.S.A.
152 B2 Pichanal Arg.
144 A2 Pichilingue Mex.
98 C2 Pickering U.K.
130 A1 Pickle Lake Can.
151 E3 Picos Brazil
153 B4 Pico Truncado Arg.
53 D2 Picton Can.
54 B2 Picton N.Z.
73 C4 Pidurutalagala mt. Sri Lanka
154 C2 Piedade Brazil
145 C3 Piedras Negras Guat.
145 B2 Piedras Negras Mex.
93 I3 Pieksämäki Fin.
92 I3 Pielinen l. Fin.
136 C2 Pierre U.S.A.
105 C3 Pierrelatte France
123 D2 Pietermaritzburg S. Africa
Pietersaari Fin. see Jakobstad
Pietersburg S. Africa see Polokwane
123 C2 Piet Retief S. Africa
110 B1 Pietrosa mt. Romania
110 C1 Pietrosu, Vârful mt. Romania
128 C2 Pigeon Lake Can.
137 F1 Pigeon River U.S.A.
153 B3 Pigüé Arg.
93 I3 Pihlajavesi l. Fin.
92 I3 Pihtipudas Fin.
145 C3 Pijijiapan Mex.
89 D2 Pikalevo Rus. Fed.
130 A2 Pikangikum Can.
136 C3 Pikes Peak U.S.A.
122 A3 Piketberg S. Africa
138 C3 Pikeville U.S.A.
103 D1 Piła Pol.
153 C3 Pilar Arg.
152 C2 Pilar Para.
75 B2 Pilibhit India
53 C2 Pilliga Austr.
154 C1 Pilões, Serra dos mts Brazil
150 C4 Pimenta Bueno Brazil
88 C3 Pina r. Belarus
153 C3 Pinamar Arg.
60 B1 Pinang i. Malaysia
80 B2 Pınarbaşı Turkey
146 B2 Pinar del Río Cuba
64 B2 Pinatubo, Mount vol. Phil.
103 E1 Pińczów Pol.
151 E3 Pindaré r. Brazil
Pindos Greece see Pindus Mountains
111 B2 Pindus Mountains Greece
140 B2 Pine Bluff U.S.A.
136 C2 Pine Bluffs U.S.A.
50 C1 Pine Creek Austr.
136 B2 Pinedale U.S.A.
86 D2 Pinega Rus. Fed.
129 D2 Pinehouse Lake Can.
111 B3 Pineios r. Greece
141 D3 Pine Islands FL U.S.A.
141 D4 Pine Islands FL U.S.A.

74 B2 Pugal India
54 B2 Pukaki, Lake N.Z.
49 K5 Pukapuka atoll Cook Is
129 D2 Pukatawagan Can.
65 B1 Pukchin N. Korea
65 B1 Pukch'ŏng N. Korea
109 C2 Pukë Albania
54 B1 Pukekohe N.Z.
65 B1 Puksubaek-san mt.
 N. Korea
108 B2 Pula China see Nyingchi
108 A3 Pula Croatia
108 A3 Pula Italy
152 B2 Pulacayo Bol.
103 E1 Puławy Pol.
92 I3 Pulkkila Fin.
134 C1 Pullman U.S.A.
64 B2 Pulog, Mount Phil.
64 B3 Pulutan Indon.
150 A3 Puná, Isla i. Ecuador
54 B2 Punakaiki N.Z.
123 D1 Punda Maria S. Africa
73 B3 Pune India
65 B1 P'ungsan N. Korea
121 C2 Púnguè r. Moz.
119 C3 Punia Dem. Rep. Congo
74 B1 Punjab state India
153 B3 Punta Alta Arg.
153 A5 Punta Arenas Chile
153 C3 Punta del Este Uru.
146 B3 Punta Gorda Belize
146 B3 Puntarenas Costa Rica
117 C4 Puntland reg. Somalia
150 B1 Punto Fijo Venez.
92 I3 Puolanka Fin.
 Puqi China see Chibi
82 G2 Pur r. Rus. Fed.
75 C3 Puri India
100 B1 Purmerend Neth.
 Purnea India see Purnia
75 C2 Purnia India
75 C2 Puruliya India
150 C3 Purus r. Brazil/Peru
61 B2 Purwakarta Indon.
61 C2 Purwodadi Indon.
65 B1 Puryŏng N. Korea
74 B3 Pusad India
65 B2 Pusan S. Korea
89 E3 Pushchino Rus. Fed.
 Pushkino Rus. see
 Biläsuvar
89 E2 Pushkino Rus. Fed.
88 C2 Pushkinskiye Gory
 Rus. Fed.
88 C2 Pustoshka Rus. Fed.
62 A1 Putao Myanmar
71 B3 Putian China
 Puting China see De'an
61 C2 Puting, Tanjung pt
 Indon.
101 F1 Putlitz Ger.
60 B1 Putrajaya Malaysia
122 B2 Putsonderwater S. Africa
102 C1 Puttgarden Ger.
150 B3 Putumayo r. Col.
61 C1 Putusibau Indon.
90 B2 Putyla Ukr.
91 C1 Putyvl' Ukr.
93 I3 Puula l. Fin.
130 C1 Puvirnituq Can.
70 B2 Puyang China
104 C3 Puylaurens France
119 C3 Pweto Dem. Rep. Congo
99 A3 Pwllheli U.K.
92 J2 Pyaozerskiy Rus. Fed.
63 A2 Pyapon Myanmar
83 G2 Pyasina r. Rus. Fed.
87 D4 Pyatigorsk Rus. Fed.
91 C2 P'yatykhatky Ukr.
62 A2 Pyè Myanmar
88 C3 Pyetrykaw Belarus
93 H3 Pyhäjärvi l. Fin.
92 H3 Pyhäjoki r. Fin.
92 I3 Pyhäsalmi Fin.
62 A1 Pyingaing Myanmar
62 A2 Pyinmana Myanmar
62 A1 Pyin-U-Lwin Myanmar
65 B2 P'yŏksŏng N. Korea
65 B2 P'yŏnggang N. Korea
65 B2 P'yŏngsan N. Korea
65 B2 P'yŏngsong N. Korea
65 B2 P'yŏng'taek S. Korea
65 B2 P'yŏngyang N. Korea
135 C2 Pyramid Lake U.S.A.
80 B3 Pyramids of Giza tourist site
 Egypt
107 D1 Pyrenees mts Europe
111 B3 Pyrgetos Greece

111 C3 Pyrgi Greece
111 B3 Pyrgos Greece
91 C1 Pyryatyn Ukr.
103 C1 Pyrzyce Pol.
88 C2 Pytalovo Rus. Fed.
111 B3 Pyxaria mt. Greece

Q

 Qaanaaq Greenland see
 Thule
 Qabqa China see Gonghe
123 C3 Qacha's Nek Lesotho
76 B3 Qā'en Iran
69 D2 Qagan Nur China
 Qahremānshahr Iran see
 Kermānshāh
68 C2 Qaidam Pendi basin China
79 C2 Qalamat Abū Shafrah
 Saudi Arabia
77 C3 Qalāt Afgh.
78 A2 Qal'at al Azlam
 Saudi Arabia
78 A2 Qal'at al Mu'azzam
 Saudi Arabia
78 B2 Qal'at Bīshah Saudi Arabia
76 C3 Qal'eh-ye Now Afgh.
129 E1 Qamanirjuaq Lake Can.
 Qamanittuaq Can. see
 Baker Lake
78 C3 Qamar, Ghubbat al b.
 Yemen
79 C3 Qamar, Ghubbat al b.
 Yemen
68 C2 Qamdo China
78 B3 Qam Hadīl Saudi Arabia
127 I2 Qaqortoq Greenland
80 A3 Qārah Egypt
 Qarkilik China see
 Ruoqiang
77 C3 Qarshi Uzbek.
78 B2 Qaryat al Ulyā Saudi Arabia
127 I2 Qasigiannguit Greenland
116 A2 Qaşr al Farāfirah Egypt
79 D2 Qaşr-e Qand Iran
81 C2 Qaşr-e Shīrīn Iran
127 I2 Qassimiut Greenland
78 B3 Qa'tabah Yemen
79 C2 Qatar country Asia
116 A2 Qattāra Depression Egypt
 Qausuittuq Can. see
 Resolute
81 C1 Qazax Azer.
81 C2 Qazımämmäd Azer.
81 C2 Qazvīn Iran
 Qena Egypt see Qinā
127 I2 Qeqertarsuaq Greenland
127 I2 Qeqertarsuaq i. Greenland
127 I2 Qeqertarsuatsiaat
 Greenland
127 I2 Qeqertarsuup Tunua b.
 Greenland
79 C2 Qeshm Iran
74 A1 Qeysār, Kūh-e mt. Afgh.
70 A3 Qiandao Hu resr China
70 A3 Qianjiang China
70 B2 Qianjiang China
69 E1 Qianjin China
65 A1 Qian Shan mts China
71 A3 Qianxi China
70 C2 Qidong China
77 E3 Qiemo China
71 A3 Qijiang China
68 C2 Qijiaojing China
127 H2 Qikiqtarjuaq Can.
74 A2 Qila Ladgasht Pak.
68 C2 Qilian Shan mts China
127 J2 Qillak i. Greenland
70 B3 Qimen China
127 H1 Qimusseriarsuaq b.
 Greenland
116 B2 Qinā Egypt
 Qincheng China see
 Nanfeng
70 A2 Qingcheng China
70 C2 Qingdao China
70 B2 Qinghai Hu salt l. China
68 C2 Qinghai Nanshan mts
 China
 Qingjiang China see
 Huai'an
 Qingjiang China see
 Zhangshu
70 B2 Qingshuihe China
70 A2 Qingtongxia China

70 A2 Qingyang China
 Qingyuan China see Yizhou
71 B3 Qingyuan Guangdong China
65 A1 Qingyuan Liaoning China
 Qingzang Gaoyuan plat.
 China see Tibet, Plateau of
70 B2 Qingzhou China
70 B2 Qinhuangdao China
 Qinjiang China see
 Shicheng
70 A2 Qin Ling mts China
 Qinting China see Lianhua
70 B2 Qinyang China
71 A3 Qinzhou China
71 B4 Qionghai China
70 A2 Qionglai Shan mts China
71 B4 Qiongshan China
71 A4 Qiongzhong China
69 E1 Qiqihar China
81 D3 Qīr Iran
 Qishan China see Qimen
79 C3 Qishn Yemen
66 B1 Qitaihe China
70 B2 Qixian Henan China
70 B2 Qixian Shanxi China
 Qogir Feng mt. China/
 Pakistan see K2
81 D2 Qom Iran
 Qomishēh Iran see
 Shāhrezā
 Qomolangma Feng mt.
 China/Nepal see
 Everest, Mount
76 B2 Qo'ng'irot Uzbek.
77 D2 Qo'qon Uzbek.
76 B2 Qoraqalpog'iston Uzbek.
80 B2 Qornet es Saouda mt.
 Lebanon
81 C2 Qorveh Iran
79 C2 Qotbābād Iran
101 C1 Quakenbrück Ger.
53 C2 Quambone Austr.
63 B2 Quang Ngai Vietnam
63 B2 Quang Tri Vietnam
62 B1 Quan Hoa Vietnam
 Quan Long Vietnam see
 Ca Mau
 Quan Phu Quoc i. Vietnam
 see Phu Quốc, Đao
71 B3 Quanzhou Fujian China
71 B3 Quanzhou Guangxi China
108 A3 Quartu Sant'Elena Italy
142 A2 Quartzsite U.S.A.
81 C1 Quba Azer.
76 B3 Qūchān Iran
53 C3 Queanbeyan Austr.
131 C3 Québec Can.
131 C2 Québec prov. Can.
101 E2 Quedlinburg Ger.
 Queen Adelaide Islands is
 Chile see Reina Adelaida,
 Archipiélago de la
128 A2 Queen Charlotte Can.
128 A2 Queen Charlotte Islands
 Can.
128 B2 Queen Charlotte Sound
 sea chan. Can.
128 B2 Queen Charlotte Strait
 Can.
126 E1 Queen Elizabeth Islands
 Can.
55 I2 Queen Mary Land reg.
 Antarctica
126 F2 Queen Maud Gulf Can.
55 E2 Queen Maud Land reg.
 Antarctica
55 P1 Queen Maud Mountains
 Antarctica
52 B3 Queenscliff Austr.
52 B1 Queensland state Austr.
51 D4 Queenstown Austr.
54 A3 Queenstown N.Z.
 Queenstown Ireland see
 Cobh
123 C3 Queenstown S. Africa
121 C2 Quelimane Moz.
153 A4 Quellón Chile
 Quelpart Island i. S. Korea
 see Cheju-do
142 B2 Quemado U.S.A.
 Que Que Zimbabwe see
 Kwekwe
154 B2 Querência do Norte Brazil
145 B2 Querétaro Mex.
101 E2 Querfurt Ger.
128 B2 Quesnel Can.
128 B2 Quesnel Lake Can.

74 A1 Quetta Pak.
146 A3 Quetzaltenango Guat.
64 B2 Quezon Phil.
64 B2 Quezon City Phil.
120 A2 Quibala Angola
150 B2 Quibdó Col.
104 B2 Quiberon France
120 A2 Quilengues Angola
104 C3 Quillan France
153 C3 Quilmes Arg.
 Quilon India see Kollam
51 D2 Quilpie Austr.
153 A3 Quilpué Chile
120 A1 Quimbele Angola
152 B2 Quimili Arg.
104 B2 Quimper France
104 B2 Quimperlé France
135 B3 Quincy CA U.S.A.
141 D2 Quincy FL U.S.A.
137 E3 Quincy IL U.S.A.
139 E2 Quincy MA U.S.A.
107 C1 Quinto Spain
121 D2 Quionga Moz.
120 A2 Quirima Angola
53 D2 Quirindi Austr.
154 B1 Quirinópolis Brazil
131 D3 Quispamsis Can.
121 C3 Quissico Moz.
120 A2 Quitapa Angola
150 B3 Quito Ecuador
151 F3 Quixadá Brazil
71 A3 Qujing China
75 D1 Qumar He r. China
123 C3 Qumrha S. Africa
115 E1 Qunayyin, Sabkhat al
 salt marsh Libya
129 E1 Quoich r. Can.
52 A2 Quorn Austr.
79 C2 Qurayat Oman
77 C3 Qŭrghonteppa Tajik.
 Quxar China see Lhazê
 Quyang China see Jingzho
63 B2 Quy Nhơn Vietnam
71 B3 Quzhou China
 Qyteti Stalin Albania see
 Kuçovë
 Qyzyltū Kazakh. see
 Kishkenekol'

R

103 D2 Raab r. Austria
92 H3 Raahe Fin.
100 C1 Raalte Neth.
61 C2 Raas i. Indon.
61 C2 Raba Indon.
114 B1 Rabat Morocco
59 E3 Rabaul P.N.G.
78 A2 Rābigh Saudi Arabia
103 D2 Rabka Pol.
 Rābniţa Moldova see
 Rîbniţa
 Rabyānah, Ramlat des. Liby
 see Rebiana Sand Sea
131 E3 Race, Cape Can.
140 B3 Raceland U.S.A.
139 E2 Race Point U.S.A.
63 B3 Rach Gia Vietnam
103 D1 Racibórz Pol.
138 C2 Racine U.S.A.
78 B3 Radaʻ Yemen
110 C1 Rădăuţi Romania
138 B3 Radcliff U.S.A.
74 B2 Radhanpur India
130 C2 Radisson Can.
103 E1 Radom Pol.
103 D1 Radomsko Pol.
90 B1 Radomyshl' Ukr.
111 B2 Radoviš Macedonia
88 B2 Radviliškis Lith.
78 A2 Raḍwá, Jabal mt.
 Saudi Arabia
90 B1 Radyvyliv Ukr.
75 C2 Rae Bareli India
100 C2 Raeren Belgium
54 C1 Raetihi N.Z.
78 A2 Rāf h. Saudi Arabia
152 B3 Rafaela Arg.
118 C2 Rafaï C.A.R.
78 B2 Rafḥaʼ Saudi Arabia
79 C1 Rafsanjān Iran
119 C2 Raga Sudan
64 B3 Ragang, Mount vol. Phil.
109 B3 Ragusa Italy
61 D2 Raha Indon.

Salajwe

130 C2 **Rupert** *r.* Can.
134 D2 **Rupert** U.S.A.
130 C2 **Rupert Bay** Can.
Rusaddir N. Africa *see* **Melilla**
121 C2 **Rusape** Zimbabwe
110 C2 **Ruse** Bulg.
137 E1 **Rush City** U.S.A.
121 C2 **Rushinga** Zimbabwe
77 D3 **Rushon** Tajik.
136 C2 **Rushville** U.S.A.
53 C3 **Rushworth** Austr.
129 D2 **Russell** Can.
54 B1 **Russell** N.Z.
137 D3 **Russell** U.S.A.
140 C2 **Russellville** *AL* U.S.A.
140 B1 **Russellville** *AR* U.S.A.
138 B3 **Russellville** *KY* U.S.A.
101 D2 **Rüsselsheim** Ger.
82 F2 **Russian Federation** *country* Asia/Europe
81 C1 **Rust'avi** Georgia
123 C2 **Rustenburg** S. Africa
140 B2 **Ruston** U.S.A.
61 D2 **Ruteng** Indon.
98 B3 **Ruthin** U.K.
139 E2 **Rutland** U.S.A.
Rutog China *see* **Dêrub**
119 C3 **Rutshuru** Dem. Rep. Congo
119 E4 **Ruvuma** *r.* Moz./Tanz.
79 C2 **Ruweis** U.A.E.
89 E2 **Ruza** Rus. Fed.
77 C1 **Ruzayevka** Kazakh.
87 D3 **Ruzayevka** Rus. Fed.
119 C3 **Rwanda** *country* Africa
89 E3 **Ryazan'** Rus. Fed.
89 F3 **Ryazhsk** Rus. Fed.
86 C2 **Rybachiy, Poluostrov** *pen.* Rus. Fed.
Rybach'ye Kyrg. *see* **Balykchy**
89 E2 **Rybinsk** Rus. Fed.
89 E2 **Rybinskoye Vodokhranilishche** *resr* Rus. Fed.
103 D1 **Rybnik** Pol.
Rybnitsa Moldova *see* **Rîbniţa**
89 E3 **Rybnoye** Rus. Fed.
99 D4 **Rye** U.K.
Rykovo Ukr. *see* **Yenakiyeve**
89 D3 **Ryl'sk** Rus. Fed.
Ryojun China *see* **Lüshunkou**
67 C3 **Ryōtsu** Japan
69 E3 **Ryukyu Islands** *is* Japan
89 D3 **Ryzhikovo** Rus. Fed.
103 E1 **Rzeszów** Pol.
91 E1 **Rzhaksa** Rus. Fed.
89 D2 **Rzhev** Rus. Fed.

S

79 C2 **Sa'ādatābād** Iran
101 E2 **Saale** *r.* Ger.
101 E2 **Saalfeld** Ger.
134 B1 **Saanich** Can.
100 C3 **Saar** *r.* Ger.
102 B2 **Saarbrücken** Ger.
88 B2 **Sääre** Estonia
88 B2 **Saaremaa** *i.* Estonia
92 I2 **Saarenkylä** Fin.
93 I3 **Saarijärvi** Fin.
100 C3 **Saarlouis** Ger.
80 B2 **Sab' Ābār** Syria
107 D1 **Sabadell** Spain
67 C3 **Sabae** Japan
61 C1 **Sabah** *state* Malaysia
61 C2 **Sabalana** *i.* Indon.
146 B2 **Sabana, Archipiélago de** *is* Cuba
150 B1 **Sabanalarga** Col.
60 A1 **Sabang** Indon.
155 D1 **Sabará** Brazil
108 B2 **Sabaudia** Italy
122 B3 **Sabelo** S. Africa
119 D2 **Sabena Desert** Kenya
115 D2 **Sabhā** Libya
123 D2 **Sabie** *r.* Moz./S. Africa
123 D2 **Sabie** S. Africa
145 B2 **Sabinas** Mex.
145 B2 **Sabinas Hidalgo** Mex.
143 E3 **Sabine** *r.* U.S.A.
131 D3 **Sable, Cape** Can.
141 D3 **Sable, Cape** U.S.A.
131 E3 **Sable Island** Can.
106 B1 **Sabugal** Port.
78 B3 **Şabyā** Saudi Arabia
76 B3 **Sabzevār** Iran
137 D2 **Sac City** U.S.A.
120 A2 **Sachanga** Angola
130 A2 **Sachigo Lake** Can.
65 B3 **Sach'on** S. Korea
126 D2 **Sachs Harbour** Can.
154 C1 **Sacramento** Brazil
135 B3 **Sacramento** U.S.A.
135 B3 **Sacramento** *r.* U.S.A.
142 B2 **Sacramento Mountains** U.S.A.
135 B2 **Sacramento Valley** U.S.A.
110 B1 **Săcueni** Romania
123 C3 **Sada** S. Africa
107 C1 **Sádaba** Spain
Sá da Bandeira Angola *see* **Lubango**
78 B3 **Şa'dah** Yemen
63 B3 **Sadao** Thai.
79 B3 **Şadārah** Yemen
63 B2 **Sa Đec** Vietnam
74 B2 **Sadiqabad** Pak.
72 D2 **Sadiya** India
67 C3 **Sadoga-shima** *i.* Japan
107 D2 **Sa Dragonera** *i.* Spain
81 D2 **Safāshahr** Iran
93 F4 **Säffle** Sweden
142 B2 **Safford** U.S.A.
99 D3 **Saffron Walden** U.K.
114 B1 **Safi** Morocco
76 C3 **Safīd Kūh** *mts* Afgh.
155 D1 **Safiras, Serra das** *mts* Brazil
86 D2 **Safonovo** Rus. Fed.
89 D2 **Safonovo** Rus. Fed.
78 B2 **Safrā' as Sark** *esc.* Saudi Arabia
75 C2 **Saga** China
67 B4 **Saga** Japan
62 A1 **Sagaing** Myanmar
67 C3 **Sagamihara** Japan
74 B2 **Sagar** India
Sagarmatha *mt.* China/Nepal *see* **Everest, Mount**
138 C2 **Saginaw** U.S.A.
138 C2 **Saginaw Bay** U.S.A.
Saglouc Can. *see* **Salluit**
106 B2 **Sagres** Port.
146 B2 **Sagua la Grande** Cuba
139 F1 **Saguenay** *r.* Can.
107 C2 **Sagunto** Spain
76 B2 **Sagyndyk, Mys** *pt* Kazakh.
106 B1 **Sahagún** Spain
114 C3 **Sahara** *des.* Africa
Şahara el Gharbîya *des.* Egypt *see* **Western Desert**
Şahara el Sharqîya *des.* Egypt *see* **Eastern Desert**
Saharan Atlas *mts* Alg. *see* **Atlas Saharien**
74 B2 **Saharanpur** India
75 C2 **Saharsa** India
114 B3 **Sahel** *reg.* Africa
74 B1 **Sahiwal** Pak.
144 B2 **Sahuayo** Mex.
78 B2 **Şāhūq** *reg.* Saudi Arabia
114 C1 **Saïda** Alg.
Saïda Lebanon *see* **Sidon**
75 C2 **Saidpur** Bangl.
67 B3 **Saigō** Japan
Saigon Vietnam *see* **Ho Chi Minh City**
75 D2 **Saiha** India
70 B1 **Saihan Tal** China
67 B4 **Saiki** Japan
93 I3 **Saimaa** *l.* Fin.
144 B2 **Sain Alto** Mex.
96 C3 **St Abb's Head** *hd* U.K.
131 E3 **St Alban's** Can.
99 C4 **St Albans** U.K.
138 C3 **St Albans** U.S.A.
St Alban's Head *hd* U.K. *see* **St Aldhelm's Head**
99 B4 **St Aldhelm's Head** *hd* U.K.
St-André, Cap *c.* Madag. *see* **Vilanandro, Tanjona**
96 C2 **St Andrews** U.K.
131 E2 **St Anthony** Can.
134 D2 **St Anthony** U.S.A.
52 B3 **St Arnaud** Austr.
131 E2 **St-Augustin** Can.
131 E2 **St Augustin** *r.* Can.
141 D3 **St Augustine** U.S.A.
99 A4 **St Austell** U.K.
104 C2 **St-Avertin** France
147 D3 **St-Barthélemy** *terr.* West Indies
98 B2 **St Bees Head** *hd* U.K.
105 D3 **St-Bonnet-en-Champsaur** France
99 A4 **St Bride's Bay** U.K.
104 B2 **St-Brieuc** France
130 C3 **St Catharines** Can.
141 D2 **St Catherines Island** U.S.A.
99 C4 **St Catherine's Point** U.K.
137 E3 **St Charles** U.S.A.
138 C2 **St Clair, Lake** Can./U.S.A.
105 D2 **St-Claude** France
99 A4 **St Clears** U.K.
137 E1 **St Cloud** U.S.A.
138 A1 **St Croix** *r.* U.S.A.
147 D3 **St Croix** Virgin Is (U.S.A.)
99 A4 **St David's** U.K.
99 A4 **St David's Head** *hd* U.K.
113 I8 **St-Denis** Can.
104 C2 **St-Denis** France
St-Denis-du-Sig Alg. *see* **Sig**
105 D2 **St-Dié** France
105 C2 **St-Dizier** France
130 C3 **Ste-Adèle** Can.
131 D3 **Ste-Anne-des-Monts** Can.
139 E1 **Ste-Foy** Can.
105 D2 **St-Égrève** France
128 A1 **St Elias Mountains** Can.
131 D2 **Ste-Marguerite** *r.* Can.
139 E1 **Ste-Marie** Can.
Ste-Marie, Cap *c.* Madag. *see* **Vohimena, Tanjona**
Sainte-Marie, Île *i.* Madag. *see* **Boraha, Nosy**
Ste-Rose-du-Dégelé Can. *see* **Dégelis**
129 E2 **Sainte Rose du Lac** Can.
104 B2 **Saintes** France
105 C2 **St-Étienne** France
104 C2 **St-Étienne-du-Rouvray** France
130 C3 **St-Félicien** Can.
97 D1 **Saintfield** U.K.
105 D3 **St-Florent** France
105 C2 **St-Flour** France
136 C3 **St Francis** U.S.A.
104 C3 **St-Gaudens** France
53 C1 **St George** Austr.
135 D3 **St George** U.S.A.
141 D3 **St George Island** U.S.A.
131 C3 **St-Georges** Can.
147 D3 **St George's** Grenada
131 E3 **St George's Bay** Can.
97 C3 **St George's Channel** Ireland/U.K.
105 D2 **St Gotthard Pass** *pass* Switz.
113 C7 **St Helena** *terr.* S. Atlantic Ocean
122 A3 **St Helena Bay** S. Africa
122 A3 **St Helena Bay** *b.* S. Africa
98 B3 **St Helens** U.K.
134 B1 **St Helens, Mount** *vol.* U.S.A.
95 C4 **St Helier** Channel Is
100 B2 **St-Hubert** Belgium
139 E1 **St-Hyacinthe** Can.
138 C1 **St Ignace** U.S.A.
130 B3 **St Ignace Island** Can.
99 A4 **St Ives** U.K.
St Jacques, Cap Vietnam *see* **Vung Tau**
128 A2 **St James, Cape** Can.
130 C3 **St-Jean, Lac** *l.* Can.
104 B2 **St-Jean-d'Angély** France
104 B3 **St-Jean-de-Luz** France
104 B2 **St-Jean-de-Monts** France
130 C3 **St-Jean-sur-Richelieu** Can.
139 E1 **St-Jérôme** Can.
134 C1 **St Joe** *r.* U.S.A.
131 D3 **Saint John** Can.
137 D3 **St John** U.S.A.
139 F1 **St John** *r.* U.S.A.
147 D3 **St John's** Antigua
131 E3 **St John's** Can.
142 B2 **St Johns** U.S.A.
141 D2 **St Johns** *r.* U.S.A.
139 E2 **St Johnsbury** U.S.A.
137 E3 **St Joseph** U.S.A.
130 A2 **St Joseph, Lake** Can.
St-Joseph-d'Alma Can. *see* **Alma**
130 B3 **St Joseph Island** Can.
139 E1 **St-Jovité** Can.
104 C2 **St-Junien** France
94 B2 **St Kilda** *is* U.K.
147 D3 **St Kitts and Nevis** *country* West Indies
151 D2 **St-Laurent-du-Maroni** Fr. Guiana
131 E3 **St Lawrence** Can.
131 D3 **St Lawrence** *inlet* Can.
131 D3 **St Lawrence, Gulf of** Can.
126 A2 **St Lawrence Island** U.S.A.
104 B2 **St-Lô** France
114 A3 **St-Louis** Senegal
137 E3 **St Louis** U.S.A.
137 E1 **St Louis** *r.* U.S.A.
147 D3 **St Lucia** *country* West Indies
147 D3 **St Lucia Channel** Martinique/St Lucia
123 D2 **St Lucia Estuary** S. Africa
96 □ **St Magnus Bay** U.K.
104 B2 **St-Malo** France
104 B2 **St-Malo, Golfe de** *g.* France
147 C3 **St-Marc** Haiti
St Mark's S. Africa *see* **Cofimvaba**
147 D3 **St-Martin** *terr.* West Indies
122 A3 **St Martin, Cape** S. Africa
129 E2 **St Martin, Lake** Can.
139 D2 **St Marys** U.S.A.
124 A3 **St Matthew Island** U.S.A.
59 D3 **St Matthias Group** *is* P.N.G.
130 C3 **St-Maurice** *r.* Can.
130 C3 **St-Michel-des-Saints** Can.
104 B2 **St-Nazaire** France
104 C1 **St-Omer** France
129 C2 **St Paul** Can.
137 E2 **St Paul** U.S.A.
156 A8 **St-Paul, Île** *i.* Indian Ocean
137 E2 **St Peter** U.S.A.
95 C4 **St Peter Port** Channel Is
89 D2 **St Petersburg** Rus. Fed.
141 D3 **St Petersburg** U.S.A.
131 E3 **St-Pierre** St Pierre and Miquelon
139 E1 **St-Pierre, Lac** *l.* Can.
131 E3 **St Pierre and Miquelon** *terr.* N. America
104 B2 **St-Pierre-d'Oléron** France
105 C2 **St-Pourçain-sur-Sioule** France
131 D3 **St Quentin** Can.
105 C2 **St-Quentin** France
105 D3 **St-Raphaël** France
122 B3 **St Sebastian Bay** S. Africa
104 B2 **St-Sébastien-sur-Loire** France
131 D3 **St-Siméon** Can.
129 E2 **St Theresa Point** Can.
130 B3 **St Thomas** Can.
105 D3 **St-Tropez** France
105 D3 **St-Tropez, Cap de** *c.* France
St Vincent, Cape *c.* Port. *see* **São Vicente, Cabo de**
52 A3 **St Vincent, Gulf** Austr.
147 D3 **St Vincent and the Grenadines** *country* West Indies
147 D3 **St Vincent Passage** St Lucia/St Vincent
100 C2 **St-Vith** Belgium
129 D2 **St Walburg** Can.
104 C2 **St-Yrieix-la-Perche** France
59 D1 **Saipan** *i.* N. Mariana Is
152 B1 **Sajama, Nevado** *mt.* Bol.
122 B2 **Sak** *watercourse* S. Africa
67 C4 **Sakai** Japan
67 B4 **Sakaide** Japan
78 B2 **Sakākah** Saudi Arabia
136 C1 **Sakakawea, Lake** U.S.A.
Sakarya Turkey *see* **Adapazarı**
111 D2 **Sakarya** *r.* Turkey
66 C3 **Sakata** Japan
65 B1 **Sakchu** N. Korea
66 D1 **Sakhalin** *i.* Rus. Fed.
123 C2 **Sakhile** S. Africa
81 C1 **Şäki** Azer.
88 B3 **Šakiai** Lith.
69 E3 **Sakishima-shotō** *is* Japan
62 B2 **Sakon Nakhon** Thai.
74 A2 **Sakrand** Pak.
122 B3 **Sakrivier** S. Africa
67 D3 **Sakura** Japan
91 C2 **Saky** Ukr.
93 G4 **Sala** Sweden
130 C3 **Salaberry-de-Valleyfield** Can.
88 B2 **Salacgrīva** Latvia
109 C2 **Sala Consilina** Italy
135 C4 **Salada, Laguna** *salt l.* Mex.
152 C2 **Saladas** Arg.
152 B3 **Salado** *r.* Arg.
153 B3 **Salado** *r.* Arg.
145 C2 **Salado** *r.* Mex.
114 B4 **Salaga** Ghana
122 B1 **Salajwe** Botswana

79 C2 **Salakh, Jabal** mt. Oman
115 D3 **Salal** Chad
78 A2 **Salâla** Sudan
79 C3 **Salālah** Oman
145 B2 **Salamanca** Mex.
106 B1 **Salamanca** Spain
139 D2 **Salamanca** U.S.A.
106 B1 **Salas** Spain
63 B2 **Salavan** Laos
59 C3 **Salawati** i. Indon.
61 D2 **Salayar** i. Indon.
157 G7 **Sala y Gómez, Isla** i. S. Pacific Ocean
Salazar Angola see **N'dalatando**
104 C2 **Salbris** France
88 C3 **Šalčininkai** Lith.
106 C1 **Saldaña** Spain
122 A3 **Saldanha** S. Africa
88 B2 **Saldus** Latvia
53 C3 **Sale** Austr.
86 F2 **Salekhard** Rus. Fed.
73 B3 **Salem** India
138 B3 **Salem** IL U.S.A.
137 E3 **Salem** MO U.S.A.
134 B2 **Salem** OR U.S.A.
137 D2 **Salem** SD U.S.A.
96 B2 **Salen** U.K.
109 B2 **Salerno** Italy
108 B2 **Salerno, Golfo di** g. Italy
98 B3 **Salford** U.K.
151 E2 **Salgado** r. Brazil
103 D2 **Salgótarján** Hungary
151 F3 **Salgueiro** Brazil
136 B3 **Salida** U.S.A.
111 C3 **Salihli** Turkey
88 C3 **Salihorsk** Belarus
121 C2 **Salima** Malawi
121 C2 **Salimo** Moz.
137 D3 **Salina** KS U.S.A.
135 D3 **Salina** UT U.S.A.
109 B3 **Salina, Isola** i. Italy
145 C3 **Salina Cruz** Mex.
155 D1 **Salinas** Brazil
144 B2 **Salinas** Mex.
135 B3 **Salinas** U.S.A.
135 B3 **Salinas** r. U.S.A.
107 D2 **Salines, Cap de ses** c. Spain
151 E3 **Salinópolis** Brazil
99 C4 **Salisbury** U.K.
139 D3 **Salisbury** MD U.S.A.
141 D1 **Salisbury** NC U.S.A.
Salisbury Zimbabwe see **Harare**
99 B4 **Salisbury Plain** U.K.
151 E3 **Salitre** r. Brazil
92 I2 **Salla** Fin.
143 E1 **Sallisaw** U.S.A.
127 G2 **Salluit** Can.
75 C2 **Sallyana** Nepal
81 C2 **Salmās** Iran
128 C3 **Salmo** Can.
134 D1 **Salmon** U.S.A.
134 C1 **Salmon** r. U.S.A.
128 C2 **Salmon Arm** Can.
134 C2 **Salmon River Mountains** U.S.A.
100 C3 **Salmtal** Ger.
118 B2 **Salo** C.A.R.
93 H3 **Salo** Fin.
105 D3 **Salon-de-Provence** France
Salonica Greece see **Thessaloniki**
110 B1 **Salonta** Romania
87 D4 **Sal'sk** Rus. Fed.
122 B3 **Salt** watercourse S. Africa
107 D1 **Salt** Spain
142 A2 **Salt** r. U.S.A.
152 B2 **Salta** Arg.
99 A4 **Saltash** U.K.
96 B3 **Saltcoats** U.K.
145 B2 **Saltillo** Mex.
134 D2 **Salt Lake City** U.S.A.
154 C2 **Salto** Brazil
152 C3 **Salto** Uru.
155 E1 **Salto da Divisa** Brazil
154 B2 **Salto del Guairá** Para.
135 C4 **Salton Sea** salt l. U.S.A.
154 B3 **Salto Osório, Represa** resr Brazil
154 B3 **Salto Santiago, Represa de** resr Brazil
141 D2 **Saluda** U.S.A.
76 B3 **Sālūk, Kūh-e** mt. Iran
108 A2 **Saluzzo** Italy
151 F4 **Salvador** Brazil
79 C2 **Salwah** Saudi Arabia
62 A2 **Salween** r. China/Myanmar

81 C2 **Salyan** Azer.
138 C3 **Salyersville** U.S.A.
122 A1 **Salzbrunn** Namibia
102 C2 **Salzburg** Austria
101 E1 **Salzgitter** Ger.
101 D2 **Salzkotten** Ger.
101 E1 **Salzwedel** Ger.
144 B1 **Samalayuca** Mex.
66 D2 **Samani** Japan
64 B2 **Samar** i. Phil.
87 E3 **Samara** Rus. Fed.
Samarahan Malaysia see **Sri Aman**
59 E3 **Samarai** P.N.G.
61 C2 **Samarinda** Indon.
77 C3 **Samarqand** Uzbek.
81 C2 **Sāmarrā'** Iraq
81 C1 **Şamaxı** Azer.
119 C3 **Samba** Dem. Rep. Congo
61 C1 **Sambaliung** mts Indon.
75 C2 **Sambalpur** India
60 C2 **Sambar, Tanjung** pt Indon.
60 B1 **Sambas** Indon.
121 □E2 **Sambava** Madag.
74 B2 **Sambhar** India
90 A2 **Sambir** Ukr.
61 C2 **Sambo** Indon.
61 C2 **Samboja** Indon.
153 C3 **Samborombón, Bahía** b. Arg.
65 B2 **Samch'ŏk** S. Korea
Samch'ŏnp'o S. Korea see **Sach'on**
81 C2 **Samdi Dağ** mt. Turkey
119 D3 **Same** Tanz.
121 B2 **Samfya** Zambia
78 B2 **Samīrah** Saudi Arabia
65 B1 **Samjiyŏn** N. Korea
Sam Neua Laos see **Xam Nua**
48 J5 **Samoa** country S. Pacific Ocean
156 E6 **Samoa Basin** S. Pacific Ocean
Samoa i Sisifo country S. Pacific Ocean see **Samoa**
109 C1 **Samobor** Croatia
110 B2 **Samokov** Bulg.
111 C3 **Samos** i. Greece
Samothrace i. Greece see **Samothraki**
111 C2 **Samothraki** Greece
111 C2 **Samothraki** i. Greece
61 C2 **Sampit** Indon.
119 C3 **Sampwe** Dem. Rep. Congo
143 E2 **Sam Rayburn Reservoir** U.S.A.
62 B2 **Sâm Sơn** Vietnam
80 B1 **Samsun** Turkey
81 C1 **Samtredia** Georgia
63 B3 **Samui, Ko** i. Thai.
63 B2 **Samut Songkhram** Thai.
114 B3 **San** Mali
78 B3 **Şan'ā'** Yemen
118 A2 **Sanaga** r. Cameroon
81 C2 **Sanandaj** Iran
146 B3 **San Andrés, Isla de** i. Caribbean Sea
106 B1 **San Andres del Rabanedo** Spain
142 B2 **San Andres Mountains** U.S.A.
145 C3 **San Andrés Tuxtla** Mex.
143 C2 **San Angelo** U.S.A.
143 D3 **San Antonio** U.S.A.
135 C4 **San Antonio, Mount** U.S.A.
152 B2 **San Antonio de los Cobres** Arg.
153 B4 **San Antonio Oeste** Arg.
108 B2 **San Benedetto del Tronto** Italy
144 A3 **San Benedicto, Isla** i. Mex.
135 C4 **San Bernardino** U.S.A.
135 C4 **San Bernardino Mountains** U.S.A.
142 B3 **San Blas** Mex.
141 C3 **San Blas, Cape** U.S.A.
146 C4 **San Blas, Punta** pt Panama
152 B1 **San Borja** Bol.
144 B2 **San Buenaventura** Mex.
64 B2 **San Carlos** Phil.
147 D4 **San Carlos** Venez.
153 A4 **San Carlos de Bariloche** Arg.
147 C4 **San Carlos del Zulia** Venez.
104 C2 **Sancerrois, Collines du** hills France
135 C4 **San Clemente** U.S.A.

135 C4 **San Clemente Island** U.S.A.
105 C2 **Sancoins** France
48 H5 **San Cristobal** i. Solomon Is
150 B2 **San Cristóbal** Venez.
145 C3 **San Cristóbal de las Casas** Mex.
146 C2 **Sancti Spíritus** Cuba
123 D1 **Sand** r. S. Africa
61 C1 **Sandakan** Malaysia
93 E3 **Sandane** Norway
111 B2 **Sandanski** Bulg.
114 A3 **Sandaré** Mali
96 C1 **Sanday** i. U.K.
143 C2 **Sanderson** U.S.A.
150 B2 **Sandia** Peru
135 C4 **San Diego** U.S.A.
111 D3 **Sandıklı** Turkey
93 E4 **Sandnes** Norway
92 F2 **Sandnessjøen** Norway
118 C3 **Sandoa** Dem. Rep. Congo
103 E1 **Sandomierz** Pol.
89 E2 **Sandovo** Rus. Fed.
94 B1 **Sandoy** i. Faroe Is
134 C1 **Sandpoint** U.S.A.
71 B3 **Sandu** China
94 B1 **Sandur** Faroe Is
138 C2 **Sandusky** U.S.A.
122 A3 **Sandveld** mts S. Africa
122 A2 **Sandverhaar** Namibia
93 F4 **Sandvika** Norway
93 G3 **Sandviken** Sweden
131 E2 **Sandwich Bay** Can.
135 D2 **Sandy** U.S.A.
129 D2 **Sandy Bay** Can.
51 E2 **Sandy Cape** Austr.
130 A2 **Sandy Lake** Can.
130 A2 **Sandy Lake** l. Can.
141 D2 **Sandy Springs** U.S.A.
144 A1 **San Felipe** Mex.
145 B2 **San Felipe** Mex.
150 C1 **San Felipe** Venez.
144 A2 **San Fernando** Mex.
145 C2 **San Fernando** Mex.
64 B2 **San Fernando** Luzon Phil.
64 B2 **San Fernando** Luzon Phil.
106 B2 **San Fernando** Spain
147 D3 **San Fernando** Trin. and Tob.
150 C2 **San Fernando de Apure** Venez.
141 D3 **Sanford** FL U.S.A.
139 E2 **Sanford** ME U.S.A.
141 E1 **Sanford** NC U.S.A.
152 B3 **San Francisco** Arg.
135 B3 **San Francisco** U.S.A.
74 B3 **Sangamner** India
83 J2 **Sangar** Rus. Fed.
108 A3 **San Gavino Monreale** Italy
101 E2 **Sangerhausen** Ger.
61 C1 **Sanggau** Indon.
118 B3 **Sangha** r. Congo
109 C3 **San Giovanni in Fiore** Italy
59 C2 **Sangir** i. Indon.
59 C2 **Sangir, Kepulauan** is Indon.
65 B2 **Sangju** S. Korea
63 A2 **Sangkhla Buri** Thai.
61 C1 **Sangkulirang** Indon.
73 B3 **Sangli** India
118 B2 **Sangmélima** Cameroon
121 C3 **Sango** Zimbabwe
San Gottardo, Passo del pass Switz. see **St Gotthard Pass**
136 B3 **Sangre de Cristo Range** mts U.S.A.
75 C2 **Sangsang** China
144 A2 **San Hipólito, Punta** pt Mex.
145 D3 **San Ignacio** Belize
152 B1 **San Ignacio** Bol.
144 A2 **San Ignacio** Mex.
130 C2 **Sanikiluaq** Can.
71 A3 **Sanjiang** China
Sanjiang China see **Jinping**
67 C3 **Sanjō** Japan
135 B3 **San Joaquin** r. U.S.A.
153 B4 **San Jorge, Golfo de** g. Arg.
146 B4 **San José** Costa Rica
64 B2 **San Jose** Phil.
64 B2 **San Jose** Phil.
135 B3 **San Jose** U.S.A.
144 A2 **San José, Isla** i. Mex.
144 B2 **San José de Bavícora** Mex.
64 B2 **San Jose de Buenavista** Phil.
144 B2 **San José de Comondú** Mex.
144 B2 **San José del Cabo** Mex.
150 B2 **San José del Guaviare** Col.
152 B3 **San Juan** Arg.

146 B3 **San Juan** r. Costa Rica/Nic...
147 C3 **San Juan** Dom. Rep.
147 D3 **San Juan** Puerto Rico
135 D3 **San Juan** r. U.S.A.
152 C2 **San Juan Bautista** Para.
145 C3 **San Juan Bautista Tuxtepe...** Mex.
147 D4 **San Juan de los Morros** Venez.
145 C2 **San Juan del Río** Mex.
134 B1 **San Juan Islands** U.S.A.
144 B2 **San Juanito** Mex.
136 B3 **San Juan Mountains** U.S.A.
153 B4 **San Julián** Arg.
75 C2 **Sankh** r. India
63 B2 **San Khao Phang Hoei** mt... Thai.
100 C2 **Sankt Augustin** Ger.
105 D2 **Sankt Gallen** Switz.
105 D2 **Sankt Moritz** Switz.
Sankt-Peterburg Rus. Fed. see **St Petersburg**
102 C2 **Sankt Veit an der Glan** Austria
100 C3 **Sankt Wendel** Ger.
80 B2 **Şanlıurfa** Turkey
142 B3 **San Lorenzo** Mex.
106 B2 **Sanlúcar de Barrameda** Spain
144 B2 **San Lucas** Mex.
153 B3 **San Luis** Arg.
145 B2 **San Luis de la Paz** Mex.
142 A2 **San Luisito** Mex.
135 B3 **San Luis Obispo** U.S.A.
135 B3 **San Luis Obispo Bay** U.S.A.
145 B2 **San Luis Potosí** Mex.
144 A1 **San Luis Río Colorado** Me...
143 D3 **San Marcos** U.S.A.
108 B2 **San Marino** country Europe
108 B2 **San Marino** San Marino
144 B2 **San Martín de Bolaños** Mex.
153 A4 **San Martín de los Andes** Arg.
135 B3 **San Mateo** U.S.A.
153 B4 **San Matías, Golfo** g. Arg.
70 B2 **Sanmenxia** China
146 B3 **San Miguel** El Salvador
152 B2 **San Miguel de Tucumán** Arg.
135 B4 **San Miguel Island** U.S.A.
145 C3 **San Miguel Sola de Vega** Mex.
71 B3 **Sanming** China
153 B3 **San Nicolás de los Arroyos** Arg.
135 C4 **San Nicolas Island** U.S.A.
110 B1 **Sânnicolau Mare** Romania
123 C2 **Sannieshof** S. Africa
114 B4 **Sanniquellie** Liberia
103 E2 **Sanok** Pol.
64 B2 **San Pablo** Phil.
144 B2 **San Pablo Balleza** Mex.
152 B2 **San Pedro** Arg.
152 B1 **San Pedro** Bol.
114 B4 **San-Pédro** Côte d'Ivoire
144 A2 **San Pedro** Mex.
142 A2 **San Pedro** watercourse U.S.A.
106 B2 **San Pedro, Sierra de** mts Spain
144 B2 **San Pedro de las Colonias** Mex.
152 C2 **San Pedro de Ycuamandyý** Para.
142 A3 **San Pedro el Saucito** Mex.
146 B3 **San Pedro Sula** Hond.
108 A3 **San Pietro, Isola di** i. Italy
96 C3 **Sanquhar** U.K.
144 A1 **San Quintín, Cabo** c. Mex.
153 B3 **San Rafael** Arg.
108 A2 **San Remo** Italy
143 D2 **San Saba** U.S.A.
147 C2 **San Salvador** i. Bahamas
146 B3 **San Salvador** El Salvador
152 B2 **San Salvador de Jujuy** Arg.
108 B2 **Sansepolcro** Italy
109 C2 **San Severo** Italy
109 C2 **Sanski Most** Bos.-Herz.
152 B1 **Santa Ana** Bol.
146 B3 **Santa Ana** El Salvador
144 A1 **Santa Ana** Mex.
135 C4 **Santa Ana** U.S.A.
152 B1 **Santa Ana de Yacuma** Bol.
144 B2 **Santa Bárbara** Mex.
135 C4 **Santa Barbara** U.S.A.
154 B2 **Santa Bárbara, Serra de** Brazil

52 B2 Santa Catalina Chile
35 C4 Santa Catalina Island U.S.A.
54 B3 Santa Catarina *state* Brazil
50 C3 Santa Clara Col.
46 C2 Santa Clara Cuba
35 B3 Santa Clara U.S.A.
35 C4 Santa Clarita U.S.A.
07 D1 Santa Coloma de Gramenet Spain
Santa Comba Angola *see* Waku-Kungo
09 C3 Santa Croce, Capo *c.* Italy
53 B5 Santa Cruz *r.* Arg.
52 B1 Santa Cruz Bol.
64 B2 Santa Cruz Phil.
35 B3 Santa Cruz U.S.A.
45 C3 Santa Cruz Barillas Guat.
55 E1 Santa Cruz Cabrália Brazil
07 C2 Santa Cruz de Moya Spain
44 A2 Santa Cruz de Tenerife Islas Canarias
52 C2 Santa Cruz do Sul Brazil
35 C4 Santa Cruz Island U.S.A.
48 H5 Santa Cruz Islands Solomon Is
07 D2 Santa Eulalia del Río Spain
52 B3 Santa Fé Arg.
42 B1 Santa Fe U.S.A.
54 B2 Santa Fé do Sul Brazil
54 B1 Santa Helena de Goiás Brazil
53 B3 Santa Isabel Arg.
Santa Isabel Equat. Guinea *see* Malabo
48 G4 Santa Isabel *i.* Solomon Is
54 B1 Santa Luisa, Serra de *hills* Brazil
51 E3 Santa Luzia Brazil
44 A2 Santa Margarita, Isla *i.* Mex.
52 C2 Santa María Brazil
44 B1 Santa María *r.* Mex.
35 B4 Santa María U.S.A.
23 D2 Santa Maria, Cabo de *c.* Moz.
06 B2 Santa Maria, Cabo de *c.* Port.
55 C1 Santa Maria, Chapadão de *hills* Brazil
51 E3 Santa Maria das Barreiras Brazil
09 C3 Santa Maria di Leuca, Capo *c.* Italy
55 D1 Santa Maria do Suaçuí Brazil
50 B1 Santa Marta Col
35 C4 Santa Monica U.S.A.
51 E4 Santana Brazil
10 B1 Sântana Romania
06 C1 Santander Spain
08 A3 Sant'Antioco Italy
08 A3 Sant'Antioco, Isola di *i.* Italy
07 D2 Sant Antoni de Portmany Spain
51 D3 Santarém Brazil
06 B2 Santarém Port.
54 B1 Santa Rita do Araguaia Brazil
53 B3 Santa Rosa Arg.
52 C2 Santa Rosa Brazil
35 B3 Santa Rosa CA U.S.A.
42 C2 Santa Rosa NM U.S.A.
46 B3 Santa Rosa de Copán Hond.
35 B4 Santa Rosa Island CA U.S.A.
40 C2 Santa Rosa Island FL U.S.A.
44 A2 Santa Rosalía Mex.
34 C2 Santa Rosa Range *mts* U.S.A.
06 B1 Santa Uxía de Ribeira Spain
54 B1 Santa Vitória Brazil
07 D1 Sant Carles de la Ràpita Spain
35 C4 Santee U.S.A.
07 D2 Sant Francesc de Formentera Spain
51 D3 Santiago Brazil
53 A3 Santiago Chile
47 C3 Santiago Dom. Rep.
44 B2 Santiago Mex.
46 B4 Santiago Panama
64 B2 Santiago Phil.
06 B1 Santiago de Compostela Spain
46 C2 Santiago de Cuba Cuba
44 B2 Santiago Ixcuintla Mex.
44 B2 Santiago Papasquiaro Mex.
06 C1 Santillana Spain

107 D2 Sant Joan de Labritja Spain
107 D1 Sant Jordi, Golf de *g.* Spain
155 D2 Santo Amaro de Campos Brazil
154 B2 Santo Anastácio Brazil
155 C2 Santo André Brazil
152 C2 Santo Angelo Brazil
154 B2 Santo Antônio da Platina Brazil
151 F4 Santo Antônio de Jesus Brazil
59 C3 Santo Antônio do Içá Brazil
107 C2 Santo Antônio do Monte Brazil
89 F3 Santo Domingo Dom. Rep.
144 A2 Santo Domingo Mex.
143 D1 Santo Domingo Pueblo U.S.A.
81 C2 Santorini *i.* Greece
81 C2 Santos Brazil
63 B2 Santos Dumont Brazil
Santos Plateau S. Atlantic Ocean
89 F3 Santo Tomé Arg.
109 C2 San Valentín, Cerro *mt.* Chile
87 E3 San Vicente El Salvador
62 A1 San Vicente Mex.
139 E2 San Vicente Phil.
109 D3 San Vicente de Cañete Peru
64 B3 San Vincenzo Italy
87 D3 San Vito, Capo *c.* Italy
87 E3 Sanya China
141 D3 São Bernardo do Campo Brazil
90 B2 São Borja Brazil
136 B2 São Carlos Brazil
139 E2 São Felipe, Serra de *hills* Brazil
61 C1 São Félix Brazil
87 D3 São Félix Brazil
61 C1 São Fidélis Brazil
79 D2 São Francisco Brazil
76 B3 São Francisco *r.* Brazil
74 B2 São Francisco, Ilha de *i.* Brazil
São Francisco do Sul Brazil
108 A2 São Gabriel Brazil
92 G2 São Gonçalo Brazil
77 C3 São Gonçalo do Abaeté Brazil
158 C3 São Gotardo Brazil
154 B1 São Jerônimo, Serra de *hills* Brazil
74 B1 São João da Barra Brazil
115 D4 São João da Boa Vista Brazil
79 D2 São João da Madeira Port.
81 D2 São João da Ponte Brazil
111 C3 São João del Rei Brazil
81 C1 São João do Paraíso Brazil
61 C1 São João Evangelista Brazil
51 D2 São João Nepomuceno Brazil
115 D2 São Joaquim da Barra Brazil
65 B2 São José Brazil
111 C2 São José do Rio Preto Brazil
77 D2 São José dos Campos Brazil
111 C2 São José dos Pinhais Brazil
104 C3 São Lourenço Brazil
59 D3 São Lourenço Brazil
153 B4 São Luís Brazil
138 C2 São Manuel Brazil
90 B1 São Marcos *r.* Brazil
60 B2 São Marcos, Baía de *b.* Brazil
111 B3 São Mateus Brazil
111 C2 São Mateus do Sul Brazil
103 E2 Saône *r.* France
87 D3 São Paulo Brazil
São Paulo *state* Brazil
105 D2 São Pedro da Aldeia Brazil
106 B1 São Raimundo Nonato Brazil
107 C1 São Romão Brazil
105 D3 São Salvador Angola *see* M'banza Congo
São Salvador do Congo Angola *see* M'banza Congo
111 C3 São Sebastião, Ilha do *i.* Brazil
103 D2 São Sebastião do Paraíso Brazil
81 D3 São Simão Brazil
77 D1 São Simão, Barragem de *resr* Brazil

59 C2 Sao-Siu Indon.
113 D5 São Tomé São Tomé and Príncipe
113 D5 São Tomé *i.* São Tomé and Príncipe
155 D2 São Tomé, Cabo de *c.* Brazil
113 D5 São Tomé and Príncipe *country* Africa
155 C2 São Vicente Brazil
106 B2 São Vicente, Cabo de *c.* Port.
59 C3 Saparua Indon.
107 C2 Sa Pobla Spain
89 F3 Sapozhok Rus. Fed.
66 D2 Sapporo Japan
109 C2 Sapri Italy
143 D1 Sapulpa U.S.A.
81 C2 Saqqez Iran
81 C2 Sarāb Iran
63 B2 Sara Buri Thai.
Saragossa Spain *see* Zaragoza
89 F3 Sarai Rus. Fed.
109 C2 Sarajevo Bos.-Herz.
87 E3 Saraktash Rus. Fed.
62 A1 Saramati *mt.* India/ Myanmar
139 E2 Saranac Lake U.S.A.
109 D3 Sarandë Albania
64 B3 Sarangani Islands Phil.
87 D3 Saransk Rus. Fed.
87 E3 Sarapul Rus. Fed.
141 D3 Sarasota U.S.A.
90 B2 Sarata Ukr.
136 B2 Saratoga U.S.A.
139 E2 Saratoga Springs U.S.A.
61 C1 Saratok Malaysia
87 D3 Saratov Rus. Fed.
79 D2 Sarāvān Iran
61 C1 Sarawak *state* Malaysia
111 C2 Saray Turkey
111 C3 Sarayköy Turkey
79 D2 Sarbāz Iran
76 B3 Sarbīsheh Iran
74 B2 Sardarshahr India
Sardegna *i.* Italy *see* Sardinia
108 A2 Sardinia *i.* Italy
92 G2 Sarektjåkkå *mt.* Sweden
77 C3 Sar-e-Pol Afgh.
158 C3 Sargasso Sea *sea* N. Atlantic Ocean
74 B1 Sargodha Pak.
115 D4 Sarh Chad
79 D2 Sarhad *reg.* Iran
81 D2 Sārī Iran
111 C3 Sarıgöl Turkey
81 C1 Sarıkamış Turkey
61 C1 Sarikei Malaysia
51 D2 Sarina Austr.
115 D2 Sarīr Tibesti *des.* Libya
65 B2 Sariwŏn N. Korea
111 C2 Sarıyer Turkey
77 D2 Sarkand Kazakh.
111 C2 Şarköy Turkey
104 C3 Sarlat-la-Canéda France
59 D3 Sarmi Indon.
153 B4 Sarmiento Arg.
138 C2 Sarnia Can.
90 B1 Sarny Ukr.
60 B2 Sarolangun Indon.
111 B3 Saronikos Kolpos *g.* Greece
111 C2 Saros Körfezi *b.* Turkey
103 E2 Sárospatak Hungary
87 D3 Sarova Rus. Fed.
Sarpan *i.* N. Mariana Is *see* Rota
105 D2 Sarrebourg France
106 B1 Sarria Spain
107 C1 Sarrión Spain
105 D3 Sartène France
Sartu China *see* Daqing
111 C3 Saruhanlı Turkey
103 D2 Sárvár Hungary
81 D3 Sarvestān Iran
77 D1 Saryarka *plain* Kazakh.
76 B2 Sarykamyshskoye Ozero *salt l.* Turkm./Uzbek.
77 D2 Saryozek Kazakh.
77 D2 Saryshagan Kazakh.
77 C2 Sarysu *watercourse* Kazakh.
77 D3 Sary-Tash Kyrg.
75 C2 Sasaram India
67 A4 Sasebo Japan
129 D2 Saskatchewan *prov.* Can.
129 D2 Saskatchewan *r.* Can.
129 D2 Saskatoon Can.
83 I2 Saskylakh Rus. Fed.

123 C2 Sasolburg S. Africa
87 D3 Sasovo Rus. Fed.
114 B4 Sassandra Côte d'Ivoire
108 A2 Sassari Italy
102 C1 Sassnitz Ger.
114 A3 Satadougou Mali
136 C3 Satanta U.S.A.
73 B3 Satara India
123 D1 Satara S. Africa
87 E3 Satka Rus. Fed.
75 C2 Satna India
77 C2 Satpayev Kazakh.
74 B2 Satpura Range *mts* India
63 B2 Sattahip Thai.
110 B1 Satu Mare Romania
63 B3 Satun Thai.
144 B2 Saucillo Mex.
93 E4 Sauda Norway
92 □B2 Sauðárkrókur Iceland
78 B2 Saudi Arabia *country* Asia
105 C3 Saugues France
137 E1 Sauk Center U.S.A.
105 C2 Saulieu France
88 B2 Saulkrasti Latvia
130 B3 Sault Sainte Marie Can.
138 C1 Sault Sainte Marie U.S.A.
77 C1 Saumalkol' Kazakh.
59 C3 Saumlaki Indon.
104 B2 Saumur France
120 B1 Saurimo Angola
109 D2 Sava *r.* Europe
49 J5 Savai'i *i.* Samoa
91 E1 Savala *r.* Rus. Fed.
114 C4 Savalou Benin
141 D2 Savannah GA U.S.A.
140 C1 Savannah TN U.S.A.
141 D2 Savannah *r.* U.S.A.
63 B2 Savannakhét Laos
130 A2 Savant Lake Can.
111 C3 Savaştepe Turkey
114 C4 Savè Benin
105 D2 Saverne France
89 F2 Savino Rus. Fed.
86 D2 Savinskiy Rus. Fed.
Savoie *reg.* France *see* Savoy
108 A2 Savona Italy
93 I3 Savonlinna Fin.
105 D2 Savoy *reg.* France
93 F4 Sävsjö Sweden
59 C3 Savu *i.* Indon.
92 I2 Savukoski Fin.
Savu Sea *sea* Indon. *see* Laut Sawu
74 B2 Sawai Madhopur India
62 A1 Sawan Myanmar
62 A2 Sawankhalok Thai.
136 B3 Sawatch Range *mts* U.S.A.
Sawhāj Egypt *see* Sūhāj
121 B2 Sawmills Zimbabwe
79 C3 Şawqirah, Ghubbat
Şawqirah Bay *b.* Oman *see* Şawqirah, Ghubbat
79 C3 Şawqirah, Ghubbat *b.* Oman
53 D2 Sawtell Austr.
134 C2 Sawtooth Range *mts* U.S.A.
68 C1 Sayano-Shushenskoye Vodokhranilishche *resr* Rus. Fed.
76 C3 Saýat Turkm.
79 C3 Sayhūt Yemen
93 I3 Säynätsalo Fin.
69 D2 Saynshand Mongolia
139 D2 Sayre U.S.A.
144 B3 Sayula Mex.
145 C3 Sayula Mex.
128 B2 Sayward Can.
Sayyod Turkm. *see* Saýat
89 C2 Sazonovo Rus. Fed.
114 B2 Sbaa Alg.
98 B2 Scafell Pike *h.* U.K.
109 C3 Scalea Italy
96 □ Scalloway U.K.
108 B2 Scandicci Italy
96 C1 Scapa Flow *inlet* U.K.
96 B2 Scarba *i.* U.K.
130 C3 Scarborough Can.
147 D3 Scarborough Trin. and Tob.
98 C2 Scarborough U.K.
64 A2 Scarborough Shoal *sea feature* S. China Sea
96 A2 Scarinish U.K.
Scarpanto *i.* Greece *see* Karpathos
100 B3 Schaerbeek Belgium
105 D2 Schaffhausen Switz.
100 B1 Schagen Neth.
102 C2 Schärding Austria
100 A2 Scharendijke Neth.

1 B3 Shangrao China
0 B2 Shangshui China
7 E2 Shangyou Shuiku resr China
0 C2 Shangyu China
9 E1 Shangzhi China
    Shangzhou China see Shangluo
4 B3 Shaniko U.S.A.
7 B2 Shannon r. Ireland
7 B2 Shannon, Mouth of the Ireland
2 A1 Shan Plateau Myanmar
    Shansi prov. China see Shanxi
1 B3 Shantou China
    Shantung prov. China see Shandong
0 B2 Shanxi prov. China
1 B3 Shaoguan China
1 B3 Shaowu China
0 C2 Shaoxing China
1 B3 Shaoyang China
6 C1 Shapinsay i. U.K.
4 C2 Shaqrā' Saudi Arabia
4 A1 Sharan Afgh.
1 D2 Shārezā Iran
0 B2 Sharhorod Ukr.
9 C2 Sharjah U.A.E.
3 C2 Sharkawshchyna Belarus
0 A2 Shark Bay Austr.
8 A2 Sharm ash Shaykh Egypt
8 C2 Sharon U.S.A.
6 D3 Shar'ya Rus. Fed.
1 B3 Shashe r. Botswana/ Zimbabwe
7 B4 Shashemenē Eth.
    Shashi China see Jingzhou
4 B2 Shasta, Mount vol. U.S.A.
4 B2 Shasta Lake U.S.A.
5 D2 Shāṭi', Wādī ash watercourse Libya
    Shatilki Belarus see Svyetlahorsk
9 E2 Shatura Rus. Fed.
9 D3 Shaunavon Can.
8 B2 Shawano U.S.A.
0 C2 Shawinigan Can.
3 D1 Shawnee U.S.A.
3 M2 Shayboveyem r. Rus. Fed.
0 B? Shay Gap (abandoned) Austr.
9 E3 Shchekino Rus. Fed.
9 E? Shchelkovo Rus. Fed.
7 C3 Shchelyayur Rus. Fed.
7 C3 Shcherbakov Rus. Fed. see Rybinsk
4 B1 Shcherbinovka Ukr. see Dzerzhyns'k
9 E3 Shchigry Rus. Fed.
1 C1 Shchors Ukr.
3 B3 Shchuchyn Belarus
  D1 Shebekino Rus. Fed.
7 C4 Shebelē Wenz, Wabē r. Ethiopia/Somalia
7 C3 Sheberghān Afgh.
8 B2 Sheboygan U.S.A.
1 D3 Shebsh r. Rus. Fed.
7 C2 Sheelin, Lough l. Ireland
4 B2 Sheep Mountain U.S.A.
9 D4 Sheerness U.K.
7 C3 Sheffield U.K.
3 C2 Sheffield U.S.A.
    Sheikh Othman Yemen see Ash Shaykh 'Uthman
    Shekhem West Bank see Nāblus
9 E2 Sheksna Rus. Fed.
9 E2 Sheksninskoye Vodokhranilishche resr Rus. Fed.
3 M2 Shelagskiy, Mys pt Rus. Fed.
1 D3 Shelburne Can.
8 B2 Shelby MI U.S.A.
4 D1 Shelby MT U.S.A.
1 D1 Shelby NC U.S.A.
8 C2 Shelbyville IN U.S.A.
0 C1 Shelbyville TN U.S.A.
3 L2 Shelikhova, Zaliv g. Rus. Fed.
6 B3 Shelikof Strait U.S.A.
9 D2 Shellbrook Can.
    Shelter Bay Can. see Port-Cartier
4 B1 Shelton U.S.A.
7 D2 Shenandoah r. U.S.A.
9 D3 Shenandoah r. U.S.A.
9 D3 Shenandoah Mountains U.S.A.
8 A2 Shendam Nigeria

    Shengli Feng mt. China/Kyrg. see Pobeda Peak
86 D2 Shenkursk Rus. Fed.
70 B2 Shenmu China
    Shensi prov. China see Shaanxi
70 C1 Shenyang China
71 B3 Shenzhen China
90 B1 Shepetivka Ukr.
53 C3 Shepparton Austr.
99 D4 Sheppey, Isle of i. U.K.
131 D3 Sherbrooke N.S. Can.
130 C3 Sherbrooke Que. Can.
97 C2 Shercock Ireland
116 B3 Shereiq Sudan
136 B2 Sheridan U.S.A.
52 A2 Sheringa Austr.
86 F2 Sherkaly Rus. Fed.
100 B2 Sherman U.S.A.
96 ☐ 's-Hertogenbosch Neth.
    Shetland Islands U.K.
76 B2 Shetpe Kazakh.
    Shevchenko Kazakh. see Aktau
120 B2 Shevchenkove Ukr.
110 C2 Sheyenne r. U.S.A.
87 F3 Shibām Yemen
88 C2 Shibata Japan
143 C3 Shibetsu Japan
    Shibotsu-jima i. Rus. Fed. see Zelenyy, Ostrov
71 B3 Shicheng China
70 C2 Shidao China
96 B2 Shiel, Loch l. U.K.
    Shigatse China see Xigazê
128 C2 Shihezi China
    Shihkiachwang China see Shijiazhuang
    Shijiao China see Fogang
70 B2 Shijiazhuang China
    Shijiusuo China see Rizhao
74 A2 Shikarpur Pak.
67 B4 Shikoku i. Japan
66 D2 Shikotsu-ko l. Japan
86 D2 Shilega Rus. Fed.
75 C2 Shiliguri India
77 D2 Shilik Kazakh.
97 C2 Shillelagh Ireland
75 D2 Shillong India
89 F3 Shilovo Rus. Fed.
69 E1 Shimanovsk Rus. Fed.
117 C3 Shimbiris mt. Somalia
67 C3 Shimizu Japan
74 B1 Shimla India
67 C4 Shimoda Japan
73 B3 Shimoga India
66 D2 Shimokita-hantō pen. Japan
67 B4 Shimonoseki Japan
89 D2 Shimsk Rus. Fed.
96 B1 Shin, Loch l. U.K.
62 A1 Shingbwiyang Myanmar
67 C4 Shingū Japan
123 D1 Shingwedzi S. Africa
123 D1 Shingwedzi r. S. Africa
66 D3 Shinjō Japan
119 D3 Shinyanga Tanz.
67 D3 Shiogama Japan
67 C4 Shiono-misaki c. Japan
71 A3 Shiping China
98 C3 Shipley U.K.
135 C3 Shiprock U.S.A.
71 A3 Shiqian China
    Shiqizhen China see Zhongshan
70 A2 Shiquan China
78 B2 Shi'r, Jabal h. Saudi Arabia
67 C3 Shirane-san mt. Japan
67 C3 Shirane-san vol. Japan
81 D3 Shīrāz Iran
66 D2 Shiretoko-misaki c. Japan
66 D2 Shiriya-zaki c. Japan
76 B3 Shīrvān Iran
74 B2 Shiv India
74 B2 Shivpuri India
70 B2 Shiyan China
77 C2 Shiyeli Kazakh.
70 A2 Shizuishan China
67 C4 Shizuoka Japan
89 D3 Shklow Belarus
109 C2 Shkodër Albania
83 H1 Shmidta, Ostrov i. Rus. Fed.
67 B4 Shōbara Japan
    Sholapur India see Solapur
158 F8 Shona Ridge S. Atlantic Ocean
77 C3 Sho'rchi Uzbek.
74 B1 Shorkot Pak.

135 C3 Shoshone CA U.S.A.
134 D2 Shoshone ID U.S.A.
135 C3 Shoshone Mountains U.S.A.
123 C1 Shoshong Botswana
91 C1 Shostka Ukr.
70 B2 Shouxian China
78 A3 Showak Sudan
142 A2 Show Low U.S.A.
91 C2 Shpola Ukr.
140 B2 Shreveport U.S.A.
99 B3 Shrewsbury U.K.
77 D2 Shu Kazakh.
    Shuangjiang China see Tongdao
62 A1 Shuangjiang China
    Shuangxi China see Shunchang
66 B1 Shuangyashan China
87 E4 Shubarkudyk Kazakh.
116 B1 Shubrā al Khaymah Egypt
89 D2 Shugozero Rus. Fed.
    Shuidong China see Dianbai
144 B2 Shumba Zimbabwe
142 A2 Shumen Bulg.
105 D2 Shumikha Rus. Fed.
116 C3 Shumilina Belarus
111 B3 Shumla U.S.A.
107 C2 Shumyachi Rus. Fed.
127 I2 Shunchang China
110 B1 Shungnak U.S.A.
110 B1 Shuqrah Yemen
60 A1 Shurugwi Zimbabwe
92 ☐B2 Shushkodom Rus. Fed.
102 B2 Shushtar Iran
100 B3 Shuswap Lake Can.
106 C1 Shuya Rus. Fed.
114 B3 Shuyskoye Rus. Fed.
88 B2 Shwebo Myanmar
63 B2 Shwedwin Myanmar
92 I3 Shwegun Myanmar
81 C2 Shwegyin Myanmar
60 B2 Shweli r. Myanmar
74 B2 Shyganak Kazakh.
74 A1 Shymkent Kazakh.
114 B3 Shyok r. India/Pak.
137 F3 Shyok India/Pak.
66 B2 Shyroke Ukr.
111 C3 Shyryayeve Ukr.
103 D2 Sia Indon.
65 A2 Siahan Range mts Pak.
88 B2 Sialkot Pak.
144 B2 Siam country Asia see Thailand
101 D1 Sian China see Xi'an
75 D2 Siargao i. Phil.
77 D1 Siasi Phil.
    Šiauliai Lith.
64 B3 Sibasa S. Africa
64 B3 Šibenik Croatia
88 B2 Siberia reg. Rus. Fed.
75 C2 Siberut i. Indon.
80 B2 Sibi Pak.
75 C1 Sibir' reg. Rus. Fed. see Siberia
110 C2 Sibiti Congo
111 C2 Sibiu Romania
93 F3 Sibolga Indon.
93 E4 Sibsagar India
88 C2 Sibu Malaysia
98 B2 Sibut C.A.R.
140 B1 Sibutu i. Phil.
123 D2 Sibuyan i. Phil.
60 B1 Sibuyan Sea Phil.
88 B2 Sicamous Can.
81 C2 Sichon Thai.
154 C1 Sichuan prov. China
74 B2 Sichuan Pendi basin China
137 E1 Sicié, Cap c. France
142 B2 Sicilia i. Italy see Sicily
136 B3 Sicilian Channel Italy/Tunisia
62 B1 Sicily i. Italy
130 C3 Sicuani Peru
111 C3 Sideros, Akrotirio pt Greece
111 C3 Sidesaviwa S. Africa
118 C2 Sidhi India
138 C2 Sidhpur India
139 D2 Sidi Aïssa Alg.
78 A3 Sidi Ali Alg.
60 A1 Sidi Bel Abbès Alg.
91 C3 Sidi Ifni Morocco
75 C2 Sidi Kacem Morocco
135 C4 Sidikalang Indon.
    Sidirokastro Greece
110 B1 Sidlaw Hills U.K.
100 C3 Sidmouth U.K.
92 I2 Sidney Can.
129 D2 Sidney MT U.S.A.

136 C2 Sidney NE U.S.A.
138 C2 Sidney OH U.S.A.
141 D2 Sidney Lanier, Lake U.S.A.
61 D1 Sidoan Indon.
80 B2 Sidon Lebanon
154 B2 Sidrolândia Brazil
103 E1 Siedlce Pol.
100 C2 Sieg r. Ger.
101 D2 Siegen Ger.
63 B2 Siĕmréab Cambodia
    Siem Reap Cambodia see Siĕmréab
108 B2 Siena Italy
103 D1 Sieradz Pol.
142 B2 Sierra Blanca U.S.A.
153 B4 Sierra Grande Arg.
114 A4 Sierra Leone country Africa
158 E4 Sierra Leone Basin N. Atlantic Ocean
158 E4 Sierra Leone Rise N. Atlantic Ocean
144 B2 Sierra Mojada Mex.
142 A2 Sierra Vista U.S.A.
105 D2 Sierre Switz.
116 C3 Sifeni Eth.
111 B3 Sifnos i. Greece
107 C2 Sig Alg.
127 I2 Sigguup Nuuaa pen. Greenland
110 B1 Sighetu Marmaţiei Romania
110 B1 Sighişoara Romania
60 A1 Sigli Indon.
92 ☐B2 Siglufjörður Iceland
102 B2 Sigmaringen Ger.
100 B3 Signy-l'Abbaye France
106 C1 Sigüenza Spain
114 B3 Siguiri Guinea
88 B2 Sigulda Latvia
63 B2 Sihanoukville Cambodia
92 I3 Siilinjärvi Fin.
81 C2 Siirt Turkey
60 B2 Sijunjung Indon.
74 B2 Sikar India
74 A1 Sikaram mt. Afgh.
114 B3 Sikasso Mali
137 F3 Sikeston U.S.A.
66 B2 Sikhote-Alin' mts Rus. Fed.
111 C3 Sikinos i. Greece
103 D2 Siklós Hungary
65 A2 Sikuaishi China
88 B2 Šilalė Lith.
144 B2 Silao Mex.
101 D1 Silberberg h. Ger.
75 D2 Silchar India
77 D1 Siletyteniz, Ozero salt l. Kazakh.
75 C2 Silgarhi Nepal
80 B2 Silifke Turkey
75 C1 Siling Co salt l. China
110 C2 Silistra Bulg.
111 C2 Silivri Turkey
93 F3 Siljan l. Sweden
93 E4 Silkeborg Denmark
88 C2 Sillamäe Estonia
98 B2 Silloth U.K.
140 B1 Siloam Springs U.S.A.
123 D2 Silobela S. Africa
60 B1 Siluas Indon.
88 B2 Šilutė Lith.
81 C2 Silvan Turkey
154 C1 Silvânia Brazil
74 B2 Silvassa India
137 E1 Silver Bay U.S.A.
142 B2 Silver City U.S.A.
136 B3 Silverton U.S.A.
62 B1 Simao China
130 C3 Simard, Lac l. Can.
111 C3 Simav Turkey
111 C3 Simav Dağları mts Turkey
118 C2 Simba Dem. Rep. Congo
138 C2 Simcoe Can.
139 D2 Simcoe, Lake Can.
78 A3 Simēn Eth.
60 A1 Simeulue i. Indon.
91 C3 Simferopol' Ukr.
75 C2 Simikot Nepal
135 C4 Simi Valley U.S.A.
    Simla India see Shimla
110 B1 Şimleu Silvaniei Romania
100 C3 Simmern (Hunsrück) Ger.
92 I2 Simo Fin.
129 D2 Simonhouse Can.
60 B2 Simpang Indon.
51 C2 Simpson Desert Austr.
93 F4 Simrishamn Sweden
60 A1 Sinabang Indon.
116 B2 Sinai pen. Egypt

Sinalunga

105	E3	Sinalunga Italy
71	A3	Sinan China
65	B2	Sinanju N. Korea
62	A1	Sinbo Myanmar
62	A1	Sinbyugyun Myanmar
150	B2	Sincelejo Col.
60	B2	Sindangbarang Indon.
111	C3	Sındırgı Turkey
86	E2	Sindor Rus. Fed.
111	C3	Sinekçi Turkey
106	B2	Sines Port.
106	B2	Sines, Cabo de c. Port.
116	B3	Singa Sudan
75	C2	Singahi India
60	B1	Singapore country Asia
61	C2	Singaraja Indon.
63	B2	Sing Buri Thai.
119	D3	Singida Tanz.
62	A1	Singkaling Hkamti Myanmar
61	D2	Singkang Indon.
60	B1	Singkawang Indon.
60	B2	Singkep i. Indon.
60	A1	Singkil Indon.
53	D2	Singleton Austr.
		Sin'gosan N. Korea see Kosan
62	A1	Singu Myanmar
		Sining China see Xining
108	A2	Siniscola Italy
109	C2	Sinj Croatia
61	D2	Sinjai Indon.
116	B3	Sinkat Sudan
151	D2	Sinnamary Fr. Guiana
		Sînnicolau Mare Romania see Sânnicolau Mare
		Sinoia Zimbabwe see Chinhoyi
80	B1	Sinop Turkey
65	B1	Sinp'o N. Korea
61	C1	Sintang Indon.
100	B3	Sint Anthonis Neth.
100	A2	Sint-Laureins Belgium
147	D3	Sint Maarten i. Neth. Antilles
100	B2	Sint-Niklaas Belgium
143	D3	Sinton U.S.A.
65	A1	Sinŭiju N. Korea
64	B3	Siocon Phil.
103	D2	Siófok Hungary
105	D2	Sion Switz.
137	D2	Sioux Center U.S.A.
137	D2	Sioux City U.S.A.
137	D2	Sioux Falls U.S.A.
130	A2	Sioux Lookout Can.
65	A1	Siping China
129	E2	Sipiwesk Lake Can.
55	P2	Siple, Mount Antarctica
55	P2	Siple Island Antarctica
		Sipolilo Zimbabwe see Guruve
60	A2	Sipura i. Indon.
64	B3	Siquijor Phil.
93	E4	Sira r. Norway
		Siracusa Italy see Syracuse
51	C1	Sir Edward Pellew Group is Austr.
110	C1	Siret Romania
110	C1	Siret r. Romania
78	A1	Sirhān, Wādī an watercourse Saudi Arabia
79	C2	Sīrīk Iran
61	C1	Sirik, Tanjung pt Malaysia
62	B2	Siri Kit, Khuan Thai.
128	B1	Sir James MacBrien, Mount Can.
79	C2	Sīrjān Iran
81	C2	Şırnak Turkey
74	B2	Sirohi India
60	A1	Sirombu Indon.
74	B2	Sirsa India
115	D1	Sirte Libya
115	D1	Sirte, Gulf of Libya
88	B2	Širvintos Lith.
109	C1	Sisak Croatia
63	B2	Sisaket Thai.
145	C2	Sisal Mex.
122	B2	Sishen S. Africa
81	C2	Sisian Armenia
127	I2	Sisimiut Greenland
129	D2	Sisipuk Lake Can.
63	B2	Sisŏphŏn Cambodia
105	D3	Sisteron France
		Sitang China see Sinan
75	C2	Sitapur India
111	C3	Siteia Greece
123	D2	Siteki Swaziland
128	A2	Sitka U.S.A.
100	B2	Sittard Neth.

62	A1	Sittaung Myanmar
62	A2	Sittaung r. Myanmar
62	A1	Sittwe Myanmar
61	C2	Situbondo Indon.
80	B2	Sivas Turkey
111	C3	Sivaslı Turkey
80	B2	Siverek Turkey
88	D2	Siverskiy Rus. Fed.
80	B2	Sivrihisar Turkey
116	A2	Sīwah Egypt
75	B1	Siwalik Range mts India/Nepal
		Siwa Oasis oasis Egypt see Wāḥāt Sīwah
105	D3	Six-Fours-les-Plages France
70	B2	Sixian China
123	C2	Siyabuswa S. Africa
		Sjælland i. Denmark see Zealand
109	D2	Sjenica Serbia
92	G2	Sjøvegan Norway
91	C2	Skadovs'k Ukr.
93	F4	Skagen Denmark
93	E4	Skagerrak str. Denmark/Norway
134	B1	Skagit r. U.S.A.
128	A2	Skagway U.S.A.
92	G2	Skaland Norway
93	F4	Skara Sweden
74	B1	Skardu Pak.
103	E1	Skarżysko-Kamienna Pol.
103	D2	Skawina Pol.
114	A2	Skaymat Western Sahara
128	B2	Skeena r. Can.
128	B2	Skeena Mountains Can.
98	D3	Skegness U.K.
92	H3	Skellefteå Sweden
92	H3	Skellefteälven r. Sweden
97	C2	Skerries Ireland
93	F4	Ski Norway
111	B3	Skiathos i. Greece
97	B3	Skibbereen Ireland
92	□B2	Skíðadals-jökull glacier Iceland
98	B2	Skiddaw h. U.K.
93	E4	Skien Norway
103	E1	Skierniewice Pol.
115	C1	Skikda Alg.
52	B3	Skipton Austr.
98	B3	Skipton U.K.
93	E4	Skive Denmark
92	H1	Skjervøy Norway
		Skobelev Uzbek. see Farg'ona
111	B3	Skopelos i. Greece
89	E3	Skopin Rus. Fed.
111	B2	Skopje Macedonia
111	C3	Skoutaros Greece
93	F4	Skövde Sweden
83	J3	Skovorodino Rus. Fed.
139	F2	Skowhegan U.S.A.
92	H2	Skröven Sweden
88	B2	Skrunda Latvia
128	A1	Skukum, Mount Can.
123	D1	Skukuza S. Africa
88	B2	Skuodas Lith.
90	B2	Skvyra Ukr.
96	A2	Skye i. U.K.
111	B3	Skyros Greece
111	B3	Skyros i. Greece
93	F4	Slagelse Denmark
60	B2	Slamet, Gunung vol. Indon.
97	C2	Slaney r. Ireland
88	C2	Slantsy Rus. Fed.
109	C1	Slatina Croatia
110	B2	Slatina Romania
143	D2	Slaton U.S.A.
129	C1	Slave r. Can.
114	C4	Slave Coast Africa
128	C2	Slave Lake Can.
77	D1	Slavgorod Rus. Fed.
88	C2	Slavkovichi Rus. Fed.
		Slavonska Požega Croatia see Požega
109	C1	Slavonski Brod Croatia
90	B1	Slavuta Ukr.
90	C1	Slavutych Ukr.
66	B2	Slavyanka Rus. Fed.
		Slavyanskaya Rus. Fed. see Slavyansk-na-Kubani
91	D2	Slavyansk-na-Kubani Rus. Fed.
89	D3	Slawharad Belarus
103	D1	Sławno Pol.
99	C3	Sleaford U.K.
97	A2	Slea Head hd Ireland
130	C1	Sleeper Islands Can.
97	D1	Slieve Donard h. U.K.

96	A2	Sligachan U.K.
		Slieve Gamph hills Ireland see Ox Mountains
		Sligeach Ireland see Sligo
97	B1	Sligo Ireland
97	B1	Sligo Bay Ireland
93	G4	Slite Sweden
110	C2	Sliven Bulg.
		Sloboda Rus. Fed. see Ezhva
110	C2	Slobozia Romania
128	C3	Slocan Can.
88	C3	Slonim Belarus
100	B1	Sloten Neth.
99	C4	Slough U.K.
103	D2	Slovakia country Europe
108	B1	Slovenia country Europe
91	D2	Slov"yans'k Ukr.
102	C1	Stubice Pol.
90	B1	Sluch r. Ukr.
100	A2	Sluis Neth.
103	D1	Słupsk Pol.
88	C3	Slutsk Belarus
97	A2	Slyne Head hd Ireland
68	C1	Slyudyanka Rus. Fed.
131	D2	Smallwood Reservoir Can.
88	C3	Smalyavichy Belarus
88	C3	Smarhon' Belarus
129	D2	Smeaton Can.
109	D2	Smederevo Serbia
109	D2	Smederevska Palanka Serbia
91	C2	Smila Ukr.
88	C2	Smilavichy Belarus
88	C2	Smiltene Latvia
137	D3	Smith Center U.S.A.
128	B2	Smithers Can.
141	E1	Smithfield NC U.S.A.
134	D2	Smithfield UT U.S.A.
139	D3	Smith Mountain Lake U.S.A.
130	C3	Smiths Falls Can.
53	D2	Smithton Austr.
53	D2	Smoky Cape Austr.
137	D3	Smoky Hills U.S.A.
92	E3	Smøla i. Norway
89	D3	Smolensk Rus. Fed.
89	D3	Smolensko-Moskovskaya Vozvyshennost' hills Belarus/Rus. Fed.
111	B2	Smolyan Bulg.
66	B2	Smolyoninovo Rus. Fed.
130	B3	Smooth Rock Falls Can.
		Smyrna Turkey see İzmir
91	D2	Smyrnove Ukr.
92	□B3	Snæfell mt. Iceland
98	A2	Snaefell h. Isle of Man
128	A1	Snag (abandoned) Can.
134	C1	Snake r. U.S.A.
134	D2	Snake River Plain U.S.A.
		Snare Lakes Can. see Wekweètì
92	F3	Snåsvatn l. Norway
100	B1	Sneek Neth.
97	B3	Sneem Ireland
122	B3	Sneeuberge mts S. Africa
		Snegurovka Ukr. see Tetiyiv
103	D1	Sněžka mt. Czech Rep.
108	B1	Snežnik mt. Slovenia
103	E1	Śniardwy, Jezioro l. Pol.
		Sniečkus Lith. see Visaginas
91	C2	Snihurivka Ukr.
93	E3	Snøhetta mt. Norway
		Snovsk Ukr. see Shchors
129	D1	Snowbird Lake Can.
99	A3	Snowdon mt. U.K.
		Snowdrift Can. see Łutselk'e
129	C1	Snowdrift r. Can.
142	A2	Snowflake U.S.A.
129	D2	Snow Lake Can.
134	C1	Snowshoe Peak U.S.A.
52	A2	Snowtown Austr.
53	C3	Snowy r. Austr.
53	C3	Snowy Mountains Austr.
143	C2	Snyder U.S.A.
121	□D2	Soalala Madag.
121	□D2	Soanierana-Ivongo Madag.
90	B2	Sob r. Ukr.
65	B2	Sobaek-sanmaek mts S. Korea
117	B4	Sobat r. Sudan
89	F2	Sobinka Rus. Fed.
151	E4	Sobradinho, Barragem de resr Brazil
151	E3	Sobral Brazil
91	D3	Sochi Rus. Fed.
65	B2	Sŏch'ŏn S. Korea
49	L5	Society Islands Fr. Polynesia

150	B2	Socorro Col.
142	B2	Socorro NM U.S.A.
142	B2	Socorro TX U.S.A.
144	A3	Socorro, Isla i. Mex.
56	B4	Socotra i. Yemen
63	B3	Soc Trăng Vietnam
106	C2	Socuéllamos Spain
92	I2	Sodankylä Fin.
134	D2	Soda Springs U.S.A.
93	G3	Söderhamn Sweden
93	G4	Södertälje Sweden
116	A3	Sodiri Sudan
117	B4	Sodo Eth.
93	G3	Södra Kvarken str. Fin./Sweden
123	C1	Soekmekaar S. Africa
		Soerabaia Indon. see Surabaya
101	D2	Soest Ger.
53	C2	Sofala Austr.
110	B2	Sofia Bulg.
121	□D2	Sofia r. Madag.
		Sofiya Bulg. see Sofia
		Sofiyevka Ukr. see Vil'nyans'k
75	D1	Sog China
93	E3	Sognefjorden inlet Norway
111	D2	Söğüt Turkey
		Sohâg Egypt see Sūhāj
		Sohar Oman see Şuḩār
100	B2	Soignies Belgium
105	C2	Soissons France
90	A1	Sokal' Ukr.
65	B2	Sokch'o S. Korea
111	C3	Söke Turkey
81	C1	Sokhumi Georgia
114	C4	Sokodé Togo
89	F2	Sokol Rus. Fed.
101	F2	Sokolov Czech Rep.
115	C3	Sokoto Nigeria
115	C3	Sokoto r. Nigeria
90	B2	Sokyryany Ukr.
73	B3	Solapur India
135	B3	Soledad U.S.A.
89	F2	Soligalich Rus. Fed.
99	C3	Solihull U.K.
86	E3	Solikamsk Rus. Fed.
87	E3	Sol'-Iletsk Rus. Fed.
100	C2	Solingen Ger.
122	M1	Solitaire Namibia
92	G3	Sollefteå Sweden
93	C4	Sollentuna Sweden
107	D2	Sóller Spain
101	D2	Solling hills Ger.
89	E2	Solnechnogorsk Rus. Fed.
60	B2	Solok Indon.
48	H4	Solomon Islands country S. Pacific Ocean
48	G4	Solomon Sea S. Pacific Ocean
61	D2	Solor, Kepulauan is Indon.
105	D2	Solothurn Switz.
81	D2	Solṭānābād Iran
101	D1	Soltau Ger.
89	D2	Sol'tsy Rus. Fed.
96	C3	Solway Firth est. U.K.
120	B2	Solwezi Zambia
111	C3	Soma Turkey
117	C4	Somalia country Africa
117	C4	Somaliland reg. Somalia
120	B1	Sombo Angola
109	C1	Sombor Serbia
144	B2	Sombrerete Mex.
138	C3	Somerset U.S.A.
123	C3	Somerset East S. Africa
126	F2	Somerset Island Can.
122	A3	Somerset West S. Africa
110	B1	Someş r. Romania
101	E2	Sömmerda Ger.
146	B3	Somoto Nic.
75	C2	Son r. India
65	C1	Sŏnbong N. Korea
93	E3	Sønderborg Denmark
101	E2	Sondershausen Ger.
		Søndre Strømfjord inlet Greenland see Kangerlussuaq
108	A1	Sondrio Italy
63	B2	Sông Câu Vietnam
62	B1	Sông Đa, Hồ resr Vietnam
119	D4	Songea Tanz.
65	B1	Sŏnggan China
65	B1	Songhua Hu resr China
		Songjianghe China
65	B2	Sŏngjin N. Korea see Kimch'aek
63	B3	Songkhla Thai.
65	B2	Sŏngnam S. Korea

65 B2 Songnim N. Korea
20 A1 Songo Angola
21 C2 Songo Moz.
Songololo Dem. Rep.
Congo see Mbanza-Ngungu
69 E1 Songyuan China
Sonid Youqi China see
Saihan Tal
74 B2 Sonipat India
89 E2 Sonkovo Rus. Fed.
62 B1 Sơn La Vietnam
74 A2 Sonmiani Pak.
74 A2 Sonmiani Bay Pak.
01 E2 Sonneberg Ger.
42 A2 Sonoita Mex.
44 A2 Sonora r. Mex.
35 B3 Sonora CA U.S.A.
43 C2 Sonora TX U.S.A.
46 B3 Sonsonate El Salvador
Soochow China see Suzhou
17 A4 Sopo watercourse Sudan
03 D2 Sopron Hungary
08 B2 Sopur India
08 B2 Sora Italy
30 C3 Sorel Can.
51 D4 Sorell Austr.
06 C1 Soria Spain
90 B2 Soroca Moldova
54 C2 Sorocaba Brazil
87 E3 Sorochinsk Rus. Fed.
Soroki Moldova see Soroca
59 D2 Sorol atoll Micronesia
59 C3 Sorong Indon.
19 D2 Soroti Uganda
92 H1 Sørøya i. Norway
08 B2 Sorrento Italy
92 G2 Sorsele Sweden
64 B2 Sorsogon Phil.
86 C2 Sortavala Rus. Fed.
92 G2 Sortland Norway
65 B2 Sŏsan S. Korea
23 C2 Soshanguve S. Africa
89 E3 Sosna r. Rus. Fed.
53 B3 Sosneado mt. Arg.
86 E2 Sosnogorsk Rus. Fed.
86 D2 Sosnovka Rus. Fed.
86 C2 Sosnovyy Bor Rus. Fed.
03 D1 Sosnowiec Pol.
91 C1 Sosnytsya Ukr.
86 F3 Sos'va Rus. Fed.
91 D2 Sosyka r. Rus. Fed.
45 C2 Soto la Marina Mex.
18 B2 Souanké Congo
11 B3 Souda Greece
04 C3 Souillac France
Sŏul S. Korea see Seoul
04 B3 Soulac-sur-Mer France
04 B3 Soulom France
Soûr Lebanon see Tyre
07 D2 Sour el Ghozlane Alg.
29 D3 Souris Man. Can.
31 D3 Souris P.E.I. Can.
29 E3 Souris r. Can.
51 F3 Sousa Brazil
15 D1 Sousse Tunisia
04 B3 Soustons France
22 B3 South Africa, Republic of
country Africa
99 C4 Southampton U.K.
29 F1 Southampton, Cape Can.
29 F1 Southampton Island Can.
73 D3 South Andaman i. India
52 A1 South Australia state Austr.
40 B2 Southaven U.S.A.
42 B2 South Baldy mt. U.S.A.
30 B3 South Baymouth Can.
38 B2 South Bend U.S.A.
41 D2 South Carolina state U.S.A.
58 B2 South China Sea
N. Pacific Ocean
South Coast Town Austr. see
Gold Coast
36 C2 South Dakota state U.S.A.
99 C4 South Downs hills U.K.
59 E6 Southeast Indian Ridge
Indian Ocean
55 O2 Southeast Pacific Basin
S. Pacific Ocean
29 D2 Southend Can.
99 D4 Southend-on-Sea U.K.
54 B2 Southern Alps mts N.Z.
50 A3 Southern Cross Austr.
29 E2 Southern Indian Lake Can.
59 D7 Southern Ocean
41 E1 Southern Pines U.S.A.
Southern Rhodesia country
Africa see Zimbabwe
96 B3 Southern Uplands hills U.K.

55 J2 South Geomagnetic Pole
(2009) Antarctica
149 G8 South Georgia terr.
S. Atlantic Ocean
149 G8 South Georgia and the
South Sandwich Islands
terr. S. Atlantic Ocean
138 B2 South Haven U.S.A.
129 E1 South Henik Lake Can.
119 D2 South Horr Kenya
54 B2 South Island N.Z.
65 B2 South Korea country Asia
135 B3 South Lake Tahoe U.S.A.
55 L3 South Magnetic Pole (2009)
Antarctica
149 F9 South Orkney Islands
S. Atlantic Ocean
136 C2 South Platte r. U.S.A.
98 B3 Southport U.K.
141 E2 Southport U.S.A.
130 C3 South River Can.
96 C1 South Ronaldsay i. U.K.
123 D3 South Sand Bluff pt
S. Africa
149 H8 South Sandwich Islands
S. Atlantic Ocean
55 C4 South Sandwich Trench
S. Atlantic Ocean
129 D2 South Saskatchewan r. Can.
129 E2 South Seal r. Can.
149 E9 South Shetland Islands
Antarctica
98 C2 South Shields U.K.
54 B1 South Taranaki Bight b.
N.Z.
156 C8 South Tasman Rise
Southern Ocean
130 C2 South Twin Island Can.
96 A2 South Uist i. U.K.
South-West Africa country
Africa see Namibia
Southwest Peru Ridge
S. Pacific Ocean see
Nazca Ridge
53 D2 South West Rocks Austr.
99 D3 Southwold U.K.
109 C3 Soverato Italy
88 B2 Sovetsk Rus. Fed.
86 F2 Sovetskiy Rus. Fed.
91 C2 Sovyets'kyy Ukr.
123 C2 Soweto S. Africa
66 D1 Sōya-misaki c. Japan
65 B2 Soyang-ho l. S. Korea
104 C2 Soyaux France
90 C1 Sozh r. Europe
110 C2 Sozopol Bulg.
100 B2 Spa Belgium
106 C1 Spain country Europe
Spalato Croatia see Split
99 C3 Spalding U.K.
135 D2 Spanish Fork U.S.A.
Spanish Guinea country
Africa see Equatorial Guinea
97 B2 Spanish Point Ireland
Spanish Sahara terr. Africa
see Western Sahara
Spanish Town Jamaica
108 B3 Sparagio, Monte mt. Italy
135 C3 Sparks U.S.A.
138 A2 Sparta U.S.A.
141 D2 Spartanburg U.S.A.
111 B3 Sparti Greece
109 C3 Spartivento, Capo c. Italy
89 D3 Spas-Demensk Rus. Fed.
89 F2 Spas-Klepiki Rus. Fed.
66 B2 Spassk-Dal'niy Rus. Fed.
89 F3 Spassk-Ryazanskiy Rus. Fed.
111 B3 Spatha, Akrotirio pt
Greece
96 B2 Spean Bridge U.K.
136 C2 Spearfish U.S.A.
143 C1 Spearman U.S.A.
Spence Bay Can. see
Taloyoak
137 D2 Spencer IA U.S.A.
134 D2 Spencer IN U.S.A.
52 A2 Spencer Gulf est. Austr.
98 C2 Spennymoor U.K.
54 B2 Spenser Mountains N.Z.
101 D3 Spessart reg. Ger.
96 C2 Spey r. U.K.
101 D3 Speyer Ger.
100 B2 Spiekeroog i. Ger.
100 B2 Spijkenisse Neth.
100 B3 Spincourt France
128 C2 Spirit River Can.
103 E2 Spišská Nová Ves Slovakia
82 C1 Spitsbergen i. Svalbard

102 C2 Spittal an der Drau Austria
93 E3 Spjelkavik Norway
109 C2 Split Croatia
129 C2 Split Lake Can.
129 E2 Split Lake l. Can.
134 C1 Spokane U.S.A.
138 A1 Spooner U.S.A.
102 C1 Spree r. Ger.
122 A2 Springbok S. Africa
131 E3 Springdale Can.
140 B1 Springdale U.S.A.
101 D1 Springe Ger.
142 C1 Springer U.S.A.
142 B2 Springerville U.S.A.
136 C3 Springfield CO U.S.A.
139 E2 Springfield IL U.S.A.
138 B3 Springfield MA U.S.A.
137 E3 Springfield MO U.S.A.
134 B2 Springfield OH U.S.A.
140 C1 Springfield OR U.S.A.
123 C3 Springfield TN U.S.A.
123 C3 Springfontein S. Africa
131 D3 Springhill Can.
141 D3 Spring Hill U.S.A.
54 B2 Springs Junction N.Z.
51 D2 Springsure Austr.
135 D2 Springville U.S.A.
98 D3 Spurn Head hd U.K.
128 B3 Squamish Can.
109 C3 Squillace, Golfo di g. Italy
Srbija country Europe see
Serbia
108 C2 Srebrenica Bos.-Herz.
110 C2 Sredets Bulg.
83 L3 Sredinnyy Khrebet mts
Rus. Fed.
83 L2 Srednekolymsk Rus. Fed.
Sredne-Russkaya
Vozvyshennost' hills
Rus. Fed. see Central
Russian Upland
Sredne-Sibirskoye
Ploskogor'ye plat. Rus. Fed.
see Central Siberian Plateau
110 B2 Srednogorie Bulg.
69 D1 Sretensk Rus. Fed.
61 C1 Sri Aman Malaysia
73 B4 Sri Jayewardenepura Kotte
Sri Lanka
73 C3 Srikakulam India
73 C4 Sri Lanka country Asia
74 B1 Srinagar India
73 B3 Srivardhan India
101 D1 Stade Ger.
101 E1 Stadensen Ger.
101 D2 Stadtallendorf Ger.
101 D2 Stadthagen Ger.
101 E2 Staffelstein Ger.
99 B3 Stafford U.K.
99 C4 Staines U.K.
91 D2 Stakhanov Ukr.
Stakhanov Rus. Fed. see
Zhukovskiy
Stalin Bulg. see Varna
Stalinabad Tajik. see
Dushanbe
Stalingrad Rus. Fed. see
Volgograd
Staliniri Georgia see
Ts'khinvali
Stalino Ukr. see Donets'k
Stalinogorsk Rus. Fed. see
Novomoskovsk
Stalinogród Pol. see
Katowice
Stalinsk Rus. Fed. see
Novokuznetsk
103 E1 Stalowa Wola Pol.
138 B1 Stambaugh U.S.A.
99 C3 Stamford U.K.
139 E2 Stamford CT U.S.A.
143 D2 Stamford TX U.S.A.
Stampalia i. Greece see
Astypalaia
122 A1 Stampriet Namibia
92 F2 Stamsund Norway
123 C2 Standerton S. Africa
138 C2 Standish U.S.A.
123 D2 Stanger S. Africa
Stanislav Ukr. see
Ivano-Frankivs'k
Stanke Dimitrov Bulg. see
Dupnitsa
153 C5 Stanley Falkland Is
136 C1 Stanley U.S.A.
Stanleyville Dem. Rep.
Congo see Kisangani

Stann Creek Belize see
Dangriga
111 B3 Stanos Greece
83 I3 Stanovoye Nagor'ye mts
Rus. Fed.
83 J3 Stanovoy Khrebet mts
Rus. Fed.
53 D1 Stanthorpe Austr.
137 E1 Staples U.S.A.
103 E1 Starachowice Pol.
Stara Planina mts
Bulg./Serbia see
Balkan Mountains
89 D2 Staraya Russa Rus. Fed.
89 D2 Staraya Toropa Rus. Fed.
110 C2 Stara Zagora Bulg.
49 L4 Starbuck Island Kiribati
103 D1 Stargard Szczeciński Pol.
89 D2 Staritsa Rus. Fed.
141 D3 Starke U.S.A.
140 C2 Starkville U.S.A.
102 C2 Starnberg Ger.
91 D2 Starobil's'k Ukr.
89 D3 Starodub Rus. Fed.
103 D1 Starogard Gdański Pol.
90 B2 Starokostyantyniv Ukr.
91 D2 Starominskaya Rus. Fed.
91 D2 Staroshcherbinovskaya
Rus. Fed.
91 D2 Starotitarovskaya Rus. Fed.
89 F3 Staroyur'yevo Rus. Fed.
89 E3 Starozhilovo Rus. Fed.
99 B4 Start Point U.K.
88 C3 Staryya Darohi Belarus
86 G2 Staryy Nadym Rus. Fed.
89 E3 Staryy Oskol Rus. Fed.
101 E2 Staßfurt Ger.
103 E1 Staszów Pol.
139 D2 State College U.S.A.
141 D2 Statesboro U.S.A.
141 D1 Statesville U.S.A.
160 L1 Station Nord Greenland
139 D3 Staunton U.S.A.
87 D4 Stavanger Norway
Stavropol' Rus. Fed.
Stavropol'-na-Volge
Rus. Fed. see Tol'yatti
87 D4 Stavropol'skaya
Vozvyshennost' hills
Rus. Fed.
52 B3 Stawell Austr.
123 C2 Steadville S. Africa
136 B2 Steamboat Springs U.S.A.
101 E2 Stedten Ger.
120 C2 Steen River Can.
134 C2 Steenkool Indon.
100 C1 Steenwijk Neth.
126 E2 Stefansson Island Can.
Stegi Swaziland see Siteki
110 B1 Stei Romania
101 E3 Steigerwald mts Ger.
100 B2 Stein Neth.
129 E3 Steinbach Can.
100 C1 Steinfurt Ger.
120 A3 Steinhausen Namibia
92 F3 Steinkjer Norway
122 A2 Steinkopf S. Africa
122 B2 Stella S. Africa
122 A3 Stellenbosch S. Africa
105 D3 Stello, Monte mt. France
105 D2 Stenay France
101 E1 Stendal Ger.
Steornabhagh U.K. see
Stornoway
Stepanakert Azer. see
Xankändi
52 B2 Stephens Creek Austr.
129 E2 Stephens Lake Can.
131 E3 Stephenville Can.
143 D2 Stephenville U.S.A.
Stepnoy Rus. Fed. see Elista
122 B3 Sterling S. Africa
136 C2 Sterling CO U.S.A.
138 B2 Sterling IL U.S.A.
136 C1 Sterling ND U.S.A.
138 C2 Sterling Heights U.S.A.
87 E3 Sterlitamak Rus. Fed.
101 E1 Sternberg Ger.
128 C2 Stettler Can.
138 C2 Steubenville U.S.A.
99 C4 Stevenage U.K.
129 E2 Stevenson Lake Can.
138 B2 Stevens Point U.S.A.
126 C2 Stevens Village U.S.A.
134 D1 Stevensville U.S.A.
128 B2 Stewart Can.
128 A1 Stewart r. Can.
54 A3 Stewart Island N.Z.

127 G2 **Stewart Lake** Can.
123 C3 **Steynsburg** S. Africa
102 C2 **Steyr** Austria
122 B3 **Steytlerville** S. Africa
128 A2 **Stikine** r. Can.
128 A2 **Stikine Plateau** Can.
122 B3 **Stilbaai** S. Africa
137 E1 **Stillwater** *MN* U.S.A.
143 D1 **Stillwater** *OK* U.S.A.
135 C3 **Stillwater Range** *mts* U.S.A.
109 D2 **Štip** Macedonia
96 C2 **Stirling** U.K.
52 A2 **Stirling North** Austr.
92 F3 **Stjørdalshalsen** Norway
103 D2 **Stockerau** Austria
93 G4 **Stockholm** Sweden
98 B3 **Stockport** U.K.
135 B3 **Stockton** U.S.A.
98 C2 **Stockton-on-Tees** U.K.
143 C2 **Stockton Plateau** U.S.A.
63 B2 **Stœng Trêng** Cambodia
96 B1 **Stoer, Point of** U.K.
99 B3 **Stoke-on-Trent** U.K.
98 C2 **Stokesley** U.K.
92 F2 **Stokmarknes** Norway
110 B2 **Stol** *mt.* Serbia
109 C2 **Stolac** Bos.-Herz.
100 C2 **Stolberg (Rheinland)** Ger.
82 E2 **Stolbovoy** Rus. Fed.
88 C3 **Stolin** Belarus
101 F2 **Stollberg** Ger.
101 D1 **Stolzenau** Ger.
96 C2 **Stonehaven** U.K.
99 C4 **Stonehenge** *tourist site* U.K.
129 E2 **Stonewall** Can.
129 D2 **Stony Rapids** Can.
92 G2 **Storavan** *l.* Sweden
Store Bælt *sea chan.* Denmark *see* **Great Belt**
92 F3 **Støren** Norway
92 I1 **Storfjordbotn** Norway
92 F2 **Storforshei** Norway
126 E2 **Storkerson Peninsula** Can.
137 D2 **Storm Lake** U.S.A.
93 E3 **Stornosa** *mt.* Norway
96 A1 **Stornoway** U.K.
86 E2 **Storozhevsk** Rus. Fed.
90 B2 **Storozhynets'** Ukr.
92 F3 **Storsjön** *l.* Sweden
92 H2 **Storslett** Norway
92 G2 **Storuman** Sweden
92 G2 **Storuman** *l.* Sweden
99 C4 **Stour** *r. England* U.K.
99 D4 **Stour** *r. England* U.K.
130 A2 **Stout Lake** Can.
88 C3 **Stowbtsy** Belarus
99 D3 **Stowmarket** U.K.
97 C1 **Strabane** U.K.
102 C2 **Strakonice** Czech Rep.
102 C1 **Stralsund** Ger.
122 A3 **Strand** S. Africa
93 E3 **Stranda** Norway
97 D1 **Strangford Lough** *inlet* U.K.
96 B3 **Stranraer** U.K.
105 D2 **Strasbourg** France
130 B3 **Stratford** Can.
54 B1 **Stratford** N.Z.
143 C1 **Stratford** U.S.A.
99 C3 **Stratford-upon-Avon** U.K.
128 C2 **Strathmore** Can.
96 C2 **Strathspey** *val.* U.K.
102 C2 **Straubing** Ger.
134 C2 **Strawberry Mountain** U.S.A.
51 C3 **Streaky Bay** Austr.
138 B2 **Streator** U.S.A.
99 B4 **Street** U.K.
110 B2 **Strehaia** Romania
94 B1 **Streymoy** *i.* Faroe Is
82 G2 **Strezhevoy** Rus. Fed.
101 F3 **Stříbro** Czech Rep.
153 B4 **Stroeder** Arg.
101 D1 **Ströhen** Ger.
109 C3 **Stromboli, Isola** *i.* Italy
96 B2 **Stromeferry** U.K.
96 C1 **Stromness** U.K.
92 G3 **Strömsund** Sweden
96 C1 **Stronsay** *i.* U.K.
53 D2 **Stroud** Austr.
99 B4 **Stroud** U.K.
100 C1 **Strücklingen (Saterland)** Ger.
111 B2 **Struga** Macedonia
88 C2 **Strugi-Krasnyye** Rus. Fed.
122 B3 **Struis Bay** S. Africa
111 B2 **Struma** *r.* Bulg.
99 A3 **Strumble Head** *hd* U.K.
111 B2 **Strumica** Macedonia
122 B2 **Strydenburg** S. Africa

111 B2 **Strymonas** *r.* Greece
93 E3 **Stryn** Norway
90 A2 **Stryy** Ukr.
90 A2 **Stryy** *r.* Ukr.
128 B2 **Stuart Lake** Can.
53 C2 **Stuart Town** Austr.
Stuchka Latvia *see* **Aizkraukle**
130 A2 **Stull Lake** Can.
89 E3 **Stupino** Rus. Fed.
55 M3 **Sturge Island** Antarctica
138 B2 **Sturgeon Bay** U.S.A.
130 C3 **Sturgeon Falls** Can.
130 A3 **Sturgeon Lake** Can.
138 B2 **Sturgis** *MI* U.S.A.
136 C2 **Sturgis** *SD* U.S.A.
52 B1 **Sturt, Mount** *h.* Austr.
50 B1 **Sturt Creek** *watercourse* Austr.
50 C1 **Sturt Plain** Austr.
52 B1 **Sturt Stony Desert** Austr.
123 C3 **Stutterheim** S. Africa
102 B2 **Stuttgart** Ger.
140 B2 **Stuttgart** U.S.A.
92 □A2 **Stykkishólmur** Iceland
90 B1 **Styr** *r.* Belarus/Ukr.
155 D1 **Suaçuí Grande** *r.* Brazil
116 B3 **Suakin** Sudan
71 C3 **Suao** Taiwan
78 A3 **Suara** Eritrea
60 B1 **Subi Besar** *i.* Indon.
109 C1 **Subotica** Serbia
110 C1 **Suceava** Romania
Suchan Rus. Fed. *see* **Partizansk**
97 B2 **Suck** *r.* Ireland
152 B1 **Sucre** Bol.
154 B2 **Sucuriú** *r.* Brazil
Suczawa Romania *see* **Suceava**
89 E2 **Suda** Rus. Fed.
91 C3 **Sudak** Ukr.
116 A3 **Sudan** *country* Africa
130 B3 **Sudbury** Can.
99 D3 **Sudbury** U.K.
117 A4 **Sudd** *swamp* Sudan
89 F2 **Sudislavl'** Rus. Fed.
89 F2 **Sudogda** Rus. Fed.
94 B1 **Suðuroy** *i.* Faroe Is
89 E3 **Sudzha** Rus. Fed.
107 C2 **Sueca** Spain
116 B2 **Suez** Egypt
116 B2 **Suez, Gulf of** Egypt
80 B2 **Suez Canal** *canal* Egypt
139 D3 **Suffolk** U.S.A.
140 A3 **Sugar Land** U.S.A.
139 E1 **Sugarloaf Mountain** U.S.A.
53 D2 **Sugarloaf Point** Austr.
70 A2 **Suhait** China
116 B2 **Sūhāj** Egypt
79 C2 **Şubār** Oman
68 D1 **Sühbaatar** Mongolia
101 E2 **Suhl** Ger.
109 C1 **Suhopolje** Croatia
70 B2 **Suide** China
66 B2 **Suifenhe** China
69 E1 **Suihua** China
71 B3 **Sui Jiang** *r.* China
70 A2 **Suining** China
70 B2 **Suiping** China
97 C2 **Suir** *r.* Ireland
Suixian China *see* **Suizhou**
70 B2 **Suiyang** China
70 B2 **Suizhou** China
74 B2 **Sujangarh** India
74 B1 **Sujanpur** India
74 A2 **Sujawal** Pak.
60 B2 **Sukabumi** Indon.
60 B2 **Sukadana** Indon.
67 D3 **Sukagawa** Japan
60 C2 **Sukaraja** Indon.
Sukarnapura Indon. *see* **Jayapura**
Sukarno, Puntjak *mt.* Indon. *see* **Jaya, Puncak**
89 E3 **Sukhinichi** Rus. Fed.
89 F2 **Sukhona** *r.* Rus. Fed.
62 A2 **Sukhothai** Thai.
74 A2 **Sukkur** Pak.
89 E2 **Sukromny** Rus. Fed.
59 C3 **Sula, Kepulauan** *is* Indon.
74 A1 **Sulaiman Range** *mts* Pak.
Sulawesi *i.* Indon. *see* **Celebes**
101 D1 **Sulingen** Ger.
76 B3 **Sullana** Peru
138 B3 **Sullivan** *IN* U.S.A.

137 E3 **Sullivan** *MO* U.S.A.
140 B2 **Sulphur** U.S.A.
143 D2 **Sulphur Springs** U.S.A.
138 C1 **Sultan** Can.
Sultanabad Iran *see* **Arāk**
64 B3 **Sulu Archipelago** *is* Phil.
64 A3 **Sulu Sea** N. Pacific Ocean
101 E3 **Sulzbach-Rosenberg** Ger.
79 C2 **Sumāil** Oman
Sumatera *i.* Indon. *see* **Sumatra**
60 A1 **Sumatra** *i.* Indon.
61 D2 **Sumba** *i.* Indon.
61 C2 **Sumba, Selat** *sea chan.* Indon.
61 C2 **Sumbawa** *i.* Indon.
61 C2 **Sumbawabesar** Indon.
119 D3 **Sumbawanga** Tanz.
120 A2 **Sumbe** Angola
96 □ **Sumburgh** U.K.
96 □ **Sumburgh Head** *hd* U.K.
119 C2 **Sumeih** Sudan
61 C2 **Sumenep** Indon.
67 D4 **Sumisu-jima** *i.* Japan
131 D3 **Summerside** Can.
138 C3 **Summersville** U.S.A.
141 D2 **Summerville** U.S.A.
137 D1 **Summit** U.S.A.
128 B2 **Summit Lake** Can.
103 D2 **Šumperk** Czech Rep.
81 C1 **Sumqayıt** Azer.
141 D2 **Sumter** U.S.A.
91 C1 **Sumy** Ukr.
75 D2 **Sunamganj** Bangl.
65 B2 **Sunan** N. Korea
79 C2 **Şunaynah** Oman
52 B3 **Sunbury** Austr.
139 D2 **Sunbury** U.S.A.
65 B2 **Sunch'ŏn** N. Korea
65 B3 **Sunch'ŏn** S. Korea
123 C2 **Sun City** S. Africa
93 H3 **Sund** Fin.
60 B2 **Sunda, Selat** *str.* Indon.
136 C2 **Sundance** U.S.A.
75 C2 **Sundarbans** *coastal area* Bangl./India
74 B1 **Sundarnagar** India
Sunda Strait *str.* Indon. *see* **Sunda, Selat**
Sunda Trench Indian Ocean *see* **Java Trench**
98 C2 **Sunderland** U.K.
128 C2 **Sundre** Can.
93 G3 **Sundsvall** Sweden
123 D2 **Sundumbili** S. Africa
60 B2 **Sungailiat** Indon.
60 B2 **Sungaipenuh** Indon.
60 B1 **Sungai Petani** Malaysia
80 B1 **Sungurlu** Turkey
75 C2 **Sun Kosi** *r.* Nepal
93 E3 **Sunndalsøra** Norway
134 C1 **Sunnyside** U.S.A.
135 B3 **Sunnyvale** U.S.A.
141 D3 **Sunrise** U.S.A.
83 I2 **Suntar** Rus. Fed.
74 A2 **Suntsar** Pak.
114 B4 **Sunyani** Ghana
92 I3 **Suomussalmi** Fin.
67 B4 **Suō-nada** *b.* Japan
86 C2 **Suoyarvi** Rus. Fed.
142 A2 **Superior** *AZ* U.S.A.
137 D2 **Superior** *NE* U.S.A.
138 A1 **Superior** *WI* U.S.A.
138 B1 **Superior, Lake** Can./U.S.A.
63 B2 **Suphan Buri** Thai.
81 C2 **Süphan Daği** *mt.* Turkey
89 D3 **Suponevo** Rus. Fed.
81 C2 **Sūq ash Shuyūkh** Iraq
70 B2 **Suqian** China
78 A2 **Sūq Suwayq** Saudi Arabia
Suqutrā *i.* Yemen *see* **Socotra**
79 C2 **Şūr** Oman
74 A2 **Surab** Pak.
61 C2 **Surabaya** Indon.
61 C2 **Surakarta** Indon.
74 B2 **Surat** India
74 B2 **Suratgarh** India
63 A3 **Surat Thani** Thai.
89 D3 **Surazh** Rus. Fed.
109 D2 **Surdulica** Serbia
100 C3 **Sûre** *r.* Lux.
74 B2 **Surendranagar** India
82 F2 **Surgut** Rus. Fed.
64 B3 **Surigao** Phil.
63 B2 **Surin** Thai.
151 D2 **Suriname** *country* S. America

75 C2 **Surkhet** Nepal
Surt Libya *see* **Sirte**
Surt, Khalīj *g.* Libya *see* **Sirte, Gulf of**
60 B2 **Surulangun** Indon.
81 C2 **Süsangerd** Iran
89 F2 **Susanino** Rus. Fed.
135 B2 **Susanville** U.S.A.
80 B1 **Suşehri** Turkey
139 D3 **Susquehanna** *r.* U.S.A.
131 D3 **Sussex** Can.
101 D1 **Süstedt** Ger.
100 C1 **Sustrum** Ger.
83 K2 **Susuman** Rus. Fed.
111 C3 **Susurluk** Turkey
74 B1 **Sutak** India
53 D2 **Sutherland** Austr.
122 B3 **Sutherland** S. Africa
136 C2 **Sutherland** U.S.A.
134 B2 **Sutherlin** U.S.A.
74 B2 **Sutlej** *r.* India/Pak.
138 C3 **Sutton** U.S.A.
99 C3 **Sutton Coldfield** U.K.
98 C3 **Sutton in Ashfield** U.K.
66 D2 **Suttsu** Japan
49 I5 **Suva** Fiji
Suvalki Pol. *see* **Suwałki**
89 E3 **Suvorov** Rus. Fed.
90 B2 **Suvorove** Ukr.
103 E1 **Suwałki** Pol.
141 D3 **Suwannee Sound** *b.* U.S.A.
63 B2 **Suwannaphum** Thai.
141 D3 **Suwanee** *r.* U.S.A.
Suways, Qanāt as *canal* Egypt *see* **Suez Canal**
Suweis, Qanâ el *canal* Egy *see* **Suez Canal**
65 B2 **Suwŏn** S. Korea
79 C2 **Sūzā** Iran
89 F2 **Suzdal'** Rus. Fed.
89 D3 **Suzemka** Rus. Fed.
70 B2 **Suzhou** *Anhui* China
70 C2 **Suzhou** *Jiangsu* China
67 C3 **Suzu** Japan
67 C3 **Suzu-misaki** *pt* Japan
82 B1 **Svalbard** *terr.* Arctic Ocea
90 A2 **Svalyava** Ukr.
92 H2 **Svappavaara** Sweden
91 D2 **Svatove** Ukr.
63 B2 **Svay Riĕng** Cambodia
93 F3 **Sveg** Sweden
88 C2 **Švenčionys** Lith.
93 F4 **Svendborg** Denmark
Sverdlovsk Rus. Fed. *see* **Yekaterinburg**
111 B2 **Sveti Nikole** Macedonia
66 C1 **Svetlaya** Rus. Fed.
88 B3 **Svetlogorsk** Rus. Fed.
87 D4 **Svetlograd** Rus. Fed.
88 B3 **Svetlyy** Rus. Fed.
93 I3 **Svetogorsk** Rus. Fed.
103 E2 **Svidník** Slovakia
111 C2 **Svilengrad** Bulg.
110 B2 **Svinecea Mare, Vârful** *mt* Romania
110 C2 **Svishtov** Bulg.
88 B3 **Svislach** Belarus
103 D2 **Svitavy** Czech Rep.
91 C2 **Svitlovods'k** Ukr.
69 E1 **Svobodnyy** Rus. Fed.
110 B2 **Svoge** Bulg.
92 F2 **Svolvær** Norway
88 C3 **Svyetlahorsk** Belarus
137 E2 **Swainsboro** U.S.A.
120 A3 **Swakopmund** Namibia
52 B3 **Swan Hill** Austr.
128 C2 **Swan Hills** Can.
129 D2 **Swan Lake** Can.
97 C1 **Swanlinbar** Ireland
129 D2 **Swan River** Can.
53 D2 **Swansea** Austr.
99 B4 **Swansea** U.K.
122 B3 **Swartkolkvloer** *salt pan* S. Africa
123 C2 **Swartruggens** S. Africa
Swatow China *see* **Shanto**
123 D2 **Swaziland** *country* Africa
93 G3 **Sweden** *country* Europe
143 C2 **Sweetwater** U.S.A.
136 B2 **Sweetwater** *r.* U.S.A.
122 B3 **Swellendam** S. Africa
103 D1 **Świdnica** Pol.
103 D1 **Świdwin** Pol.
103 E1 **Świebodzin** Pol.
103 D1 **Świecie** Pol.
129 D2 **Swift Current** Can.
97 C1 **Swilly, Lough** *inlet* U.K.
99 C4 **Swindon** U.K.

Tarkwa

114	B4	Tarkwa Ghana
64	B2	Tarlac Phil.
105	C3	Tarn r. France
92	G2	Tärnaby Sweden
77	C3	Tarnak r. Afgh.
110	B1	Târnăveni Romania
103	E1	Tarnobrzeg Pol.
		Tarnopol Ukr. see Ternopil'
114	H2	Tarnów Pol.
103	E1	Tarnów Pol.
51	D2	Taroom Austr.
114	B1	Taroudannt Morocco
141	D3	Tarpon Springs U.S.A.
69	L1	Tarqi China
108	B2	Tarquinia Italy
107	D1	Tarragona Spain
107	D1	Tàrrega Spain
80	B2	Tarsus Turkey
152	B2	Tartagal Arg.
104	B3	Tartas France
88	C2	Tartu Estonia
80	B2	Ţarţūs Syria
155	D1	Tarumirim Brazil
89	E3	Tarusa Rus. Fed.
90	B2	Tarutyne Ukr.
108	B1	Tarvisio Italy
		Tashauz Turkm. see Daşoguz
		Tashauz Turkm. see Dashkhovuz
75	D2	Tashigang Bhutan
81	D3	Tashk, Daryācheh-ye l. Iran
		Tashkent Uzbek. see Toshkent
130	C2	Tasialujjuaq, Lac l. Can.
130	C2	Tasiat, Lac l. Can.
131	D2	Tasiujaq Can.
76	B1	Taskala Kazakh.
77	E2	Taskesken Kazakh.
54	B2	Tasman Bay N.Z.
51	D4	Tasmania state Austr.
54	B2	Tasman Mountains N.Z.
156	D8	Tasman Sea S. Pacific Ocean
61	D2	Tataba Indon.
103	D2	Tatabánya Hungary
90	B2	Tatarbunary Ukr.
83	K3	Tatarskiy Proliv str. Rus. Fed.
		Tatar Strait str. Rus. Fed. see Tatarskiy Proliv
67	C4	Tateyama Japan
128	C1	Tathlina Lake Can.
78	B3	Tathlīth Saudi Arabia
78	B2	Tathlīth, Wādī watercourse Saudi Arabia
53	C3	Tathra Austr.
62	A1	Tatkon Myanmar
128	B2	Tatla Lake Can.
		Tatra Mountains mts Pol./Slovakia see Tatry
103	D2	Tatry Pol./Slovakia
154	C2	Tatuí Brazil
143	C2	Tatum U.S.A.
81	C2	Tatvan Turkey
151	E3	Taua Brazil
155	C2	Taubaté Brazil
101	D3	Tauberbischofsheim Ger.
54	C1	Taumarunui N.Z.
122	B2	Taung S. Africa
62	A1	Taunggyi Myanmar
62	A2	Taung-ngu Myanmar
62	A2	Taungup Myanmar
74	B1	Taunsa Pak.
99	B4	Taunton U.K.
101	C2	Taunus hills Ger.
54	C1	Taupo N.Z.
54	C1	Taupo, Lake N.Z.
88	B2	Tauragė Lith.
54	C1	Tauranga N.Z.
139	E1	Taureau, Réservoir resr Can.
80	B2	Taurus Mountains mts Turkey
111	C3	Tavas Turkey
86	F3	Tavda Rus. Fed.
106	B2	Tavira Port.
99	A4	Tavistock U.K.
63	A2	Tavoy Myanmar
111	C3	Tavşanlı Turkey
99	A4	Taw r. U.K.
138	C2	Tawas City U.S.A.
61	C1	Tawau Malaysia
64	A3	Tawi-Tawi i. Phil.
71	C3	Tawu Taiwan
145	C3	Taxco Mex.
76	B2	Taxiatosh Uzbek.
77	D3	Taxkorgan China
96	C2	Tay r. U.K.
76	C3	Tāybād Iran

96	C2	Tay, Firth of est. U.K.
96	B2	Tay, Loch l. U.K.
128	B2	Taylor Can.
138	C2	Taylor MI U.S.A.
143	D2	Taylor TX U.S.A.
138	B3	Taylorville U.S.A.
78	A2	Taymā' Saudi Arabia
83	H2	Taymura r. Rus. Fed.
83	H2	Taymyr, Ozero l. Rus. Fed.
		Taymyr, Poluostrov pen. Rus. Fed. see Taymyr Peninsula
83	G2	Taymyr Peninsula pen. Rus. Fed.
63	B2	Tây Ninh Vietnam
96	C2	Tayport U.K.
64	A2	Taytay Phil.
82	G2	Taz r. Rus. Fed.
114	B1	Taza Morocco
129	D2	Tazin Lake Can.
86	G2	Tazovskaya Guba sea chan. Rus. Fed.
81	C1	T'bilisi Georgia
91	E2	Tbilisskaya Rus. Fed.
118	B3	Tchibanga Gabon
115	C3	Tchin-Tabaradene Niger
118	B2	Tcholliré Cameroon
103	D1	Tczew Pol.
144	B2	Teacapán Mex.
54	A3	Te Anau N.Z.
54	A3	Te Anau, Lake N.Z.
145	C3	Teapa Mex.
54	C1	Te Awamutu N.Z.
115	C1	Tébessa Alg.
60	B2	Tebingtinggi Indon.
60	A1	Tebingtinggi Indon.
144	A1	Tecate Mex.
114	B4	Techiman Ghana
144	B3	Tecomán Mex.
144	B2	Tecoripa Mex.
145	B3	Técpan Mex.
144	B2	Tecuala Mex.
110	C1	Tecuci Romania
68	C1	Teeli Rus. Fed.
98	C2	Tees r. U.K.
111	C3	Tefenni Turkey
60	B2	Tegal Indon.
146	B3	Tegucigalpa Hond.
115	C3	Teguidda-n-Tessoumt Niger
129	E1	Tehek Lake Can.
		Teheran Iran see Tehrān
114	B4	Téhini Côte d'Ivoire
81	D2	Tehrān Iran
145	C3	Tehuacán Mex.
		Tehuantepec, Golfo de g. Mex. see Tehuantepec, Gulf of
145	C3	Tehuantepec, Gulf of g. Mex.
145	C3	Tehuantepec, Istmo de isth. Mex.
114	A2	Teide, Pico del vol. Islas Canarias
99	A3	Teifi r. U.K.
		Teixeira de Sousa Angola see Luau
76	C3	Tejen Turkm.
76	C3	Tejen r. Turkm.
		Tejo r. Port. see Tagus
145	B3	Tejupilco Mex.
54	B2	Tekapo, Lake N.Z.
145	D2	Tekax Mex.
116	B3	Tekezē Wenz r. Eritrea/Eth.
111	C2	Tekirdağ Turkey
54	C1	Te Kuiti N.Z.
75	C2	Tel r. India
81	C1	T'elavi Georgia
80	B2	Tel Aviv-Yafo Israel
145	D2	Telchac Puerto Mex.
128	A2	Telegraph Creek Can.
154	B2	Telêmaco Borba Brazil
61	C1	Telen r. Indon.
50	B2	Telfer Mining Centre Austr.
99	B3	Telford U.K.
128	B2	Telkwa Can.
60	A1	Telo Indon.
86	E2	Telpoziz, Gora mt. Rus. Fed.
88	B2	Telšiai Lith.
60	B2	Telukbatang Indon.
60	A1	Telukdalam Indon.
60	B1	Teluk Intan Malaysia
114	C4	Tema Ghana
130	C3	Temagami Lake Can.
61	C2	Temanggung Indon.
123	C2	Temba S. Africa
83	H2	Tembenchi r. Rus. Fed.
60	B2	Tembilahan Indon.
123	C2	Tembisa S. Africa

120	A1	Tembo Aluma Angola
		Tembué Moz. see Chifunde
99	B3	Teme r. U.K.
135	C4	Temecula U.S.A.
63	B3	Temengor, Tasik resr Malaysia
60	B1	Temerluh Malaysia
77	D1	Temirtau Kazakh.
139	D1	Témiscamingue, Lac l. Can.
53	C2	Temora Austr.
142	A2	Tempe U.S.A.
143	D2	Temple U.S.A.
97	C2	Templemore Ireland
102	C1	Templin Ger.
145	C2	Tempoal Mex.
120	A2	Tempué Angola
91	D2	Temryuk Rus. Fed.
91	D2	Temryukskiy Zaliv b. Rus. Fed.
153	A3	Temuco Chile
54	B2	Temuka N.Z.
145	C2	Tenabo Mex.
143	E2	Tenaha U.S.A.
73	C3	Tenali India
63	A2	Tenasserim Myanmar
99	A4	Tenby U.K.
117	C3	Tendaho Eth.
105	D3	Tende France
105	D3	Tende, Col de pass France/Italy
73	D4	Ten Degree Channel India
114	A3	Te-n-Dghâmcha, Sebkhet salt marsh Maur.
67	D3	Tendō Japan
114	B3	Ténenkou Mali
115	C2	Ténéré reg. Niger
115	D3	Ténéré, Erg du des. Niger
115	D2	Ténéré du Tafassâsset des. Niger
114	A2	Tenerife i. Islas Canarias
107	D2	Ténès Alg.
61	C2	Tengah, Kepulauan is Indon.
		Tengcheng China see Tengxian
62	A1	Tengchong China
61	C2	Tenggarong Indon.
70	A2	Tengger Shamo des. China
77	C1	Tengiz, Ozero salt l. Kazakh.
71	B3	Tengxian China
119	C4	Tenke Dem. Rep. Congo
114	B3	Tenkodogo Burkina
51	C1	Tennant Creek Austr.
140	C1	Tennessee r. U.S.A.
140	C1	Tennessee state U.S.A.
61	C1	Tenom Malaysia
145	C3	Tenosique Mex.
61	D2	Tenteno Indon.
53	D1	Tenterfield Austr.
141	D3	Ten Thousand Islands U.S.A.
154	B2	Teodoro Sampaio Brazil
155	D1	Teófilo Otoni Brazil
145	C3	Teopisca Mex.
59	C3	Tepa Indon.
144	B2	Tepache Mex.
54	B1	Te Paki N.Z.
144	B3	Tepalcatepec Mex.
144	B2	Tepatitlán Mex.
144	B2	Tepehuanes Mex.
109	D2	Tepelenë Albania
102	C1	Teplice Czech Rep.
89	E3	Teploye Rus. Fed.
90	B2	Teplyk Ukr.
54	C1	Te Puke N.Z.
144	B2	Tequila Mex.
49	K3	Teraina i. Kiribati
108	B2	Teramo Italy
52	B3	Terang Austr.
89	E3	Terbuny Rus. Fed.
90	B2	Terebovlya Ukr.
87	D4	Terek r. Rus. Fed.
154	B2	Terenos Brazil
151	E3	Teresina Brazil
155	D2	Teresópolis Brazil
63	A3	Teressa Island India
100	A3	Tergnier France
80	B1	Terme Turkey
108	B3	Termini Imerese Italy
145	C3	Términos, Laguna de lag. Mex.
77	C3	Termiz Uzbek.
109	B2	Termoli Italy
59	C2	Ternate Indon.
100	A2	Terneuzen Neth.
66	C1	Terney Rus. Fed.
108	B2	Terni Italy

90	B2	Ternopil' Ukr.
52	A2	Terowie Austr.
69	F1	Terpeniya, Mys c. Rus. Fed.
69	F1	Terpeniya, Zaliv g. Rus. Fed.
128	B2	Terrace Can.
130	B3	Terrace Bay Can.
122	B2	Terra Firma S. Africa
140	B3	Terrebonne Bay U.S.A.
138	B3	Terre Haute U.S.A.
131	E3	Terrenceville Can.
100	B1	Terschelling i. Neth.
108	A3	Tertenia Italy
107	C1	Teruel Spain
92	H2	Tervola Fin.
109	C2	Tešanj Bos.-Herz.
116	B3	Teseney Eritrea
66	D2	Teshio Japan
66	D2	Teshio-gawa r. Japan
128	A1	Teslin Can.
128	A1	Teslin Lake Can.
154	B1	Tesouro Brazil
115	C3	Tessaoua Niger
99	C4	Test r. U.K.
121	C2	Tete Moz.
90	C1	Teteriv r. Ukr.
101	F1	Teterow Ger.
90	B2	Tetiyiv Ukr.
114	B1	Tétouan Morocco
110	B2	Tetovo Macedonia
		Tetyukhe Rus. Fed. see Dal'negorsk
		Teuchezhsk Rus. Fed. see Adygeysk
152	B2	Teuco r. Arg.
144	B2	Teul de González Ortega Mex.
101	D1	Teutoburger Wald hills Ger.
		Tevere r. Italy see Tiber
54	A3	Teviot N.Z.
96	C3	Teviot r. U.K.
96	C3	Teviothead U.K.
61	C2	Tewah Indon.
51	E2	Tewantin Austr.
54	C2	Te Wharau N.Z.
99	B4	Tewkesbury U.K.
143	E2	Texarkana U.S.A.
53	D1	Texas Austr.
143	D2	Texas state U.S.A.
143	E3	Texas City U.S.A.
145	C3	Texcoco Mex.
100	B1	Texel i. Neth.
143	D2	Texoma, Lake U.S.A.
123	C2	Teyateyaneng Lesotho
89	F2	Teykovo Rus. Fed.
89	F2	Teza r. Rus. Fed.
75	D2	Tezpur India
72	D2	Tezu India
129	E1	Tha-anne r. Can.
123	C2	Thabana-Ntlenyana mt. Lesotho
123	C2	Thaba Nchu S. Africa
123	C2	Thaba Putsoa mt. Lesotho
123	C1	Thaba-Tseka Lesotho
123	C1	Thabazimbi S. Africa
123	C2	Thabong S. Africa
63	A2	Thagyettaw Myanmar
62	B1	Thai Binh Vietnam
63	B2	Thailand country Asia
63	B2	Thailand, Gulf of Asia
62	B1	Thai Nguyên Vietnam
62	B2	Thakhèk Laos
74	B1	Thal Pak.
63	A3	Thalang Thai.
74	B1	Thal Desert Pak.
101	E2	Thale (Harz) Ger.
62	B2	Tha Li Thai.
53	C1	Thallon Austr.
123	C1	Thamaga Botswana
78	B3	Thamar, Jabal mt. Yemen
79	C3	Thamarīt Oman
130	B3	Thames r. Can.
54	C1	Thames N.Z.
99	D4	Thames est. U.K.
99	D4	Thames r. U.K.
79	B3	Thamūd Yemen
		Thana India see Thane
63	A2	Thanbyuzayat Myanmar
62	A1	Thandwè Myanmar
74	B3	Thane India
62	B1	Thanh Hoa Vietnam
73	B3	Thanjavur India
62	B1	Thanlyin Myanmar
74	A2	Thano Bula Khan Pak.
62	B1	Than Uyên Vietnam
74	A2	Thar Desert India/Pak.
52	B1	Thargomindah Austr.
81	C2	Tharthār, Buḩayrat ath l. Iraq

Tomazina

U

United Arab Emirates

79 C2 United Arab Emirates country Asia
United Arab Republic country Africa see Egypt
95 C3 United Kingdom country Europe
United Provinces state India see Uttar Pradesh
133 B3 United States of America country N. America
129 D2 Unity Can.
100 C2 Unna Ger.
96 □ Unst i. U.K.
101 E2 Unstrut r. Ger.
89 E3 Upa r. Rus. Fed.
119 C3 Upemba, Lac l. Dem. Rep. Congo
122 B2 Upington S. Africa
74 B2 Upleta India
49 J5 'Upolu i. Samoa
134 B2 Upper Alkali Lake U.S.A.
128 C2 Upper Arrow Lake Can.
54 C2 Upper Hutt N.Z.
134 B2 Upper Klamath Lake U.S.A.
128 B1 Upper Liard Can.
97 C1 Upper Lough Erne l. U.K.
137 E1 Upper Red Lake U.S.A.
Upper Tunguska r. Rus. Fed. see Angara
Upper Volta country Africa see Burkina
93 G4 Uppsala Sweden
78 B2 'Uqlat aş Şuqūr Saudi Arabia
Urad Qianqi China see Xishanzui
76 B2 Ural r. Kazakh./Rus. Fed.
53 D2 Uralla Austr.
87 E3 Ural Mountains Rus. Fed.
76 B1 Ural'sk Kazakh.
Ural'skiy Khrebet mts Rus. Fed. see Ural Mountains
119 D3 Urambo Tanz.
53 C3 Urana Austr.
129 D2 Uranium City Can.
86 F2 Uray Rus. Fed.
98 C2 Ure r. U.K.
86 D3 Uren' Rus. Fed.
82 G2 Urengoy Rus. Fed.
144 A2 Ures Mex.
Urfa Turkey see Şanlıurfa
76 C2 Urganch Uzbek.
100 B1 Urk Neth.
111 C3 Urla Turkey
110 C2 Urlaţi Romania
81 C2 Urmia Iran
81 C2 Urmia, Lake salt l. Iran
Uroševac Kosovo see Ferizaj
144 B2 Uruáchic Mex.
151 E4 Uruaçu Brazil
144 B3 Uruapan Mex.
150 B4 Urubamba r. Peru
151 D3 Urucara Brazil
151 E3 Uruçuí Brazil
151 E3 Uruçuí, Serra do hills Brazil
151 D3 Urucurituba Brazil
152 C2 Uruguaiana Brazil
153 C3 Uruguay country S. America
Urumchi China see Ürümqi
68 B2 Ürümqi China
Urundi country Africa see Burundi
53 D2 Urunga Austr.
119 D3 Uruwira Tanz.
110 C2 Urziceni Romania
67 B4 Usa Japan
86 E2 Usa r. Rus. Fed.
111 C3 Uşak Turkey
120 A3 Usakos Namibia
88 C2 Ushachy Belarus
82 G1 Ushakova, Ostrov i. Rus. Fed.
77 E2 Usharal Kazakh.
77 D2 Ushtobe Kazakh.
Ush-Tyube Kazakh. see Ushtobe
153 B5 Ushuaia Arg.
86 E2 Usinsk Rus. Fed.
99 B4 Usk r. U.K.
88 C3 Uskhodni Belarus
89 E3 Usman' Rus. Fed.
86 D2 Usogorsk Rus. Fed.
104 C2 Ussel France
66 C1 Ussuri r. China/Rus. Fed.
66 B2 Ussuriysk Rus. Fed.
Ust'-Abakanskoye Rus. Fed. see Abakan

Ust'-Balyk Rus. Fed. see Nefteyugansk
91 E2 Ust'-Donetskiy Rus. Fed.
108 B3 Ustica, Isola di i. Italy
83 H3 Ust'-Ilimsk Rus. Fed.
86 E2 Ust'-Ilych Rus. Fed.
102 C1 Ústí nad Labem Czech Rep.
Ustinov Rus. Fed. see Izhevsk
103 D1 Ustka Pol.
83 L3 Ust'-Kamchatsk Rus. Fed.
77 E2 Ust'-Kamenogorsk Kazakh.
86 F2 Ust'-Kara Rus. Fed.
86 F2 Ust'-Kulom Rus. Fed.
83 I3 Ust'-Kut Rus. Fed.
91 D2 Ust'-Labinsk Rus. Fed.
Ust'-Labinskaya Rus. Fed. see Ust'-Labinsk
88 C2 Ust'-Luga Rus. Fed.
86 E2 Ust'-Nem Rus. Fed.
83 K2 Ust'-Nera Rus. Fed.
83 I2 Ust'-Olenek Rus. Fed.
83 K2 Ust'-Omchug Rus. Fed.
83 H3 Ust'-Ordynskiy Rus. Fed.
103 E2 Ustrzyki Dolne Pol.
86 E2 Ust'-Tsil'ma Rus. Fed.
86 D2 Ust'-Ura Rus. Fed.
76 B2 Ustyurt Plateau Kazakh./Uzbek.
89 E2 Ustyuzhna Rus. Fed.
146 B3 Usulután El Salvador
Usumbura Burundi see Bujumbura
89 D2 Usvyaty Rus. Fed.
135 D3 Utah state U.S.A.
135 D2 Utah Lake U.S.A.
88 C2 Utena Lith.
119 D3 Utete Tanz.
63 B2 Uthai Thani Thai.
74 A2 Uthal Pak.
139 D2 Utica U.S.A.
107 C2 Utiel Spain
128 C2 Utikuma Lake Can.
93 G4 Utländan i. Sweden
100 B1 Utrecht Neth.
123 D2 Utrecht S. Africa
106 B2 Utrera Spain
92 I2 Utsjoki Fin.
67 C3 Utsunomiya Japan
87 D4 Utta Rus. Fed.
62 B2 Uttaradit Thai.
75 B1 Uttarakhand state India
75 B2 Uttar Pradesh state India
Uummannaq Greenland see Dundas
127 I2 Uummannaq Greenland
127 I2 Uummannaq Fjord inlet Greenland
93 H3 Uusikaupunki Fin.
120 A2 Uutapi Namibia
143 D3 Uvalde U.S.A.
119 D3 Uvinza Tanz.
123 D3 Uvongo S. Africa
68 C1 Uvs Nuur salt l. Mongolia
67 B4 Uwajima Japan
78 A2 'Uwayriḍ, Ḥarrat al lava field Saudi Arabia
116 A2 Uweinat, Jebel mt. Sudan
83 H3 Uyar Rus. Fed.
115 C4 Uyo Nigeria
152 B2 Uyuni Bol.
152 B2 Uyuni, Salar de salt flat Bol.
76 C2 Uzbekistan country Asia
Uzbekskaya S.S.R. country Asia see Uzbekistan
Uzbek S.S.R. country Asia see Uzbekistan
104 C2 Uzerche France
105 C3 Uzès France
90 C1 Uzh r. Ukr.
90 A2 Uzhhorod Ukr.
Uzhorod Ukr. see Uzhhorod
109 C2 Užice Serbia
89 E3 Uzlovaya Rus. Fed.
111 C3 Üzümlü Turkey
111 C2 Uzunköprü Turkey

V

123 B2 Vaal r. S. Africa
92 I3 Vaala Fin.
123 C2 Vaal Dam S. Africa
123 C1 Vaalwater S. Africa
92 H3 Vaasa Fin.
103 D2 Vác Hungary

152 C2 Vacaria Brazil
154 B2 Vacaria, Serra hills Brazil
135 B3 Vacaville U.S.A.
74 B2 Vadodara India
92 I1 Vadsø Norway
105 D2 Vaduz Liechtenstein
94 B1 Vágar i. Faroe Is
94 B1 Vágur Faroe Is
103 D2 Váh r. Slovakia
49 I4 Vaiaku Tuvalu
88 B2 Vaida Estonia
136 B3 Vail U.S.A.
77 C3 Vakhsh Tajik.
Vakhstroy Tajik. see Vakhsh
79 C2 Vakīlābād Iran
108 B1 Valdagno Italy
Valdai Hills hills Rus. Fed. see Valdayskaya Vozvyshennost'
89 D2 Valday Rus. Fed.
89 D2 Valdayskaya Vozvyshennost' hills Rus. Fed.
106 B2 Valdecañas, Embalse de resr Spain
93 G4 Valdemarsvik Sweden
106 C2 Valdepeñas Spain
153 B4 Valdés, Península pen. Arg.
126 C2 Valdez U.S.A.
153 A3 Valdivia Chile
130 C3 Val-d'Or Can.
141 D2 Valdosta U.S.A.
128 C2 Valemount Can.
152 E1 Valença Brazil
105 C3 Valence France
107 C2 Valencia Spain
107 C2 Valencia reg. Spain
150 C1 Valencia Venez.
107 D2 Valencia, Golfo de g. Spain
106 B1 Valencia de Don Juan Spain
97 A3 Valencia Island Ireland
105 C1 Valenciennes France
136 C2 Valentine U.S.A.
64 B2 Valenzuela Phil.
150 B2 Valera Venez.
88 C2 Valga Estonia
109 C2 Valjevo Serbia
88 C2 Valka Latvia
93 H3 Valkeakoski Fin.
100 B2 Valkenswaard Neth.
91 D2 Valky Ukr.
55 G2 Valkyrie Dome Antarctica
145 C2 Valladolid Mex.
106 C1 Valladolid Spain
93 E4 Valle Norway
145 C2 Vallecillos Mex.
150 C2 Valle de la Pascua Venez.
150 B1 Valledupar Col.
145 C2 Valle Hermoso Mex.
135 B3 Vallejo U.S.A.
152 A2 Vallenar Chile
84 F5 Valletta Malta
137 D1 Valley City U.S.A.
134 B2 Valley Falls U.S.A.
128 C2 Valleyview Can.
107 D1 Valls Spain
129 D3 Val Marie Can.
88 C2 Valmiera Latvia
104 B2 Valognes France
88 C3 Valozhyn Belarus
154 B2 Valparaíso Brazil
153 A3 Valparaíso Chile
105 C3 Valréas France
59 D3 Vals, Tanjung c. Indon.
74 B2 Valsad India
122 B2 Valspan S. Africa
91 D1 Valuyki Rus. Fed.
106 B2 Valverde del Camino Spain
81 C2 Van Turkey
81 C2 Van, Lake salt l. Turkey
81 C1 Vanadzor Armenia
83 H2 Vanavara Rus. Fed.
140 B1 Van Buren AR U.S.A.
139 F1 Van Buren ME U.S.A.
Van Buren U.S.A. see Kettering
128 B3 Vancouver Can.
134 B1 Vancouver U.S.A.
128 B3 Vancouver Island Can.
138 B3 Vandalia IL U.S.A.
138 C3 Vandalia OH U.S.A.
123 C2 Vanderbijlpark S. Africa
128 B2 Vanderhoof Can.
122 B3 Vanderkloof Dam dam S. Africa
50 C1 Van Diemen Gulf Austr.
88 C2 Vandra Estonia
Väner, Lake l. Sweden see Vänern

93 F4 Vänern l. Sweden
93 F4 Vänersborg Sweden
121 □D3 Vangaindrano Madag.
Van Gölü salt l. Turkey see Van, Lake
142 C2 Van Horn U.S.A.
59 D3 Vanimo P.N.G.
83 K3 Vanino Rus. Fed.
104 B2 Vannes France
Vannovka Kazakh. see Turar Ryskulov
59 D3 Van Rees, Pegunungan m Indon.
122 A3 Vanrhynsdorp S. Africa
93 H3 Vantaa Fin.
49 I5 Vanua Levu i. Fiji
48 H5 Vanuatu country S. Pacific Ocean
138 C2 Van Wert U.S.A.
122 B3 Vanwyksvlei S. Africa
122 B2 Van Zylsrus S. Africa
75 C2 Varanasi India
92 I1 Varangerfjorden sea chan. Norway
92 I1 Varangerhalvøya pen. Norway
109 C1 Varaždin Croatia
93 F4 Varberg Sweden
111 B3 Varda Greece
111 B2 Vardar r. Macedonia
93 E4 Varde Denmark
92 J1 Vardø Norway
101 J1 Varel Ger.
88 B3 Varėna Lith.
139 E1 Varennes Can.
108 A1 Varese Italy
155 C2 Varginha Brazil
93 I3 Varkaus Fin.
110 C2 Varna Bulg.
93 F4 Värnamo Sweden
155 D1 Várzea da Palma Brazil
86 C2 Varzino Rus. Fed.
Vasa Fin. see Vaasa
88 C3 Vasilyevichy Belarus
88 C2 Vasknarva Estonia
110 C1 Vaslui Romania
93 G4 Västerås Sweden
93 G3 Västerdalälven r. Sweden
88 A2 Västerhaninge Sweden
93 G4 Västervik Sweden
108 B2 Vasto Italy
91 D2 Vasylivka Ukr.
90 C1 Vasyl'kiv Ukr.
91 D2 Vasyl'kivka Ukr.
104 C2 Vatan France
111 B3 Vatheia Greece
108 B2 Vatican City Europe
92 □B3 Vatnajökull Iceland
110 C1 Vatra Dornei Romania
Vätter, Lake l. Sweden see Vättern
93 F4 Vättern l. Sweden
142 B2 Vaughn U.S.A.
105 C3 Vauvert France
121 □D2 Vavatenina Madag.
49 J5 Vava'u Group is Tonga
88 C3 Vawkavysk Belarus
93 F4 Växjö Sweden
Vayenga Rus. Fed. see Severomorsk
86 E1 Vaygach, Ostrov i. Rus. Fe
154 C1 Vazante Brazil
101 D1 Vechta Ger.
110 C2 Vedea r. Romania
100 C1 Veendam Neth.
100 B1 Veenendaal Neth.
92 F2 Vega i. Norway
128 C2 Vegreville Can.
106 B2 Vejer de la Frontera Spain
93 E4 Vejle Denmark
110 B2 Velbŭzhdki Prokhod pass Bulg./Macedonia
122 A3 Velddrif S. Africa
100 B2 Veldhoven Neth.
109 C2 Velebit mts Croatia
100 C2 Velen Ger.
109 C1 Velenje Slovenia
109 D2 Veles Macedonia
106 C2 Vélez-Málaga Spain
155 D1 Velhas r. Brazil
109 D2 Velika Plana Serbia
88 C2 Velikaya r. Rus. Fed.
89 D2 Velikiye Luki Rus. Fed.
89 D2 Velikiy Novgorod Rus. Fed.
86 D2 Velikiy Ustyug Rus. Fed.
110 C2 Veliko Tŭrnovo Bulg.
108 B2 Veli Lošinj Croatia

89 D2 Velizh Rus. Fed.
108 B2 Velletri Italy
73 B3 Vellore India
86 D2 Vel'sk Rus. Fed.
101 F1 Velten Ger.
91 D1 Velykyy Burluk Ukr.
Velykyy Tokmak Ukr. see Tokmak
59 E4 Vema Trench Indian Ocean
108 B2 Venafro Italy
154 C2 Venceslau Bráz Brazil
104 C2 Vendôme France
108 B1 Veneta, Laguna lag. Italy
89 E3 Venev Rus. Fed.
Venezia Italy see Venice
150 C2 Venezuela country S. America
150 B1 Venezuela, Golfo de g. Venez.
108 B1 Venice Italy
141 D3 Venice U.S.A.
108 B1 Venice, Gulf of Europe
100 C2 Venlo Neth.
93 E4 Vennesla Norway
100 B2 Venray Neth.
88 B2 Venta r. Latvia/Lith.
88 B2 Venta Lith.
123 C2 Ventersburg S. Africa
123 C3 Venterstad S. Africa
108 A2 Ventimiglia Italy
99 C4 Ventnor U.K.
88 B2 Ventspils Latvia
135 C4 Ventura U.S.A.
143 C3 Venustiano Carranza, Presa resr Mex.
107 C2 Vera Spain
154 C2 Vera Cruz Brazil
45 C3 Veracruz Mex.
74 B2 Veraval India
108 A1 Verbania Italy
108 A1 Vercelli Italy
105 D3 Vercors reg. France
92 F3 Verdalsøra Norway
154 B1 Verde r. Brazil
154 B1 Verde r. Brazil
144 B2 Verde r. Mex.
55 D1 Verde Grande r. Brazil
101 D1 Verden (Aller) Ger.
154 B1 Verdinho, Serra do mts Brazil
105 D2 Verdon r. France
105 D2 Verdun France
23 C2 Vereeniging S. Africa
106 B1 Verín Spain
91 D3 Verkhnebakanskiy Rus. Fed.
89 D3 Verkhnedneprovskiy Rus. Fed.
92 J2 Verkhnetulomskiy Rus. Fed.
92 J2 Verkhnetulomskoye Vodokhranilishche resr Rus. Fed.
87 D4 Verkhniy Baskunchak Rus. Fed.
91 E1 Verkhniy Mamon Rus. Fed.
86 D2 Verkhnyaya Toyma Rus. Fed.
89 E3 Verkhov'ye Rus. Fed.
90 A2 Verkhovyna Ukr.
83 J2 Verkhoyanskiy Khrebet mts Rus. Fed.
29 C2 Vermilion Can.
37 D2 Vermillion U.S.A.
30 A3 Vermillion Bay Can.
39 E2 Vermont state U.S.A.
35 E2 Vernal U.S.A.
22 B2 Verneuk Pan salt pan S. Africa
28 C2 Vernon Can.
43 D2 Vernon U.S.A.
41 D3 Vero Beach U.S.A.
11 B2 Veroia Greece
108 B1 Verona Italy
104 C2 Versailles France
104 B2 Vertou France
23 C2 Verulam S. Africa
105 C2 Verviers Belgium
105 C2 Vervins France
105 D3 Vescovato France
87 E3 Veselaya, Gora mt. Rus. Fed.
91 E2 Vesele Ukr.
91 E2 Veselyy Rus. Fed.
102 C2 Vesoul France
40 C2 Vestavia Hills U.S.A.
92 F2 Vesterålen is Norway
92 F2 Vestfjorden sea chan. Norway

94 B1 Vestmanna Faroe Is
92 □A3 Vestmannaeyjar Iceland
92 □A3 Vestmannaeyjar is Iceland
93 E3 Vestnes Norway
Vesuvio vol. Italy see Vesuvius
108 B2 Vesuvius vol. Italy
89 E2 Ves'yegonsk Rus. Fed.
103 D2 Veszprém Hungary
93 G4 Vetlanda Sweden
86 D3 Vetluga Rus. Fed.
86 D3 Vetluzhskiy Rus. Fed.
100 A2 Veurne Belgium
119 D2 Veveno r. Sudan
105 D2 Vevey Switz.
91 D1 Veydelevka Rus. Fed.
80 B1 Vezirköprü Turkey
Vialar Alg. see Tissemsilt
152 C3 Viamao Brazil
151 E3 Viana Brazil
106 B1 Viana do Castelo Port.
Viangchan Laos see Vientiane
62 B1 Viangphoukha Laos
111 C3 Viannos Greece
154 C1 Vianópolis Brazil
108 B2 Viareggio Italy
93 E4 Viborg Denmark
Viborg Rus. Fed. see Vyborg
109 C3 Vibo Valentia Italy
107 D1 Vic Spain
144 A1 Vicente Guerrero Mex.
108 B1 Vicenza Italy
89 F2 Vichuga Rus. Fed.
105 C2 Vichy France
140 B2 Vicksburg U.S.A.
155 D2 Viçosa Brazil
52 A3 Victor Harbor Austr.
50 C1 Victoria r. Austr.
52 B3 Victoria state Austr.
Victoria Cameroon see Limbe
128 B3 Victoria Can.
153 A3 Victoria Chile
Victoria Malaysia see Labuan
113 I6 Victoria Seychelles
143 D3 Victoria U.S.A.
119 D3 Victoria, Lake Africa
52 B2 Victoria, Lake Austr.
62 A1 Victoria, Mount Myanmar
59 D3 Victoria, Mount P.N.G.
154 B3 Victoria, Sierra de la hills Arg.
120 B2 Victoria Falls waterfall Zambia/Zimbabwe
120 B2 Victoria Falls Zimbabwe
126 E2 Victoria Island Can.
55 M2 Victoria Land coastal area Antarctica
50 C1 Victoria River Downs Austr.
139 E1 Victoriaville Can.
122 B3 Victoria West S. Africa
135 C4 Victorville U.S.A.
141 D2 Vidalia U.S.A.
110 C2 Videle Romania
92 □A2 Víðidalsá Iceland
110 B2 Vidin Bulg.
74 B2 Vidisha India
140 B2 Vidor U.S.A.
153 B4 Viedma Arg.
153 A4 Viedma, Lago l. Arg.
102 C2 Viehberg mt. Austria
107 D1 Vielha Spain
100 B2 Vielsalm Belgium
101 E2 Vienenburg Ger.
103 D2 Vienna Austria
138 C3 Vienna U.S.A.
105 C2 Vienne France
104 C2 Vienne r. France
62 B2 Vientiane Laos
100 C2 Viersen Ger.
104 C2 Vierzon France
144 B2 Viesca Mex.
109 C2 Vieste Italy
62 B2 Vietnam country Asia
62 B1 Viêt Tri Vietnam
64 B2 Vigan Phil.
108 A1 Vigevano Italy
106 B1 Vigo Spain
Viipuri Rus. Fed. see Vyborg
73 C3 Vijayawada India
92 □B3 Vík Iceland
111 B2 Vikhren mt. Bulg.
128 C2 Viking Can.
92 F3 Vikna i. Norway
Vila Alferes Chamusca Moz. see Guija

Vila Arriaga Angola see Bibala
Vila Bugaço Angola see Camanongue
Vila Cabral Moz. see Lichinga
Vila da Ponte Angola see Kuvango
Vila de Aljustrel Angola see Cangamba
Vila de Almoster Angola see Chiange
Vila de João Belo Moz. see Xai-Xai
Vila de Trego Morais Moz. see Chókwé
106 B2 Vila Fontes Moz. see Caia
106 B1 Vila Franca de Xira Port.
123 D1 Vilagarcía de Arousa Spain
106 B1 Vila Gomes da Costa Moz.
Vilalba Spain see
Vila Luísa Moz. see Marracuene
Vila Marechal Carmona Angola see Uíge
Vila Miranda Moz. see Macaloge
121 □D2 Vilanandro, Tanjona c. Madag.
88 C2 Vilāni Latvia
106 B1 Vila Nova de Gaia Port.
107 D1 Vilanova i la Geltrú Spain
Vila Paiva de Andrada Moz. see Gorongosa
Vila Pery Moz. see Chimoio
106 B1 Vila Real Port.
106 B1 Vilar Formoso Port.
Vila Salazar Angola see N'dalatando
Vila Salazar Zimbabwe see Sango
Vila Teixeira de Sousa Angola see Luau
155 D2 Vila Velha Brazil
150 B4 Vilcabamba, Cordillera mts Peru
82 F1 Vil'cheka, Zemlya i. Rus. Fed.
92 G3 Vilhelmina Sweden
150 C4 Vilhena Brazil
88 C2 Viljandi Estonia
123 C2 Viljoenskroon S. Africa
88 B3 Vilkaviškis Lith.
83 H1 Vil'kitskogo, Proliv str. Rus. Fed.
144 B1 Villa Ahumada Mex.
106 B1 Villablino Spain
102 C2 Villach Austria
Villa Cisneros Western Sahara see Ad Dakhla
144 B2 Villa de Cos Mex.
152 B3 Villa Dolores Arg.
145 C3 Villa Flores Mex.
153 C3 Villa Gesell Arg.
145 C2 Villagrán Mex.
145 C3 Villahermosa Mex.
144 A2 Villa Insurgentes Mex.
107 C2 Villajoyosa-La Vila Joiosa Spain
152 B3 Villa María Arg.
153 B3 Villa Mercedes Arg.
152 B2 Villa Montes Bol.
144 B2 Villanueva Mex.
106 B2 Villanueva de la Serena Spain
106 C2 Villanueva de los Infantes Spain
152 B3 Villa Ocampo Arg.
142 B3 Villa Ocampo Mex.
108 A3 Villaputzu Italy
152 C2 Villarrica Para.
106 C2 Villarrobledo Spain
Villasalazar Zimbabwe see Sango
152 B2 Villa Unión Arg.
144 B2 Villa Unión Mex.
144 B2 Villa Unión Mex.
150 B2 Villavicencio Col.
152 B2 Villazon Bol.
104 C3 Villefranche-de-Rouergue France
105 C2 Villefranche-sur-Saône France
107 C2 Villena Spain
100 A2 Villeneuve-d'Ascq France
104 C3 Villeneuve-sur-Lot France

140 B2 Ville Platte U.S.A.
100 A3 Villers-Cotterêts France
105 C2 Villeurbanne France
123 C2 Villiers S. Africa
102 B2 Villingen Ger.
137 E2 Villisca U.S.A.
88 C3 Vilnius Lith.
91 C2 Vil'nohirs'k Ukr.
91 D2 Vil'nyans'k Ukr.
100 B2 Vilvoorde Belgium
88 C3 Vilyeyka Belarus
83 J2 Vilyuy r. Rus. Fed.
93 G4 Vimmerby Sweden
153 A3 Viña del Mar Chile
107 D1 Vinaròs Spain
138 B3 Vincennes U.S.A.
55 J3 Vincennes Bay Antarctica
139 D3 Vineland U.S.A.
62 B2 Vinh Vietnam
63 B2 Vinh Long Vietnam
143 D1 Vinita U.S.A.
90 B2 Vinnytsya Ukr.
55 R2 Vinson Massif mt. Antarctica
93 E3 Vinstra Norway
105 E2 Vipiteno Italy
64 B2 Virac Phil.
74 B2 Viramgam India
80 B2 Viranşehir Turkey
129 D3 Virden Can.
104 B2 Vire France
120 A2 Virei Angola
155 D1 Virgem da Lapa Brazil
142 A1 Virgin r. U.S.A.
123 C2 Virginia S. Africa
137 E1 Virginia U.S.A.
139 D3 Virginia state U.S.A.
139 D3 Virginia Beach U.S.A.
135 C3 Virginia City U.S.A.
147 D3 Virgin Islands (U.K.) terr. West Indies
147 D3 Virgin Islands (U.S.A.) terr. West Indies
63 B2 Viróchey Cambodia
109 C1 Virovitica Croatia
100 B3 Virton Belgium
88 B2 Virtsu Estonia
73 B4 Virudhunagar India
109 C2 Vis i. Croatia
88 C2 Visaginas Lith.
135 C3 Visalia U.S.A.
74 B2 Visavadar India
64 B2 Visayan Sea Phil.
93 G4 Visby Sweden
126 E2 Viscount Melville Sound sea chan. Can.
151 E3 Viseu Brazil
106 B1 Viseu Port.
110 B1 Vişeu de Sus Romania
73 C3 Vishakhapatnam India
88 C2 Viški Latvia
103 D1 Vistula r. Pol.
Vitebsk Belarus see Vitsyebsk
108 B2 Viterbo Italy
49 I5 Viti Levu i. Fiji
83 I3 Vitim r. Rus. Fed.
155 D2 Vitória Brazil
151 E4 Vitória da Conquista Brazil
106 C1 Vitoria-Gasteiz Spain
104 B2 Vitré France
105 C2 Vitry-le-François France
89 D2 Vitsyebsk Belarus
105 D2 Vittel France
108 B3 Vittoria Italy
108 B1 Vittorio Veneto Italy
106 B1 Viveiro Spain
136 C2 Vizagapatam India see Vishakhapatnam
142 A3 Vizcaíno, Desierto de des. Mex.
144 A2 Vizcaíno, Sierra mts Mex.
111 C2 Vize Turkey
73 C3 Vizianagaram India
100 B2 Vlaardingen Neth.
87 D4 Vladikavkaz Rus. Fed.
89 F2 Vladimir Rus. Fed.
66 B2 Vladivostok Rus. Fed.
109 D2 Vlasotince Serbia
100 B1 Vlieland i. Neth.
100 A2 Vlissingen Neth.
109 C2 Vlorë Albania
102 C1 Vltava r. Czech Rep.
102 C2 Vöcklabruck Austria
109 C2 Vodice Croatia

41 D2	Waynesboro GA U.S.A.
39 D3	Waynesboro VA U.S.A.
37 E3	Waynesville MO U.S.A.
41 D1	Waynesville NC U.S.A.
74 B1	Wazirabad Pak.
50 A1	We, Pulau i. Indon.
98 C2	Wear r. U.K.
43 D1	Weatherford OK U.S.A.
43 D2	Weatherford TX U.S.A.
34 B2	Weaverville U.S.A.
43 D3	Webb U.S.A.
30 B2	Webequie Can.
37 D1	Webster U.S.A.
37 E2	Webster City U.S.A.
55 C3	Weddell Abyssal Plain Southern Ocean
55 B3	Weddell Sea Antarctica
23 D2	Weenen S. Africa
90 B2	Weert Neth.
53 C2	Weethalle Austr.
53 C2	Wee Waa Austr.
00 C2	Wegberg Ger.
03 E1	Węgorzewo Pol.
03 E1	Węgrów Pol.
70 B2	Weichang China Qianjin
70 B2	Weifang China
70 C2	Weihai China
70 B2	Wei He r. China
53 C1	Weilmoringle Austr.
01 E2	Weimar Ger.
62 B1	Weinan China
43 D3	Weiner U.S.A.
70 A2	Weinan China
71 A3	Weining China
51 D1	Weipa Austr.
53 C1	Weir r. Austr.
38 C2	Weirton U.S.A.
62 B1	Weishan China
01 E2	Weiße Elster r. Ger.
01 E2	Weißenfels Ger.
02 C2	Weißkugel mt. Austria/Italy
71 A3	Weixin China Weizhou China see Wenchuan
03 D1	Wejherowo Pol.
28 C1	Wekweètì Can.
38 C3	Welch U.S.A.
17 B3	Weldiya Eth.
23 C2	Welkom S. Africa
99 C3	Welland r. U.K.
51 C1	Wellesley Islands Austr.
99 C3	Wellingborough U.K.
53 C2	Wellington Austr.
54 B2	Wellington N.Z.
22 A3	Wellington S. Africa
36 B2	Wellington CO U.S.A.
37 D3	Wellington KS U.S.A.
35 D3	Wellington UT U.S.A.
53 A4	Wellington, Isla i. Chile
53 C3	Wellington, Lake Austr.
28 B2	Wells Can.
99 B4	Wells U.K.
34 D2	Wells U.S.A.
50 B2	Wells, Lake imp. l. Austr.
54 B1	Wellsford N.Z.
99 D3	Wells-next-the-Sea U.K.
42 A2	Wellton U.S.A.
02 C2	Wels Austria
99 B3	Welshpool U.K.
23 C2	Wembesi S. Africa
30 C2	Wemindji Can.
34 B1	Wenatchee U.S.A.
71 B4	Wenchang China
44 B4	Wenchi Ghana Wenchow China see Wenzhou
70 A2	Wenchuan China
70 C2	Wendeng China
01 E1	Wendisch Evern Ger.
17 B4	Wendo Eth.
35 D2	Wendover U.S.A.
71 B3	Wengyuan China Wenhua China see Weishan Wenlan China see Mengzi Wenlin China see Renshou
71 C3	Wenling China Wenquan China see Yingshan
71 A3	Wenshan China
52 B2	Wentworth Austr.
71 C3	Wenxian China
23 C2	Wepener S. Africa
22 B2	Werda Botswana
101 F2	Werdau Ger.
101 F1	Werder Ger.
101 F3	Wernberg-Köblitz Ger.
101 E2	Wernigerode Ger.
101 D2	Werra r. Ger.
52 B2	Werrimull Austr.
53 D2	Werris Creek Austr.
101 D3	Wertheim Ger.
100 C2	Wesel Ger.
101 E1	Wesendorf Ger.
101 D1	Weser r. Ger.
101 D1	Weser sea chan. Ger.
51 C1	Wessel, Cape Austr.
51 C1	Wessel Islands Austr.
123 C2	Wesselton S. Africa
55 P2	West Antarctica reg. Antarctica
156 B7	West Australian Basin Indian Ocean
80 B2	West Bank terr. Asia
138 B2	West Bend U.S.A.
75 C2	West Bengal state India
99 C3	West Bromwich U.K.
139 E2	Westbrook U.S.A.
137 E2	West Des Moines U.S.A.
100 C2	Westerburg Ger.
100 C1	Westerholt Ger.
139 E2	Westerly U.S.A.
50 B2	Western Australia state Austr.
122 B3	Western Cape prov. S. Africa
116 A2	Western Desert Egypt Western Dvina r. Europe see Zapadnaya Dvina
73 B3	Western Ghats mts India
114 A2	Western Sahara terr. Africa Western Samoa country S. Pacific Ocean see Samoa Western Sayan Mountains reg. Rus. Fed. see Zapadnyy Sayan
100 A2	Westerschelde est. Neth.
100 C1	Westerstede Ger.
101 C2	Westerwald hills Ger.
153 D5	West Falkland i. Falkland Is
138 B3	West Frankfort U.S.A.
100 B1	West Frisian Islands Neth.
96 C2	Westhill U.K.
55 I3	West Ice Shelf Antarctica
147 D2	West Indies is Caribbean Sea
100 A2	Westkapelle Neth.
96 A1	West Loch Roag b. U.K.
128 C2	Westlock Can.
100 B2	Westmalle Belgium Westman Islands is Iceland see Vestmannaeyjar
53 C1	Westmar Austr.
156 C4	West Mariana Basin N. Pacific Ocean
140 B1	West Memphis U.S.A.
138 C3	Weston U.S.A.
99 B4	Weston-super-Mare U.K.
141 D3	West Palm Beach U.S.A.
137 E3	West Plains U.S.A.
137 D2	West Point U.S.A.
54 B2	Westport N.Z.
97 B2	Westport Ireland
129 D2	Westray Can.
96 C1	Westray i. U.K.
82 G2	West Siberian Plain plain Rus. Fed.
100 B1	West-Terschelling Neth.
136 A2	West Thumb U.S.A. West Town Ireland see An Baile Thiar
135 D2	West Valley City U.S.A.
138 C3	West Virginia state U.S.A.
53 C2	West Wyalong Austr.
134 D2	West Yellowstone U.S.A.
59 C3	Wetar i. Indon.
128 C2	Wetaskiwin Can.
119 D3	Wete Tanz.
101 D2	Wetzlar Ger.
59 D3	Wewak P.N.G.
97 C2	Wexford Ireland
97 C2	Wexford Harbour b. Ireland
129 D3	Weyakwin Can.
129 D3	Weyburn Can.
101 D1	Weyhe Ger.
99 B4	Weymouth U.K.
54 C1	Whakatane N.Z.
129 E1	Whale Cove Can.
96 □	Whalsay i. U.K.
54 B1	Whangamomona N.Z.
54 B1	Whangaparaoa N.Z.
54 B1	Whangarei N.Z.
98 C3	Wharfe r. U.K.
143 D3	Wharton U.S.A.
128 C1	Whati Can.
136 B2	Wheatland U.S.A.
138 B2	Wheaton U.S.A.
140 C2	Wheeler Lake resr U.S.A.
142 B1	Wheeler Peak NM U.S.A.
135 D3	Wheeler Peak NV U.S.A.
138 C2	Wheeling U.S.A.
98 B2	Whernside h. U.K.
128 B2	Whistler Can.
98 C2	Whitby U.K.
128 A1	White r. Can./U.S.A.
140 B2	White r. AR U.S.A.
138 B3	White r. IN U.S.A.
50 B2	White, Lake imp. l. Austr.
131 E3	White Bay Can.
136 C1	White Butte mt. U.S.A.
52 B2	White Cliffs Austr.
128 C2	Whitecourt Can.
134 D1	Whitefish U.S.A.
98 B2	Whitehaven U.K.
97 D1	Whitehead U.K.
128 A1	Whitehorse Can.
140 B3	White Lake U.S.A.
51 D4	Whitemark Austr.
135 C3	White Mountain Peak U.S.A.
116 B3	White Nile r. Sudan/Uganda White Russia country Europe see Belarus
86 C2	White Sea Rus. Fed.
134 D1	White Sulphur Springs U.S.A.
141 E2	Whiteville U.S.A.
114 B3	White Volta r. Burkina/Ghana
136 B3	Whitewater U.S.A.
142 B2	Whitewater Baldy mt. U.S.A.
130 B2	Whitewater Lake Can.
129 D2	Whitewood Can.
96 B3	Whithorn U.K.
54 C1	Whitianga N.Z.
135 C3	Whitney, Mount U.S.A.
99 D4	Whitstable U.K.
51 D2	Whitsunday Island Austr.
53 C3	Whittlesea Austr.
52 A2	Whyalla Austr.
62 A2	Wiang Pa Pao Thai.
100 A2	Wichelen Belgium
137 D3	Wichita U.S.A.
143 D2	Wichita Falls U.S.A.
143 D2	Wichita Mountains U.S.A.
96 C1	Wick U.K.
142 A2	Wickenburg U.S.A.
97 C2	Wicklow Ireland
97 D2	Wicklow Head hd Ireland
97 C2	Wicklow Mountains Ireland
98 B3	Widnes U.K.
101 D1	Wiehengebirge hills Ger.
100 C2	Wiehl Ger.
103 D1	Wieluń Pol. Wien Austria see Vienna
103 D2	Wiener Neustadt Austria
100 B1	Wieringerwerf Neth.
101 D2	Wiesbaden Ger.
100 C1	Wiesmoor Ger.
103 D1	Wieżyca h. Pol.
98 B3	Wigan U.K.
99 C3	Wight, Isle of i. U.K.
96 B3	Wigtown U.K.
100 B2	Wijchen Neth.
130 B3	Wikwemikong Can.
52 B2	Wilcannia Austr. Wilczek Land i. Rus. Fed. see Vil'cheka, Zemlya
123 C3	Wild Coast S. Africa
101 D1	Wildeshausen Ger.
136 C2	Wild Horse Hill mt. U.S.A.
123 C2	Wilge r. S. Africa
48 F4	Wilhelm, Mount P.N.G.
101 D1	Wilhelmshaven Ger.
139 D2	Wilkes-Barre U.S.A.
55 K3	Wilkes Land reg. Antarctica
129 D2	Wilkie Can.
134 B1	Willamette r. U.S.A.
134 B1	Willapa Bay U.S.A.
142 B2	Willcox U.S.A.
100 B2	Willebroek Belgium
147 D3	Willemstad Neth. Antilles
52 B3	William, Mount Austr.
52 A1	William Creek Austr.
142 A1	Williams U.S.A.
138 C3	Williamsburg KY U.S.A.
139 D3	Williamsburg VA U.S.A.
128 B2	Williams Lake Can.
138 C3	Williamson U.S.A.
139 D2	Williamsport U.S.A.
141 E1	Williamston U.S.A.
122 B3	Williston S. Africa
136 C1	Williston U.S.A.
128 B2	Williston Lake Can.
135 B3	Willits U.S.A.
137 D1	Willmar U.S.A.
122 B3	Willowmore S. Africa
135 B3	Willows U.S.A.
123 C3	Willowvale S. Africa
50 B2	Wills, Lake imp. l. Austr.
52 A3	Willunga Austr.
52 A2	Wilmington Austr.
139 D3	Wilmington DE U.S.A.
141 E2	Wilmington NC U.S.A.
138 C3	Wilmington OH U.S.A.
141 D2	Wilmington Island U.S.A. Wilno Lith. see Vilnius
101 D2	Wilsdorf Ger.
101 D1	Wilseder Berg h. Ger.
141 E1	Wilson U.S.A.
53 C3	Wilson's Promontory pen. Austr.
100 B3	Wiltz Lux.
50 B2	Wiluna Austr.
99 D4	Wimereux France
123 C2	Winburg S. Africa
99 B4	Wincanton U.K.
99 C4	Winchester U.K.
138 C3	Winchester KY U.S.A.
139 D3	Winchester VA U.S.A.
98 B2	Windermere l. U.K.
122 A1	Windhoek Namibia
137 D2	Windom U.S.A.
51 D2	Windorah Austr.
136 B2	Wind River Range mts U.S.A.
53 D2	Windsor Austr.
130 B3	Windsor Can.
147 D3	Windward Islands Caribbean Sea
147 C3	Windward Passage Cuba/Haiti
137 D3	Winfield U.S.A.
100 A2	Wingene Belgium
53 D2	Wingham Austr.
130 B2	Winisk r. Can.
130 B2	Winisk (abandoned) Can.
130 B2	Winisk Lake Can.
63 A2	Winkana Myanmar
129 E3	Winkler Can.
114 B4	Winneba Ghana
138 B2	Winnebago, Lake U.S.A.
134 C2	Winnemucca U.S.A.
136 C2	Winner U.S.A.
140 B2	Winnfield U.S.A.
137 E1	Winnibigoshish, Lake U.S.A.
129 E3	Winnipeg Can.
129 E2	Winnipeg r. Can.
129 E2	Winnipeg, Lake Can.
129 D2	Winnipegosis, Lake Can.
139 E2	Winnipesaukee, Lake U.S.A.
140 B2	Winnsboro U.S.A.
137 E2	Winona MN U.S.A.
140 C2	Winona MS U.S.A.
100 C1	Winschoten Neth.
101 D1	Winsen (Aller) Ger.
101 E1	Winsen (Luhe) Ger.
142 A1	Winslow U.S.A.
141 D1	Winston-Salem U.S.A.
101 D2	Winterberg Ger.
141 D3	Winter Haven U.S.A.
100 C2	Winterswijk Neth.
105 D2	Winterthur Switz.
51 D2	Winton Austr.
54 A3	Winton N.Z.
52 A2	Wirrabara Austr.
52 A2	Wirraminna Austr.
99 D3	Wisbech U.K.
138 A2	Wisconsin r. U.S.A.
138 B2	Wisconsin state U.S.A.
138 B2	Wisconsin Rapids U.S.A. Wisła r. Pol. see Vistula
101 E1	Wismar Ger.
123 C2	Witbank S. Africa
122 A2	Witbooisvlei Namibia
98 D3	Witham r. U.K.
98 D3	Witham U.K.
100 B1	Witmarsum Neth.
99 C4	Witney U.K.
123 C2	Witrivier S. Africa
123 C3	Witteberg mts S. Africa
101 F2	Wittenberg, Lutherstadt Ger.
101 E1	Wittenberge Ger.
101 E1	Wittenburg Ger.
50 A2	Wittenoom Austr.
101 E1	Wittingen Ger.

Wittlich

119 C2 Zémio C.A.R.
107 D2 Zemmora Alg.
145 C3 Zempoaltépetl, Nudo de mt. Mex.
109 D2 Zemun Serbia
65 B1 Zengfeng Shan mt. China
109 C2 Zenica Bos.-Herz.
107 D2 Zenzach Alg.
101 F2 Zerbst Ger.
76 C4 Zereh, Gowd-e depr. Afgh.
105 D2 Zermatt Switz.
91 E2 Zernograd Rus. Fed.
Zernovoy Rus. Fed. see Zernograd
101 E2 Zeulenroda Ger.
101 D1 Zeven Ger.
100 C2 Zevenaar Neth.
100 B2 Zevenbergen Neth.
83 J3 Zeya Rus. Fed.
79 C2 Zeydābād Iran
79 C2 Zeynālābād Iran
83 J3 Zeyskoye Vodokhranilishche resr Rus. Fed.
103 D1 Zgierz Pol.
88 B3 Zhabinka Belarus
Zhabye Ukr. see Verkhovyna
Zhaksy Sarysu watercourse Kazakh. see Sarysu
76 A2 Zhalpaktal Kazakh.
77 C1 Zhaltyr Kazakh.
Zhambyl Kazakh. see Taraz
76 B2 Zhanakala Kazakh.
76 B2 Zhanaozen Kazakh.
Zhangaqazaly Kazakh. see Ayteke Bi
Zhangde China see Anyang
Zhangdian China see Zibo
66 A1 Zhangguangcai Ling mts China
71 B3 Zhangjiajie China
70 B1 Zhangjiakou China
71 B3 Zhangping China
71 B3 Zhangshu China
65 A1 Zhangwu China
70 A2 Zhangxian China
68 C2 Zhangye China
71 B3 Zhangzhou China
76 A2 Zhanibek Kazakh.
71 B3 Zhanjiang China
71 B3 Zhao'an China
69 E1 Zhaodong China
Zhaoge China see Qixian
71 B3 Zhaoqing China

71 A3 Zhaotong China
75 C1 Zhari Namco salt l. China
77 E2 Zharkent Kazakh.
89 D2 Zharkovskiy Rus. Fed.
77 E2 Zharma Kazakh.
90 C2 Zhashkiv Ukr.
Zhaxi China see Weixin
77 D2 Zhayrem Kazakh.
Zhdanov Ukr. see Mariupol'
71 C3 Zhejiang prov. China
82 F1 Zhelaniya, Mys c. Rus. Fed.
Zheleznodorozhnyy Rus. Fed. see Yemva
Zheleznodorozhnyy Uzbek. see Qo'ng'irot
89 E3 Zheleznogorsk Rus. Fed.
Zheltye Vody Ukr. see Zhovti Vody
76 B2 Zhem r. Kazakh.
70 A2 Zhen'an China
70 A2 Zhenba China
71 A3 Zheng'an China
71 B3 Zhenghe China
70 B2 Zhengzhou China
70 B2 Zhenjiang China
Zhenjiang China see Zhenjiang
71 A3 Zhenyuan China
91 E1 Zherdevka Rus. Fed.
86 D2 Zheshart Rus. Fed.
77 D2 Zhetysuskiy Alatau mts China/Kazakh.
77 C2 Zhezkazgan Kazakh.
77 C2 Zhezkazgan Kazakh.
83 J2 Zhigansk Rus. Fed.
70 B2 Zhijiang China
76 C1 Zhitikara Kazakh.
89 D3 Zhizdra Rus. Fed.
88 D3 Zhlobin Belarus
90 B2 Zhmerynka Ukr.
74 A1 Zhob Pak.
88 C3 Zhodzina Belarus
83 L1 Zhokhova, Ostrov i. Rus. Fed.
Zholkva Ukr. see Zhovkva
Zhongba China see Jiangyou
75 C2 Zhongba China
Zhongduo China see Youyang
Zhonghe China see Xiushan
70 A2 Zhongning China
Zhongping China see Huize

71 B3 Zhongshan China
Zhongshan China see Lupanshui
70 A2 Zhongwei China
Zhongxin China see Xangyi'nyilha
76 C2 Zhosaly Kazakh.
70 B2 Zhoukou China
70 C2 Zhoushan China
90 A1 Zhovkva Ukr.
91 C2 Zhovti Vody Ukr.
65 A2 Zhuanghe China
70 B2 Zhucheng China
89 D3 Zhukovka Rus. Fed.
89 E2 Zhukovskiy Rus. Fed.
70 B2 Zhumadian China
Zhuoyang China see Suiping
71 B3 Zhuzhou Hunan China
71 B3 Zhuzhou Hunan China
90 A2 Zhydachiv Ukr.
88 C3 Zhytkavichy Belarus
90 B1 Zhytomyr Ukr.
103 D2 Žiar nad Hronom Slovakia
70 B2 Zibo China
103 D1 Zielona Góra Pol.
100 A2 Zierikzee Neth.
62 A1 Zigaing Myanmar
115 E2 Zighan Libya
71 A3 Zigong China
Zigui China see Guojiaba
114 A3 Ziguinchor Senegal
144 B3 Zihuatanejo Mex.
103 D2 Žilina Slovakia
115 D2 Zillah Libya
83 H3 Zima Rus. Fed.
145 C2 Zimapán Mex.
121 B2 Zimbabwe country Africa
114 A4 Zimmi Sierra Leone
110 C2 Zimnicea Romania
86 C2 Zimniy Bereg coastal area Rus. Fed.
115 C3 Zinder Niger
78 B3 Zinjibār Yemen
91 C1 Zin'kiv Ukr.
Zinoyevsk Ukr. see Kirovohrad
150 B2 Zipaquirá Col.
103 D2 Zirc Hungary
75 D2 Ziro India
79 C2 Zīr Rūd Iran
103 D2 Zistersdorf Austria
145 B3 Zitácuaro Mex.
103 C1 Zittau Ger.

87 E3 Zlatoust Rus. Fed.
103 D2 Zlín Czech Rep.
115 D1 Zlīţan Libya
103 D1 Złotów Pol.
89 D3 Zlynka Rus. Fed.
89 E3 Zmiyevka Rus. Fed.
91 D2 Zmiyiv Ukr.
89 E3 Znamenka Rus. Fed.
91 E1 Znamenka Rus. Fed.
91 C2 Znam"yanka Ukr.
103 D2 Znojmo Czech Rep.
122 B3 Zoar S. Africa
70 A2 Zoigê China
91 D1 Zolochiv Ukr.
90 A2 Zolochiv Ukr.
91 C2 Zolotonosha Ukr.
89 E3 Zolotukhino Rus. Fed.
121 C2 Zomba Malawi
118 B2 Zongo Dem. Rep. Congo
80 B1 Zonguldak Turkey
105 D3 Zonza France
114 B3 Zorgho Burkina
114 B4 Zorzor Liberia
115 D2 Zouar Chad
114 A2 Zouérat Maur.
109 D1 Zrenjanin Serbia
89 D2 Zubtsov Rus. Fed.
105 D2 Zug Switz.
81 C1 Zugdidi Georgia
Zuider Zee l. Neth. see IJsselmeer
106 B2 Zújar r. Spain
100 C2 Zülpich Ger.
100 A2 Zulte Belgium
121 C2 Zumbo Moz.
145 C3 Zumpango Mex.
142 B1 Zuni Mountains U.S.A.
71 A3 Zunyi China
109 C1 Županja Croatia
105 D2 Zürich Switz.
Zürichsee l. Switz.
100 C1 Zutphen Neth.
115 D1 Zuwārah Libya
90 C2 Zvenyhorodka Ukr.
121 C3 Zvishavane Zimbabwe
103 D2 Zvolen Slovakia
109 C2 Zvornik Bos.-Herz.
114 B4 Zwedru Liberia
123 C3 Zwelitsha S. Africa
103 D2 Zwettl Austria
101 F2 Zwickau Ger.
100 C1 Zwolle Neth.
83 L2 Zyryanka Rus. Fed.
77 E2 Zyryanovsk Kazakh.

Acknowledgements

pages 34–5
Köppen classification map: Kottek, M., J. Grieser, C. Beck, B. Rudolf, and F. Rubel, 2006: World Map of the Köppen-Geiger climate classification updated. Meteorol. Z., **15**, 259–263. http://koeppen-geiger.vu-wien.ac.at

pages 36–37
Land cover map: © ESA / ESA GlobCover Project, led by MEDIAS-France.
Olivier Arino, Patrice Bicheron, Frederic Achard, John Latham, Ron Witt, Jean-Louis Weber, November 2008, GLOBCOVER. The most detailed portrait of Earth. ESA *Bulletin* **136**, pp25–31
Available at
www.esa.int/esapub/bulletin/bulletin136/bul136d_arino.pdf

pages 38–39
Population map data:
Gridded Population of the World (GPW), Version 3.
Palisades, NY: CIESN, Columbia University. Available at http://sedac.ciesin.columbia.edu/plue/gpw

Cover
Lake Rukwa, Tanzania.
Image courtesy of the Image Science and Analysis Laboratory, NASA-Johnson Space Center.
http://eol.jsc.nasa.gov

KEY TO MAP PAGES AFRICA, NORTH AMERICA, SOUTH AMERICA
(see front endpapers for Oceania, Asia and Europe)

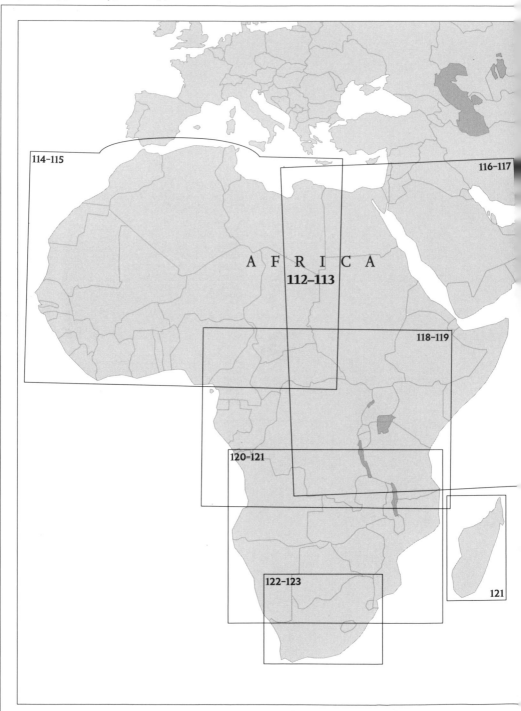

114–115

116–117

A F R I C A
112–113

118–119

120–121

121

122–123